JACARANDA MATHS QUEST
FOUNDATION MATHEMATICS 11

VCE UNITS 1 AND 2 | SECOND EDITION

JACARANDA MATHS QUEST

FOUNDATION MATHEMATICS 11

VCE UNITS 1 AND 2 | SECOND EDITION

MARK BARNES

CONTRIBUTING AUTHORS

Pauline Holland

Christine Utber

Second edition published 2023 by
John Wiley & Sons Australia, Ltd
155 Cremorne Street, Cremorne, Vic 3121

First edition published 2010

Typeset in 10.5/13 pt TimesLTStd

ISBN: 978-1-119-87605-2

The covers of the *Jacaranda Maths Quest VCE Mathematics* series are the work of Victorian artist Lydia Bachimova.

Lydia is an experienced, innovative and creative artist with over 10 years of professional experience, including five years of animation work with Walt Disney Studio in Sydney. She has a passion for hand drawing, painting and graphic design.

Illustrated by diacriTech and Wiley Composition Services

Typeset in India by diacriTech

A catalogue record for this book is available from the National Library of Australia

Printed in Singapore
M121034_260822

Contents

About this resource

Everything you need for your students to succeed

JACARANDA MATHS QUEST
FOUNDATION
MATHEMATICS 11 VCE UNITS 1 AND 2 | SECOND EDITION

Developed by expert Victorian teachers for VCE students

Tried, tested and trusted. The NEW Jacaranda VCE Mathematics series continues to deliver curriculum-aligned material that caters to students of all abilities.

Completely aligned to the VCE Mathematics Study Design

Our expert author team of practising teachers ensures 100 per cent coverage of the new VCE Mathematics Study Design (2023–2027).

Everything you need for your students to succeed, including:

- **NEW!** Access carefully scaffolded question sets, to ensure that all students can experience success and take the next steps. Ensure assessment preparedness with practice Mathematical investigations.

- **NEW!** Be confident your students can get unstuck and progress, in class or at home. For every question online they receive immediate feedback and fully worked solutions.

- **NEW!** Teacher-led videos to unpack concepts and worked examples to fill learning gaps after COVID-19 disruptions.

Learn online with Australia's most

Everything you need for each of your lessons in one simple view

- Trusted, curriculum-aligned theory
- Engaging, rich multimedia
- All the teacher support resources you need
- Deep insights into progress
- Immediate feedback for students
- Create custom assignments in just a few clicks.

Practical teaching advice and ideas for each lesson provided in teachON

Each lesson linked to the Key Knowledge (and Key Skills) from the VCE Mathematics Study Design

Reading content and rich media including embedded videos and interactivities

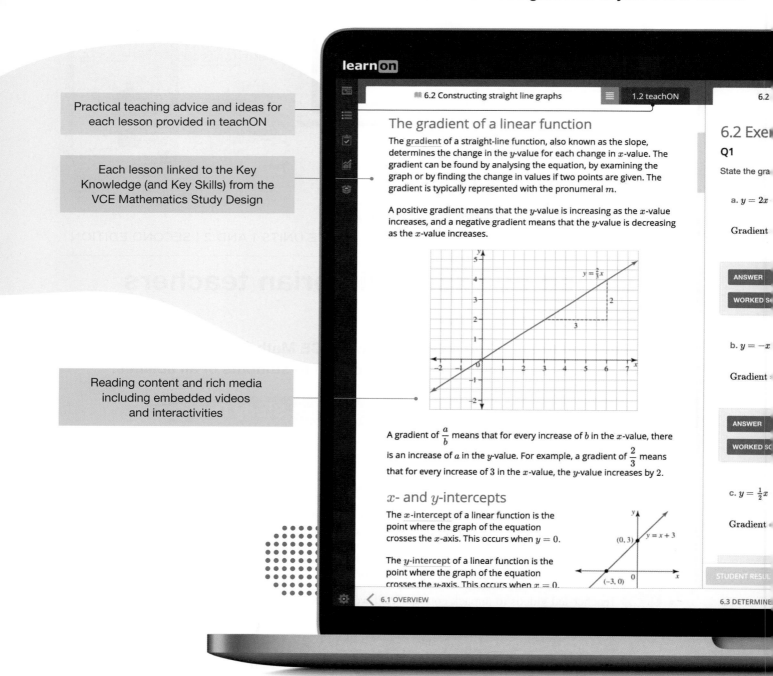

learnon

6.2 Constructing straight line graphs — 1.2 teachON — 6.2

The gradient of a linear function

The gradient of a straight-line function, also known as the slope, determines the change in the y-value for each change in x-value. The gradient can be found by analysing the equation, by examining the graph or by finding the change in values if two points are given. The gradient is typically represented with the pronumeral m.

A positive gradient means that the y-value is increasing as the x-value increases, and a negative gradient means that the y-value is decreasing as the x-value increases.

$y = \frac{2}{3}x$

A gradient of $\frac{a}{b}$ means that for every increase of b in the x-value, there is an increase of a in the y-value. For example, a gradient of $\frac{2}{3}$ means that for every increase of 3 in the x-value, the y-value increases by 2.

x- and y-intercepts

The x-intercept of a linear function is the point where the graph of the equation crosses the x-axis. This occurs when $y = 0$.

The y-intercept of a linear function is the point where the graph of the equation crosses the y-axis. This occurs when $x = 0$.

$y = x + 3$
$(0, 3)$
$(-3, 0)$

6.1 OVERVIEW — 6.3 DETERMINE

6.2 Exe

Q1

State the gra

a. $y = 2x$

Gradient

ANSWER

WORKED S

b. $y = -x$

Gradient

ANSWER

WORKED S

c. $y = \frac{1}{2}x$

Gradient

STUDENT RESUL

powerful learning tool, learnON

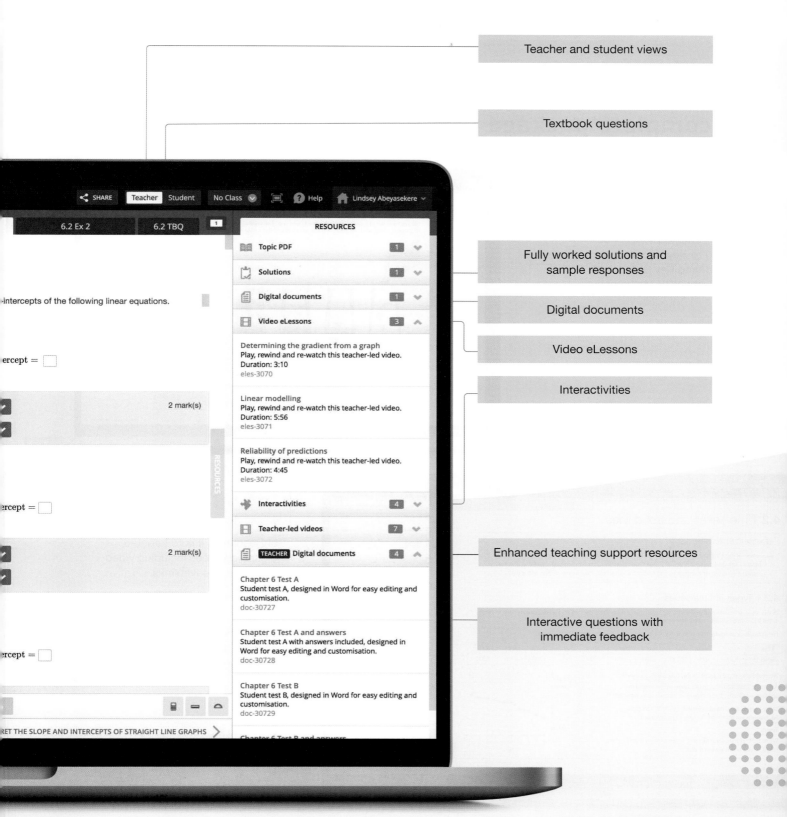

Teacher and student views

Textbook questions

Fully worked solutions and sample responses

Digital documents

Video eLessons

Interactivities

Enhanced teaching support resources

Interactive questions with immediate feedback

Get the most from your online resources

Online, these new editions are the complete package

Trusted Jacaranda theory, plus tools to support teaching and make learning more engaging, personalised and visible.

Each topic is linked to Key Knowledge (and Key Skills) from the VCE Mathematics Study Design.

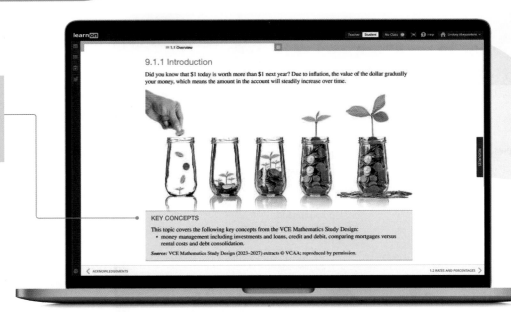

9.1.1 Introduction

Did you know that $1 today is worth more than $1 next year? Due to inflation, the value of the dollar gradually your money, which means the amount in the account will steadily increase over time.

KEY CONCEPTS

This topic covers the following key concepts from the VCE Mathematics Study Design:
• money management including investments and loans, credit and debit, comparing mortgages versus rental costs and debt consolidation.

Source: VCE Mathematics Study Design (2023–2027) extracts © VCAA; reproduced by permission.

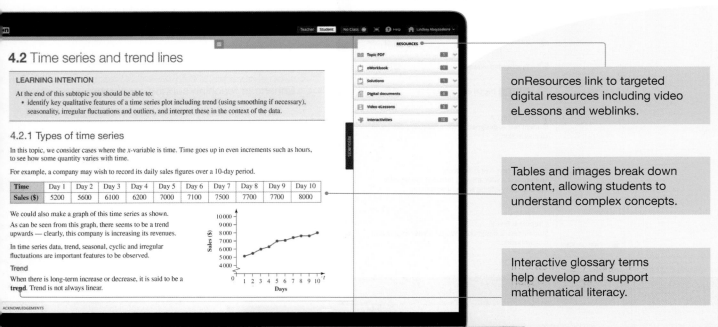

4.2 Time series and trend lines

LEARNING INTENTION

At the end of this subtopic you should be able to:
• identify key qualitative features of a time series plot including trend (using smoothing if necessary), seasonality, irregular fluctuations and outliers, and interpret these in the context of the data.

4.2.1 Types of time series

In this topic, we consider cases where the x-variable is time. Time goes up in even increments such as hours, to see how some quantity varies with time.

For example, a company may wish to record its daily sales figures over a 10-day period.

Time	Day 1	Day 2	Day 3	Day 4	Day 5	Day 6	Day 7	Day 8	Day 9	Day 10
Sales ($)	5200	5600	6100	6200	7000	7100	7500	7700	7700	8000

We could also make a graph of this time series as shown.

As can be seen from this graph, there seems to be a trend upwards — clearly, this company is increasing its revenues.

In time series data, trend, seasonal, cyclic and irregular fluctuations are important features to be observed.

Trend

When there is long-term increase or decrease, it is said to be a **trend**. Trend is not always linear.

onResources link to targeted digital resources including video eLessons and weblinks.

Tables and images break down content, allowing students to understand complex concepts.

Interactive glossary terms help develop and support mathematical literacy.

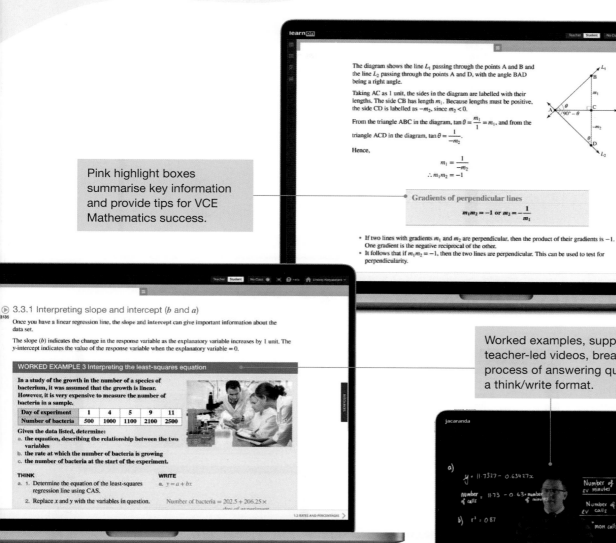

Pink highlight boxes summarise key information and provide tips for VCE Mathematics success.

Worked examples, supported by teacher-led videos, break down the process of answering questions using a think/write format.

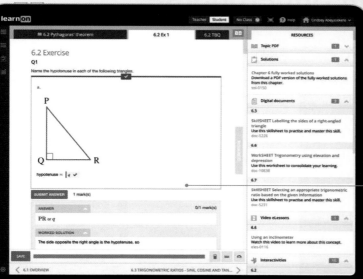

- Online and offline question sets contain practice questions, with exemplary responses.
- Every question has immediate, corrective feedback to help students to overcome misconceptions as they occur and to study independently — in class and at home.

Topic reviews

Topic reviews include online summaries and topic level review exercises that cover multiple concepts.

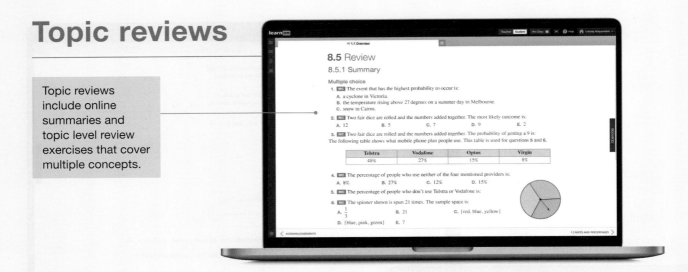

Get exam-ready!

Topic level review questions are structured just like the exams — with multiple choice, short answer and extended response questions.

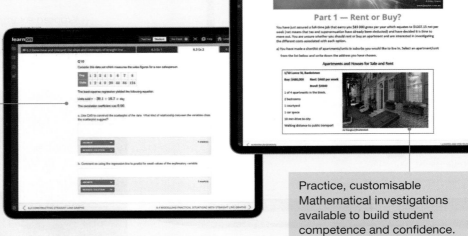

Practice, customisable Mathematical investigations available to build student competence and confidence.

Combine units flexibly with the Jacaranda Supercourse

Build the course you've always wanted with the Jacaranda Supercourse. You can combine all Foundation Mathematics Units 1 to 4, so students can move backwards and forwards freely. Or General and Foundation Units 1 & 2 for when students switch courses. The possibilities are endless!

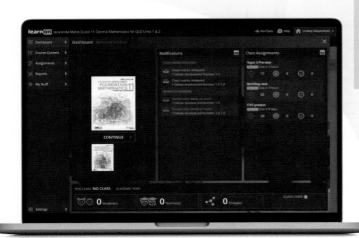

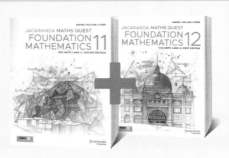

A wealth of teacher resources

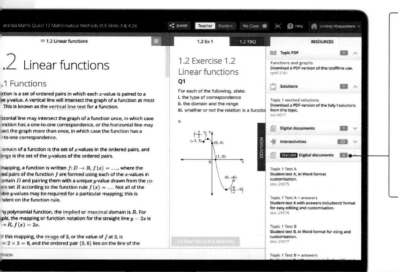

Enhanced teacher support resources, including:

- work programs and curriculum grids
- teaching advice and additional activities
- quarantined topic tests (with solutions)
- quarantined Mathematical investigations (with worked solutions and marking rubrics)

Customise and assign

A testmaker enables you to create custom tests from the complete bank of thousands of questions.

Reports and results

Data analytics and instant reports provide data-driven insights into performance across the entire course.

Show students (and their parents or carers) their own assessment data in fine detail. You can filter their results to identify areas of strength and weakness.

Acknowledgements

The authors and publisher would like to thank the following copyright holders, organisations and individuals for their assistance and for permission to reproduce copyright material in this book

Selected extracts from the VCE Mathematics Study Design (2023–2027) are copyright Victorian Curriculum and Assessment Authority (VCAA), reproduced by permission. VCE® is a registered trademark of the VCAA. The VCAA does not endorse this product and makes no warranties regarding the correctness and accuracy of its content. To the extent permitted by law, the VCAA excludes all liability for any loss or damage suffered or incurred as a result of accessing, using or relying on the content. Current VCE Study Designs and related content can be accessed directly at www.vcaa.vic.edu.au. Teachers are advised to check the VCAA Bulletin for updates.

Images

- © GREG WOOD/AFP/Getty Images: **165** • © INDRANIL MUKHERJEE/Getty Images: **236**
- © PeopleImages/Getty Images: **424** • © Getty Images: **228, 705** • © Mark Dadswell/Getty Images Sport/Getty Images: **237** • © Ryan Pierse/Getty Images: **268** • © domonabike/Alamy Images: **477** • © Hugh PETERSWALD/Alamy Stock Photo: **163** • © Independent/Alamy Stock Photo: **78** • © Neil Tingle/Alamy Images: **491** • © Anja Koppitsch/Getty Images: **435** • © Shutterstock: **158, 180, 695** • © Africa Studio/Shutterstock: **2, 202, 239, 247, 295, 308, 443, 516, 733** • DaisyCooil/Shutterstock: **524** • © Dragon Images/Shutterstock: **423, 436, 446, 556** • FMStox/Shutterstock: **707** • © hddigital/Shutterstock: **92, 101** • Maksym Dykha/Shutterstock: **515** • Pixel Embargo/Shutterstock: **515, 648** • ridersuperone/Shutterstock: **43** • Stefano Chiacchia/Shutterstock: **515** • © Aaron Amat/Shutterstock: **106** • © Adamlee01/Shutterstock: **50** • © Aleksandar Todorovic/Shutterstock: **460** • © Aleksandr Markin/Shutterstock: **546** • © Aleksandra Gigowska/Shutterstock: **411** • © Alex Bogatyrev/Shutterstock: **268** • © Alexa Mat/Shutterstock: **58** • © Alexandra Lande/Shutterstock: **185** • © Alik Drew/Shutterstock: **546** • © Andrey Izvolinin/Shutterstock: **264** • © Andrey Sayfutdinov/Shutterstock: **477** • © Andrey_Popov/Shutterstock: **67** • © AndreySkat/Shutterstock: **417** • © Anna Kucherova/Shutterstock: **44** • © Anterovium/Shutterstock: **196** • © Anton Balazh/Shutterstock: **168** • © antoniodiaz/Shutterstock: **417, 517** • © aodaodaodaod/Shutterstock: **195** • © arek_malang/Shutterstock: **421** • © Ariwasabi/Shutterstock: **263** • © astudio/Shutterstock: **313** • BARRI/Shutterstock: **242** • © BearFotos/Shutterstock: **17** • © Budimir Jevtic/Shutterstock: **96** • © BushAlex/Shutterstock: **272** • © canbedone/Shutterstock: **295** • © Chatham172/Shutterstock: **317** • © Chekyravaa/Shutterstock: **27** • © Chris D573/Shutterstock: **475** • © chrisdorney/Shutterstock: **62** • © chuanpis/Shutterstock: **412** • © cigdem/Shutterstock: **275** • © cowardlion/Shutterstock: **223** • © cybrain/Shutterstock: **459** • © Daria Nipot/Shutterstock: **656** • © David Gyung/Shutterstock: **523** • © David Papazian/Shutterstock: **95** • © DAVIDE CERATI/Shutterstock: **234** • © Denis Raev/Shutterstock: **534** • © Denis Tabler/Shutterstock: **485** • © DenPhotos/Shutterstock: **274** • © Dima Zel/Shutterstock: **287** • © DisobeyArt/Shutterstock: **20** • © Dmitry Kalinovsky/Shutterstock: **423** • © dotshock.com/Shutterstock: **459** • © Double Brain/Shutterstock: **210** • © Dr. Victor Wong/Shutterstock: **696** • © Eladstudio/Shutterstock: **192** • © Elena Elisseeva/Shutterstock: **163** • © epic_images/Shutterstock: **288, 292** • © Ermolaev Alexander/Shutterstock: **416** • © Eugene Partyzan/Shutterstock: **268** • © Evgeniya Uvarova/Shutterstock: **268** • © fongbeerredhot/Shutterstock: **659** • © Franck Boston/Shutterstock: **628** • © frantic00/Shutterstock: **16** • © Freedomz/Shutterstock: **413** • © Galina Petrova/Shutterstock: **262** • © gangnavigator/Shutterstock: **532** • © Gater Images/Shutterstock: **256** • © gmstockstudio/Shutterstock: **680** • © Golubovy/Shutterstock: **662** • © Goncharov_Artem/Shutterstock: **58** • © Gordon Bell/Shutterstock: **500** • © gpointstudio/Shutterstock: **147** • © gvictoria/Shutterstock: **112** • © Hans Christiansson/Shutterstock: **99** • © happymay/Shutterstock: **194** • © Heleen Van Assche/Shutterstock: **677** • © hidesy/Shutterstock: **410** • © HomeStudior/Shutterstock: **508** • © Hrytsiv Oleksandr/Shutterstock: **662** • © Igisheva Maria/Shutterstock: **114** • © ILIA BLIZNYUK/Shutterstock: **51, 215** • © Imagentle/Shutterstock: **255** • © inspiredbyartr/Shutterstock: **508** • © insta_photos/Shutterstock: **643, 659** • © interstid/Shutterstock: **634** • © Iryna Imago/Shutterstock: **79** • © Isogood_patrick/Shutterstock: **166** • © Ivanko80/Shutterstock: **36** • © Jacob Crook - JBC Studios/

Shutterstock: **165** • © Jacob Lund/Shutterstock: **412** • © Javier Catano Gonzalez/Shutterstock: **453** • © Jfunk/Shutterstock: **45** • © JM-Design/Shutterstock: **201** • © js59/Shutterstock: **92** • © Juice Dash/Shutterstock: **77** • © Juice Flair/Shutterstock: **39** • © kai keisuke/Shutterstock: **21** • © Keitma/Shutterstock: **35** • © kenary820/Shutterstock: **672** • © kitzcorner/Shutterstock: **477** • © Kozak Bohdan/Shutterstock: **515** • © Krakenimages.com/Shutterstock: **675** • © Kritchanut/Shutterstock: **411** • © Kuchina/Shutterstock: **213** • © kudla/Shutterstock: **519** • © Leonard Zhukovsky/Shutterstock: **165, 273, 275** • © LightField Studios/Shutterstock: **446** • © Lisa-S/Shutterstock: **216** • © LittlePanda29/Shutterstock: **263** • © Ljupco Smokovski/Shutterstock: **18** • © Lucky Business/Shutterstock: **448** • © lunamarina/Shutterstock: **318** • © M-Production/Shutterstock: **426** • © M. Unal Ozmen/Shutterstock: **177, 193** • © Markus Gebauer/Shutterstock: **264** • © maxpro/Shutterstock: **274** • © maxstockphoto/Shutterstock: **246** • © Maxx-Studio/Shutterstock: **529, 683** • © Mega Pixel/Shutterstock: **163** • © Mehaniq/Shutterstock: **322** • © mentatdgt/Shutterstock: **125** • © Michael Stokes/Shutterstock: **249** • © MichaelHahnr/Shutterstock: **515** • © michaeljung/Shutterstock: **148, 433** • © Michele Morrone/Shutterstock: **266** • © Mike Flippo/Shutterstock: **264, 534** • © Minerva Studio/Shutterstock: **633** • © Monkey Business Images/Shutterstock: **69, 443, 447, 631, 650, 669** • © Murni/Shutterstock: **301** • © NASA images/Shutterstock: **237, 454** • © Nata-Lia/Shutterstock: **282** • © Nattawit Khomsanit/Shutterstock: **114** • © Neale Cousland/Shutterstock: **39, 100, 161** • © Neil Balderson/Shutterstock: **485** • © Nestor Rizhniak/Shutterstock: **273** • © New Africa/Shutterstock: **69** • © Ok-product studio/Shutterstock: **457** • © Oksana Kuzmina/Shutterstock: **658** • © OksankaFra/Shutterstock: **588** • © Oliver Hoffmann/Shutterstock: **202** • © Olleg/Shutterstock: **645** • © oneSHUTTER oneMEMORY/Shutterstock: **246** • © OSTILL is Franck Camhi/Shutterstock: **235, 248** • © PabloBenii/Shutterstock: **125** • © Pakhnyushchy/Shutterstock: **178, 229** • © Paulharding00/Shutterstock: **459** • © Pavaphon Supanantananont/Shutterstock: **135** • © Pavel L Photo and Video/Shutterstock: **495** • © Pavel Skopets/Shutterstock: **235** • © pedica018/Shutterstock: **97** • © Peter Hermes Furian/Shutterstock: **316** • © Photographee.eu/Shutterstock: **249, 670** • © Pixel-Shot/Shutterstock: **241** • © Ralf Kleemann/Shutterstock: **634** • © Rashevskyi Viacheslav/Shutterstock: **457, 512** • © Rawpixel.com/Shutterstock: **630, 633, 657** • © RaymondZ/Shutterstock: **35** • © Reamolko/Shutterstock: **294** • © Robyn Mackenzie/Shutterstock: **7** • © rodimov/Shutterstock: **238** • © Romolo Tavani/Shutterstock: **57** • © Rtimages/Shutterstock: **200** • © Ruslan Ivantsov/Shutterstock: **417** • © S-F/Shutterstock: **67** • © s_bukley/Shutterstock: **516** • © Sandratsky Dmitriy/Shutterstock: **207** • © sandystifler/Shutterstock: **153** • © seagames50 images/Shutterstock: **246** • © SeventyFour/Shutterstock: **70, 676** • © Shane White/Shutterstock: **178** • © CHEN WS/Shutterstock: **228** • © LoopAll/Shutterstock: **194** • © Sorbis/Shutterstock: **193** • © Source: El Nariz/Shutterstock: **418** • © steamroller_blues/Shutterstock: **228, 274** • © Steve Cukrov/Shutterstock: **114** • © stockcreations/Shutterstock: **191** • © stockfour/Shutterstock: **261** • © StockImageFactory.com/Shutterstock: **319** • © stockyimages/Shutterstock: **571** • © STRINGER Image/Shutterstock: **260** • © Studio Barcelona/Shutterstock: **190** • © Sunflowerey/Shutterstock: **409** • © Suvorov_Alex/Shutterstock: **280** • © SvartKat/Shutterstock: **137** • © szefei/Shutterstock: **668** • © theshots.co/Shutterstock: **416** • © ThomasLENNE/Shutterstock: **192** • © Tienuskin/Shutterstock: **175** • © TippaPatt/Shutterstock: **629** • © tishomir/Shutterstock: **225** • © Toasted Pictures/Shutterstock: **195** • © TorriPhoto/Shutterstock: **214** • © trabantos/Shutterstock: **459** • © Tyler Olson/Shutterstock: **94** • © urfin/Shutterstock: **17** • © Vadim Ovchinnikov/Shutterstock: **723** • © vchal/Shutterstock: **157** • © virtu studio/Shutterstock: **687** • © Visual Collective/Shutterstock: **461** • © Vixit/Shutterstock: **7** • © Vlad Teodor/Shutterstock: **431** • © VladFree/Shutterstock: **418** • © Vladi333/Shutterstock: **517** • © vlasque/Shutterstock: **233** • © World_of_Textiles/Shutterstock: **415** • © Yeti studio/Shutterstock: **229** • © Yuganov Konstantin/Shutterstock: **461** • © Zagardinova Anastasiya/Shutterstock: **185** • © zentilia/Shutterstock: **201** • © Zryzner/Shutterstock: **224** • © zstock/Shutterstock: **188** • Cliff Day/Shutterstock: **293** • motorolka/Shutterstock: **290, 291, 293** • © Vetta/Getty Images: **638** • © Folio Images/Alamy Stock Photo: **441** • © GL Archive/Alamy Stock Photo: **186** • © NDP/Alamy Stock Photo: **148** • © 9-Kilo/Shutterstock: **512** • © A__N/Shutterstock: **274** • © Africa Studior/Shutterstock: **507** • © Aleks Melnik/Shutterstock: **208** • © alle12/iStock/Getty Images: **510** • © AlonaPhoto/Shutterstock: Volodymyr Krasyuk/Shutterstock, Maksim Denisenko/Shutterstock: **304** • © Aprilphoto/Shutterstock: Sashkin/Shutterstock: **194** • © artellia/Shutterstock: **192** • © Bardocz Peter/Shutterstock: **187** • © creatOR76/Shutterstock: **193** • © cTermit/Shutterstock: **198** • © Daria Krav/Shutterstock: **748** • © David Svetlik/Shutterstock: **200** • © Dawid Rojek/Shutterstock: **518** • © Delpixel/Shutterstock: **233** • © Denis Belitsky/Shutterstock: **120** • © Denis Shitikoff/Shutterstock: **304** • © Dorling Kindersley ltd/Alamy Stock Photo: **721**

• © EPSTOCK/Shutterstock: **435** • © Eric Isselee/Shutterstock: **233** • © Fine Art/Shutterstock: **279** • © Golden Pixels LLC/Shutterstock: **153** • © Gts/Shutterstock: **217** • © inception-D/Shutterstock: **633** • © jkerrigan/Shutterstock: **207** • © John Wiley & Sons: **269, 281** • © Jon Tyson/Unsplash: **1** • © Lightspring/Shutterstock: **26** • © local_doctor/Shutterstock: **499** • © maratr/Shutterstock: **170** • © marina_ua/Shutterstock: **219** • © Maurizio De Mattei/Shutterstock: **229** • © MilanTomazin/Shutterstock: **201** • © Nataliia G/Shutterstock: **216** • © Nils Z/Shutterstock: **232** • © ONYXprj/Shutterstock: **171** • © Patrick Fore/Unsplash: **144** • © PeopleImages/E+/Getty Images: Rido/Shutterstock: **419** • © PhotoDisc: Inc: **510** • © photosync/Shutterstock: **202** • © photowind/Shutterstock: **198** • © Pyty/Shutterstock: **201** • © Robert Plociennik/Shutterstock: **182** • © sabri deniz kizil/Shutterstock: **181** • © Scott Prokop/Adobe Stock: **510** • © Sergey Nivens/Shutterstock: **176** • © Source: Based on 2008 and 2014-15 National Aboriginal and Torres Strait Islander Social Survey (NATSISS): 2012-13 and 2018-19 National Aboriginal and Torres Strait Islander Health Survey (NATSIHS): **169** • © Source: Map drawn by MAPgraphics Pvt Ltd: Bribane: **205** • © SS1001/Shutterstock: **183** • © TanjaJovicic/Shutterstock: **199** • © Tatiana Popova/Shutterstock: **282** • © teen boy in blue T-shirt being measured: **226** • © Teia/Shutterstock: **208** • © Tupungato/Shutterstock: **152** • © urbanbuzz/Shutterstock: **164** • © vectorisland/Shutterstock: **204** • © Vitaliy Hrabar/Shutterstock: **186** • © vlastas/Shutterstock: **632** • © Wiley art: **711** • © YesPhotographers/Shutterstock: **25** • © Yoko Design/Shutterstock: **182** • © Andrey Arkusha/Shutterstock: **455** • © Anthony Hall/Shutterstock: **264** • © Benny Marty/Shutterstock: **455** • © Bildagentur Zoonar GmbH/Shutterstock: **457** • © Carsten Reisinger/Shutterstock: **34** • © CurrywurstmitPommes/Shutterstock: **456** • © Ddanilovicr/Shutterstock: **513** • © Duda Vasilii/Shutterstock: **457** • © fckncg/Shutterstock: **34** • © Julenochek/Shutterstock: **456** • © Lenscap Photography/Shutterstock: **516** • © Lipowski Milan/Shutterstock: **516** • © losw/Shutterstock: **295** • © Monthirar/Shutterstock: **513** • © Nonnakrit/Shutterstock: **197** • © Oleksiy Mark/Shutterstock: **264** • © photogal/Shutterstock: **516** • © Sofia Zhuravetcr/Shutterstock: **513** • © Tom Oliveira/Shutterstock: **236** • © Vereshchagin Dmitryr/Shutterstock: **513** • © Volosovich Igor/Shutterstock: **457** • © XAOC/Shutterstock: **236** • © Jim Barber/Shutterstock: **337** • © Australian Bureau of Statistics: **331** • © 4zevar/Shutterstock: **342** • © Pixsooz/Shutterstock: **341** • © reezuan/Shutterstock: **340** • © ESB Professional/Shutterstock: **350** • © Julija Sapic/Shutterstock: **355** • © Dean Drobot/Shutterstock: **372** • © Alexey Boldin/Shutterstock: **339** • © Pandur/Shutterstock: **339** • © DGLimages/Shutterstock: **338** • © Brian A Jackson/Shutterstock: **345** • © Eduard Zhukov/Shutterstock: **345** • © Evgeny Karandaev/Shutterstock: **344** • © Tony J Tan/Shutterstock: **352** • © Fedorov Oleksiy/Shutterstock: **352** • © Halfpoint/Shutterstock: **351** • © Billion Photos/Shutterstock: **368** • © Shift Drive/Shutterstock: **367** • © Margarita Borodina/Shutterstock: **367** • © Lekhawattana/Shutterstock: **383** • © Boule/Shutterstock: **382** • © Auscape/Getty Images: **381** • © wavebreakmedia/Shutterstock: **381** • © ALPA PROD/Shutterstock: **380** • © Dave Clark Digital Photo/Shutterstock: **397** • © ESB Professional/Shutterstock: **395** • © Stefano Chiacchiarini '74/Shutterstock: **515** • © buradaki/Shutterstock: **119** • © Evgeny Atamanenko/Shutterstock: **330** • © Source: Based on information taken from The Age, Money, 27 March 2013: **363** • © R-O-M-A/Shutterstock: **695** • © C Squared Studios/Stockbyte/Getty Images: **705** • © FlashStudio/Shutterstock: **143** • © austinding/Shutterstock: **158** • © Inpho Photography/Getty Images Sport/Getty Images: **228** • © Kampol Taepanich/Shutterstock: **295** • © alexandre zveiger/Shutterstock: **317** • © Nicola Katie/Shutterstock: **436** • © MARGRIT HIRSCH/Shutterstock: **531** • © rangizzz/Adobe Stock: **349** • © Denis Tabler/Adobe Stock: **349** • © pattang/Shutterstock: **349** • © Yellow duck/Adobe Stock: **349** • © Source: Australian Bureau of Statistics, http://www.abs.gov.au/ausstats/abs@.nsf/Lookup/4221.0main+features22017. License under CC- BY 4.0: **369** • © Source: Oreopoulos, P., Salvanes, K. G., 'Priceless: the nonpecuniary benefits of schooling', Journal of Economic Perspectives, vol. 25, no. 1, Winter 2011, pp. 159–84: **370**

Text

Source: Data from TAC road safety statistical summary October 2012, page 6: **383**

1 Integers

LEARNING SEQUENCE

Fully worked solutions for this topic are available online.

1.1 Overview

1.1.1 Introduction

An integer is a whole number (e.g. 100, 1, 0, −1 and −100), with no fractions or parts of a whole included. Integers can be positive (above zero), negative (below zero) or equal to zero.

Being able to add, subtract, multiply and divide integers is important in many parts of everyday life. It helps with budgeting and knowing what you can and can't afford to buy — or how much of something to buy. It also helps to keep you safe; for example, being able to read the integers on the speedometer helps you stay within the speed limit. It is also useful for understanding temperatures, weights and measures when you're cooking or you need to store food safely.

In finance, positive numbers are used to represent the amount of money in someone's bank account, while negative numbers are used to represent how much money someone owes (for example, how much they have to pay back after borrowing money or taking out a loan from a bank).

KEY CONCEPTS

This topic covers the following key concepts from the VCE Mathematics Study Design:
- application of integers, fractions and decimals, their properties and related operations
- estimation, approximation and reasonableness of calculations and results.

Note: Concepts shown in grey are covered in other topics.

Source: VCE Mathematics Study Design (2023–2027) extracts © VCAA; reproduced by permission.

1.2 Adding and subtracting integers

1.2.1 Addition and subtraction of integers using a number line

An **integer** is a whole number. It can be positive (2, 4, 89, 1035) or negative (-2, -4, -89, -1035). Zero is also an integer because it is a whole number.

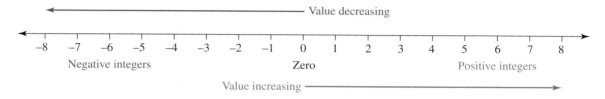

The positive numbers do not usually have a symbol to show that they are positive ($+2$ is the same as 2). However, the negative numbers must include the negative sign to show that they are negative (-2, -27).

A whole number is a number that does not include any fractions or parts of a number. For example, the following are *not* whole numbers or integers:

- Fractions $\left(\dfrac{1}{2}, -12\dfrac{3}{4}, 5\dfrac{1}{2}\right)$ because they are or include parts of a whole number
- Numbers that continue after a decimal point (0.5, -12.75, 5.5)

Number lines can be used to remember the rules for adding and subtracting integers.

> **Addition and subtraction using a number line**
>
> **Start at the first number.**
>
> **To add a:**
> - **positive integer, move to the right**
> - **negative integer, move to the left.**
>
> **To subtract a:**
> - **positive integer, move to the left**
> - **negative integer, move to the right.**

To show $-3 + (+2)$ on a number line, start at the first number (-3) and place a pointer on the number line at -3. To add the positive integer ($+2$), move 2 places to the right.

Therefore, using the number line, we can see that $-3 + (+2) = -1$.

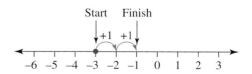

WORKED EXAMPLE 1 Adding and subtracting integers using a number line

▶ tlvd-3549

Use a number line showing intervals from −6 to 3 to calculate each of the following.

a. $2 + (+1)$ b. $3 + (-2)$ c. $-4 - (+2)$ d. $-3 - (-5)$

THINK	WRITE
a. 1. Start at 2 and move 1 unit to the right, as this is the addition of a positive number.	**a.** 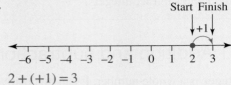
2. Write the answer.	$2 + (+1) = 3$
b. 1. Start at 3 and move 2 units to the left, as this is the addition of a negative number.	**b.** 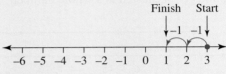
2. Write the answer.	$3 + (-2) = 1$
c. 1. Start at −4 and move 2 units to the left, as this is the subtraction of a positive number.	**c.** 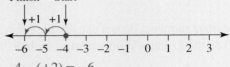
2. Write the answer.	$-4 - (+2) = -6$
d. 1. Start at −3 and move 5 units to the right, as this is the subtraction of a negative number.	**d.** 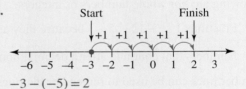
2. Write the answer.	$-3 - (-5) = 2$

1.2.2 Addition and subtraction of integers using symbols

Another method of remembering the rules for adding and subtracting integers is to imagine positive and negative symbols or counters representing each of the numbers in an equation. A pink positive symbol represents +1 and a blue negative symbol represents −1.

When calculating:

- 2 is written as:

- −2 is written as:

- positive and negative symbols cancel each other out: $-1 + 1 = 0$

WORKED EXAMPLE 2 Adding and subtracting integers using symbols

Calculate the value of each of the following.
a. $2 + 2$
b. $-2 + 2$
c. $-2 - (3)$
d. $-4 - (-2)$

THINK	WRITE
a. 1. Set the equation up using symbols.	**a.** $2 + 2 = $ ⊕ ⊕ + ⊕ ⊕
2. Add all the positives.	4 positives $= +4$
3. Write the answer.	$2 + 2 = 4$
b. 1. Set the equation up using symbols.	**b.** $-2 + 2 = $ ⊖ ⊖ + ⊕ ⊕
2. Cancel any pairs of positives and negatives.	⊗ ⊗ + ⊗ ⊗
3. Count how many symbols are left.	Zero symbols remain.
4. Write the answer.	$-2 + 2 = 0$
c. 1. Set the equation up using symbols.	**c.** $-2 - (3) = $ ⊖ ⊖ − ⊕ ⊕ ⊕
2. When subtracting positive integers, change the symbols' sign and then add the symbols.	⊖ ⊖ + ⊖ ⊖ ⊖
3. Count all the negatives.	5 negatives $= -5$
4. Write the answer.	$-2 - (3) = -5$
d. 1. Set the equation up using symbols.	**d.** $-4 - (-2) = $ ⊖ ⊖ ⊖ ⊖ − ⊖ ⊖
2. When subtracting negative integers, change the symbol's sign and then add the symbols.	⊖ ⊖ ⊖ ⊖ + ⊕ ⊕
3. Cancel any pairs of positive and negative symbols.	⊗ ⊗ ⊖ ⊖ + ⊗ ⊗
4. Count how many positives or negatives remain.	Two negatives remain.
5. Write the answer.	$-4 - (-2) = -2$

1.2.3 Addition and subtraction of integers by applying rules

You can also follow the rules shown in this table to add and subtract positive and negative numbers.

Rules for addition and subtraction of integers

Rule		Example
Adding a positive integer is the same as adding.	$+ + = +$	$+2 + (+5) = +7$
Adding a negative integer is the same as subtracting.	$+ - = -$	$+2 + (-5) = -3$
Subtracting a positive integer is the same as subtracting.	$- + = -$	$+2 - (+5) = -3$
Subtracting a negative integer is the same as adding.	$- - = +$	$-2 - (-5) = 3$

WORKED EXAMPLE 3 Adding and subtracting integers by applying rules

tlvd-3551

Calculate the value of each of the following.

a. $2+2$ b. $2+(-2)$ c. $2-(+3)$ d. $-4-(-2)$

THINK	WRITE
a. 1. Adding a positive integer is the same as adding.	**a.** $2+2$
2. Apply the rule $++=+$ and calculate the value.	$2+2=4$
b. 1. Adding a negative integer is the same as subtracting.	**b.** $2+(-2)$
2. Apply the rule $+-=-$ and calculate the value.	$2+(-2)=2-2$
3. Write the answer.	$2-2=0$
c. 1. Subtracting a positive integer is the same as subtracting.	**c.** $2-(+3)$
2. Apply the rule $+-=-$ and calculate the value.	$2-(+3)=2-3$
3. Write the answer.	$2-3=-1$
d. 1. Subtracting a negative integer is the same as adding.	**d.** $-4-(-2)$
2. Apply the rule $--=+$ and calculate the value.	$-4--2=-4+2$
3. Write the answer.	$-4+2=-2$

on Resources

 Video eLesson Integers on the number line (eles-0040)

Interactivities Direct number target (int-0074)
Addition of positive integers (int-3922)
Subtraction of positive integers (int-3924)

1.2 Exercise

Students, these questions are even better in jacPLUS

Receive immediate feedback and access sample responses

Access additional questions

Track your results and progress

Find all this and MORE in jacPLUS

1. Select the integers from the following numbers.

$$5, -2, \frac{3}{4}, 212, 12.3, -2.5, -33, -2\frac{1}{2}$$

2. **WE1** Calculate each of the following using the number line method.

a. $(-5)+2$ b. $(-6)+1$ c. $5+(-3)$
d. $(-8)+(-3)$ e. $21+(+9)$ f. $18+(-5)$

3. **WE2** Calculate each of the following using the cancelling signs method.

a. $3+(+1)$ b. $3+(-2)$ c. $4-(+5)$
d. $2+(-2)$ e. $3-(-2)$ f. $(-1)-(-3)$

4. **WE3** Calculate the value of each of the following using rules.
   a. $(-7) + (-3)$
   b. $8 - (-7)$
   c. $(-23) + (+15)$
   d. $(-18) - (-17)$
   e. $26 - (-13)$
   f. $(-72) - (-26)$

5. Calculate the value of each of the following using the method of your choice.
   a. $(-3) + (-2)$
   b. $(-7) + (-3)$
   c. $8 + (-2)$
   d. $15 - (+6)$
   e. $(-15) - (-4)$
   f. $(-23) + (-14)$

6. Calculate the value of each of the following using the method of your choice.
   a. $(-37) + (-12)$
   b. $42 - (+7)$
   c. $(-14) - (+18)$
   d. $(-27) - (-15)$
   e. $37 - (+12)$
   f. $135 - (-37)$

7. Calculate the value of each of the following using the method of your choice.
   a. $12 + (-4) + (+6)$
   b. $28 + (-7) - (-10)$
   c. $(-15) + (+5) - (+8)$
   d. $28 - (+15) - (+4)$
   e. $18 - (-12) + (-5)$
   f. $(-42) - (-21) - (-21)$

8. Write out these equations, filling in the missing numbers.
   a. $-5 + \square = 2$
   b. $12 - \square = 8$
   c. $(-8) - \square = -20$
   d. $30 + \square = 25$
   e. $-2 - \square - (-8) = 2$
   f. $\square - 3 + 5 = 7$

9. State whether the following expressions are True or False.
   a. $(-5) + (-10) = -15$
   b. $(-7) - (-6) = -13$
   c. $-4 - 3 = 7$
   d. $8 - (-6) + (-10) = 4$
   e. $50 + 12 - (-12) = 50$
   f. $-23 + 10 + 6 = 7$

10. Evaluate and compare the following pairs of expressions.
    a. $-7 + 3$ and $3 - 7$
    b. $-8 + 6$ and $6 - 8$
    c. $-15 + 5$ and $5 - 15$

11. Evaluate and compare the following pairs of expressions.
    a. $-3 + (-4)$ and $-(3 + 4)$
    b. $-8 + (-3)$ and $-(8 + 3)$
    c. $-10 - (+3)$ and $-(10 + 3)$

12. Layla is standing in a park. She runs 20 m to the right, then 5 m to the left, before running another 45 m to the right. Use a number line to show where Layla finishes compared to where she started.

13. Bryce decides to cook meat pies for lunch. He takes the pies out of the freezer that is set at $-18\,°C$ to defrost on the bench where the room temperature is $21\,°C$. He sets the oven to $180\,°C$ to cook the pies.

    a. State the difference between the temperature of the freezer and the room temperature.
       (*Hint:* difference = largest number − smallest number)
    b. Calculate the difference between the room temperature and the temperature of the oven.
    c. Calculate the difference of the freezer temperature and the temperature of the oven.

14. Calculate the difference between the two temperatures on Mount Everest shown in the image.

1.3 Multiplying integers

1.3.1 Multiplication of integers

When multiplying integers, the following rules apply.

> **Rules for the multiplication of integers**
>
> - **When multiplying two integers with the same sign, the answer is positive.**
>
> $$+ \times + = +$$
> $$- \times - = +$$
>
> - **When multiplying two integers with different signs, the answer is negative.**
>
> $$+ \times - = -$$
> $$- \times + = -$$

WORKED EXAMPLE 4 Multiplying integers by applying rules

Evaluate each of the following.

a. $(-4) \times +3$

b. $(-7) \times (-6)$

THINK	WRITE
a. The two numbers have different signs, so the answer is negative.	a. $(-4) \times (+3)$ $= -12$
b. The two numbers have the same sign, so the answer is positive.	b. $(-7) \times (-6)$ $= 42$ (or $+42$)

1.3.2 Powers, squares and square roots of integers

Powers

A **power** of a number is how many times the number has been multiplied by itself. For example, 8 to the power of 2 is written as $8^2 = 8 \times 8$ and -6 to the power of 3 is written as $(-6)^3 = (-6) \times (-6) \times (-6)$.

A number written in power form is represented with a base number and an exponent. When a negative number is raised to a power, the sign of the answer will be:

- positive, if the power is even;
  e.g. $(-5)^2 = (-5) \times (-5) = +25$
- negative, if the power is odd;
  e.g. $(-5)^3 = (-5) \times (-5) \times (-5) = (+25) \times (-5) = -125$

Squares

A square number is any whole number multiplied by itself. When written in index form, it will have a power of 2.

A square number can be illustrated by looking at the area of a square with a whole number as its side length. Consider the image shown. We can say that 5^2 or 25 is a square number, since 5^2 or $25 = 5 \times 5$.

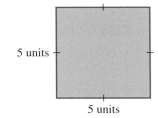

5 units

5 units

Square roots

The square root of a number is a positive value that, when multiplied by itself, gives the original number. The symbol for the square root is $\sqrt{}$. Calculating the square root of a number is the opposite of squaring the number.

For example, if $5^2 = 25$ then the square root of 25 is written as $\sqrt{25} = 5$.

Note: When solving certain equations, we may be required to take both the positive and the negative of a square root. We use the notation $\pm\sqrt{}$ when we are considering both.

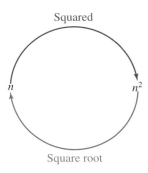

Squared

n n^2

Square root

tlvd-3552

WORKED EXAMPLE 5 Calculating powers and square roots of integers

Calculate each of the following.

a. $(-3)^3$

b. The square root of 64

THINK

a. 1. Write the expression in expanded form.

    2. Evaluate by working from left to right, remembering that the answer will be negative because the power is odd.

b.   Look for the number that, when squared, result in 64 ($8 \times 8 = 64$).

WRITE

a. $(-3)^3 = (-3) \times (-3) \times (-3)$

    $= (+9) \times (-3)$

    $= -27$

b. $\sqrt{64} = 8$

1. **WE4** Evaluate each of the following.
   a. $(-3) \times 5$
   b. $3 \times (-7)$
   c. $(-6) \times (-5)$
   d. $2 \times (-10)$
   e. $(-8) \times (-5)$
   f. $(-7) \times 8$

2. Evaluate each of the following.
   a. $(-10) \times 25$
   b. $(-125) \times 10$
   c. $(-6) \times 9$
   d. $(+9) \times (-7)$
   e. $(-11) \times (-7)$
   f. $250 \times (-2)$

3. Use an appropriate method to evaluate the following.
   a. $(-2) \times 3 \times (-5) \times (-10)$
   b. $6 \times (-1) \times 5 \times (-2) \times 1$
   c. $8 \times (-3) \times (-1) \times (-2) \times 4$
   d. $(-3) \times 5 \times (-2) \times (-1) \times (-1) \times (-1)$
   e. $(-5) \times 6 \times (-2) \times (-2)$
   f. $6 \times (-1) \times 3 \times (-2)$

4. Complete the following equations.
   a. $6 \times \square = -42$
   b. $(-5) \times \square = -30$
   c. $15 \times \square = -30$
   d. $\square \times (-4) = 24$
   e. $\square \times (-9) = 36$
   f. $\square \times 11 = -77$

5. Complete the following equations.
   a. $\square \times (-12) = 108$
   b. $\square \times (-11) = -66$
   c. $\square \times 4 = -48$
   d. $(-10) \times \square = 80$
   e. $(-9) \times \square = -81$
   f. $(-12) \times \square = 144$

6. **WE5a** Evaluate each of the following.
   a. $(-2)^3$
   b. $(-3)^2$
   c. $(-2)^4$
   d. $(-3)^4$
   e. $(-2)^5$
   f. $(-9)^2$

7. Evaluate each of the following.
   a. $(-4)^2$
   b. $(-5)^3$
   c. $(-4)^4$
   d. $(-5)^4$
   e. $(-6)^3$
   f. $(-8)^3$

8. Complete the following statements.
   a. If a negative number is raised to an even power, the answer is (positive/negative).
   b. If a negative number is raised to an odd power, the answer is (positive/negative).

9. **WE5b** Evaluate the square root of the following numbers.
   a. 121
   b. 100
   c. 36
   d. 16
   e. 9
   f. 144

10. For each of the following, write three possible sets of integers that can be placed in the boxes to make the equation a true statement.
    a. $\square \times \square \times \square = -16$
    b. $\square \times \square \times \square = 24$

11. For each of the following, determine whether the result is a positive or negative value. You do not have to work out the value.
    a. $(-47) \times (-52) \times 100$
    b. $(-56) \times (-8) \times (-66)$
    c. $21 \times (-21) \times 42 \times (-32)$
    d. $40 \times (-5) \times 10 \times (-2)$

12. Explain what happens when a number is multiplied by -1, using examples to justify your answer.

13. Explain why the answer to this question is negative.

$$(-2) \times (-4) \times (+3) \times (-6) \times (+4) \times (+3)$$

14. Evaluate $(-1)^n \times (+1)^{n+1}$ if:

    a. n is even
    b. n is odd.

15. In a Year 11 Mathematics examination, there are 30 multiple choice questions. Students are given 2 marks for a correct answer, -1 mark for an incorrect answer and zero marks for an unanswered question. A student scored a total of 33 marks in the multiple choice section. Explain how they could have reached this total.

1.4 Dividing integers

LEARNING INTENTION

At the end of this subtopic you should be able to:
- divide integers.

1.4.1 Division of integers

Division is the inverse or opposite operation of multiplication. We can use the multiplication facts for integers to discover the division facts for integers.

Multiplication fact	Division fact	Multiplication pattern	Division pattern
$4 \times 5 = 20$	$20 \div 5 = 4$ and $20 \div 4 = 5$	positive × positive = positive	$\dfrac{\text{positive}}{\text{positive}} = \text{positive}$
$(-4) \times (-5) = 20$	$20 \div (-5) = -4$ and $20 \div (-4) = -5$	negative × negative = positive	$\dfrac{\text{positive}}{\text{negative}} = \text{negative}$
$(-4) \times 5 = -20$	$(-20) \div 5 = -4$ and $(-20) \div (-4) = 5$	negative × positive = negative and positive × negative = negative	$\dfrac{\text{negative}}{\text{positive}} = \text{negative}$ and $\dfrac{\text{negative}}{\text{negative}} = \text{positive}$

Determining the sign of the answer when dividing integers

- **When dividing two integers with the same sign, the answer is positive.**

$$+ \div + = +$$
$$- \div - = +$$

- **When dividing two integers with different signs, the answer is negative.**

$$+ \div - = -$$
$$- \div + = -$$

Remember that division statements can be written as fractions and then simplified.

For example:

$$(-12) \div (-4) = \frac{-12}{-4}$$
$$= \frac{12 \times \cancel{-1}}{4 \times \cancel{-1}}$$
$$= 3$$

WORKED EXAMPLE 6 Dividing integers

Evaluate each of the following.

a. $(-48) \div 6$

b. $\dfrac{-54}{-9}$

c. $144 \div (-6)$

d. $(-240) \div (-16)$

THINK	WRITE
a. The two numbers have different signs, so the answer is negative.	a. $(-48) \div 6$ $= -8$
b. The two numbers have the same sign, so the answer is positive.	b. $\dfrac{-54}{-9} = \dfrac{(-1) \times 54}{(-1) \times 9}$ $= \dfrac{54}{9}$ $= 6$
c. 1. Complete the division as if both numbers were positive numbers.	c. $\begin{array}{r} 24 \\ 6\overline{)14^24} \end{array}$
2. Determine the sign of the answer. The two numbers have different signs, so the answer is negative.	$144 \div (-6) = -24$
d. 1. Complete the division as if both numbers were positive numbers.	d. $\begin{array}{r} 15 \\ 16\overline{)24^80} \end{array}$
2. Determine the sign of the answer. The two numbers have the same sign, so the answer is positive.	$(-240) \div (-16) = 15$

1.4 Exercise

Students, these questions are even better in jacPLUS

 Receive immediate feedback and access sample responses

 Access additional questions

 Track your results and progress

Find all this and MORE in jacPLUS ▶

1. **WE6** Evaluate each of the following.
   a. $(-54) \div 9$
   b. $10 \div (-2)$
   c. $(-8) \div (-2)$
   d. $(-5) \div (-1)$
   e. $99 \div (-11)$
   f. $0 \div (-9)$

2. Evaluate each of the following.
   a. $42 \div (-3)$
   b. $(-130) \div 5$
   c. $(-56) \div (-8)$
   d. $(+88) \div (-4)$
   e. $(-66) \div (-6)$
   f. $168 \div (-8)$

3. Evaluate each of the following.
   a. $\dfrac{-132}{-12}$
   b. $\dfrac{-16}{4}$
   c. $\dfrac{-40}{-8}$
   d. $\dfrac{28}{-7}$
   e. $\dfrac{-250}{-50}$
   f. $\dfrac{144}{-12}$

4. Evaluate the following.
   a. $184 \div (-8)$
   b. $(-189) \div 9$
   c. $(-161) \div (-7)$
   d. $(-132) \div (-2)$
   e. $(-204) \div 6$
   f. $1080 \div (-9)$

5. Evaluate the following.
   a. $216 \div (-12)$
   b. $(-345) \div 15$
   c. $(-1536) \div (-24)$
   d. $(-1764) \div (-49)$
   e. $4096 \div 64$
   f. $(-2695) \div 55$

6. Write three different division statements, each of which has an answer of -6.

7. Copy and complete the following by placing the correct integer in the box.
   a. $(-36) \div \square = -9$
   b. $(-72) \div \square = -9$
   c. $72 \div \square = -36$
   d. $(-24) \div \square = 4$
   e. $\square \div 6 = -5$
   f. $\square \div 4 = -12$

8. Calculate the value of each of the following by working from left to right.
   a. $(-48) \div 6 \div (-4)$
   b. $(-240) \div 12 \div (-5)$
   c. $400 \div (-5) \div 8 \div (-2)$

9. Copy and complete these tables.

a.

×			− 6	+ 8
			18	
−10		−40		
	10		30	
−7				−56

b.

×			− 9	
6	30			−42
		36		
	−55		99	
		− 6	−18	

10. Copy and complete these tables. Divide the number on the top by the number on the side.

a.

÷	4	−10	12	−8
−2				
7				
−3				
−10				

b.

÷				−4
		−2		
−8	−4	3		
6			−6	
				1

11. Given $\dfrac{(-1)^n}{(-1)^m}$, state the values of m and n that would make this fraction positive.

12. Ross played in a recent golf tournament. He scored $-3, +1, -4, +2$ compared to par. Calculate his average score compared to par.
    (*Hint:* To calculate the average score, divide the sum of the scores by the number of scores.)

1.5 Combining operations on integers

LEARNING INTENTION

At the end of this subtopic you should be able to:
- apply the order of operations to solve an equation
- check the reasonableness of answers.

1.5.1 Order of operations

There are rules that determine which parts of an equation you need to calculate first, when the equation contains multiple operations. The term BIDMAS is used to remember the correct order in which to complete operations within an equation.

The order of operations

BIDMAS helps us to remember the correct order in which we should perform the various operations, working from left to right.

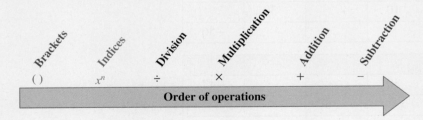

Order	Operations	What does it mean?
First	Brackets	Calculate any parts of the expression that are shown in brackets. For example: $(3 + 1) \times 4 = 4 \times 4$
Second	Indices	Multiply or divide out any indices. Indices are: • powers (3^2), where a number is multiplied by itself $3^2 = 3 \times 3$ $= 9$ $3^3 = 3 \times 3 \times 3$ $= 27$ • roots ($\sqrt{9}$), the opposite of a power. $\sqrt{9} = 3$ $\sqrt[3]{27} = 3$
Third	Division and Multiplication	Calculate any parts of the expression that involve division or multiplication. If the expression contains both multiplication and division, start from the left and work across to the right. For example: $3 \times 4 + 36 \div 6 = 12 + 6$ $= 18$
Last	Addition and Subtraction	Calculate any parts of the expression that involve addition or subtraction. If the expression contains both addition and subtraction, start from the left and work across to the right. For example (remember division comes before addition or subtraction): $4 + 24 \div 4 - 2 = 4 + 6 - 2$ $= 10 - 2$ $= 8$

tlvd-3553

WORKED EXAMPLE 7 Applying the order of operations

Calculate the value of each of the following.

a. $23 - 6 \times 4$

b. $(12 - 8) + 5^2 - (10 + 2^2)$

c. $\dfrac{3(4 + 8) + 4}{4 + 2(3^2 - 1)}$

THINK

a. 1. Apply BIDMAS to determine the first step (in this instance multiplication).

2. Complete the next step in the calculation (in this instance subtraction) and write the answer.

b. 1. Apply BIDMAS to determine the first step (perform the calculations in brackets first, then remove the brackets).

2. Complete the next steps in the calculation (resolve the powers, then carry out addition and subtraction from left to right).

3. Complete the final step and write the answer.

WRITE

a. $23 - 6 \times 4$

$= 23 - 24$

$= -1$

b. $(12 - 8) + 5^2 - (10 + 2^2)$

$= 4 + 5^2 - (10 + 4)$

$= 4 + 5^2 - 14$

$= 4 + 25 - 14$

$= 29 - 14$

$= 15$

c. 1. Apply BIDMAS to determine the first step (brackets).

   2. Complete the next step in the calculation (multiplication).

   3. Perform the calculations on the numerator and denominator separately.

   4. Complete the division and write the answer.

c.
$$\frac{3(4+8)+4}{4+2(3^2-1)} = \frac{3\times 12+4}{4+2(9-1)}$$
$$= \frac{3\times 12+4}{4+2\times 8}$$
$$= \frac{36+4}{4+16}$$
$$= \frac{40}{20}$$
$$= 2$$

1.5.2 Reasonableness

When calculating mathematical answers, it is always important to understand the question so you have an idea of what a reasonable answer would be. Checking the reasonableness of answers can indicate possible mistakes in your working.

WORKED EXAMPLE 8 Checking for reasonableness of an answer

Cathy goes shopping to purchase an outfit for a hike. She buys a $145 pair of waterproof pants and a $180 jacket, and gets $30 off the combined price by purchasing the two together. She also buys a pair of boots for $120 and a pair of socks for $15.
Calculate how much Cathy spent in total and check your answer for reasonableness.

THINK

1. Read the question carefully to understand what it is about.

2. Have an idea of what sort of answer you expect.

3. Write a mathematical expression to calculate the total amount that Cathy spent, including the discount.

4. Check the answer for reasonableness.

WRITE

The question asks you to add up Cathy's total shopping bill including the discount.

Cathy purchased three items, each over $100 dollars, so the answer should be over $300.

Cathy spent $= \$145 + \$180 - \$30 + \$120 + \$15$
           $= \$430$

This answer is above $300, so it seems reasonable.

1.5 Exercise

1. **WE7** Calculate the values of the following.
   a. $(-4) \times 2 + 1$
   b. $8 \div (2-4) + 4$
   c. $9 \times (8-3)$
   d. $(-3) - 40 \div 8 + 2$
   e. $4 + 12 \times (-5)$
   f. $(-5) \times 12 + 2$

2. Calculate the values of the following.
   a. $12 - 6 \div 3$
   b. $45 \div (27 \div (-3))$
   c. $(17 - +7) \div (-5)$
   d. $-12 + 8 \times 7$
   e. $100 \div ((-50) \div (-2)) + 10$
   f. $9 + \dfrac{24}{-6} \times 3$

3. Calculate the values of the following.
   a. $(-7) + 4 \times -4$
   b. $((-63) \div (-7)) \times (-3) + (-2)$
   c. $(-5)^2 - 3 \times (-5)$
   d. $(-6) \times (-8) - (3 + (-6)^2)$
   e. $52 \div ((-9) - 4) - 8$
   f. $-6 - 64 \div (-16) + 8$

4. **WE8** Kyle went to a sports shop to get some clothes for the gym. He purchased a $120 tracksuit and a pair of $150 runners as a package, and got $35 off the combined price. Kyle also purchased a pair of running shorts for $30 and a singlet top for $25.
   Determine how much Kyle spent at the sports shop. Check your answer for reasonableness.

5. Bob goes shopping for some party food. He buys 6 packets of chips at $2.50 each and 2 boxes of soft drink at $8.50 each. Since he purchases 2 boxes of soft drink, he gets a $2 discount. He also buys 5 packets of biscuits at $1 each and 3 dips at $2 each.
   Determine how much Bob spends in total and check your answer for reasonableness.

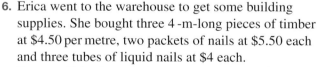

6. Erica went to the warehouse to get some building supplies. She bought three 4-m-long pieces of timber at $4.50 per metre, two packets of nails at $5.50 each and three tubes of liquid nails at $4 each.
   Because Erica is a regular customer, she received a $15 discount off the total purchase price.
   Determine how much Erica spent in total and check your answer for reasonableness.

7. Taiki headed north on a bike ride, initially travelling 25 km. He then turned around and travelled 15 km south before stopping for a drink.
After his drink, Taiki continued to ride south for another 20 km, before again turning around and travelling north for a further 25 km.

   a. Calculate how many kilometres Taiki covered on his ride.

   b. Determine how far north he finished from where he started.

8. Students were given the following question to evaluate.

$$4 + 8 \div (-2)^2 - 7 \times 2$$

   a. A number of different answers were obtained, including $-8, -12$ and -17. Determine which one of these is correct.

   b. Using only brackets, change the question in two ways so that the other answers would be correct.

9. Calculate the number required to make the following equation true:

$$13 \times (15 - 1) = 180 + \square$$

10. Calculate the number required to make the following equation true:

$$\left(10^2 + 12 \div 3\right) \div (-8) = \square$$

1.6 Review

1.6.1 Summary

doc-38042

Hey students! Now that it's time to revise this topic, go online to:

Access the topic summary

Review your results

Watch teacher-led videos

Practise questions with immediate feedback

Find all this and **MORE** in jacPLUS

1.6 Exercise

Multiple choice

1. **MC** State which of the following is an integer.

   A. -1.3

   B. $\dfrac{3}{4}$

   C. 23

   D. 0.25

   E. 1.234

2. **MC** State which of the following is incorrect.

   A. $5 > 1$

   B. $-5 < -1$

   C. $-25 < 0$

   D. $-25 > 1$

   E. $25 < 1$

3. **MC** The integers between -11 and -7 are

   A. $-10, -9, -8$

   B. $10, 9, 8$

   C. $-12, -13, -14$

   D. $-10.5, -9.5, -8.5$

   E. $-8\dfrac{1}{2}, -9\dfrac{1}{2}, -11\dfrac{1}{2}$

4. **MC** Arrange the following from lowest to highest: $-12, 34, 0, -3, 7$

   A. $0, -3, 7, -12, 34$

   B. $-12, 34, 0, -3, 7$

   C. $-3, 7, -12, 34, 0$

   D. $34, 7, 0, -3, -12$

   E. $-12, -3, 0, 7, 34$

5. **MC** The value of the expression $17 + 3 \times 7$ is

   A. 27

   B. 140

   C. 38

   D. 28

   E. 357

6. **MC** The value of the expression $-3 \times (5 - (-10)) + 50$ is

   A. 95

   B. 5

   C. 65

   D. -5

   E. 10

7. **MC** State which of the following statements is True.

   A. Multiplying an even number of negative numbers together gives a negative answer.
   B. The only square root of 25 is $+5$.
   C. Subtracting a positive number from another positive number always gives a negative number.
   D. Adding two negative numbers together gives a positive answer.
   E. Dividing a negative number by another negative number gives a positive answer.

8. **MC** Determine which of the following has an answer of -4.

    A. $2+2$ **B.** $2-(+2)$ **C.** $(-2)+2$

    D. $(-2)+(-2)$ **E.** 2^2

9. **MC** Determine which of the following has an answer of -3.

    A. $\dfrac{-24}{-8}$ **B.** $\dfrac{-8}{24}$ **C.** $\dfrac{27}{-9}$

    D. $\dfrac{-27}{-9}$ **E.** $\dfrac{-30}{-10}$

10. **MC** Determine which of the following has an answer of $+36$.

    A. $(-3)\times(-12)$ **B.** $(-9)\times(+4)$ **C.** $(-3)\times(-6)$

    D. $(+6)\times(-6)$ **E.** $(-3)\times(-11)$

Short answer

11. Calculate the following.

    a. $(-7)+(-13)$ **b.** $23-(-18)$ **c.** $(-4)+(-5)-(+6)$

    d. $(-2)-(-7)+(-5)$

12. Calculate the following.

    a. $(-4)\times(-9)$ **b.** $(-7)\times+8$ **c.** $\sqrt{81}$

    d. $-(-3)^3$

13. Calculate the following.

    a. $(-24)\div(+6)$ **b.** $72\div(-12)$ **c.** $\dfrac{+56}{-7}$ **d.** $\dfrac{-144}{-12}$

14. Calculate the following.

    a. $36\div9\times8-12$ **b.** $45-15\div3$ **c.** $30-10\times2+7$ **d.** $64-(16+8)+26$

15. Fill in the missing number.

    a. $13-\square=18$ **b.** $(-23)+\square=7$ **c.** $\square-4-(-5)=3$ **d.** $(-32)-\square=-18$

Extended response

16. Scott bought some pipe to complete the downpipes on his new carport. He purchased 34.2 m lengths at $5.15 per metre and 23.6 m lengths at $6.75 per metre. He also purchased some 5 fittings at $1.25 each. Calculate how much Scott spent on the downpipes.

17. Yi Rong is saving for her end-of-year holiday with her school friends. Her parents said if she did work around the house they would help contribute to the holiday expenses. They agreed to pay her $25 a week for doing the dishes and $15 a week for doing the washing; for extra incentive, they said they would double her pay if she did each of the jobs for the next 4 weeks without missing a day.

Assuming Yi Rong didn't miss a day, calculate how much money she earned in the 4 weeks.

18. On a test, each correct answer scores 5 points, each incorrect answer scores −2 points and each unanswered question scores 0 points. Calculate the following students' test scores.

 a. Student 1: 17 correct answers, 2 incorrect answers and 1 question unanswered.
 b. Student 2: 13 correct answers, 5 incorrect answers and 2 questions unanswered.

19. Answer the following questions.

 a. You have $75 and you spend $15 on soft drink and another $8 on chips. A friend gives you $6 to buy lunch, which cost $16 for the two of you. You decide to also buy an ice cream for $3. Calculate how much money you have left.
 b. Two friends saved $895 together for their holiday. They spend $350 on accommodation and then $125 each to go whale watching. Due to bad weather, the trip was cut short, so they got back $50 each. They then decided to go out to dinner and together spent $148. After the holiday, they split the remaining money in half. Calculate how much each friend got back.

20. Write the integers from −6 to +2 in the circles on this figure so that each line of three circles has each of the following totals.

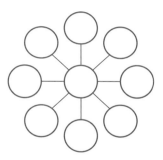

 a. −6   b. −3   c. −9

Answers

Topic 1 Integers

1.2 Adding and subtracting integers

1.2 Exercise

1. $5, -2, 212, -33$
2. a. -3 b. -5 c. 2
   d. -11 e. 30 f. 13
3. a. 4 b. 1 c. -1
   d. 0 e. 5 f. 2
4. a. -10 b. 15 c. -8
   d. -1 e. 39 f. -46
5. a. -5 b. -10 c. 6
   d. 9 e. -11 f. -37
6. a. -49 b. 35 c. -32
   d. -12 e. 25 f. 172
7. a. 14 b. 31 c. -18
   d. 9 e. 25 f. 0
8. a. 7 b. 4 c. 12
   d. -5 e. 4 f. 5
9. a. True   b. False   c. False
   d. True   e. False   f. False
10. a. -4 and -4 b. -2 and -2 c. -10 and -10
11. a. -7 and -7 b. -11 and -11 c. -13 and -13
12. 60 m to the right of where she started.
13. a. $39\,°C$ b. $159\,°C$ c. $198\,°C$
14. $7\,°C$

1.3 Multiplying integers

1.3 Exercise

1. a. -15 b. -21 c. 30
   d. -20 e. 40 f. -56
2. a. -250 b. -1250 c. -54
   d. -63 e. 77 f. -500
3. a. -300 b. 60 c. -192
   d. -30 e. -120 f. 36
4. a. -7 b. 6 c. -2
   d. -6 e. -4 f. -7
5. a. -9 b. 6 c. -12
   d. -8 e. 9 f. -12
6. a. -8 b. 9 c. 16
   d. 81 e. -32 f. 81
7. a. 16 b. -125 c. 256
   d. 625 e. -216 f. -512
8. a. Positive   b. Negative
9. a. 11 b. 10 c. 6
   d. 4 e. 3 f. 12
10. a. $2 \times 2 \times (-4) = -16$
    $(-4) \times 2 \times 2 = -16$
    $2 \times (-4) \times 2 = -16$
    There are more options.

b. $3 \times 2 \times 4 = 24$
$3 \times (-2) \times (-4) = 24$
$(-3) \times (-2) \times 4 = 24$
There are more options.

11. a. Positive   b. Negative   c. Positive
    d. Positive
12. When multiplying by 1, the number stays the same, so when multiplying by -1, the number becomes negative.
13. The answer will be negative, since there are three negative numbers (odd number).
14. a. Positive   b. Negative
15. One option could be 20 correct answers, 7 incorrect answers and 3 unanswered.
$$(20 \times 2) + (7 \times (-1)) + (3 \times 0) = 40 + (-7) + 0$$
$$= 40 - 7$$
$$= 33$$

1.4 Dividing integers

1.4 Exercise

1. a. -6 b. -5 c. 4
   d. 5 e. -9 f. 0
2. a. -14 b. -26 c. 7
   d. -22 e. 11 f. -21
3. a. 11 b. -4 c. 5
   d. -4 e. 5 f. -12
4. a. -23 b. -21 c. 23
   d. 66 e. -34 f. -120
5. a. -18 b. -23 c. 64
   d. 36 e. 64 f. -49
6. There are many answers. Sample response:
   $36 \div (-6) = -6$
   $(-54) \div 9 = -6$
   $72 \div (-12) = -6$
7. a. 4 b. 8 c. -2
   d. -6 e. -30 f. -48
8. a. 2 b. 4 c. 5
9. a.

×	−2	4	−6	+8
3	−6	12−	−18	24
−10	20	−40	60	−80
−5	10	−20	30	−40
−7	14	−28	42	−56

b.

×	5	−3	−9	−7
6	30	−18	−54	−42
−12	−60	36	108	84
−11	−55	33	99	77
2	10	−6	−18	−14

10. a.

÷	4	−10	12	−8
−2	−2	5	−6	4
7	$\frac{4}{7}$	$-\frac{10}{7}$	$\frac{12}{7}$	$-\frac{8}{7}$
−3	$-\frac{4}{3}$	$\frac{10}{3}$	−4	$\frac{8}{3}$
−10	$-\frac{2}{5}$	1	$-\frac{6}{5}$	$\frac{4}{5}$

b.

÷	32	−24	−36	−4
12	$\frac{8}{3}$	−2	−3	$-\frac{1}{3}$
−8	−4	3	$\frac{9}{2}$	$\frac{1}{2}$
6	$\frac{16}{3}$	−4	−6	$-\frac{2}{3}$
−4	−8	6	9	1

11. For the numerator (top) and the denominator (bottom), positive n and m need to be even numbers.
    For the numerator (top) and the denominator (bottom), negative n and m need to be odd numbers.

12. −1

1.5 Combining operations on integers

1.5 Exercise

1. a. −7   b. 0   c. 45
   d. −6   e. −56   f. −58

2. a. 10   b. −5   c. −2
   d. 44   e. 14   f. −3

3. a. −23   b. −29   c. 40
   d. 9   e. −12   f. 6

4. $290

5. $41
   The answer is reasonable, since chips ≈ $10, soft drink ≈ $20, biscuits ≈ $5 and dips ≈ $5.
   This totals $40, which reduces to $38 after the discount.
   This value is close to $41, so it is reasonable.

6. $62
   The answer is reasonable since timber ≈ $60, nails ≈ $10 and liquid nails ≈ $10.
   This totals $40, which reduces to $25 after the discount.
   This value is close to $26, so it is reasonable.

7. a. 85 km   b. 15 km north

8. a. −8
   b. $4 + 8 \div -(2)^2 - 7 \times 2 = -12$
      $(4 + 8) \div -(2)^2 - 7 \times 2 = -17$

9. 2

10. −13

1.6 Review

1.6 Exercise

Multiple choice

1. C   2. D   3. A   4. E   5. C
6. B   7. E   8. D   9. C   10. A

Short answer

11. a. −20   b. 41   c. −15   d. 0
12. a. 36   b. −56   c. 9   d. 27
13. a. −4   b. −6   c. −8   d. 12
14. a. 20   b. 40   c. 17   d. 66
15. a. −5   b. 30   c. 2   d. −14

Extended response

16. $341.68
17. $320
18. a. 81   b. 55
19. a. $39   b. $123.50
20. a.

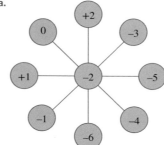

b.

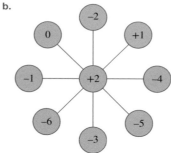

c.

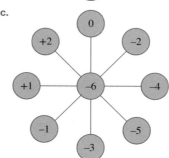

2 Fractions

Fully worked solutions for this topic are available online.

2.1 Overview

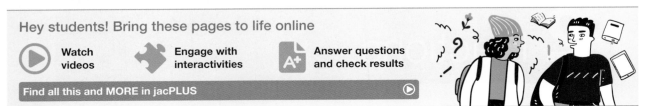
2.1.1 Introduction

Whole numbers cannot be used to measure everything. Instead, fractions and decimals must be used when you need to divide things into equal parts, such as an apple or the bill at a restaurant. How will you share anything equally among friends if you do not know about fractions? Fractions are a way of expressing parts of whole things, amounts or quantities. This concept helps us in cooking, shopping and telling the time. Fractions are used extensively in many different professions, including hospitality, finance, statistics and journalism. The shutter speed of a camera is calculated using fractions. A doctor prescribes different dosages for patients according to their weight using fractions.

KEY CONCEPTS

This topic covers the following key concepts from the VCE Mathematics Study Design:
- application of integers, fractions and decimals, their properties and related operations.
- estimation, approximation and reasonableness of calculations and results.

Note: Concepts shown in grey are covered in other topics.

Source: VCE Mathematics Study Design (2023–2027) extracts @VCAA; reproduced by permission.

2.2 Introducing fractions

LEARNING INTENTION

At the end of this subtopic you should be able to:
- understand the concept of fractions
- convert between equivalent fractions
- place fractions on number lines
- compare fractions by using fraction walls, number lines and comparison symbols.

2.2.1 Understanding fractions

Fractions are used to describe parts of a whole.

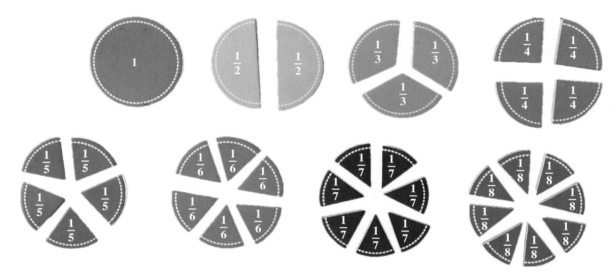

The **numerator**, or top number of the fraction, shows how many parts are required. The numerator must be an integer. The **denominator**, or bottom number, shows the number of parts into which the whole can be divided. The denominator must be an integer, but cannot be zero. The horizontal bar separating the numerator from the denominator is called the **vinculum**.

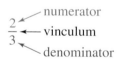

$\dfrac{2}{3}$ ← numerator ← vinculum ← denominator

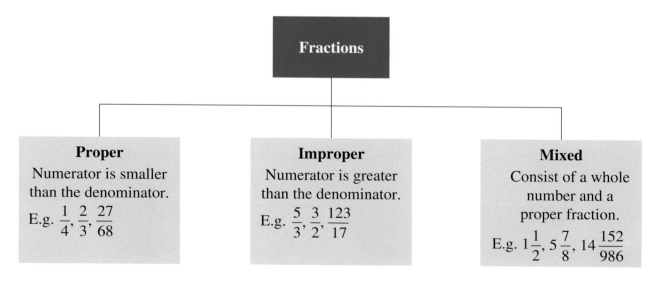

Fractions		
Proper	**Improper**	**Mixed**
Numerator is smaller than the denominator.	Numerator is greater than the denominator.	Consist of a whole number and a proper fraction.
E.g. $\dfrac{1}{4}, \dfrac{2}{3}, \dfrac{27}{68}$	E.g. $\dfrac{5}{3}, \dfrac{3}{2}, \dfrac{123}{17}$	E.g. $1\dfrac{1}{2}, 5\dfrac{7}{8}, 14\dfrac{152}{986}$

Understanding fractions

A fraction can be understood as numerator *out of* denominator.

For example: the fraction $\dfrac{6}{10}$ means 6 *out of* 10.

When the numerator is equal to the denominator, we have a 'whole', which is equivalent to 1.

For example: $\dfrac{10}{10}$ means 10 *out of* 10 and is one whole.

WORKED EXAMPLE 1 Identifying fractions of a quantity

a. Identify the fraction of the rectangle that has been shaded.
b. Express the number of unshaded squares as a fraction of the number of squares in total.

THINK

a. 1. Count how many equal parts the rectangle has been divided into.
   2. State how many parts are shaded.
   3. Write the number of shaded parts as a fraction of the total number of parts.

b. 1. Count the number of unshaded squares.
   2. State the number of squares in total.
   3. Write the number of unshaded squares as a fraction of the total number of squares.

WRITE

a. Total number of parts $= 8$

   5 parts are shaded.

   $\dfrac{5}{8}$ of the rectangle has been shaded.

b. There are 3 unshaded squares.
   There are 8 squares in total.

   $\dfrac{3}{8}$ of the squares are unshaded.

2.2.2 Equivalent fractions

Equivalent fractions are equal fractions.

For example, the fractions $\dfrac{1}{2}, \dfrac{2}{4}, \dfrac{5}{10}, \dfrac{8}{16}$ are all equivalent because they are all equal to a half.

Equivalent fractions for $\dfrac{1}{2}$ are shown in the following diagrams.

Note that the same portion $\dfrac{1}{2}$ of the circle, has been shaded in each case.

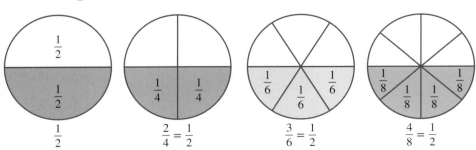

Equivalent fractions

Equivalent fractions are equal fractions. Equivalent fractions can be found by multiplying or dividing the numerator and denominator by the same number.

For example:

$$\frac{2}{3} = \frac{4}{6} \quad \text{(×2)}$$

$$\frac{10}{20} = \frac{1}{2} \quad \text{(÷10)}$$

WORKED EXAMPLE 2 Determining the multiplier

For the equivalent fractions $\dfrac{3}{5} = \dfrac{6}{10}$, determine the number (multiplier) that has been used to multiply the numerator and the denominator of the first fraction in order to get the second fraction.

THINK	WRITE
1. What number is 3 multiplied by to equal 6? $(3 \times \_ = 6)$ What number is 5 multiplied by to equal 10? $(5 \times \_ = 10)$	$\dfrac{3}{5} = \dfrac{3 \times 2}{5 \times 2} = \dfrac{6}{10}$
2. Write the answer.	The number used is 2.

---

tlvd-3555

WORKED EXAMPLE 3 Determining equivalent fractions

From the following list, determine which fractions are equivalent to $\dfrac{1}{3}$.

$$\frac{3}{9}, \frac{3}{6}, \frac{6}{18}, \frac{10}{15}, \frac{5}{15}, \frac{17}{51}$$

THINK	WRITE
1. Multiply the numerator and denominator of $\dfrac{1}{3}$ by the numerator of the first fraction in the list (3) to check whether the new fraction is in the list.	$\dfrac{1 \times 3}{3 \times 3} = \dfrac{3}{9}$
2. Multiply the numerator and denominator of $\dfrac{1}{3}$ by the next different numerator in the list (6) and check whether the new fraction is in the list.	$\dfrac{1 \times 6}{3 \times 6} = \dfrac{6}{18}$

3. Continue until all fractions have been considered.

$$\frac{1 \times 10}{3 \times 10} = \frac{10}{30}$$

$$\frac{1 \times 5}{3 \times 5} = \frac{5}{15}$$

$$\frac{1 \times 17}{3 \times 17} = \frac{17}{51}$$

4. Write the equivalent fractions.

From the given list, the equivalent fractions of $\frac{1}{3}$ are

$$\frac{3}{9}, \frac{6}{18}, \frac{5}{15} \text{ and } \frac{17}{51}.$$

WORKED EXAMPLE 4 Writing a sequence of equivalent fractions

Write the sequence of the first three equivalent fractions for $\frac{3}{4}$.

THINK

1. Write the first three equivalent fractions in the sequence by multiplying both the numerator and denominator by 2, 3 and 4.

2. Write the equivalent fractions.

3. Write the answer.

WRITE

$$\frac{3}{4} = \frac{3 \times 2}{4 \times 2} = \frac{6}{8}$$

$$= \frac{3 \times 3}{4 \times 3} = \frac{9}{12}$$

$$= \frac{3 \times 4}{4 \times 4} = \frac{12}{16}$$

$$\frac{3}{4} = \frac{6}{8} = \frac{9}{12} = \frac{12}{16}$$

The first three equivalent fractions of $\frac{3}{4}$ are $\frac{6}{8}, \frac{9}{12}$ and $\frac{12}{16}$.

2.2.3 Fractions on a number line

The space between each whole number on a number line can be divided into equal parts. The number of parts is equal to the denominator of the fraction, as shown.

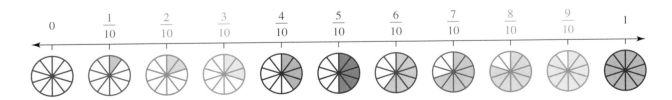

WORKED EXAMPLE 5 Placing fractions on a number line

Show the positions of $\dfrac{5}{6}$ and $1\dfrac{1}{6}$ on a number line.

THINK

1. Draw a number line from 0 to 2.

2. The denominator of both fractions is 6. Divide the sections between 0 and 1 and between 1 and 2 into 6 equal parts each.

3. To place $\dfrac{5}{6}$, count 5 marks from 0. To place $1\dfrac{1}{6}$, count 1 mark from the whole number 1.

WRITE/DRAW

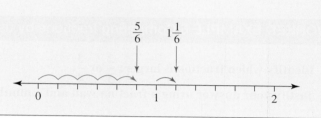

When choosing a scale for your number lines, take into account the number of equal parts each whole number has been divided into. For example, when plotting thirds, it is easier to use a scale of 3 cm than of 1 cm for each whole number (each part).

2.2.4 Comparing fractions by using a fraction wall and number line

Fraction walls, or number lines, are useful tools when comparing fractions. The following figure shows an example of a fraction wall.

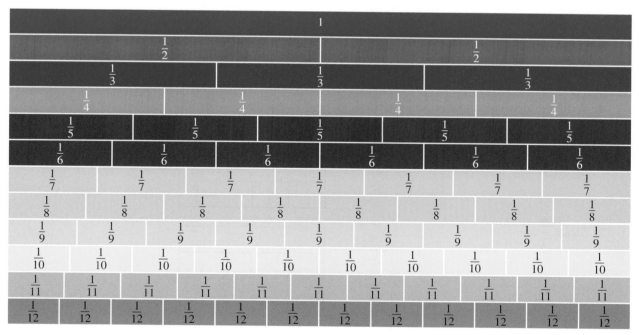

With a diagram such as a fraction wall or a number line, we can see that $\dfrac{1}{4}$ is less than $\dfrac{1}{3}$. That is, if we divided a block into 4 equal parts, each part would be smaller than if we divided it into 3 equal parts.

If the numerators are the same, the smaller fraction is the one with the larger denominator.

For example, $\dfrac{1}{7}$ is less than $\dfrac{1}{6}$, and $\dfrac{3}{10}$ is less than $\dfrac{3}{7}$.

Comparing fractions

If the fractions you wish to compare do not have the same denominator:
- use equivalent fractions to convert both fractions so that they have the same denominator
- use the lowest common multiple of the two denominators (which is the best choice for the denominator).

Once the fractions have the same denominator, compare the numerators. The fraction with the larger numerator is bigger.

WORKED EXAMPLE 6 Comparing fractions by using a fraction wall and number line

a. Identify which fraction is larger: $\dfrac{2}{3}$ or $\dfrac{3}{4}$.

b. Justify your answer using a fraction wall and a number line.

THINK

a. 1. Determine the lowest common multiple of the denominators. First, list the multiples of 3 and 4. Identify the lowest number that is common to both lists.

2. Write each fraction as an equivalent fraction using the lowest common multiple (12) as the denominator.

3. Decide which is larger by comparing the numerators of the equivalent fractions.

4. Write the answer using words.

b. 1. Create a fraction wall showing thirds and quarters.

2. Shade $\dfrac{2}{3}$ and $\dfrac{3}{4}$.

3. Compare the lengths of the shaded areas to compare the fractions. Answer the question using words.

WRITE/DRAW

a. Multiples of 3 are 3, 6, 9, ⑫, 15, 18, ...
Multiples of 4 are 4, 8, ⑫, 16, ...
The lowest common multiple is 12.

$$\frac{2}{3} = \frac{2 \times 4}{3 \times 4} \quad \text{and} \quad \frac{3}{4} = \frac{3 \times 3}{4 \times 3}$$

$$= \frac{8}{12} \qquad\qquad = \frac{9}{12}$$

$\dfrac{8}{12} = \dfrac{2}{3}$ is less than $\dfrac{9}{12} = \dfrac{3}{4}$.

$\dfrac{3}{4}$ is larger than $\dfrac{2}{3}$.

b.

1		
$\frac{1}{3}$	$\frac{1}{3}$	$\frac{1}{3}$
$\frac{1}{4}$	$\frac{1}{4}$	$\frac{1}{4}$	$\frac{1}{4}$

1		
$\frac{1}{3}$	$\frac{1}{3}$	$\frac{1}{3}$
$\frac{1}{4}$	$\frac{1}{4}$	$\frac{1}{4}$	$\frac{1}{4}$

$\dfrac{2}{3}$ is less than $\dfrac{3}{4}$.

4. Draw a number line showing 0 to 1 in intervals of $\frac{1}{12}$ (found by using the lowest common multiple of the denominators).

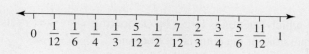

5. Mark $\frac{2}{3}$ and $\frac{3}{4}$.

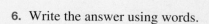

6. Write the answer using words.

$\frac{3}{4}$ is larger than $\frac{2}{3}$.

2.2.5 Comparing fractions by using comparison symbols

The symbol '>' means *is larger than* or *is greater than*.

The symbol '<' means *is smaller than* or *is less than*.

When we write $\frac{1}{2} < \frac{3}{4}$, it means that $\frac{1}{2}$ is less than $\frac{3}{4}$.

tlvd-3557

WORKED EXAMPLE 7 Comparing fractions by using comparison symbols

Insert the appropriate symbol, < or >, between each pair of fractions to make a true statement.

a. $\frac{5}{8}$ $\frac{6}{9}$ b. $\frac{5}{12}$ $\frac{4}{9}$

THINK

a. 1. Determine the lowest common multiple of the denominators.

2. Write each fraction as an equivalent fraction using the lowest common multiple (72) as the denominator.

3. Decide which fraction is larger by comparing the numerators of the equivalent fractions.

4. Write the answer.

b. 1. Determine the lowest common multiple of the denominators.

2. Write each fraction as an equivalent fraction using the lowest common multiple (36) as the denominator.

WRITE

a. Multiples of 9 are:
9, 18, 27, 36, 45, 54, 63, ⑦②, 80, ...

Multiples of 8 are:
8, 16, 24, 32, 40, 48, 56, 64, ⑦②, 80, 88, ...

The lowest common multiple is 72.

$\frac{5}{8} = \frac{5 \times 9}{8 \times 9} = \frac{45}{72}$ and $\frac{6}{9} = \frac{6 \times 8}{9 \times 8} = \frac{48}{72}$

$\frac{45}{72} = \frac{5}{8}$ is less than $\frac{48}{72} = \frac{6}{9}$.

$\frac{5}{8} < \frac{6}{9}$

b. Multiples of 12 are: 12, 24, ㉟, 48, ...

Multiples of 9 are: 9, 18, 27, ㉟, 45, ...

The lowest common multiple is 36.

$\frac{5}{12} = \frac{5 \times 3}{12 \times 3} = \frac{15}{36}$ and $\frac{4}{9} = \frac{4 \times 4}{9 \times 4} = \frac{16}{36}$

3. Decide which fraction is larger by comparing the numerators of the equivalent fractions.

$\frac{15}{36} = \frac{5}{12}$ is less than $\frac{16}{36} = \frac{4}{9}$.

4. Write the answer.

$\frac{5}{12} < \frac{4}{9}$

 Resources

 Interactivities Understanding fractions (int-3938)

Expressing one quantity as a fraction of another (int-3939)

Equivalent fractions (int-3940)

Fractions on a number line (int-3941)

Comparing fractions (int-3942)

2.2 Exercise

Students, these questions are even better in jacPLUS

Receive immediate feedback and access sample responses

Access additional questions

Track your results and progress

Find all this and MORE in jacPLUS

1. **WE1** Determine what fraction of each of the following rectangles has been shaded.

a.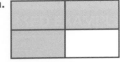

b.

c.

d.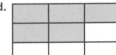

2. Determine what fraction of each of these flags is coloured red.

a.

b.

3. **a.** Determine what fraction of the total number of pieces of fruit shown below is made up of:

    i. bananas   **ii.** apples.

  b. If one finger of the KitKat is eaten, determine what fraction remains.

4. **WE2** Draw a number line from 0 to 2. Show the position of each of the following fractions on the line.

  a. $\dfrac{1}{4}$ **b.** $\dfrac{3}{4}$ **c.** $1\dfrac{1}{4}$ **d.** $\dfrac{7}{4}$

5. Draw a number line from 0 to 3. Show the position of each of the following fractions on the number line.

  a. $\dfrac{1}{3}$ **b.** $\dfrac{2}{3}$ **c.** $\dfrac{7}{3}$ **d.** $\dfrac{6}{3}$

6. **WE3** For the following equivalent fractions, determine the number that has been used to multiply both the numerator and the denominator.

  a. $\dfrac{2}{3} = \dfrac{6}{9}$ **b.** $\dfrac{1}{4} = \dfrac{2}{8}$ **c.** $\dfrac{2}{5} = \dfrac{6}{15}$ **d.** $\dfrac{3}{8} = \dfrac{12}{32}$

7. For the following equivalent fractions, determine the number that has been used to multiply both the numerator and the denominator.

  a. $\dfrac{5}{6} = \dfrac{25}{30}$ **b.** $\dfrac{7}{10} = \dfrac{28}{40}$ **c.** $\dfrac{7}{8} = \dfrac{63}{72}$ **d.** $\dfrac{5}{8} = \dfrac{55}{88}$

8. **WE4** From the list, identify which fractions are equivalent to the fraction marked in red.

  a. $\dfrac{2}{3}, \dfrac{7}{9}, \dfrac{20}{30}, \dfrac{5}{8}, \dfrac{12}{16}, \dfrac{14}{21}, \dfrac{40}{60}$ **b.** $\dfrac{4}{5}, \dfrac{12}{15}, \dfrac{15}{20}, \dfrac{36}{45}, \dfrac{16}{20}, \dfrac{28}{35}, \dfrac{80}{100}$

  c. $\dfrac{7}{8}, \dfrac{17}{18}, \dfrac{40}{45}, \dfrac{56}{64}, \dfrac{14}{18}, \dfrac{18}{19}, \dfrac{21}{24}$ **d.** $\dfrac{7}{10}, \dfrac{18}{25}, \dfrac{35}{50}, \dfrac{14}{21}, \dfrac{21}{30}, \dfrac{14}{20}, \dfrac{140}{200}$

9. Fill in the gaps.

  a. $\dfrac{2}{5} = \dfrac{12}{\square} = \dfrac{\square}{30} = \dfrac{\square}{45}$ **b.** $\dfrac{3}{4} = \dfrac{\square}{44} = \dfrac{15}{\square} = \dfrac{\square}{36}$ **c.** $\dfrac{2}{3} = \dfrac{\square}{27} = \dfrac{10}{\square} = \dfrac{8}{\square}$ **d.** $\dfrac{5}{6} = \dfrac{25}{\square} = \dfrac{\square}{42} = \dfrac{60}{\square}$

10. **WE5** Write a sequence of the first three equivalent fractions for each of the following.

  a. $\dfrac{4}{5}$ **b.** $\dfrac{3}{8}$ **c.** $\dfrac{4}{10}$ **d.** $\dfrac{1}{7}$ **e.** $\dfrac{5}{8}$ **f.** $\dfrac{2}{3}$

11. **WE6** Determine which fraction is bigger. Justify your answer using a fraction wall or number line.

  a. $\dfrac{2}{5}$ or $\dfrac{3}{5}$ **b.** $\dfrac{5}{8}$ or $\dfrac{7}{8}$ **c.** $\dfrac{1}{5}$ or $\dfrac{1}{6}$ **d.** $\dfrac{1}{8}$ or $\dfrac{1}{10}$ **e.** $\dfrac{1}{2}$ or $\dfrac{3}{10}$ **f.** $\dfrac{3}{4}$ or $\dfrac{3}{5}$

12. **WE7** Insert the appropriate symbol, $<$ or $>$, between each pair of fractions to make a true statement.

  a. $\dfrac{3}{8}$ $\dfrac{2}{7}$ **b.** $\dfrac{1}{4}$ $\dfrac{2}{9}$ **c.** $\dfrac{3}{10}$ $\dfrac{2}{5}$ **d.** $\dfrac{5}{12}$ $\dfrac{1}{2}$ **e.** $\dfrac{1}{8}$ $\dfrac{2}{15}$ **f.** $\dfrac{3}{7}$ $\dfrac{10}{21}$

13. **MC** a. Determine which of the following fractions is smaller than $\frac{5}{8}$.

A. $\frac{7}{8}$ B. $\frac{17}{24}$ C. $\frac{11}{12}$ D. $\frac{3}{5}$ E. $\frac{13}{16}$

    MC b. Determine which of the following fractions is equivalent to $\frac{3}{4}$.

A. $\frac{3}{8}$ B. $\frac{6}{8}$ C. $\frac{9}{16}$ D. $\frac{12}{24}$ E. $\frac{15}{24}$

14. Write the following fractions with the same denominator, and then write them in ascending order (from smallest to largest).

a. $\frac{3}{10}, \frac{1}{2}, \frac{1}{5}$ b. $\frac{3}{8}, \frac{1}{2}, \frac{1}{3}$ c. $\frac{2}{3}, \frac{4}{5}, \frac{6}{15}$ d. $\frac{3}{4}, \frac{2}{3}, \frac{7}{15}$

15. Four friends ordered a pizza each, but none could eat the whole pizza. Chloe ate $\frac{6}{8}$ of her seafood pizza. Layla ate $\frac{1}{2}$ of her Hawaiian pizza. Scott ate $\frac{3}{4}$ of his special pizza. Bryce ate $\frac{5}{6}$ of his vegetarian pizza.

    a. Draw the four pizzas and shade the amount that each person ate.
    b. Determine who ate the most pizza.
    c. Determine whether Chloe or Scott ate more pizza.

16. Stephanie puts four equally spaced points on a number line. The first point is at the $\frac{1}{4}$ mark and the last point is at the $\frac{5}{8}$ mark. Determine where she has placed the two points in between.

2.3 Simplifying fractions

LEARNING INTENTION

At the end of this subtopic you should be able to:
• simplify fractions and mixed numbers.

2.3.1 Simplifying fractions

Simplifying fractions means to write them in the simplest form possible.

For example, $\frac{4}{8}$ can be simplified to $\frac{2}{4}$, which can be further simplified to $\frac{1}{2}$. This is illustrated in the following diagram.

$$\frac{4}{8} \qquad = \qquad \frac{2}{4} \qquad = \qquad \frac{1}{2}$$
$$\text{(four-eighths)} \qquad \text{(two-quarters)} \qquad \text{(one-half)}$$

Generally, fractions are written in their simplest form. In other words, we reduce the fraction to its lowest equivalent form. A fraction is in simplest form when it cannot be simplified further. This occurs when the numerator and denominator share no common factor. To simplify a fraction, divide the numerator and the denominator by the same number. The highest common factor (HCF) of two numbers is the largest factor of both numbers.

Simplifying fractions

There are two ways to convert a fraction into its simplest form:
1. **Divide the numerator by the highest common factor (HCF) of the numerator and denominator.**
2. **Simplify repeatedly, by whichever common factor you like, until there are no common factors between the numerator and denominator.**

$$\frac{10}{50} = \frac{1}{5} \quad \div 10$$

WORKED EXAMPLE 8 Simplifying a fraction in its simplest form

Write $\dfrac{7}{14}$ in simplest form.

THINK

1. Write the fraction and determine the highest common factor (HCF), or the largest number that is a factor of both the numerator and the denominator.
2. Divide the numerator and denominator by this factor ($7 \div 7 = 1, 14 \div 7 = 2$).

3. Write the answer in simplest form.

WRITE

$\dfrac{7}{14}$; HCF $= 7$

$$\frac{7}{14} = \frac{1}{2} \quad \div 7$$

$= \dfrac{1}{2}$

2.3.2 Simplifying mixed numbers

A **mixed number** consists of a whole number and a **proper fraction**. To simplify a mixed number, leave the whole number part and simplify the fraction part.

tlvd-3558

WORKED EXAMPLE 9 Simplifying mixed numbers

Simplify $5\dfrac{28}{49}$.

THINK

1. Write the mixed number and determine the HCF of the numerator and denominator of the fraction part.

WRITE

$5\dfrac{28}{49}$; HCF $= 7$

2. Divide both numerator and denominator
   by this factor ($28 \div 7 = 4, 49 \div 7 = 7$).

$$\frac{2}{49} = \frac{4}{7}$$

$$\div 7$$

3. Write the answer as a mixed number in simplest form.

$$= 5\frac{4}{7}$$

 Resources

 Interactivities Simplifying fractions (int-3943)

2.3 Exercise

Students, these questions are even better in jacPLUS

Receive immediate
feedback and access
sample responses

Access
additional
questions

Track your
results and
progress

Find all this and MORE in jacPLUS ▶

1. **WE8** Write the following fractions in simplest form.

   a. $\dfrac{81}{90}$

   b. $\dfrac{8}{24}$

   c. $\dfrac{48}{64}$

   d. $\dfrac{21}{56}$

   e. $\dfrac{49}{70}$

   f. $\dfrac{63}{90}$

2. Write the following fractions in simplest form.

   a. $\dfrac{50}{80}$

   b. $\dfrac{22}{77}$

   c. $\dfrac{21}{63}$

   d. $\dfrac{18}{36}$

   e. $\dfrac{15}{40}$

   f. $\dfrac{12}{108}$

3. **WE9** Simplify.

   a. $2\dfrac{3}{9}$

   b. $5\dfrac{10}{20}$

   c. $4\dfrac{7}{42}$

   d. $8\dfrac{21}{49}$

4. Simplify.

   a. $6\dfrac{16}{48}$

   b. $3\dfrac{16}{64}$

   c. $1\dfrac{33}{99}$

   d. $2\dfrac{72}{144}$

5. **MC** Select the equivalent fraction to $\dfrac{3}{4}$.

   A. $\dfrac{2}{3}$

   B. $\dfrac{2}{6}$

   C. $\dfrac{3}{8}$

   D. $\dfrac{9}{12}$

   E. $\dfrac{13}{14}$

6. Annie's netball team scored 36 goals. Annie scored 30 of her team's goals. Calculate the fraction of the team's goals scored by Annie. Simplify the answer.

7. Noah's basketball team scored 56 points. Noah scored 24 of the team's points.

   a. Calculate the fraction of the team's points scored by Noah. Simplify the answer.

   b. Calculate the fraction of the team's points scored by the rest of the team. Simplify the answer.

8. Year 11 students at Springfield High School held a sausage sizzle to raise money for the Royal Children's Hospital. They raised a total of $1200 and drew up a table showing how much money each class raised.

11A	11B	11C	1D	1E
$300	$180	$350	$120	$250

Express as a simple fraction how much of the total was raised by each class.

9. The table shows the goal scorers of a football team.

Kyle	25 goals
Anthony	5 goals
Callum	15 goals
Harry	20 goals

   a. Determine how many goals were scored in total.

   b. For each player, record the number of goals as a fraction of the total number of goals scored. Where possible, reduce the fractions to their simplest forms.

10. The Geelong to Melbourne train runs on time three out of every five trains. The Melbourne to Geelong train runs on time four out of every seven trains. Determine which train is more reliable.

11. Answer the following questions.

    a. Draw three lines to divide the triangle shown into four equal parts.

    b. State what fraction of the original triangle is one of the new triangles.

    c. Halve each smaller triangle.

    d. State what fraction of the smaller triangle is one of these even smaller triangles.

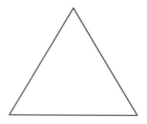

12. The fraction $\dfrac{17}{n}$ is not written in its simplest form. Explain what number the denominator could be.

2.4 Mixed numbers and improper fractions

LEARNING INTENTION

At the end of this subtopic you should be able to:
- convert between mixed numbers and improper fractions.

2.4.1 Converting improper fractions to mixed numbers

An **improper fraction** has a numerator larger than the denominator; for example, $\frac{3}{2}$. It can be changed to a **mixed number** by dividing the denominator into the numerator and writing the remainder over the same denominator.

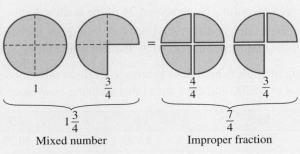

Mixed numbers and improper fractions

Mixed numbers and improper fractions are different ways of writing the same quantity.

$1 \qquad \frac{3}{4}$ = $\frac{4}{4} \qquad \frac{3}{4}$

$1\frac{3}{4}$ $\frac{7}{4}$

Mixed number   Improper fraction

---

tlvd-3559

WORKED EXAMPLE 10 Converting improper fractions to mixed numbers

Draw a diagram to show $\frac{7}{4}$ as parts of a circle, then write the improper fraction as a mixed number.

THINK	WRITE/DRAW
1. Draw a whole circle and divide it into the number of parts shown by the denominator.	This is 4 quarters or $\frac{4}{4}$.
2. Determine the number of parts left over and draw them.	 Extra $\frac{3}{4}$ is needed.
3. Write the improper fraction as a mixed number.	$\frac{7}{4} = 1\frac{3}{4}$

Express $\dfrac{13}{5}$ as a mixed number.

THINK

1. Write the improper fraction.

2. Determine how many times the denominator can be divided into the numerator and what the remainder is. The whole number is part of the answer and the remainder becomes the numerator of the fractional part with the same denominator as the original improper fraction.

3. Write the answer.

WRITE

$$\dfrac{13}{5}$$

$$= 13 \div 5$$
$$= 2 \text{ remainder } 3$$

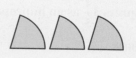

$$= 2\dfrac{3}{5}$$

2.4.2 Converting mixed numbers to improper fractions

A mixed number can be changed to an improper fraction by first multiplying the whole number by the denominator and then adding the numerator. The denominator stays the same. A diagram can help visualise the conversion.

Mixed numbers to improper fractions

$$2\dfrac{1}{4} = \dfrac{(2 \times 4) + 1}{4} = \dfrac{8 + 1}{4} = \dfrac{9}{4}$$

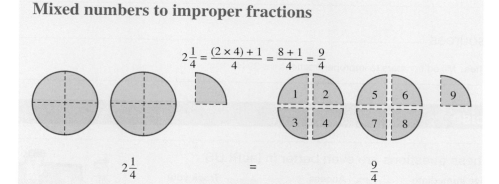

$$2\dfrac{1}{4} \qquad = \qquad \dfrac{9}{4}$$

tlvd-3560

Draw a diagram to illustrate $2\dfrac{1}{3}$ as pieces of a circle, then write $2\dfrac{1}{3}$ as an improper fraction.

THINK

1. Draw two whole circles and $\dfrac{1}{3}$ of a circle.

WRITE/DRAW

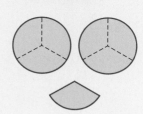

2. Divide the whole circles into thirds.

3. Count the number of thirds and write the mixed number as an improper fraction.

$$2\frac{1}{3} = \frac{7}{3}$$

WORKED EXAMPLE 13 Converting mixed numbers to improper fractions

Express $3\frac{4}{5}$ as an improper fraction.

THINK	WRITE
1. Write the mixed number.	$3\frac{4}{5}$
2. Multiply the whole number by the denominator, then add the numerator. The result becomes the numerator, and the denominator stays the same.	$= \frac{3 \times 5 + 4}{5}$
3. Evaluate the top line of the fraction.	$= \frac{15 + 4}{5}$
4. Write the answer.	$= \frac{19}{5}$

 Resources

 Interactivities Mixed numbers to improper fractions (int-3944)

2.4 Exercise

Students, these questions are even better in jacPLUS

 Receive immediate feedback and access sample responses

Access additional questions

Track your results and progress

Find all this and MORE in jacPLUS ▶

1. **WE10** Draw a diagram to show the following improper fractions as pieces of a circle, then write each improper fraction as a mixed number.

   a. $\frac{5}{2}$

   b. $\frac{4}{3}$

   c. $\frac{11}{6}$

   d. $\frac{13}{8}$

   e. $\frac{12}{4}$

   f. $\frac{7}{6}$

2. a. Eve ate $\frac{4}{3}$ of an orange. Express $\frac{4}{3}$ as a mixed number.

b. Finn ate $\frac{9}{4}$ of a pizza. Express $\frac{9}{4}$ as a mixed number.

3. `WE11` Express these improper fractions as mixed numbers.

a. $\frac{7}{3}$ **b.** $\frac{9}{5}$ **c.** $\frac{21}{6}$

d. $\frac{51}{7}$ **e.** $\frac{27}{5}$ **f.** $\frac{32}{9}$

4. Express these improper fractions as mixed numbers.

a. $\frac{10}{3}$ **b.** $\frac{33}{7}$ **c.** $\frac{100}{11}$ **d.** $\frac{22}{7}$ **e.** $\frac{35}{8}$ **f.** $\frac{69}{11}$

5. Change these improper fractions to mixed numbers.

a. $\frac{89}{6}$ **b.** $\frac{113}{10}$ **c.** $\frac{25}{12}$ **d.** $\frac{7}{5}$

6. Change these improper fractions to mixed numbers.

a. $\frac{57}{7}$ **b.** $\frac{97}{9}$ **c.** $\frac{65}{9}$ **d.** $\frac{13}{9}$

7. `MC` Select the equivalent fraction to $\frac{53}{8}$.

A. $5\frac{5}{8}$ **B.** $6\frac{5}{8}$ **C.** $7\frac{5}{8}$ **D.** $8\frac{5}{8}$ **E.** $9\frac{5}{8}$

8. `MC` Select the equivalent fraction to $\frac{77}{10}$.

A. $3\frac{7}{10}$ **B.** $7\frac{2}{5}$ **C.** $7\frac{7}{10}$ **D.** $7\frac{4}{5}$ **E.** $7\frac{4}{10}$

9. `MC` Select the equivalent fraction to $\frac{44}{5}$.

A. $8\frac{4}{5}$ **B.** $5\frac{5}{9}$ **C.** $9\frac{5}{9}$ **D.** $9\frac{3}{5}$ **E.** $9\frac{4}{9}$

10. **WE12** Draw a diagram to show the following mixed numbers as pieces of a circle and then write each one as an improper fraction.
   a. $1\frac{1}{8}$
   b. $3\frac{3}{4}$
   c. $2\frac{1}{6}$
   d. $1\frac{1}{6}$
   e. $5\frac{1}{4}$
   f. $1\frac{3}{5}$

11. **WE13** Express the following mixed numbers as improper fractions.
   a. $1\frac{2}{3}$
   b. $1\frac{3}{5}$
   c. $2\frac{2}{3}$
   d. $2\frac{3}{4}$

12. Express the following mixed numbers as improper fractions.
   a. $3\frac{5}{6}$
   b. $5\frac{4}{5}$
   c. $6\frac{7}{8}$
   d. $8\frac{7}{9}$

13. Change the following mixed numbers to improper fractions.
   a. $3\frac{1}{3}$
   b. $4\frac{2}{5}$
   c. $5\frac{4}{7}$
   d. $2\frac{8}{10}$

14. Change the following mixed numbers to improper fractions.
   a. $5\frac{7}{12}$
   b. $9\frac{3}{8}$
   c. $3\frac{2}{11}$
   d. $9\frac{5}{6}$

15. **MC** Select the fraction that is the same as $10\frac{3}{5}$.
   A. $\frac{53}{10}$
   B. $\frac{53}{3}$
   C. $\frac{50}{3}$
   D. $\frac{30}{5}$
   E. $\frac{53}{5}$

16. **MC** Select the fraction that is the same as $9\frac{4}{7}$.
   A. $\frac{67}{9}$
   B. $\frac{67}{7}$
   C. $\frac{67}{4}$
   D. $\frac{94}{7}$
   E. $\frac{97}{4}$

17. **MC** Select the fraction that is the same as $6\frac{4}{9}$.
   A. $\frac{58}{6}$
   B. $\frac{64}{9}$
   C. $\frac{58}{9}$
   D. $\frac{50}{9}$
   E. $\frac{69}{4}$

18. Colleen and Narelle were making some apple juice. Colleen used 13 quarters of an apple and Narelle used 17 quarters. Determine how many apples each person used.

19. The local bakery is well known for its lemon meringue pie. If the bakery sold 63 pieces of pie in one day and each piece is one-eighth of a pie, determine how many pies were sold.

20. Belinda supplies oranges for her son's football team. If 23 players all ate $\frac{1}{4}$ of an orange each, determine how many oranges were eaten.

21. Insert the appropriate < or > sign between each pair of fractions to make a true statement.

a. $\frac{5}{2}$ $3\frac{1}{2}$

b. $\frac{7}{3}$ $1\frac{2}{3}$

c. $\frac{12}{5}$ $2\frac{1}{5}$

d. $\frac{9}{4}$ $2\frac{3}{4}$

22. Insert the appropriate < or > sign between each pair of fractions to make a true statement.

a. $2\frac{1}{3}$ $\frac{8}{3}$

b. $5\frac{2}{3}$ $\frac{16}{3}$

c. $4\frac{1}{7}$ $\frac{30}{7}$

d. $6\frac{1}{5}$ $\frac{33}{5}$

23. Answer the following questions.

a. Jim made family-size pies for a family lunch. He divided each pie into 6 pieces and after lunch he noticed that $5\frac{5}{6}$ pies had been eaten. Calculate how many pieces had been eaten.

b. Marion supplied the drinks for lunch, and she calculated that she could get 15 drinks from each bottle. At the end of the lunch, $4\frac{13}{15}$ bottles had been used. Calculate how many drinks had been consumed.

24. Ava ordered a birthday cake for 20 people and took it to school to celebrate her birthday. Unexpectedly, other teachers joined her class, so 30 people ended up sharing the cake. Calculate what fraction of the original slice each person will receive if they get an equal amount each.

25. Answer the following questions.

a. Nasira made apple pies for her Grandma's 80th birthday party. She divided each pie into 6 pieces, and after the party she noted that $4\frac{1}{6}$ pies had been eaten. Calculate how many pieces were eaten.

b. Nasira's cousin Ria provided cordial for the same party and calculated that she could make 20 drinks from each bottle. At the end of the party, $3\frac{17}{20}$ bottles had been used. Calculate how many drinks were consumed.

2.5 Adding and subtracting fractions

2.5.1 Adding and subtracting proper fractions with equal denominators

Fractions can be added and subtracted if they have the same denominator. Addition and subtraction of fractions can be visualised using areas of shapes.

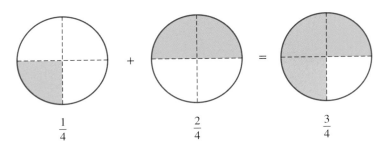

$$\frac{1}{4} \qquad\qquad + \qquad\qquad \frac{2}{4} \qquad\qquad = \qquad\qquad \frac{3}{4}$$

Adding and subtracting fractions with equal denominators

To add or subtract fractions of the same denominator, simply add or subtract the numerators.
Leave the denominator unchanged.

For example:
$$\frac{1}{5} + \frac{2}{5} = \frac{1+2}{5} = \frac{3}{5} \qquad\qquad \frac{5}{8} - \frac{2}{8} = \frac{5-2}{8} = \frac{3}{8}$$

WORKED EXAMPLE 14 Adding proper fractions with equal denominators

Evaluate $\dfrac{3}{8} + \dfrac{1}{8}$.

THINK	WRITE
1. Write the question.	$\dfrac{3}{8} + \dfrac{1}{8}$
2. Since the denominators are the same, add the numerators.	$= \dfrac{4}{8}$
3. Simplify the answer, if possible, by cancelling. Divide by the highest common factor $(4 \div 4 = 1, 8 \div 4 = 2)$.	$= \dfrac{4^1}{8^2}$ $= \dfrac{1}{2}$

2.5.2 Adding and subtracting fractions with different denominators

To add or subtract fractions with different denominators, convert them to equivalent fractions with the lowest common denominator (LCD). The lowest common denominator (LCD) is the lowest number that is a multiple of the denominators. This is also known as the lowest common multiple (LCM) of the denominators. Let us visualise the previous example using areas of circles. We can divide both circles into the same number of parts and then add the parts.

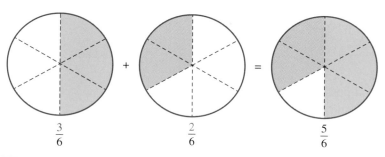

$$\frac{3}{6} \qquad \frac{2}{6} \qquad \frac{5}{6}$$

Adding and subtracting fractions with different denominators

To add or subtract fractions with different denominators, convert to equivalent fractions with the lowest common denominator (LCD), and then add the numerators.

For example:

$$\frac{1}{2} + \frac{1}{3}$$

The LCD of 2 and 3 is 6.

Add the intermediate step to the equation and calculate the answer:

$$\frac{1}{2} + \frac{1}{3} = \frac{3}{6} + \frac{2}{6} = \frac{3+2}{6} = \frac{5}{6}$$

tlvd-3561

WORKED EXAMPLE 15 Subtracting proper fractions with different denominators

Evaluate $\dfrac{5}{6} - \dfrac{1}{12}$, expressing the answer in simplest form.

THINK

1. List multiples of each denominator.

2. Identify the lowest common denominator.

3. Rewrite the fractions as equivalent fractions with 12 as the LCD.

WRITE

Multiples of 6 are $6, 12, \ldots$
Multiples of 12 are $12, \ldots$

LCD is 12.

$$\frac{5}{6} - \frac{1}{12} = \frac{5 \times 2}{6 \times 2} - \frac{1}{12} = \frac{10}{12} - \frac{1}{12}$$

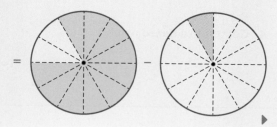

4. Subtract the numerators.

$$= \frac{9}{12}$$

5. Simplify the answer if possible by cancelling. Divide by the highest common factor ($9 \div 3 = 3$, $12 \div 3 = 4$).

$$= \frac{\cancel{9}^{3}}{\cancel{12}^{4}}$$

$$= \frac{3}{4}$$

WORKED EXAMPLE 16 Subtracting a fraction from a whole number

Evaluate $3 - \dfrac{1}{3}$, **expressing the answer as a mixed number.**

THINK

WRITE

1. Express the whole number as a fraction over 1.

$$3 - \frac{1}{3} = \frac{3}{1} - \frac{1}{3}$$

2. Equate the denominators.

$$= \frac{9}{3} - \frac{1}{3}$$

3. Perform the subtraction.

$$= \frac{8}{3}$$

4. Convert to a mixed number.

$$= 2\frac{2}{3}$$

WORKED EXAMPLE 17 Application of adding and subtracting fractions

Tahlia eats $\frac{2}{3}$ of her apple and her friend Sarah eats $\frac{1}{6}$ of hers. Calculate how much of the apples they ate together.

THINK	WRITE
1. Add the fractions to calculate the total fraction of apples eaten.	$\dfrac{2}{3} + \dfrac{1}{6} = \dfrac{4}{6} + \dfrac{1}{6}$ 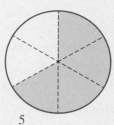 + 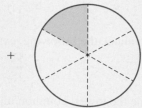 $= \dfrac{5}{6}$ 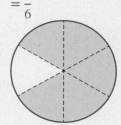
2. Write the answer.	$\dfrac{5}{6}$ of the apples were eaten by Tahlia and Sarah.

 Resources

▶ **Video eLesson** Addition and subtraction of fractions (eles-1866)

✦ **Interactivities** Addition and subtraction of proper fractions (int-3945)

2.5 Exercise

Students, these questions are even better in jacPLUS

 Receive immediate feedback and access sample responses

 Access additional questions

 Track your results and progress

Find all this and MORE in jacPLUS

1. **WE14** Evaluate:

   a. $\dfrac{1}{4} + \dfrac{2}{4}$ b. $\dfrac{2}{7} + \dfrac{3}{7}$ c. $\dfrac{2}{6} + \dfrac{3}{6}$ d. $\dfrac{4}{9} + \dfrac{1}{9}$

2. Evaluate:

   a. $\dfrac{4}{9} - \dfrac{2}{9}$ b. $\dfrac{3}{5} - \dfrac{1}{5}$ c. $\dfrac{16}{25} - \dfrac{12}{25}$ d. $\dfrac{13}{15} - \dfrac{6}{15}$

3. Evaluate the following, expressing the answers in simplest form.

   a. $\dfrac{7}{12} + \dfrac{3}{12}$ b. $\dfrac{3}{10} + \dfrac{2}{10}$ c. $\dfrac{15}{18} - \dfrac{3}{18}$ d. $\dfrac{7}{8} - \dfrac{5}{8}$

4. Evaluate the following, expressing the answers in simplest form.

   a. $\dfrac{8}{14} + \dfrac{6}{14}$ b. $\dfrac{13}{20} - \dfrac{7}{20}$ c. $\dfrac{21}{24} - \dfrac{3}{24}$ d. $\dfrac{6}{15} + \dfrac{4}{15}$

5. Evaluate the following, expressing the answers in simplest form.

   a. $\dfrac{1}{3} + \dfrac{1}{6}$ b. $\dfrac{1}{4} + \dfrac{2}{3}$ c. $\dfrac{1}{8} + \dfrac{3}{4}$ d. $\dfrac{1}{6} + \dfrac{1}{4}$

6. **WE15** Evaluate the following, expressing the answers in simplest form.

   a. $\dfrac{1}{2} - \dfrac{1}{4}$ b. $\dfrac{12}{15} - \dfrac{2}{3}$ c. $\dfrac{6}{10} - \dfrac{1}{4}$ d. $\dfrac{7}{11} - \dfrac{3}{5}$

7. Determine the lowest common multiple of each of the following pairs of numbers.

   a. 6 and 8   b. 4 and 6   c. 5 and 15   d. 8 and 3   e. 7 and 3   f. 5 and 6

8. Add or subtract these fractions.

   a. $\dfrac{11}{12} - \dfrac{3}{4}$ b. $\dfrac{13}{15} - \dfrac{1}{3}$ c. $\dfrac{1}{7} + \dfrac{1}{4}$ d. $\dfrac{2}{9} + \dfrac{1}{5}$ e. $\dfrac{3}{5} - \dfrac{3}{8}$ f. $\dfrac{5}{6} - \dfrac{3}{8}$

9. Answer the following by first identifying the lowest common denominator. Simplify your answer if necessary.

   a. $\dfrac{7}{8} - \dfrac{3}{5}$ b. $\dfrac{2}{3} - \dfrac{1}{13}$ c. $\dfrac{5}{12} + \dfrac{2}{9}$ d. $\dfrac{2}{10} - \dfrac{1}{30}$ e. $\dfrac{2}{3} - \dfrac{2}{9}$ f. $\dfrac{7}{8} - \dfrac{3}{5}$

10. **MC** The lowest common denominator of $\dfrac{1}{3}, \dfrac{1}{4}$ and $\dfrac{1}{6}$ is:

    A. 16   B. 24   C. 18   D. 12   E. 72

11. **WE16** Evaluate the following, expressing the answer in simplest form.

    a. $1 - \dfrac{1}{3}$ b. $2 - \dfrac{1}{4}$ c. $3 - \dfrac{1}{5}$ d. $4 - \dfrac{4}{5}$

12. Answer the following:

    a. $\dfrac{5}{6} - \dfrac{7}{18} - \dfrac{1}{9}$ b. $\dfrac{2}{3} - \dfrac{2}{6} - \dfrac{1}{4}$ c. $\dfrac{1}{2} + \dfrac{3}{4} - \dfrac{2}{3}$ d. $\dfrac{11}{12} - \dfrac{8}{15} - \dfrac{3}{20}$

13. Ross and Joanne took the family camping by the river. Ross collected $\dfrac{2}{3}$ of a bucket of water from the river, while Joanne collected $\dfrac{1}{4}$ of a bucket.

    a. Calculate how much of a bucket of water they collected together.
    b. Determine what fraction they still need to fill in order to fill the entire bucket.

14. **WE17** To get good crackling on the pork roast, Scott initially cooks the pork for $\dfrac{2}{3}$ of an hour on low heat and then turns it up to a high heat for $\dfrac{1}{2}$ of an hour.
    Calculate the total cooking time in hours.

15. Kyle, Finn and Dan play together in a basketball team. In their most recent game, Kyle scores $\frac{3}{8}$ of the team's score, and Finn scores $\frac{1}{4}$ and Dan scores $\frac{1}{6}$.

Calculate the fraction of the team's score that was scored by:
   a. Kyle and Finn
   b. Finn and Dan
   c. The three boys together.

16. A concrete mix requires cement, sand and stones. Darcy makes a concrete mix of $\frac{1}{2}$ stones and $\frac{1}{6}$ cement.

   a. Calculate the fraction of the concrete mix made of both stones and cement.
   b. Calculate the fraction of the concrete mix made of sand.

2.6 Multiplying fractions

LEARNING INTENTION

At the end of this subtopic you should be able to:
- multiply fractions by a whole number
- multiply fractions by another fraction
- calculate fractions of a fraction
- calculate fractions of an amount.

2.6.1 Multiplying fractions

Multiplication by a whole number can be thought of as repeated addition.

For example, $3 \times 4 = 4 + 4 + 4 = 12$.

The same principle holds when multiplying a fraction by a whole number.

For example, $3 \times \frac{1}{4} = \frac{1}{4} + \frac{1}{4} + \frac{1}{4} = \frac{3}{4}$.

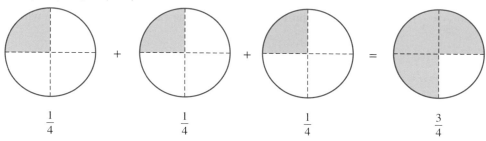

An efficient way of multiplying a fraction by a whole number is to multiply the numerator by the whole number.

Multiplying a fraction by a whole number

To multiply a fraction by a whole number, simply multiply the numerator by the whole number and keep the denominator the same. For example:

$$\frac{1}{7} \times 5 = \frac{1 \times 5}{7} = \frac{5}{7}$$

WORKED EXAMPLE 18 Multiplying a fraction by a whole number

Calculate $6 \times \dfrac{1}{3}$.

THINK	WRITE

1. Write the question.

$$6 \times \frac{1}{3}$$

2. Multiply by a whole number using repeated addition. Add $\dfrac{1}{3}$ six times.

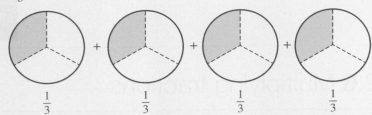

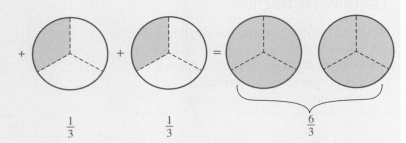

3. Alternatively, multiply the numerator by the whole number.

$$= \frac{6 \times 1}{3}$$

$$= \frac{6}{3}$$

4. Simplify and write the answer.

$$= 2$$

The multiplication $\dfrac{1}{3} \times \dfrac{1}{2}$ is the same as one-third of a half; this is the same as one-sixth of a whole unit, as shown here.

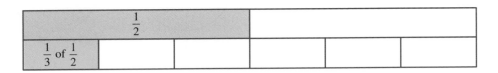

$$\frac{1}{3} \times \frac{1}{2} = \frac{1}{6} \text{ (of a whole unit)}$$

Multiplying a fraction by another fraction

To multiply a fraction by another fraction, simply multiply the numerators by the numerators and the denominators by the denominators. For example:

$$\frac{1}{3} \times \frac{2}{5} = \frac{1 \times 2}{3 \times 5} = \frac{2}{15}$$

Fractions don't need to have the same denominator to be multiplied together.

WORKED EXAMPLE 19 Multiplying a fraction by another fraction

Calculate $\frac{1}{7} \times \frac{2}{3}$.

THINK	WRITE
1. Write the question.	$\frac{1}{7} \times \frac{2}{3}$
2. Multiply the numerators and the denominators.	$= \frac{1 \times 2}{7 \times 3}$
3. Write the answer.	$= \frac{2}{21}$

Before being multiplied, fractions can be simplified by dividing both the numerator and denominator by their highest common factor. This is often called *cancelling*.

Applying cancelling before multiplying fractions

When multiplying fractions, cancelling can only occur:

- vertically: for example, $\dfrac{3^1}{24^8} \times \dfrac{5}{7} = \dfrac{1 \times 5}{8 \times 7}$

$$= \frac{5}{56}$$

- diagonally: for example, $\dfrac{2^1}{3} \times \dfrac{5}{6^3} = \dfrac{1 \times 5}{3 \times 3}$

$$= \frac{5}{9}$$

tlvd-3563

WORKED EXAMPLE 20 Multiplying fractions by applying cancelling

Calculate $\frac{3}{4} \times \frac{12}{15}$.

THINK	WRITE
1. Write the question.	$\frac{3}{4} \times \frac{12}{15}$

2. Simplify by dividing the numbers in the numerator and denominator by their highest common factor. The HCF of 3 and 15 is 3, and the HCF of 4 and 12 is 4.
Note: This may be done vertically or diagonally.

$$= \frac{\cancel{3}^{1} \times \cancel{12}^{3}}{\cancel{4}^{1} \times \cancel{15}^{5}}$$

3. Multiply the numerators, then the denominators, and simplify the answer if appropriate.

$$= \frac{1 \times 3}{1 \times 5}$$
$$= \frac{3}{5}$$

2.6.2 Using the word *of*

Sometimes the word *of* means 'multiply'. For example, $\frac{1}{2}$ of $\frac{1}{3}$ can be visualised as:

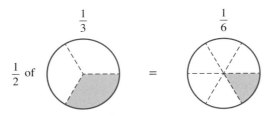

$$\text{Half} \quad \text{of} \quad \frac{1}{3} = \frac{1}{2} \times \frac{1}{3}$$
$$= \frac{1}{6}$$

A third of an hour, or $\frac{1}{3}$ of 60 minutes, can be written as $\frac{1}{3} \times \frac{60}{1} = 20$ minutes.

A diagram can also be used to display multiplication of a fraction by another fraction. The following model shows how to calculate $\frac{2}{3}$ of $\frac{4}{5}$.

The rectangle is divided vertically into fifths; $\frac{4}{5}$ is shaded blue.

The rectangle is then divided horizontally into thirds; $\frac{2}{3}$ of the blue rectangles are now shaded pink.
The pink shading represents $\frac{2}{3}$ of $\frac{4}{5}$.
$$\frac{2}{3} \times \frac{4}{5} = \frac{8}{15}$$

WORKED EXAMPLE 21 Calculating a fraction of a fraction

Calculate $\dfrac{2}{3}$ of $\dfrac{1}{4}$.

THINK	WRITE
1. Write the question.	$\dfrac{2}{3}$ of $\dfrac{1}{4}$

	$\dfrac{1}{4}$							
$\dfrac{2}{3}$ of $\dfrac{1}{4}$								

THINK	WRITE
2. Change 'of' to × and cancel if appropriate.	$= \dfrac{2^1}{3} \times \dfrac{1}{4^2}$
3. Perform the multiplication and write the answer.	$= \dfrac{1}{6}$

WORKED EXAMPLE 22 Calculating fractions of an amount

A 2-litre carton of pineapple juice is in the refrigerator. After coming home from school, Cara drank one-quarter of the juice, Bridget drank one-half of it and Phoenix drank one-sixth of it. Calculate how much pineapple juice, in litres, was drunk by each person.

THINK	WRITE
1. Write the fraction of the pineapple juice that Cara drank and simplify the answer.	Cara: $\dfrac{1}{4}$ of $2 = \dfrac{1}{4} \times 2$
2. Perform the multiplication.	$= \dfrac{1}{2\!\!\!/4} \times \dfrac{2^1}{1}$
	$= \dfrac{1}{2}$
3. Write the fraction of the juice that Bridget drank and simplify the answer.	Bridget: $\dfrac{1}{2}$ of $2 = \dfrac{1}{2} \times 2$
4. Perform the multiplication.	$= \dfrac{1}{{}^1\!2} \times \dfrac{2^1}{1}$
	$= 1$
5. Write the fraction of the juice that Phoenix drank and simplify the answer.	Phoenix: $\dfrac{1}{6}$ of $2 = \dfrac{1}{6} \times 2$
6. Perform the multiplication.	$= \dfrac{1}{{}^3\!6} \times \dfrac{2^1}{1}$
	$= \dfrac{1}{3}$
7. Write the answer in litres.	Cara drank half a litre, Bridget drank 1 litre and Phoenix drank a third of a litre of pineapple juice.

WORKED EXAMPLE 23 Calculating a fraction of a fraction of an amount

Before a road trip John completely fills up his car's fuel tank with 72 litres of petrol. On the first day of the trip, the car uses $\frac{3}{8}$ of the full tank. On the second day, the car uses $\frac{2}{5}$ of the remaining amount. Determine how much petrol was used on the second day.

THINK

1. Determine the fraction left after $\frac{3}{8}$ of the petrol is used on the first day.

2. Calculate the total fraction of the tank of petrol that has been used on day 2.

3. Calculate the amount of petrol used on day 2. Remember to write the unit.

WRITE

$1 - \frac{3}{8} = \frac{5}{8}$

$\frac{5}{8}$ of the tank remains after the first day.

$\frac{2}{5}$ of $\frac{5}{8}$ of a tank $= \frac{2 \times 5}{5 \times 8}$

$\qquad\qquad\qquad\qquad\quad = \frac{1}{4}$

A quarter of the full tank is used on day 2.

$\frac{1}{4} \times 72 = \frac{72}{4}$

$\qquad\qquad = 18$

18 litres of petrol were used on the second day.

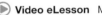 Resources

 Video eLesson Multiplication and division of fractions (eles-1867)

 Interactivities Repeated addition (int-2738)

Multiplication of rational numbers (int-3946)

Using the word *of* (int-3947)

2.6 Exercise

Students, these questions are even better in jacPLUS

 Receive immediate feedback and access sample responses

Access additional questions

Track your results and progress

Find all this and MORE in jacPLUS

1. **WE18** Calculate the following.

a. $\frac{2}{5} \times 5$

b. $\frac{3}{14} \times 7$

c. $\frac{15}{33} \times 11$

d. $\frac{7}{16} \times 4$

2. **WE19** Calculate the following, then check your answers with a calculator.

a. $\frac{1}{3} \times \frac{1}{4}$

b. $\frac{2}{3} \times \frac{1}{3}$

c. $\frac{2}{3} \times \frac{5}{9}$

d. $\frac{9}{13} \times \frac{1}{2}$

3. Calculate the following, then check your answers with a calculator.

   a. $\dfrac{3}{5} \times \dfrac{7}{11}$

   b. $\dfrac{11}{12} \times \dfrac{11}{12}$

   c. $\dfrac{5}{7} \times \dfrac{5}{9}$

   d. $\dfrac{7}{8} \times \dfrac{1}{3}$

4. Calculate the following, then check your answers with a calculator.

   a. $\dfrac{5}{9} \times \dfrac{9}{11}$

   b. $\dfrac{5}{11} \times \dfrac{5}{12}$

   c. $\dfrac{2}{3} \times \dfrac{6}{7}$

   d. $\dfrac{3}{7} \times \dfrac{4}{5}$

5. **WE20** Calculate the following, then check your answers with a calculator.

   a. $\dfrac{5}{10} \times \dfrac{3}{6}$

   b. $\dfrac{2}{5} \times \dfrac{1}{4}$

   c. $\dfrac{3}{9} \times \dfrac{5}{6}$

   d. $\dfrac{5}{8} \times \dfrac{4}{15}$

6. Calculate the following, then check your answers with a calculator.

   a. $\dfrac{20}{11} \times \dfrac{33}{40}$

   b. $\dfrac{32}{35} \times \dfrac{7}{8}$

   c. $\dfrac{18}{32} \times \dfrac{64}{72}$

   d. $\dfrac{42}{64} \times \dfrac{16}{70}$

7. Calculate the following, then check your answers with a calculator.

   a. $\dfrac{15}{54} \times \dfrac{36}{60}$

   b. $\dfrac{6}{42} \times \dfrac{14}{24}$

   c. $\dfrac{4}{7} \times \dfrac{14}{20}$

   d. $\dfrac{5}{25} \times \dfrac{12}{48}$

8. **WE21** Calculate each of the following.

   a. $\dfrac{1}{2}$ of $\dfrac{1}{5}$

   b. $\dfrac{3}{4}$ of $\dfrac{1}{3}$

   c. $\dfrac{5}{6}$ of $\dfrac{6}{8}$

   d. $\dfrac{1}{4}$ of $\dfrac{8}{9}$

9. Calculate each of the following.

   a. $\dfrac{7}{8}$ of 20

   b. $\dfrac{9}{10}$ of 70

   c. $\dfrac{3}{5}$ of 35

   d. $\dfrac{4}{5}$ of 30

10. **WE22** Peter's dad made 8 litres of cordial for his basketball friends. John drank $\dfrac{1}{4}$ of the cordial, Anthony drank $\dfrac{1}{8}$ and Will drank $\dfrac{1}{12}$. Calculate how much each person drank in litres.

11. Charlie and Kristy went to the beach on a nice summer's day. They were at the beach for 1 hour and 20 minutes. Charlie spent $\dfrac{1}{8}$ of the time in the water, and Kristy spent $\dfrac{1}{10}$ of the time in the water. Calculate in minutes how much time:

    a. Charlie spent in the water

    b. Kristy spent in the water.

12. For Easter Jessica got a box of chocolates that contained 48 chocolates. Jessica shared some of them with her friends. Emily had $\frac{1}{12}$, Ethan had $\frac{1}{6}$ and Erica had $\frac{1}{8}$ of the box. The remaining chocolates were eaten by Jessica and her family. Determine how many chocolates:

   a. Emily ate
   b. Ethan ate
   c. Erica ate
   d. Jessica and her family ate.

13. **WE23** A car's petrol tank holds 48 litres of fuel. The tank was full at the start of a trip.

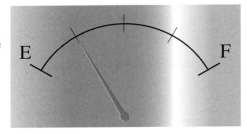

   a. By examining the gauge shown, determine what fraction of the tank of petrol has been used.
   b. Determine how many litres of petrol have been used.

14. If pies are cut into quarters, determine how many quarter pieces are in:

   a. 1 pie   b. 3 pies   c. 5 pies.

15. Ibrahim takes home $3600 a month. From this he spends a third on rent, a quarter on food and one-sixth on clothing. He tries to save money each month, so he puts one-half of what is left over into his savings account. Calculate how much money he puts into his savings account each month.

16. Bob the builder is building a house 18 m by 12 m. If the plans are $\frac{1}{150}$ of the actual size of the house, calculate, in cm:
   a. the length of the house on the plans
   b. the width of the house on the plans.

2.7 Dividing fractions

LEARNING INTENTION

At the end of this subtopic you should be able to:
• divide fractions.

2.7.1 Dividing fractions

To divide fractions, you need to know their reciprocal. The **reciprocal** of a fraction is found by turning the fraction upside down. So $\frac{3}{5}$ is the reciprocal of $\frac{5}{3}$.

A whole number can be written as a fraction by putting it over 1. Therefore, the reciprocal of a whole number is 1 over the number.

For example, since $4 = \dfrac{4}{1}$, the reciprocal of 4 is $\dfrac{1}{4}$.

WORKED EXAMPLE 24 Reciprocal of a proper fraction

Identify the reciprocal of $\dfrac{3}{4}$.

THINK	WRITE
Turn the fraction upside down and write the answer as a sentence.	The reciprocal of $\dfrac{3}{4}$ is $\dfrac{4}{3}$.

To determine the reciprocal of a mixed number, first express it as an improper fraction and then turn the improper fraction upside down.

WORKED EXAMPLE 25 Determining the reciprocal of a mixed number

Determine the reciprocal of $2\dfrac{3}{4}$.

THINK	WRITE
1. Write $2\dfrac{3}{4}$ as an improper fraction.	$2\dfrac{3}{4} = \dfrac{11}{4}$
2. Turn the improper fraction upside down and write the answer as a sentence.	The reciprocal of $2\dfrac{3}{4}$ is $\dfrac{4}{11}$.

Consider $10 \div 4$. This can be interpreted as 'How many groups of 4 are there in 10?'

$$10 \div 4 = 2\dfrac{1}{2} \text{ (groups of 4)}$$

← 10 ones

← Groups of 4 ones

Division of fractions follows the same logic.

For example, $3 \div \dfrac{1}{2}$ can be interpreted as 'How many halves are there in 3?'

$$3 \div \dfrac{1}{2} = 6 \left(\text{groups of } \dfrac{1}{2}\right)$$

1	1	1

← 3 ones

$\frac{1}{2}$	$\frac{1}{2}$	$\frac{1}{2}$	$\frac{1}{2}$	$\frac{1}{2}$	$\frac{1}{2}$

← Groups of 1 half

This method is great for understanding the concept of division of fractions, but an efficient way of dividing fractions is to multiply the first fraction by the reciprocal of the second fraction.

Dividing fractions

To divide fractions, multiply the first fraction by the reciprocal of the second fraction.

E.g. $\dfrac{1}{3} \div \dfrac{2}{5} = \dfrac{1}{3} \times \dfrac{5}{2}$

$\qquad\qquad = \dfrac{5}{6}$

This process can be remembered easily by the saying KEEP, CHANGE, FLIP.

$$\dfrac{3}{4} \qquad \div \qquad \dfrac{2}{7}$$

$$\text{KEEP} \qquad \text{CHANGE} \qquad \text{FLIP}$$

$$\dfrac{3}{4} \qquad\quad \times \qquad\quad \dfrac{7}{2}$$

tlvd-3565

WORKED EXAMPLE 26 Dividing fractions

Evaluate $\dfrac{3}{5} \div \dfrac{9}{10}$.

THINK

1. Write the question.

2. Keep the first fraction the same. Change the division sign to a multiplication sign and flip the second fraction.

3. Cancel diagonally.

4. Perform the multiplication.

WRITE

$\dfrac{3}{5} \div \dfrac{9}{10}$

$= \dfrac{3}{5} \times \dfrac{10}{9}$

$= \dfrac{\cancel{3}^{1}}{\cancel{5}^{1}} \times \dfrac{\cancel{10}^{2}}{\cancel{9}^{3}}$

$= \dfrac{2}{3}$

 Resources

Interactivities   Division of rational numbers (int-3948)

2.7 Exercise

1. **WE24** Identify the reciprocals of each of the following.

   a. $\dfrac{2}{3}$ b. $\dfrac{5}{7}$ c. $\dfrac{8}{5}$ d. $\dfrac{7}{3}$ e. 3

2. Identify the reciprocals of each of the following.

   a. $\dfrac{5}{11}$ b. $\dfrac{11}{2}$ c. $\dfrac{6}{5}$ d. $\dfrac{12}{7}$ e. 18

3. Identify the reciprocals of each of the following.

   a. $\dfrac{1}{5}$ b. $\dfrac{4}{13}$ c. $\dfrac{3}{25}$ d. $\dfrac{1}{8}$ e. 1

4. **WE25** Determine the reciprocals of these mixed numbers.

   a. $1\dfrac{3}{5}$ b. $3\dfrac{3}{8}$ c. $4\dfrac{3}{7}$ d. $4\dfrac{3}{8}$ e. $7\dfrac{7}{10}$

5. Determine the reciprocals of these mixed numbers.

   a. $8\dfrac{6}{7}$ b. $5\dfrac{4}{5}$ c. $10\dfrac{1}{4}$ d. $12\dfrac{5}{8}$ e. $4\dfrac{5}{9}$

6. Multiply each of these numbers by its reciprocal.

   a. $\dfrac{5}{9}$ b. $\dfrac{3}{7}$ c. $3\dfrac{2}{5}$

7. **WE26** Evaluate the following, then check your answers with a calculator.

   a. $\dfrac{1}{5} \div \dfrac{2}{3}$ b. $\dfrac{9}{10} \div \dfrac{7}{9}$ c. $\dfrac{1}{9} \div \dfrac{2}{3}$

   d. $\dfrac{3}{11} \div \dfrac{4}{7}$ e. $\dfrac{10}{7} \div \dfrac{20}{28}$

8. Evaluate the following, then check your answers with a calculator.

   a. $\dfrac{6}{7} \div \dfrac{18}{21}$ b. $\dfrac{5}{7} \div \dfrac{5}{7}$ c. $\dfrac{3}{5} \div 10$

   d. $\dfrac{5}{3} \div 12$ e. $\dfrac{20}{9} \div 5$

9. Evaluate the following, then check your answers with a calculator.

   a. $\dfrac{7}{9} \div \dfrac{3}{2}$ b. $\dfrac{5}{9} \div 15$ c. $\dfrac{6}{7} \div \dfrac{3}{14}$

   d. $\dfrac{25}{16} \div \dfrac{5}{4}$ e. $\dfrac{60}{7} \div 15$

10. Linda says that the following statement is correct.

$$3\frac{1}{2} \div \frac{4}{6} > \frac{4}{3} \div \frac{2}{15}$$

Her friend Lisa does not agree. Explain who is correct.

11. Sophie is making a cake. The recipe requires $\frac{1}{4}$ cup of butter. She has a 500 g tub of butter. If 1 cup of butter weighs 100 g, calculate the number of batches of the recipe she can make.

12. Adam bought $2\frac{1}{2}$ kg of Freddo Frogs for his birthday party. He is going to give each friend attending the party $\frac{1}{10}$ kg.
Calculate how many of his friends will be attending.

13. A carpenter has a length of timber that is $\frac{16}{5}$ cm long. She divides the timber into lengths that are $\frac{4}{5}$ cm long.
Determine how many lengths of timber she has now.

14. Calculate the average of $\frac{1}{4}, \frac{1}{5}, \frac{1}{10}$ and $\frac{1}{20}$.

15. Determine the value of the following expression.

$$\frac{1}{3 + \frac{1}{\frac{1}{4}}}$$

2.8 Working with mixed numbers

LEARNING INTENTION

At the end of this subtopic you should be able to:
- add, subtract, multiply and divide mixed numbers
- check answers by estimation.

2.8.1 Adding and subtracting mixed numbers

To add or subtract mixed numbers, first convert to improper fractions, and then add and subtract as usual.

To perform operations on mixed numbers, first convert them to improper fractions, and then use the rules for operations on fractions to evaluate.

Working with mixed numbers

To add, subtract, multiply or divide mixed numbers, convert them to improper fractions and then perform the operation as usual.

WORKED EXAMPLE 27 Adding mixed numbers

Evaluate $3\dfrac{1}{3} + 1\dfrac{1}{6}$.

THINK	WRITE
1. Write the question.	$3\dfrac{1}{3} + 1\dfrac{1}{6}$
2. Convert each mixed number into an improper fraction.	$= \dfrac{10}{3} + \dfrac{7}{6}$
3. Write both fractions with the lowest common denominator (LCD).	$= \dfrac{10 \times 2}{3 \times 2} + \dfrac{7}{6}$
	$= \dfrac{20}{6} + \dfrac{7}{6}$
4. Add the numerators.	$= \dfrac{27}{6}$
5. Convert to a mixed number, if appropriate, and write the answer in simplified form.	$= 4\dfrac{3}{6}$
	$= 4\dfrac{1}{2}$

WORKED EXAMPLE 28 Subtracting mixed numbers

Evaluate $4\dfrac{1}{3} - 2\dfrac{3}{4}$.

THINK	WRITE
1. Write the question.	$4\dfrac{1}{3} - 2\dfrac{3}{4}$
2. Convert each mixed number into an improper fraction.	$= \dfrac{13}{3} - \dfrac{11}{4}$
3. Write both fractions with the lowest common denominator (LCD).	$= \dfrac{13 \times 4}{3 \times 4} - \dfrac{11 \times 3}{4 \times 3}$
	$= \dfrac{52}{12} - \dfrac{33}{12}$
4. Subtract the numerators.	$= \dfrac{19}{12}$
5. Convert to a mixed number, if appropriate, and write the answer.	$= 1\dfrac{7}{12}$

2.8.2 Checking answers by estimation

Estimation can be used to determine approximate answers to addition and subtraction of mixed numbers.

This allows you to check easily whether your answer to a calculation is reasonable. To estimate, add or subtract the whole number parts. The answer should be close to the result obtained by this method.

tlvd-3567

WORKED EXAMPLE 29 Checking answers by estimation

Evaluate $4\dfrac{1}{2} + 3\dfrac{1}{3}$ and check the answer by estimation.

THINK	WRITE
1. Write the question.	$4\dfrac{1}{2} + 3\dfrac{1}{3}$
2. Convert each mixed number into an improper fraction.	$= \dfrac{9}{2} + \dfrac{10}{3}$
3. Write both fractions with the lowest common denominator (LCD).	$= \dfrac{9 \times 3}{2 \times 3} + \dfrac{10 \times 2}{3 \times 2}$
	$= \dfrac{27}{6} + \dfrac{20}{6}$
4. Add the numerators.	$= \dfrac{47}{6}$
5. Convert to a mixed number if appropriate.	$= 7\dfrac{5}{6}$
6. Check the answer by adding the whole numbers to determine an approximation.	$4\dfrac{1}{2} + 3\dfrac{1}{3} \simeq 4 + 3$
	$\simeq 7$
	The approximation shows that the answer should be close to 7.

2.8.3 Multiplying and dividing mixed numbers

To multiply mixed numbers, first convert to improper fractions, and then multiply as usual.

To divide mixed numbers, first convert them into improper fractions, and then divide as usual.

WORKED EXAMPLE 30 Multiplying mixed numbers

Evaluate $1\dfrac{2}{3} \times 2\dfrac{1}{4}$.

THINK	WRITE
1. Write the question.	$1\dfrac{2}{3} \times 2\dfrac{1}{4}$
2. Change mixed numbers to improper fractions and cancel if possible.	$= \dfrac{5}{\cancel{3}^{1}} \times \dfrac{\cancel{9}^{3}}{4}$

3. Multiply the numerators and then the denominators. $= \dfrac{15}{4}$

4. Convert to a mixed number, if appropriate, and write the answer. $= 3\dfrac{3}{4}$

WORKED EXAMPLE 31 Dividing mixed numbers

Evaluate $2\dfrac{1}{2} \div 2\dfrac{3}{5}$.

THINK	WRITE
1. Write the question.	$2\dfrac{1}{2} \div 2\dfrac{3}{5}$
2. Convert the mixed numbers to improper fractions.	$= \dfrac{5}{2} \div \dfrac{13}{5}$
3. Change $\div$ to $\times$ and flip the second fraction.	$= \dfrac{5}{2} \times \dfrac{5}{13}$
4. Multiply the numerators and the denominators, and simplify the answer if necessary.	$= \dfrac{25}{26}$

 Resources

🧩 **Interactivities** Addition and subtraction of mixed numbers (int-3951)

2.8 Exercise

Students, these questions are even better in jacPLUS

💬 Receive immediate feedback and access sample responses

🔒 Access additional questions

⭐ Track your results and progress

Find all this and MORE in jacPLUS

1. **WE27, WE28** Evaluate the following, giving answers as mixed numbers in simplest form.

   a. $1\dfrac{1}{4} + 2\dfrac{1}{3}$ b. $4\dfrac{2}{3} - 1\dfrac{1}{3}$ c. $3\dfrac{2}{3} - 2\dfrac{1}{2}$ d. $7\dfrac{3}{8} + 3$

2. Evaluate the following, giving answers as mixed numbers in simplest form.

   a. $3\dfrac{3}{5} - 1\dfrac{1}{5}$ b. $4\dfrac{1}{2} + 2\dfrac{1}{4}$

   c. $5 - 3\dfrac{2}{3}$ d. $3\dfrac{1}{3} - \dfrac{5}{6}$

3. Evaluate the following, giving answers as mixed numbers in simplest form.

   a. $4\dfrac{1}{4} - 1\dfrac{5}{8}$

   b. $5\dfrac{3}{5} - 3\dfrac{1}{3}$

   c. $6\dfrac{1}{5} - 2\dfrac{1}{4}$

   d. $5\dfrac{1}{4} - 2\dfrac{7}{12}$

4. **WE29** Evaluate the following and check the answers by estimation.

   a. $5\dfrac{1}{3} + 2\dfrac{1}{4}$

   b. $1\dfrac{1}{6} + 2\dfrac{1}{2}$

   c. $3\dfrac{1}{3} + 1\dfrac{1}{6}$

   d. $6\dfrac{1}{2} - 2\dfrac{1}{4}$

5. **WE30** Evaluate each of the following.

   a. $1\dfrac{1}{4} \times \dfrac{2}{3}$

   b. $1\dfrac{1}{2} \times 1\dfrac{1}{2}$

   c. $2\dfrac{5}{8} \times 1\dfrac{3}{4}$

   d. $2\dfrac{1}{5} \times 3\dfrac{1}{2}$

6. Evaluate each of the following.

   a. $2\dfrac{5}{9} \times 7\dfrac{4}{7}$

   b. $5\dfrac{2}{5} \times 6\dfrac{2}{3}$

   c. $2\dfrac{3}{5} \times 3\dfrac{1}{3}$

   d. $\dfrac{7}{8} \times 5\dfrac{1}{4}$

7. **WE31** Evaluate each of the following.

   a. $1\dfrac{3}{4} \div \dfrac{5}{8}$

   b. $\dfrac{9}{4} \div 3\dfrac{1}{2}$

   c. $\dfrac{10}{3} \div 3\dfrac{1}{3}$

   d. $6\dfrac{5}{9} \div \dfrac{2}{3}$

8. Evaluate each of the following.

   a. $2\dfrac{1}{2} \div \dfrac{1}{2}$

   b. $8\dfrac{7}{18} \div 3$

   c. $2\dfrac{2}{7} \div 3\dfrac{1}{7}$

   d. $4\dfrac{5}{6} \div 2\dfrac{2}{3}$

9. To make apple-raspberry cordial, Callum uses $\dfrac{3}{8}$ of a litre of cordial with $1\dfrac{3}{4}$ litres of water. Calculate how many litres of cordial Callum made.

10. Heather is dividing up $4\dfrac{1}{4}$ litres of drink for 8 people. Calculate how much drink each person gets in litres.

11. If Toby takes $\dfrac{1}{3}$ of an hour to complete a soccer game on PlayStation, determine how many games he can complete in $4\dfrac{2}{3}$ hours.

12. Doris is organising a 5-month holiday around England. She plans to spend $\frac{5}{8}$ of a month in Liverpool, $1\frac{1}{4}$ months in Manchester, $1\frac{1}{2}$ months in London and the rest of the time in Birmingham.
Calculate how long Doris will spend in Birmingham.

13. Cory is planning on building a new pool. He is told it takes $1\frac{1}{4}$ days to dig the hole, $1\frac{1}{2}$ days to prepare the steel, $3\frac{5}{8}$ days to pour the concrete and let it set, then $3\frac{3}{4}$ days to tile and paint.
Calculate how long it takes to complete the whole process.

14. Fiona has $5\frac{3}{4}$ kg of rice that she wants to store in containers that can hold $1\frac{1}{2}$ kg or rice each.

   a. Calculate how many containers she can fill.
   b. Determine how much rice Fiona has left after she fills as many containers as possible.

15. Scott heads out on his $7\frac{1}{2}$-km run in the morning. He gets $\frac{3}{4}$ of the way through his run before he needs to stop to stretch his calves.

   a. Calculate how far Scott ran before he needed to stretch his calves.
   b. Calculate how much further Scott needed to run after stretching his calves.

2.9 Review

2.9.1 Summary

doc-38043

2.9 Exercise

Multiple choice

1. **MC** Determine what fraction of eggs are white in the diagram.

   A. $\dfrac{1}{10}$

   B. $\dfrac{3}{20}$

   C. $\dfrac{3}{24}$

   D. $\dfrac{3}{22}$

   E. $\dfrac{3}{10}$

2. **MC** The fraction $\dfrac{12}{48}$ simplifies to:

   A. $\dfrac{1}{2}$ B. $\dfrac{1}{3}$ C. $\dfrac{1}{4}$ D. $\dfrac{1}{5}$ E. $\dfrac{1}{6}$

3. **MC** The mixed number $4\dfrac{3}{7}$ as an improper fraction is:

   A. $\dfrac{13}{7}$ B. $\dfrac{43}{7}$ C. $\dfrac{24}{7}$ D. $\dfrac{19}{7}$ E. $\dfrac{31}{7}$

4. **MC** The fraction $\dfrac{24}{5}$ as a mixed number is:

   A. $5\dfrac{1}{5}$ B. $5\dfrac{4}{5}$ C. $4\dfrac{3}{5}$ D. $4\dfrac{4}{5}$ E. $2\dfrac{4}{5}$

5. **MC** The reciprocal of $\dfrac{7}{8}$ is:

   A. $\dfrac{7}{8}$ B. $\dfrac{3}{4}$ C. $\dfrac{8}{7}$ D. $\dfrac{4}{3}$ E. $\dfrac{5}{6}$

6. **MC** $\dfrac{3}{6} + \dfrac{5}{3}$ evaluates to:

   A. $2\dfrac{2}{3}$ B. $2\dfrac{1}{3}$ C. $\dfrac{8}{9}$ D. $\dfrac{6}{7}$ E. $2\dfrac{1}{6}$

7. MC $3\dfrac{1}{4} - 1\dfrac{2}{3}$ evaluates to:

   A. $1\dfrac{7}{12}$ **B.** $2\dfrac{1}{4}$ **C.** $2\dfrac{1}{3}$ **D.** $2\dfrac{3}{4}$ **E.** $3\dfrac{3}{4}$

8. MC $\dfrac{3}{4} \times \dfrac{8}{9}$ evaluates to:

   A. $\dfrac{3}{4}$ **B.** $\dfrac{3}{8}$ **C.** $\dfrac{3}{2}$ **D.** $\dfrac{2}{3}$ **E.** $\dfrac{1}{3}$

9. MC $\dfrac{18}{30} \div \dfrac{24}{10}$ evaluates to:

   A. $\dfrac{1}{4}$ **B.** $\dfrac{2}{3}$ **C.** $\dfrac{1}{2}$ **D.** $\dfrac{1}{8}$ **E.** $\dfrac{1}{6}$

10. MC $\dfrac{3}{4}$ of 48 is:

   A. 24   **B.** 30   **C.** 36   **D.** 40   **E.** 44

Short answer

11. Ricky, Leo and Gabriel were the highest goal scorers for their local soccer team throughout the season. Ricky scored $\dfrac{2}{7}$ of the goals, Leo scored $\dfrac{1}{3}$ of the goals and Gabriel scored $\dfrac{1}{4}$ of the goals.

   a. Determine who scored the most goals.
   b. State whether Ricky or Gabriel scored more goals.

12. Determine what fraction of months of the year begin with the letter J. Express your answer in simplest form.

13. Anthony's monthly take-home pay is $2800. He spends a quarter on his home loan payments, one-half on food and drink, and one-seventh on clothing. One-half of the remainder goes into his savings account. Calculate how much money Anthony saves each month.

14. The temperature rose from $21\dfrac{1}{2}$ °C to $21\dfrac{3}{4}$ °C. Calculate the temperature rise.

15. Marcella was in charge of baking and selling lemon slices at the school fete. Each tray she baked contained 15 lemon slices. If Marcella sold 68 lemon slices, determine how many empty trays there were.

Extended response

16. Tomas buys a 24-piece pizza and eats 8 pieces. Drew buys a 16-piece pizza of the same size, and eats 5 pieces. Determine who ate more pizza and explain your choice.

17. Six containers, each holding $5\dfrac{1}{3}$ litres, are filled from a large tank of chemicals. If the large tank initially contained 50 litres of chemicals, determine how much is left in the tanks after six containers are filled.

18. A skydiver opens her parachute at 846 m above ground level. She has already fallen five-sevenths of the distance to the ground. Determine the altitude of the plane when she jumped.

19. Mr Thompson was earning \$2400 a month. Recently, he had a pay increase. He now earns $1\frac{1}{4}$ times his old salary. Calculate his new salary.

Mr Thompson's colleague, Mr Goody, was earning \$3000 per month and has had a pay cut. He now gets $\frac{1}{6}$ less than his old salary.

Calculate Mr Goody's new salary.

20. Byron needs $\frac{2}{5}$ kg of catalyst mixed with $4\frac{1}{2}$ kg of resin to create one surfboard.

a. If he has $2\frac{1}{4}$ kg of catalyst and $20\frac{1}{4}$ kg of resin, determine how many whole surfboards Byron can build.

b. Calculate how much more catalyst and resin Byron needs to build one more surfboard. Explain your reasoning.

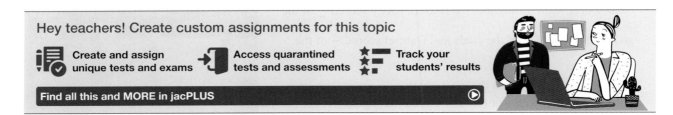

Answers

Topic 2 Fractions

2.2 Introducing fractions

2.2 Exercise

1. a. $\dfrac{3}{4}$ b. $\dfrac{15}{16}$ c. $\dfrac{7}{8}$ d. $\dfrac{5}{9}$

2. a. $\dfrac{1}{3}$ b. $\dfrac{3}{4}$

3. a. i. $\dfrac{4}{7}$ ii. $\dfrac{3}{7}$

   b. $\dfrac{3}{4}$

4.

5.

6. a. 3   b. 2   c. 3   d. 4

7. a. 5   b. 4   c. 9   d. 11

8. a. Equivalent fractions to $\dfrac{2}{3} = \dfrac{20}{30}, \dfrac{14}{21}, \dfrac{40}{60}$

   b. Equivalent fractions to $\dfrac{4}{5} = \dfrac{12}{15}, \dfrac{36}{45}, \dfrac{16}{20}, \dfrac{28}{35}, \dfrac{80}{100}$

   c. Equivalent fractions to $\dfrac{7}{8} = \dfrac{56}{64}, \dfrac{21}{24}$

   d. Equivalent fractions to $\dfrac{7}{10} = \dfrac{35}{50}, \dfrac{21}{30}, \dfrac{14}{20}, \dfrac{140}{200}$

9. a. $\dfrac{2}{5} = \dfrac{12}{30} = \dfrac{12}{30} = \dfrac{18}{45}$ b. $\dfrac{3}{4} = \dfrac{33}{44} = \dfrac{15}{20} = \dfrac{27}{36}$

   c. $\dfrac{2}{3} = \dfrac{18}{27} = \dfrac{10}{15} = \dfrac{8}{12}$ d. $\dfrac{5}{6} = \dfrac{25}{30} = \dfrac{35}{42} = \dfrac{60}{72}$

10. a. $\dfrac{4}{5} = \dfrac{8}{10} = \dfrac{12}{15} = \dfrac{16}{20}$ b. $\dfrac{3}{8} = \dfrac{6}{16} = \dfrac{9}{24} = \dfrac{12}{32}$

    c. $\dfrac{4}{10} = \dfrac{8}{20} = \dfrac{12}{30} = \dfrac{16}{40}$ d. $\dfrac{1}{7} = \dfrac{2}{14} = \dfrac{3}{21} = \dfrac{4}{28}$

    e. $\dfrac{5}{8} = \dfrac{10}{16} = \dfrac{15}{24} = \dfrac{20}{32}$ f. $\dfrac{2}{3} = \dfrac{4}{6} = \dfrac{6}{9} = \dfrac{8}{12}$

11. a. Larger fraction: $\dfrac{3}{5}$

    b. Larger fraction: $\dfrac{7}{8}$

c. Larger fraction: $\dfrac{1}{5}$

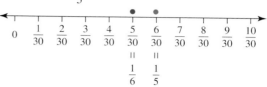

d. Larger fraction: $\dfrac{1}{2}$

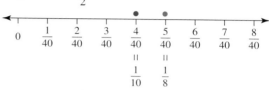

e. Larger fraction: $\dfrac{1}{2}$

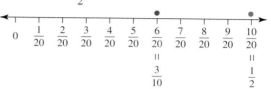

f. Larger fraction: $\dfrac{3}{4}$

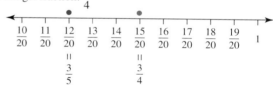

12. a. >   b. >   c. <   d. <   e. <   f. <

13. a. D   b. B

14. a. $\dfrac{1}{5}, \dfrac{3}{10}, \dfrac{1}{2}$ b. $\dfrac{1}{3}, \dfrac{3}{8}, \dfrac{1}{2}$

    c. $\dfrac{6}{15}, \dfrac{2}{3}, \dfrac{4}{5}$ d. $\dfrac{7}{15}, \dfrac{2}{3}, \dfrac{3}{4}$

15. a.

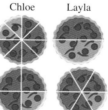

Chloe   Layla

Scott   Bryce

    b. Bryce
    c. The same

16.

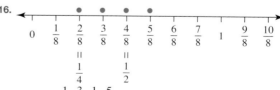

Points at $\dfrac{1}{4}, \dfrac{3}{8}, \dfrac{1}{2}, \dfrac{5}{8}$

2.3 Simplifying fractions

2.3 Exercise

1. a. $\dfrac{9}{10}$ b. $\dfrac{1}{3}$ c. $\dfrac{3}{4}$

   d. $\dfrac{3}{8}$ e. $\dfrac{7}{10}$ f. $\dfrac{7}{10}$

2. a. $\dfrac{5}{8}$ b. $\dfrac{2}{7}$ c. $\dfrac{1}{3}$

   d. $\dfrac{1}{2}$ e. $\dfrac{3}{8}$ f. $\dfrac{1}{9}$

3. a. $2\dfrac{1}{3}$ b. $5\dfrac{1}{2}$ c. $4\dfrac{1}{6}$ d. $8\dfrac{3}{7}$

4. a. $6\dfrac{1}{3}$ b. $3\dfrac{1}{4}$ c. $1\dfrac{1}{3}$ d. $2\dfrac{1}{2}$

5. D

6. $\dfrac{5}{6}$

7. a. $\dfrac{3}{7}$ b. $\dfrac{4}{7}$

8. 11A: $\dfrac{1}{4}$

   11B: $\dfrac{3}{20}$

   11C: $\dfrac{7}{24}$

   11D: $\dfrac{1}{10}$

   11E: $\dfrac{5}{24}$

9. a. 65

   b. Kyle: $\dfrac{5}{13}$; Anthony: $\dfrac{1}{13}$; Callum: $\dfrac{3}{13}$; Harry: $\dfrac{4}{13}$

10. The Geelong to Melbourne train is more reliable.

11. a. 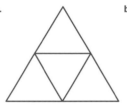 b. $\dfrac{1}{4}$

    c. 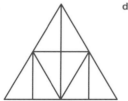 d. $\dfrac{1}{8}$

12. For $\dfrac{17}{n}$ not to be in simplest form, the numerator and denominator would need to have a common factor. Therefore, n would need to be a factor of 17. For example:

If $n = 34\,(2 \times 17 = 34)$, then $\dfrac{17}{34} = \dfrac{1}{2}$.

If $n = 51\,(3 \times 17 = 51)$, then $\dfrac{17}{51} = \dfrac{1}{3}$.

2.4 Mixed numbers and improper fractions

2.4 Exercise

1. a.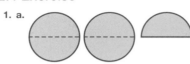

   $2\dfrac{1}{2}$

   b.

   $1\dfrac{1}{3}$

   c.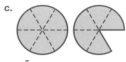

   $1\dfrac{5}{6}$

   d.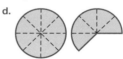

   $1\dfrac{5}{8}$

   e.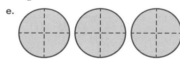

   3

   f.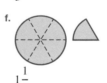

   $1\dfrac{1}{6}$

2. a. $1\dfrac{1}{3}$ b. $2\dfrac{1}{4}$

3. a. $2\dfrac{1}{3}$ b. $1\dfrac{4}{5}$ c. $3\dfrac{1}{2}$

   d. $7\dfrac{2}{7}$ e. $5\dfrac{2}{5}$ f. $3\dfrac{5}{9}$

4. a. $3\dfrac{1}{3}$ b. $4\dfrac{5}{7}$ c. $9\dfrac{1}{11}$

   d. $3\dfrac{1}{7}$ e. $4\dfrac{3}{8}$ f. $6\dfrac{3}{11}$

5. a. $14\frac{5}{6}$ **b.** $11\frac{3}{10}$ **c.** $2\frac{1}{12}$ **d.** $1\frac{2}{5}$

6. a. $8\frac{1}{7}$ **b.** $10\frac{7}{9}$ **c.** $7\frac{2}{9}$ **d.** $1\frac{4}{9}$

7. B

8. C

9. A

10. a.

b.

c.

d.

e.

f.

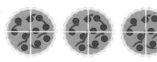

11. a. $\frac{5}{3}$ **b.** $\frac{8}{5}$ **c.** $\frac{8}{3}$ **d.** $\frac{11}{4}$

12. a. $\frac{23}{6}$ **b.** $\frac{29}{5}$ **c.** $\frac{55}{8}$ **d.** $\frac{79}{9}$

13. a. $\frac{10}{3}$ **b.** $\frac{22}{5}$ **c.** $\frac{39}{7}$ **d.** $\frac{28}{10}$

14. a. $\frac{67}{12}$ **b.** $\frac{75}{8}$ **c.** $\frac{35}{11}$ **d.** $\frac{59}{6}$

15. E

16. B

17. C

18. $7\frac{1}{2}$ apples

19. $7\frac{7}{8}$ pies

20. $5\frac{3}{4}$ oranges

21. a. $\frac{5}{2} < 3\frac{1}{2}$ **b.** $\frac{7}{3} > 1\frac{2}{3}$ **c.** $\frac{12}{5} > 2\frac{1}{5}$ **d.** $\frac{9}{4} < 2\frac{3}{4}$

22. a. $2\frac{1}{3} < \frac{8}{3}$ **b.** $5\frac{2}{3} > \frac{16}{3}$ **c.** $4\frac{1}{7} < \frac{30}{7}$ **d.** $6\frac{1}{5} < \frac{33}{5}$

23. a. 35   **b.** 73

24. $\frac{2}{3}$

25. a. 25   **b.** 77

2.5 Adding and subtracting fractions

2.5 Exercise

1. a. $\frac{3}{4}$ **b.** $\frac{5}{7}$ **c.** $\frac{5}{6}$ **d.** $\frac{5}{9}$

2. a. $\frac{2}{9}$ **b.** $\frac{2}{5}$ **c.** $\frac{4}{25}$ **d.** $\frac{7}{15}$

3. a. $\frac{5}{6}$ **b.** $\frac{1}{2}$ **c.** $\frac{2}{3}$ **d.** $\frac{1}{4}$

4. a. 1   **b.** $\frac{3}{10}$ **c.** $\frac{3}{4}$ **d.** $\frac{2}{3}$

5. a. $\frac{1}{2}$ **b.** $\frac{11}{12}$ **c.** $\frac{7}{8}$ **d.** $\frac{5}{12}$

6. a. $\frac{1}{4}$ **b.** $\frac{2}{15}$ **c.** $\frac{7}{20}$ **d.** $\frac{2}{55}$

7. a. 24   **b.** 12   **c.** 15   **d.** 24
   e. 21   **f.** 30

8. a. $\frac{1}{6}$ **b.** $\frac{8}{15}$ **c.** $\frac{11}{28}$ **d.** $\frac{19}{45}$
   e. $\frac{9}{40}$ **f.** $\frac{11}{24}$

9. a. $\frac{11}{40}$ **b.** $\frac{1}{3}$ **c.** $\frac{23}{36}$ **d.** $\frac{1}{6}$
   e. $\frac{4}{9}$ **f.** $\frac{11}{40}$

10. D

11. a. $\frac{2}{3}$ **b.** $1\frac{3}{4}$ **c.** $2\frac{4}{5}$ **d.** $3\frac{1}{5}$

12. a. $\frac{1}{3}$ **b.** $\frac{1}{12}$ **c.** $\frac{7}{12}$ **d.** $\frac{7}{30}$

13. a. $\frac{11}{12}$ **b.** $\frac{1}{12}$

14. $1\frac{1}{6}$ hour

15. a. $\frac{5}{8}$ **b.** $\frac{5}{12}$ **c.** $\frac{19}{24}$

16. a. $\frac{2}{3}$ **b.** $\frac{1}{3}$

2.6 Multiplying fractions

2.6 Exercise

1. a. 2   b. $\dfrac{3}{2}$ c. 5   d. $\dfrac{7}{4}$

2. a. $\dfrac{1}{12}$ b. $\dfrac{2}{9}$ c. $\dfrac{10}{27}$ d. $\dfrac{9}{26}$

3. a. $\dfrac{21}{55}$ b. $\dfrac{121}{144}$ c. $\dfrac{25}{63}$ d. $\dfrac{7}{24}$

4. a. $\dfrac{5}{11}$ b. $\dfrac{25}{132}$ c. $\dfrac{4}{7}$ d. $\dfrac{12}{35}$

5. a. $\dfrac{1}{4}$ b. $\dfrac{1}{10}$ c. $\dfrac{5}{18}$ d. $\dfrac{1}{6}$

6. a. $\dfrac{3}{2}$ b. $\dfrac{4}{5}$ c. $\dfrac{1}{2}$ d. $\dfrac{3}{20}$

7. a. $\dfrac{1}{6}$ b. $\dfrac{1}{12}$ c. $\dfrac{2}{5}$ d. $\dfrac{1}{20}$

8. a. $\dfrac{1}{10}$ b. $\dfrac{1}{4}$ c. $\dfrac{5}{8}$ d. $\dfrac{2}{9}$

9. a. $\dfrac{35}{2}$ b. 63   c. 21   d. 24

10. John = 2 litres

    Anthony = 1 litre

    Will = $\dfrac{2}{3}$ litre

11. a. 10 minutes   b. 8 minutes

12. a. 4   b. 8   c. 6   d. 30

13. a. $\dfrac{3}{4}$ b. 36 litres

14. a. 4   b. 12   c. 20

15. $450

16. a. 12 cm   b. 8 cm

2.7 Dividing fractions

2.7 Exercise

1. a. $\dfrac{3}{2}$ b. $\dfrac{7}{5}$ c. $\dfrac{5}{8}$

   d. $\dfrac{3}{7}$ e. $\dfrac{1}{3}$

2. a. $\dfrac{11}{5}$ b. $\dfrac{2}{11}$ c. $\dfrac{5}{6}$

   d. $\dfrac{7}{12}$ e. $\dfrac{1}{18}$

3. a. 5   b. $\dfrac{13}{4}$ c. $\dfrac{25}{3}$

   d. 8   e. 1

4. a. $\dfrac{5}{8}$ b. $\dfrac{8}{27}$ c. $\dfrac{7}{31}$

   d. $\dfrac{8}{35}$ e. $\dfrac{10}{77}$

5. a. $\dfrac{7}{62}$ b. $\dfrac{5}{29}$ c. $\dfrac{4}{41}$

   d. $\dfrac{8}{101}$ e. $\dfrac{9}{41}$

6. a. 1   b. 1   c. 1

7. a. $\dfrac{3}{10}$ b. $\dfrac{81}{70}$ c. $\dfrac{1}{6}$ d. $\dfrac{21}{44}$ e. 2

8. a. 1   b. 1   c. $\dfrac{3}{50}$ d. $\dfrac{5}{36}$ e. $\dfrac{4}{9}$

9. a. $\dfrac{14}{27}$ b. $\dfrac{1}{27}$ c. 4   d. $\dfrac{5}{4}$ e. $\dfrac{4}{7}$

10. Lisa is correct.

11. 20 batches

12. 25

13. 4

14. $\dfrac{3}{20}$

15. $\dfrac{1}{7}$

2.8 Working with mixed numbers

2.8 Exercise

1. a. $3\dfrac{7}{12}$ b. $3\dfrac{1}{3}$ c. $1\dfrac{1}{6}$ d. $10\dfrac{3}{8}$

2. a. $2\dfrac{2}{5}$ b. $6\dfrac{3}{4}$ c. $1\dfrac{1}{3}$ d. $2\dfrac{1}{2}$

3. a. $2\dfrac{5}{8}$ b. $2\dfrac{4}{15}$ c. $3\dfrac{19}{20}$ d. $2\dfrac{8}{12}$

4. a. $7\dfrac{7}{12}$; estimate = 7

   b. $3\dfrac{2}{3}$; estimate = 3

   c. $4\dfrac{1}{2}$; estimate = 4

   d. $4\dfrac{1}{4}$; estimate = 4

5. a. $\dfrac{5}{6}$ b. $2\dfrac{1}{4}$ c. $4\dfrac{19}{32}$ d. $7\dfrac{7}{10}$

6. a. $19\dfrac{22}{63}$ b. 36   c. $8\dfrac{2}{3}$ d. $4\dfrac{19}{32}$

7. a. $2\dfrac{4}{5}$ b. $\dfrac{9}{14}$ c. 1   d. $9\dfrac{5}{6}$

8. a. 5   b. $2\dfrac{43}{54}$ c. $\dfrac{8}{11}$ d. $1\dfrac{13}{16}$

9. $2\dfrac{1}{8}$ litres

10. $\dfrac{17}{32}$ litres

11. 14 games

12. $1\dfrac{5}{8}$ months

13. $10\dfrac{1}{8}$ days

14. **a.** 3 containers   **b.** $1\dfrac{1}{4}$ kg left over

15. **a.** $7\dfrac{1}{2} \times \dfrac{3}{4} = \dfrac{15}{2} \times \dfrac{3}{4}$

$$= \dfrac{45}{8}$$

$$= 5\dfrac{5}{8} \text{ km}$$

   b. $1\dfrac{7}{8}$ km

2.9 Review

2.9 Exercise

Multiple choice

1. B
2. C
3. E
4. D
5. C
6. E
7. A
8. D
9. A
10. C

Short answer

11. **a.** Leo   **b.** Ricky

12. $\dfrac{1}{4}$

13. $150

14. $\dfrac{1}{4}$ °C

15. 4

Extended response

16. Tomas

17. 18 litres

18. $1184\dfrac{2}{5}$ m

19. New salary of Mr Thompson = $3000
New salary of Mr Goody = $2500

20. **a.** 4   **b.** $\dfrac{1}{2}$ kg of resin

3 Decimals

LEARNING SEQUENCE

Fully worked solutions for this topic are available online.

3.1 Overview

3.1.1 Introduction

You have already seen that whole numbers cannot be used to measure everything. How much did you pay for that burger and chips? $12.60? Our money system is in dollars and cents and uses decimals to represent parts of the dollar. Decimals play an important part in our daily lives. You may have been using decimals without realising their importance. Since our lives are so dependent on money, knowing how to work with decimals is very important. We all need to be skilled in the basic operations of addition, subtraction, multiplication and division with decimals.

Decimals are used not only with money but also with time and measurement. They are also used as a way of expressing a fraction or percentage of something. Did you know that Cathy Freeman won the gold medal in the 400-metre event at the 2000 Sydney Olympics in a time of 49.11 seconds, a winning margin of just 0.02 seconds? Usain Bolt set the men's 100-metre world record in 2009 with a time of just 9.58 seconds. Every person will use decimals in their lives, including when checking their pay or their bank balance. Understanding and using decimals is vital in retail, hospitality, design, commerce, construction, the health industry and all areas of business.

KEY CONCEPTS

This topic covers the following key concepts from the VCE Mathematics Study Design:
- application of integers, fractions and decimals, their properties and related operations.

Note: Concepts shown in grey are covered in other topics.

Source: VCE Mathematics Study Design (2023–2027) extracts © VCAA; reproduced by permission.

3.2 Place value and comparing decimals

LEARNING INTENTION

At the end of this subtopic you should be able to:
- understand the place values used in decimal numbers
- determine the value of digits in decimal numbers
- compare the sizes of decimal numbers.

3.2.1 Whole numbers and decimal parts

Each position in a number has its own place value.

Hundred thousands	Ten thousands	Thousands	Hundreds	Tens	Units
100 000	10 000	1000	100	10	1

Each place to the left of another position has a value that is 10 times larger. Each place to the right of another position has a value that is $\frac{1}{10}$ of the previous position.

The position of a digit in a number gives the value of the digit.

The following table shows the value of the digit 3 in some numbers.

Number	Value of 3 in the number
132	3 tens or 30
3217	3 thousands or 3000
4103	3 units (ones) or 3

Decimal numbers

In a decimal number, the whole number part and the fractional part are separated by a decimal point.

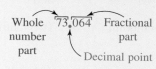

Whole number part   73.064   Fractional part

Decimal point

A place value table can be extended to include decimal place values. It can be continued to an infinite number of **decimal places**.

Thousands	Hundreds	Tens	Units	.	Tenths	Hundredths	Thousandths	Ten thousandths
1000	100	10	1	.	$\frac{1}{10}$	$\frac{1}{100}$	$\frac{1}{1000}$	$\frac{1}{10\,000}$

The following table shows the value of the digit 3 in some decimal numbers.

Number	Value of 3 in the number
14.32	3 tenths or $\dfrac{3}{10}$
106.013	3 thousandths or $\dfrac{3}{1000}$
0.000 03	3 hundred thousandths or $\dfrac{3}{100\,000}$

The number of decimal places in a decimal is the number of digits after the decimal point. The number 73.064 has 3 decimal places. The zero (0) in 73.064 means that there are no tenths. The zero must be written to hold the place value; otherwise the number would be written as 73.64, which does not have the same value.

WORKED EXAMPLE 1 Stating the place value

State the value of 5 in each of the following.

a. 12.57　　　　　　　　**b. 2.154**　　　　　　　　**c. 424.075**

THINK	WRITE
a. The value of the first place to the right of the decimal point is tenths, so the value of 5 is five tenths.	a. $\dfrac{5}{10}$
b. The second place after the decimal point is hundredths, so the value of 5 is five hundredths.	b. $\dfrac{5}{100}$
c. The third place after the decimal point is thousandths, so the value of 5 is five thousandths.	c. $\dfrac{5}{1000}$

WORKED EXAMPLE 2 Expressing place values in words

For the number 76.204, write the value of each digit in words and numbers.

THINK	WRITE
1. 7 is in the tens position.	Seventy, 70
2. 6 is in the units position.	Six, 6
3. 2 is in the first position after the decimal point, so it is tenths.	Two tenths, $\dfrac{2}{10}$
4. 0 is in the hundredths position.	Zero hundredths, $\dfrac{0}{100}$
5. 4 is in the thousandths position.	Four thousandths, $\dfrac{4}{1000}$

Expanded notation

A number can be written in expanded notation by adding the values of each digit.

The number 76.204 can be written in expanded notation as:

$$(7 \times 10) + (6 \times 1) + \left(2 \times \frac{1}{10}\right) + \left(4 \times \frac{1}{1000}\right)$$

WORKED EXAMPLE 3 Writing in expanded notation

Write 5.3407 in expanded notation.

THINK	WRITE
1. Write the decimal.	5.3407
2. Determine the place value of each digit. 5: 5 units 3: 3 tenths 4: 4 hundredths 0: 0 thousandths 7: 7 ten thousandths	$= 5$ units $+ 3$ tenths $+ 4$ hundredths $+ 0$ thousandths $+ 7$ ten thousandths
3. Write the number in expanded notation.	$= (5 \times 1) + \left(3 \times \frac{1}{10}\right) + \left(4 \times \frac{1}{100}\right) + \left(7 \times \frac{1}{10\,000}\right)$

3.2.2 Comparing decimals

Decimals are compared using digits with the same place value.

Comparing decimals

The decimal with the largest number in the highest place-value column is the largest decimal, *regardless of the number of decimal places.*

For example, 15.71 is larger than 15.702 because the first place value with different digits (moving from left to right) is hundredths, and 1 is greater than 0; that is, 15.71 > 15.702.

Tens	Units (ones)	.	Tenths	Hundredths	Thousandths
1	5	.	7	1	
1	5	.	7	0	2

tlvd-3675

WORKED EXAMPLE 4 Determining the largest number with a decimal

State the largest number in each of the following.

a. 0.126, 0.216, 0.122   b. 2.384, 2.388, 2.138   c. 0.506, 0.605, 0.612

THINK

WRITE

a. 1. As the units digit is 0 in each number, compare the tenths. The number 0.216 has 2 tenths while the others have 1 tenth, so 0.216 is the largest number.

  2. Answer the question.

b. 1. As the units digits are the same, compare the tenths. 2.138 has 1 tenth while the others have 3 tenths, so 2.138 is the smallest.

  2. The hundredths are the same in the two larger numbers, so compare the thousandths. 2.388 has 8 thousandths and 2.384 has 4 thousandths, so 2.388 is larger.

  3. Answer the question.

c. 1. As the units digits are the same, compare the tenths. 0.506 has 5 tenths while the others have 6 tenths, so 0.506 is the smallest number.

  2. Compare the hundredths in the two larger numbers. 0.612 has 1 hundredth and 0.605 has 0 hundredths, so 0.612 is larger.

  3. Answer the question.

a. 0.216 is larger than 0.126 and 0.122.

  The largest number is 0.216.

b. 2.384 and 2.388 are both larger than 2.138.

  2.388 is larger than 2.384 and 2.138.

  The largest number is 2.388.

c. 0.605 and 0.612 are larger than 0.506.

  0.612 is larger than 0.605 and 0.506.

  The largest number is 0.612.

When comparing two numbers, it is easier to use symbols instead of words, as shown in Topic 2 Fractions. In Worked example 4a, the result '0.216 is larger than 0.126' could be written as $0.216 > 0.126$. We could also say that 0.126 is less than 0.216, or write it as $0.126 < 0.216$.

WORKED EXAMPLE 5 Comparing numbers with decimals

Insert the appropriate < or > sign between the following pairs of numbers to make true statements.

a. 0.524   0.526
b. 0.0351   0.0299

THINK

WRITE

a. Compare the numbers. Both numbers have the same number of tenths and the same number of hundredths, so compare thousandths and insert the correct sign.

b. Compare the numbers. Neither number has a tenth, so compare hundredths and insert the correct sign.

a. $0.524 < 0.526$

b. $0.0351 > 0.0299$

 Resources

Interactivities Whole numbers (int-3974)
Decimal parts (int-3975)
Comparing decimals (int-3976)
Comparing decimals using the area model (int-3977)

3.2 Exercise

Students, these questions are even better in jacPLUS

 Receive immediate feedback and access sample responses

 Access additional questions

 Track your results and progress

Find all this and MORE in jacPLUS

1. **WE1** State the value of 2 in each of the following.
   a. 3.2
   b. 57.32
   c. 875.21
   d. 0.762 058

2. State the value of 9 in each of the following.
   a. 1.9
   b. 22.93
   c. 54.39
   d. 16.092

3. For each of the following numbers write the value of each digit in *numbers*.
   a. 0.6
   b. 2.3
   c. 6.25
   d. 0.982

4. For each of the following numbers write the value of each digit in *numbers*.
   a. 0.2121
   b. 3.874
   c. 0.6733
   d. 200.09

5. **WE2** For the following numbers write the value of each digit in *words* and *numbers*.
   a. 2.95
   b. 0.379
   c. 12.003
   d. 3.4027

6. **WE3** Write the following numbers in expanded notation.
   a. 1.75
   b. 55.03
   c. 25.01
   d. 7.987

7. Write the following numbers in expanded notation.
   a. 23.27
   b. 0.3284
   c. 0.7001
   d. 5.65

8. **MC** Five hundredths, 2 thousandths and 7 ten thousandths equals:
   A. 527
   B. 52.7
   C. 5.27
   D. 0.0527
   E. 0.527

9. Copy and complete the following table by putting only one digit in each box.

		Tens	Units	.	Tenths	Hundredths	Thousandths
Example	37.684	3	7	.	6	8	4
a.	0.205			.			
b.	1.06			.			
c.	74.108			.			
d.	0.108			.			
e.	50.080			.			

10. **WE4** State the largest number in each of the following.
    a. 0.35, 0.33, 0.29
    b. 0.67, 0.69, 0.65
    c. 2.001, 2.101, 2.110

11. State the largest number in each of the following.
    a. 0.0392, 0.039 90, 0.0039   b. 2.506, 2.305, 2.559   c. 0.110 43, 0.110 49, 0.110 40

12. **WE5** Insert the appropriate < or > sign between each of the following pairs of numbers to make true statements.
    a. 3.2   2.9   b. 8.6   8.9   c. 0.64   0.67
    d. 0.29   0.39   e. 13.103   13.112

13. Insert the appropriate < or > sign between each of the following pairs of numbers to make true statements.
    a. 0.427   0.424   b. 0.580   0.508   c. 0.0101   0.0120
    d. 0.048 01   0.4801 e. 1.3830   1.3824

14. Answer the following questions.
    a. State how many decimal places we use in our currency.
    b. Write the amount of money shown here:
       i. in words
       ii. in numbers.

15. Write the following in ascending order (from smallest to largest).
    a. 0.21, 0.39, 0.17, 0.45, 0.33
    b. 0.314, 0.413, 0.420, 0.391, 0.502
    c. 0.821, 0.803, 0.811, 0.807, 0.902

16. Write the following in descending order (from largest to smallest).
    a. 0.36, 0.31, 0.39, 0.48, 0.19
    b. 0.91, 0.97, 0.90, 0.95, 0.99
    c. 1.264, 1.279, 1.273, 1.291, 1.288

17. **MC** The largest number in the list 0.4261, 0.4265, 0.4273, 0.4199, 0.3999 is:
    A. 0.4261   B. 0.4199   C. 0.4265   D. 0.3999   E. 0.4273

18. **MC** The smallest number in the list 0.4261, 0.4265, 0.4273, 0.4199, 0.3999 is:
    A. 0.4261   B. 0.4199   C. 0.4265   D. 0.3999   E. 0.4273

19. **MC** The list 0.4261, 0.4265, 0.4273, 0.4199, 0.3999 arranged from smallest to largest is:
    A. 0.4273, 0.4265, 0.4261, 0.4199, 0.3999
    B. 0.4273, 0.4261, 0.4265, 0.4199, 0.3999
    C. 0.3999, 0.4199, 0.4265, 0.4261, 0.4273
    D. 0.3999, 0.4199, 0.4261, 0.4273, 0.4265
    E. 0.3999, 0.4199, 0.4261, 0.4265, 0.4273

20. Write True (T) or False (F) for each of the following and justify your response.
    a. 76.34 has 4 decimal places.
    b. $\dfrac{6}{10} + \dfrac{3}{100} + \dfrac{4}{10\,000}$ is the same as 0.6304.
    c. 4.03 has the same value as 4.3.
    d. 29.60 has the same value as 29.6.
    e. 1.2804 could be written as $1 + \dfrac{2}{10} + \dfrac{8}{100} + \dfrac{4}{1000}$.
    f. 1090.264 51 has 5 decimal places.

21. For each of the numbers listed from **a** to **e**:

 i. state the place value of the zero
 ii. state whether the value of the number would change if the zero wasn't there. (Write yes or no.)

   a. 5.03
   b. 6.450
   c. 0.75
   d. 50
   e. 302.58

22. Year 11 students competing in their school swimming sports recorded the following times in the 50-metre freestyle, backstroke and breaststroke events.

Time (seconds) recorded by contestants							
Event	**Carolyn**	**Jessica**	**Mara**	**Jenika**	**Robyn**	**Shelley**	**Kyah**
Freestyle	37.23	39.04	40.90	38.91	37.45	37.02	37.89
Backstroke	40.23	43.87	44.08	42.65	41.98	40.29	41.05
Breaststroke	41.63	42.70	41.10	41.21	42.66	41.33	41.49

 a. State who won the freestyle event.
 b. State who won the backstroke event.
 c. State who won the breaststroke event.
 d. List the first 3 placings of the freestyle event.
 e. List the first 3 placings of the backstroke event.
 f. List the first 3 placings of the breaststroke event.
 g. Determine whether any students obtained a placing in all three events.

3.3 Converting decimals to fractions and fractions to decimals

LEARNING INTENTION

At the end of this subtopic you should be able to:
• convert decimals to fractions and fractions to decimals.

3.3.1 Converting decimals to fractions

Decimals can be written as fractions by using place values.

Number	Ones	·	Tenths	Hundredths
0.3	0	·	3	0
1.25	1	·	2	5

The decimal 0.3 can be written as $\dfrac{3}{10}$. The decimal 1.25 can be thought of as $1 + \dfrac{2}{10} + \dfrac{5}{100} = 1 + \dfrac{20}{100} + \dfrac{5}{100}$

$$= 1\dfrac{25}{100}$$
$$= 1\dfrac{1}{4}.$$

Converting decimals to fractions

- **The digits after the decimal point become the numerator, and the place value of the last digit gives the denominator.**
- **Simplify fractions by dividing the denominator and numerator by a common factor; e.g. $1\dfrac{25}{100} = 1\dfrac{1}{4}$.**

tlvd-3676

WORKED EXAMPLE 6 Converting decimals to fractions

Write the following decimals as fractions, then simplify where appropriate.

a. **0.4** b. **0.78** c. **0.5031**

THINK	WRITE
a. 1. Write the decimal.	a. 0.4
2. The numerator is 4 and the last decimal place is tenths, so the denominator is 10.	$= \dfrac{4}{10}$
3. Simplify the fraction and write the answer.	$= \dfrac{2}{5}$
b. 1. Write the decimal.	b. 0.78
2. The numerator is 78. The last decimal place is hundredths, so the denominator is 100.	$= \dfrac{78}{100}$
3. Simplify the fraction and write the answer.	$= \dfrac{39}{50}$
c. 1. Write the decimal.	c. 0.5031
2. The numerator is 5031. The last place is ten thousandths, so the denominator is 10 000.	$= \dfrac{5031}{10\,000}$

WORKED EXAMPLE 7 Converting mixed numbers to fractions

Write each of the following as a mixed number in simplest form.

a. **5.037** b. **3.192**

THINK	WRITE
a. 1. Write the decimal.	a. 5.037
2. Write the whole number part and change the decimal part to a fraction. The numerator is 37. The last decimal place is thousandths, so the denominator is 1000.	$= 5\dfrac{37}{1000}$
b. 1. Write the decimal.	b. 3.192
2. Write the whole number part and change the decimal part to a fraction. The numerator is 192 and the denominator is 1000.	$= 3\dfrac{192}{1000}$

3. Simplify the fraction.	$= 3\dfrac{192 \div 8}{1000 \div 8}$
4. Write the answer in simplest form.	$= 3\dfrac{24}{125}$

3.3.2 Converting fractions to decimals

Fractions can be written as decimals by dividing the numerator by the denominator.

For example, to convert $\dfrac{1}{4}$ into a decimal, divide 1 by 4: $4\overline{)1.00}^{\,0.25}$.

Note: Add trailing zeros to the numerator where required.

tlvd-3677

WORKED EXAMPLE 8 Converting fractions to decimals

Change the following fractions into decimals.

a. $\dfrac{2}{5}$

b. $\dfrac{1}{8}$

THINK

WRITE

a. **1.** Set out the question as for division of whole numbers, adding a decimal point and the required number of zeros.

a. $5\overline{)2.0}$

2. Divide, writing the answer with the decimal points aligned.

$5\overline{)2.0}^{\,0.4}$

3. Write the answer.

$\dfrac{2}{5} = 0.4$

b. **1.** Set out the question as for division of whole numbers, adding a decimal point and the required number of zeros.

Note: $\dfrac{1}{8} = 1 \div 8$

b. $8\overline{)1.000}^{\,0.125}$

2. Divide, writing the answer with the decimal point exactly in line with the decimal point in the question, and write the answer.

$\dfrac{1}{8} = 0.125$

By knowing the decimal equivalent of any fraction, it is possible to determine the equivalent of any multiple of that fraction.

WORKED EXAMPLE 9 Converting multiples of a known fraction to decimals

Use the results of Worked example 8 to determine decimal equivalents for these numbers.

a. $\dfrac{3}{8}$

b. $4\dfrac{5}{8}$

THINK	WRITE
a. 1. Write the decimal equivalent for the fraction with 1 as the numerator.	**a.** $\dfrac{1}{8} = 0.125$
2. Multiply both sides of this equation by the appropriate multiple (3 in this case).	$\dfrac{1}{8} \times 3 = 0.125 \times 3$
3. Write the answer.	$\dfrac{3}{8} = 0.375$
b. 1. Consider only the fraction part of the mixed number. Write the decimal equivalent of this fraction with 1 as the numerator.	**b.** $\dfrac{1}{8} = 0.125$
2. Multiply both sides of this equation by the appropriate multiple (5 in this case).	$\dfrac{1}{8} \times 5 = 0.125 \times 5$
3. Simplify both sides.	$\dfrac{5}{8} = 0.625$
4. Combine with the whole number and write the answer.	$4\dfrac{5}{8} = 4.625$

 Resources

Interactivities Conversion of decimals to fractions (int-3978)

Conversion of fractions to decimals (int-3979)

3.3 Exercise

Students, these questions are even better in jacPLUS

 Receive immediate feedback and access sample responses

 Access additional questions

 Track your results and progress

Find all this and MORE in jacPLUS

1. **WE6** Write the following decimals as fractions, then simplify where appropriate.

   a. 0.1   **b.** 0.5   **c.** 0.37   **d.** 0.6   **e.** 0.44

2. Write the following decimals as fractions, then simplify where appropriate.

   a. 0.63   **b.** 0.622   **c.** 0.719   **d.** 0.70   **e.** 0.48

3. Write the following decimals as fractions, then simplify where appropriate.

   a. 0.36   **b.** 0.9456   **c.** 0.9209   **d.** 0.4621   **e.** 0.140

4. **WE7** Write the following decimals as mixed numbers in simplest form.

   a. 1.7   b. 2.3   c. 8.4   d. 1.2   e. 4.2

5. Write the following decimals as mixed numbers in simplest form.

   a. 6.5   b. 7.27   c. 19.182   d. 6.15   e. 7.25

6. **MC** 0.17 as a fraction is:

   A. $\dfrac{17}{10}$ B. $\dfrac{17}{100}$ C. $\dfrac{17}{1000}$ D. $\dfrac{1.7}{100}$ E. $1\dfrac{7}{10}$

7. **MC** 0.207 as a fraction is:

   A. $\dfrac{207}{1000}$ B. $\dfrac{207}{100}$ C. $2\dfrac{7}{10}$ D. $20\dfrac{7}{10}$ E. 207

8. **MC** 0.48 as a fraction in simplest form is:

   A. $\dfrac{48}{100}$ B. $\dfrac{24}{50}$ C. $\dfrac{12}{25}$ D. $\dfrac{24}{100}$ E. $\dfrac{12}{50}$

9. **MC** 0.716 as a fraction in simplest form is:

   A. $\dfrac{716}{10\,000}$ B. $\dfrac{368}{560}$ C. $\dfrac{716}{1000}$ D. $\dfrac{179}{250}$ E. $\dfrac{358}{1000}$

10. **MC** 9.125 as a fraction in simplest form is:

    A. $\dfrac{9125}{1000}$ B. $\dfrac{125}{1000}$ C. $9\dfrac{125}{1000}$ D. $9\dfrac{25}{200}$ E. $9\dfrac{1}{8}$

11. **WE8** Change the following fractions to decimals.

    a. $\dfrac{3}{4}$ b. $\dfrac{1}{2}$ c. $\dfrac{3}{5}$ d. $\dfrac{1}{10}$

12. Change the following fractions to decimals.

    a. $\dfrac{9}{12}$ b. $\dfrac{1}{50}$ c. $\dfrac{6}{25}$ d. $\dfrac{3}{15}$

13. **WE9** Write $\dfrac{1}{8}$ as a decimal. Using this value, write:

    a. $\dfrac{3}{8}$ as a decimal   b. $\dfrac{7}{8}$ as a decimal   c. $\dfrac{1}{16}$ as a decimal.

14. You are competing in a long jump contest. Your first jump was 6.78 m, your second jump was $6\dfrac{8}{9}$ m and your third jump is $6\dfrac{4}{5}$ m. You are congratulated on your third jump and told it was your best. Explain whether you agree.

15. A student received the following scores in his last three tests:
    Maths $\dfrac{12}{15}$, English $\dfrac{7}{10}$ and Science $\dfrac{75}{100}$. Place his test scores in order from highest to lowest (by first converting each score to a decimal).

16. Place the following fractions in order from smallest to largest by first converting the fractions to decimals:

$$\dfrac{3}{4}, \dfrac{6}{15}, \dfrac{8}{25}, \dfrac{30}{100}, \dfrac{5}{8}$$

3.4 Rounding and repeating decimals

LEARNING INTENTION

At the end of this subtopic you should be able to:
- round decimal numbers to the required number of decimal places.

3.4.1 Rounding

When rounding decimals, look at the first digit after the number of decimal places required.

Rules for rounding

- **If the first digit after the number of decimal places required is 0, 1, 2, 3 or 4, write the number without any change.**
- **If the first digit after the number of decimal places required is 5, 6, 7, 8 or 9, add 1 to the digit in the last required decimal place.**

We can use the symbol $\approx$ to represent approximation when rounding has occurred, as shown in Worked example 10.

tlvd-3678

WORKED EXAMPLE 10 Rounding decimal numbers to a given number of decimal places

Round the following to the number of decimal places shown in the brackets.

a. 7.562 432 (2)
b. 27.875 327 (2)
c. 35.324 971 (4)
d. 0.129 91 (3)

THINK	WRITE
a. 1. Write the number and underline the required decimal place.	a. 7.56̲2 432
2. Circle the next digit and round according to the rule. *Note:* Since the circled digit is less than 5, we leave the number as it is.	= 7.56②432 ≈ 7.56
b. 1. Write the number and underline the required decimal place.	b. 27.87̲5 327
2. Circle the next digit and round according to the rule. *Note:* Since the circled digit is greater than or equal to 5, add 1 to the last decimal place that is being kept.	= 27.87⑤327 ≈ 27.88
c. 1. Write the number and underline the required decimal place.	c. 35.324 9̲71

2. Circle the next digit and round according to the rule.
   Note: Since the circled digit is greater than 5, add 1 to the last decimal place that is being kept. As 1 is being added to 9, write 0 in the last place and add 1 to the previous digit.

$= 35.3249\enclose{circle}{7}1$

≈ 35.3250

d. 1. Write the number and underline the required decimal place.

d. $0.129\underline{9}\,91$

2. Circle the next digit and round according to the rule.
   Note: Since the circled digit is greater than 5, add 1 to the last decimal place that is being kept. As 1 is being added to 9, write 0 in the last place and add 1 to the previous digit.

$= 0.129\enclose{circle}{9}1$

≈ 0.130

Note: If you need to add 1 to the last decimal place and the digit in this position is 9, the result is 10. Zero is put in the last required place and 1 is added to the digit in the next place to the left.
For example, 0.298 rounded to 2 decimal places is 0.30.

Truncation

Truncation is another method of rounding where all decimal places are dropped off past a certain point. For example, if 3.141 592 65 is truncated to 4 decimal places, it would be approximately 3.1415. If 3.141 592 65 is rounded to 4 decimal places, it would be 3.1416.

Truncation can be a useful means of limiting or removing decimal places in computer programs.

WORKED EXAMPLE 11 Truncating decimals to a given number of decimal places

Truncate the following to the number of decimal places shown in the brackets.
a. **4.3851 (3)**   b. **2.58 (0)**

THINK	WRITE
a. 1. Write the number and underline the required decimal place.	a. $4.38\underline{5}1$
2. Delete any decimal places past this point.	≈ 4.385
b. 1. Write the number and underline the required decimal place.	b. $\underline{2}.58$
2. Delete any decimal places past this point.	≈ 2

WORKED EXAMPLE 12 Rounding decimals to the nearest unit

Round 5.783 to the nearest unit.

THINK	WRITE
1. Write the decimal.	5.783
2. Look at the first digit after the decimal point. Since it is greater than 5, add 1 to the whole number (you can think of the question this way: 'Is 5.783 closer to 5 or 6?').	≈ 6

3.4.2 Rounding to the nearest 5 cents

As the smallest denomination of physical money in Australia is the
5-cent coin, prices must be rounded to the nearest 5 cents upon payment in cash.
Rounding to the nearest 5 cents works in the same way as rounding to the nearest 10:
determine whether the number is closest to 0, 5 or 10, then round accordingly.

Rounding guidelines for cash transactions	
Final cash amount	**Round to:**
1 and 2 cents	nearest 10
3 and 4 cents	nearest 5
6 and 7 cents	nearest 5
8 and 9 cents	nearest 10

Source: www.accc.gov.au

WORKED EXAMPLE 13 Rounding to the nearest 5 cents

Danica had $51.67 in her bank account. She wanted to withdraw all her money in cash, so the bank rounded the amount to the nearest 5 cents. State how much money the teller gave Danica.

THINK	WRITE
1. Write the actual amount she had in her account.	$51.67
2. Determine whether the last digit is closer to 5 or to 10, then rewrite the approximate value. *Note:* Alternatively, it can be seen that 67 cents is closer to 65 cents than 70 cents.	$\approx$ $51.65
3. Write the answer as a sentence.	Danica will receive $51.65 from the bank.

3.4.3 Repeating decimals

A **terminating decimal** is a decimal number that ends (terminates) after any number of decimal places; for example, $\frac{9}{20} = 0.45$ is a terminating decimal.

A **recurring decimal**, or a repeating decimal, occurs when we divide the denominator into the numerator and the answer keeps repeating.

For example, $\frac{1}{3} = 0.333\ldots$

$\frac{1}{6} = 0.1666\ldots$

tlvd-3680

WORKED EXAMPLE 14 Repeating decimals

Convert $\frac{1}{11}$ to a decimal. Continue dividing until a pattern emerges, then round the answer to 2 decimal places.

THINK	WRITE
1. Set out the question as for division of whole numbers, adding a decimal point and enough zeros to see a pattern emerging.	$11\overline{)1.0000}$

2. Divide, writing the answer with the decimal
   point exactly in line with the decimal point in
   the question. (The amount left over each time is
   10, then 1, then 10, then 1 again. The decimal
   answer is also repeating.)

$$11\overline{\smash)1.{}^1 0\, {}^{10}0\, {}^1 0\, {}^{10}0\, {}^1 0\, {}^{10}0\, {}^1 0\, {}^{10}0}$$
$$0.0\ 9\ 0\ 9\ 0\ 9\ 0\ 9\ ...$$

3. Write the approximate answer rounded to
   2 decimal places.

$$\frac{1}{11} \approx 0.09$$

Shorter ways of writing recurring decimals

Recurring decimals can be written in one of the following shorter ways for an exact answer.
- 4.6666 … could be written as $4.\dot{6}$ (with a dot above the repeating part of the decimal).
- 3.512 512 … could be written as $3.\dot{5}1\dot{2}$ (with a dot above the first and last digits of the repeating
  part), or as $3.\overline{512}$ (with a line above the repeating part of the decimal).
- Like terminating decimals, the decimal equivalent of a fraction can be used to determine the
  decimal equivalent of any multiple of that fraction.

WORKED EXAMPLE 15 Equivalent repeating fractions

Use the result from Worked example 14 to calculate the decimal equivalent of $\dfrac{6}{11}$.

THINK

1. Write the decimal equivalent for the fraction
   with 1 as the numerator. In this case, it is an
   infinite recurring decimal.

2. Multiply both sides of this equation by the
   appropriate multiple (6 in this case).

3. Simplify and write the answer.

WRITE

$$\frac{1}{11} = 0.090\,909 \dots$$

$$\frac{1}{11} \times 6 = 0.090\,909 \dots \times 6$$
$$= 0.545\,454 \dots$$

$$\frac{6}{11} = 0.\overline{54}$$

on Resources

Interactivities   Rounding (int-3980)

Repeating decimals (int-3981)

3.4 Exercise

1. **WE10** Round the following to the number of decimal places shown in the brackets.
   a. 0.3342 (2)   b. 0.874 (2)   c. 3.247 30 (2)   d. 17.039 91 (3)   e. 3.0702 (1)

2. Round the following to 2 decimal places.
   a. 3.128 74   b. 110.7247   c. 67.370 49   d. 0.0073   e. 7.176 221

3. Round the following to 1 decimal place.
   a. 0.520   b. 0.69   c. 12.16   d. 32.35   e. 5.83

4. **WE11** Truncate the following to the number of decimal places shown in the brackets.
   a. 100.07 (1)   b. 15.63 (0)   c. 99.94 (1)   d. 93.994 (2)   e. 0.8989 (3)

5. **MC** 13.179 rounded to 2 decimal places is:
   A. 13.17   B. 13.20   C. 13.18   D. 13.27   E. 13.19

6. **MC** 2.998 rounded to 1 decimal place is:
   A. 3.0   B. 2.9   C. 2.8   D. 3.1   E. 3.9

7. **WE12** Round the following to the nearest unit.
   a. 12.7   b. 6.2   c. 5.7   d. 87.5   e. 121.1

8. Round the following to the nearest unit.
   a. 31.67   b. 43.0489   c. 3237.50   d. 0.18   e. 0.507

9. Write the following infinite recurring decimals using one of the short forms.
   a. 3.555 ...   b. 0.333 ...   c. 47.111 ...   d. 0.353 535 ...

10. Write the following infinite recurring decimals using one of the short forms.
    a. 0.125 125 ...   b. 7.431 243 12 ...   c. 121.921 921 ...   d. 0.127 7777 ...

11. **MC** 1.8888 ... written as an exact answer is:
    A. $1.\dot{8}$ B. 1.888   C. 1.88   D. 1.9   E. 1.889

12. **MC** $\dfrac{3}{7}$ as a decimal in exact form is:
    A. $0.\dot{4}285 7\dot{1}$ B. $0.\overline{428}$ C. 0.4   D. 0.428 517 4   E. $2.\dot{3}$

13. **WE13** In the supermarket, Chloe's shopping bill came to $58.78. As there are no 1- or 2-cent pieces, this amount must be rounded to the nearest 5 cents.
    State how much Chloe will pay in cash for her shopping.

14. **WE14** Convert each of the following to a decimal. Continue dividing until a pattern emerges, then round the answer to the number of decimal places indicated in the brackets.

    a. $\frac{1}{3}$ (1)
    b. $\frac{1}{9}$ (1)
    c. $\frac{5}{12}$ (2)

15. Convert each of the following to a decimal. Continue dividing until a pattern emerges, then round the answer to the number of decimal places indicated in the brackets.

    a. $\frac{1}{6}$ (2)
    b. $\frac{2}{11}$ (3)
    c. $\frac{7}{15}$ (2)

16. **WE15** Determine decimal equivalents for the following fractions.

    a. $\frac{9}{2}$
    b. $7\frac{1}{3}$
    c. $\frac{1}{4}$

17. Round the following to the nearest ten.

    a. 12
    b. 67
    c. 158
    d. 362
    e. 375

18. Round the following to the nearest hundred.

    a. 530
    b. 480
    c. 238
    d. 5762
    e. 8660.082

19. Round the following to the nearest thousand.

    a. 2437
    b. 13 200
    c. 21 730
    d. 9374
    e. 2098

20. Rank the following decimals from smallest to largest:

$$0.295, 0.29\dot{5}, 0.2\dot{9}5, 0.\dot{2}9\dot{5}$$

21. Using a calculator, Warren worked out that the piece of timber required for a door frame should have a length of 2.426 78 metres. Realistically, the timber can be measured only to the nearest millimetre (nearest thousandth of a metre). Explain what measurement should be used for the length of the timber and why 2.426 78 m is an unreasonable measurement for timber.

22. The maximum daily temperature was recorded as 32.8 °C. In the news broadcast, the presenter quoted this to the nearest degree. Write the temperature quoted.

23. a. Round the decimal number 0.7286 to the nearest:

    i. tenth
    ii. hundredth
    iii. thousandth.

    b. Explain whether the rules for rounding a decimal up or down depend on the number of decimal places.

24. Sasha went to the supermarket to grab a few items. The items she purchased cost $1.85, $5.65, $0.75, $6.40 and $3.55. By rounding each item's cost to the nearest unit, approximate the total cost.

3.5 Adding and subtracting decimals

3.5.1 Adding decimals

Decimals can be added by hand. Set out the addition in vertical columns and line up the decimal points so that the digits with the same place value are underneath each other. If the question is not written in columns, it is necessary to rewrite it with the decimal points lined up.

Note: Fill the empty places with zeros.

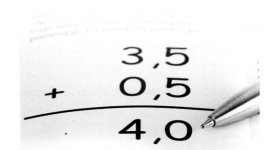

WORKED EXAMPLE 16 Adding numbers with decimals

Calculate each of the following.

a. 1.6
   +0.2
   ─────

b. 15.76
   +3.25
   ─────

c. 2.25
   7.154
   +4.0
   ─────

THINK

a. Copy the question exactly and add the digits as you would for whole numbers, working from right to left. Write the decimal point directly below the decimal points in the question.

b. Copy the question exactly and add the digits as you would for whole numbers, working from right to left. Write the decimal point directly below the decimal points in the question.

c. Write the question, replacing the spaces with zeros. Add the digits as you would for whole numbers, working from right to left. Write the decimal point directly below the decimal points in the question.

WRITE

a. 1.6
   +0.2
   ─────
    1.8

b. 15.76
   +3.25
   ─────
   19.01

c. 2.250
    7.154
   +4.000
   ──────
   13.404

WORKED EXAMPLE 17 Adding numbers with decimals

Rewrite in columns and then add 0.38 + 2.7 + 10.324.

THINK	WRITE
1. Write the question in columns with the decimal points directly beneath each other, with the zeros included.	$\begin{array}{r} 0.380 \\ 2.700 \\ +10.324 \\ \hline 13.404 \end{array}$
2. Add the digits as you would for whole numbers. Write the decimal point directly below the decimal points in the question.	

3.5.2 Subtracting decimals

Decimals can be subtracted by hand. Set out the subtraction in vertical columns and line up the decimal points so that the digits with the same place value are underneath each other. If the question is not written in columns, it is necessary to rewrite it with the decimal points lined up.

Note: Fill the empty places with zeros.

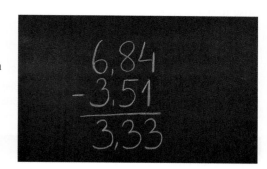

WORKED EXAMPLE 18 Subtracting numbers with decimals

Calculate:
$$\begin{array}{r} 0.68 \\ -0.24 \\ \hline \end{array}$$

THINK	WRITE
Copy the question exactly and subtract the digits as you would for whole numbers, working from right to left. Write the decimal point directly below the decimal points in the question.	$\begin{array}{r} 0.68 \\ -0.24 \\ \hline 0.44 \end{array}$

WORKED EXAMPLE 19 Subtracting numbers with decimals

Rewrite the following in columns, then subtract.

a. 2.64 − 0.37   b. 4.682 − 0.75

THINK	WRITE
a. Write in columns with the decimal points directly underneath each other. Subtract, and insert the decimal point directly below the other decimal points in the question.	a. $\begin{array}{r} \overset{5}{2}.\overset{1}{6}4 \\ -0.37 \\ \hline 2.27 \end{array}$

▶

b. Write in columns with the decimal points directly underneath each other, adding zeros as appropriate. Subtract as you would for whole numbers and insert the decimal point directly below the other decimal points.

b. $\begin{array}{r}{}^{3}\!4.{}^{1}682\\-0.750\\\hline 3.932\end{array}$

Checking by estimating

Answers to decimal addition and subtraction can be checked by estimating. Round each decimal to the nearest whole number or to a similar number of decimal places and then add or subtract them.

For example, in Worked example 17, you could check the answer by rounding to get an estimate by adding $0.4 + 3 + 11 = 14.4$, which is close to 14.274.

In Worked example 18, you could check the answer by rounding to get an estimate of $0.7 - 0.2 = 0.5$, which is close to 0.42.

 Resources

 Interactivities Addition of decimals (int-3982)
Checking by estimating (int-3983)
Subtraction of decimals (int-3984)

3.5 Exercise

Students, these questions are even better in jacPLUS

 Receive immediate feedback and access sample responses

 Access additional questions

 Track your results and progress

Find all this and MORE in jacPLUS

1. **WE16a, b** Calculate the following.

   a. $\begin{array}{r}1.6\\+2.2\\\hline\end{array}$ **b.** $\begin{array}{r}2.57\\+1.01\\\hline\end{array}$ **c.** $\begin{array}{r}9.064\\+2.757\\\hline\end{array}$ **d.** $\begin{array}{r}36.058\\+48.753\\\hline\end{array}$

2. **WE16c** Calculate the following, after filling the blank spaces.

   a. $\begin{array}{r}8.25\\+0.7\\\hline\end{array}$ **b.** $\begin{array}{r}3.26\\+18.6460\\\hline\end{array}$ **c.** $\begin{array}{r}4.2\\62.013\\+1946.12\\\hline\end{array}$ **d.** $\begin{array}{r}54.1584\\3.2\\+102.43\\\hline\end{array}$

3. **WE17** Rewrite the following in columns, then add. Check your answer by rounding to get an estimate.

   a. $2.1 + 3.7$
   c. $2.863 + 0.18$
   b. $8.2 + 0.7$
   d. $12.864 + 2.274$

4. Rewrite the following sums, then add. Check your answer by rounding to get an estimate.

   a. $10.8271 + 6.5$
   c. $0.24 + 3.16 + 8.29$
   b. $1.8 + 18.6329$
   d. $14.23 + 1.06 + 86.29 + 3.64$

5. **WE18** Calculate the following, filling the spaces with zeros as required. Check your answer by rounding to get an estimate.

   a. 8.87
   -8.35
   _____

   b. 15.462
   -7.034
   _____

   c. 0.7312
   -0.5473
   _____

   d. 5.0082
   -2.5471
   _____

6. Calculate the following, filling the spaces with zeros as required. Check your answer by rounding to get an estimate.

   a. 43.05
   -17
   _____

   b. 98.26
   -9.07
   _____

   c. 232
   -64.2
   _____

   d. 3.2
   -0.467
   _____

7. **WE19** Rewrite the following in columns, then subtract. Check your answer by rounding to get an estimate.

   a. $6.48 - 2.5$ b. $18.08 - 5.22$ c. $0.687 - 0.36$ d. $15.226 - 11.08$

8. Rewrite the following in columns, then subtract. Check your answer by rounding to get an estimate.

   a. $8.742 - 3.8$
   b. $12.2 - 8.911$
   c. $200.482 - 72.0462$
   d. $38 - 21.234$

9. **MC** $1.6 + 4.8$ equals:

   A. 5.4   B. 6.4   C. 0.54   D. 0.64   E. 64

10. **MC** $1.84 + 0.61 + 4.07$ equals:

    A. 6.52   B. 6.42   C. 5.42   D. 5.52   E. 0.652

11. **MC** $0.39 - 0.15$ equals:

    A. 0.0024   B. 0.024   C. 0.24   D. 2.4   E. 24

12. **MC** $1.4 - 0.147$ can be rewritten as:

    A. 1.4
    -0.417
    _____

    B. 1.400
    -0.417
    _____

    C. 1.40
    -1.47
    _____

    D. 1.004
    -0.147
    _____

    E. 1.040
    -0.147
    _____

13. **MC** $150.278 - 0.99$ equals:

    A. 150.728   B. 149.288   C. 1.492 88   D. 159.388   E. 1.593 88

14. James deposited $52.60 into his bank account. If his balance before the deposit was $252.80, determine his new bank balance.

15. A triathlon consists of a 0.8-kilometre swim, a 24.56-kilometre bike ride and an 8.2-kilometre run. Calculate how far the competitors travel altogether.

16. Mia walked 5.4 kilometres to school, 1.2 kilometres from school to the shops, 1.8 kilometres from the shops to a friend's house and finally 3.6 kilometres from her friend's house to her home. Calculate how far Mia walked.

17. For lunch Jill ordered 1 potato cake, 1 dim sim, minimum chips and a milkshake from the menu shown below.
Calculate how much Jill spent on her lunch.

Menu			
Flake	$3.50	Coffee	$2.20
Whiting	$3.50	Tea	$2.20
Dim sims	$0.60	Soft drinks	$1.80
Potato cakes	$0.50	Milkshakes	$3.00
Minimum chips	$2.50	Water	$1.80

18. Raffa works at McDonald's. A customer buys $24.75 worth of food and gives Raffa a $50 note. Calculate the change Raffa gives the customer.

19. A jockey has a mass of 52.3 kilograms. After exercising and training for 2 days and spending time in a sauna, the jockey has lost 1.82 kilograms. Determine the jockey's mass now.

20. If 2.48 metres is cut from a piece of pipe that is 4.2 metres long, calculate how much pipe is left.

21. Cathy Freeman won a particular 400-metre race in 51.35 seconds. In her next race, her time was 2.97 seconds faster than this. Determine Cathy's time in the latter race.

22. The following table shows the times recorded for each swimmer in the under-17, 50-metre freestyle relay for 6 teams.

Time for each swimmer (in seconds)				
Team	Swimmer 1	Swimmer 2	Swimmer 3	Swimmer 4
1	36.7	41.3	39.2	35.8
2	38.1	46.5	38.8	35.9
3	34.6	39.2	39.9	35.2
4	41.6	40.8	43.7	40.5
5	37.9	40.2	38.6	39.2
6	38.3	39.1	40.8	37.6

a. Determine the total time for each team. Put your results in a table.
b. Identify the team that won the relay.
c. Determine the difference in time between the first- and second-placed teams.

23. Australian coins, which are minted by the Royal Australian Mint, have the following masses:

5c	2.83 g	10c	5.65 g	20c	11.3 g
50c	15.55 g	$1	9.0 g	$2	6.6 g

I have 6 coins in my pocket, with a total value of $2.45. Determine what the total mass of these coins could be.

24. You purchased some shares over a 6-month period. The price fell by $21.55 in the first month, rose by $18.97 in the second month, rose by $15.62 in the third month and fell by $5.34 in the fourth month. The price remained the same for the next two months.
Explain whether the shares increased or decreased in value over the six months, and by how much.

3.6 Multiplying decimals (including powers of 10)

LEARNING INTENTION

At the end of this subtopic you should be able to:
• multiply decimals, including by multiples of 10
• square decimal numbers.

3.6.1 Multiplying decimals

The following calculation shows the multiplication 1.7×2.3. The diagram is a visual representation of each step in the calculation. There are 1.7 rows of 2.3, or 1.7 groups of 2.3.

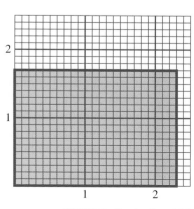

$$
\begin{array}{r}
1.7 \\
\times\, 2.3 \\
\hline
0.51 \leftarrow 0.3 \times 1.7 = 0.51 \\
3.40 \leftarrow 2.0 \times 1.7 = 3.40 \\
\hline
3.91 \leftarrow \textbf{Total}
\end{array}
$$

The smallest place value in the answer is determined by multiplying the smallest place values from each of the decimal numbers. The first multiplication is $\frac{3}{10} \times \frac{7}{10} = \frac{21}{100}$, so the smallest place value in the answer will be hundredths.

> **Multiplying decimals**
>
> **The number of decimal places in the answer is equal to the total number of decimal places in the numbers being multiplied.**

WORKED EXAMPLE 20 Multiplying numbers with decimals

Calculate the following.

a. 13.4×6

b. $\begin{array}{r} 2.37 \\ \times\, 0.2 \\ \hline \end{array}$

c. $\begin{array}{r} 0.4629 \\ \times\, 2.6 \\ \hline \end{array}$

THINK	WRITE
a. 1. Rewrite and multiply the digits, ignoring the decimal places.	a. $\begin{array}{r} 134 \\ \times 6 \\ \hline 804 \end{array}$
2. Count the number of decimal places altogether (1) and put in the decimal point.	$13.4 \times 6 = 80.4$
3. Check the answer by rounding: $10 \times 6 = 60$, which is close to 80.4.	
b. 1. Multiply, ignoring the decimal places.	b. $\begin{array}{r} 237 \\ \times 2 \\ \hline 474 \end{array}$
2. Count the number of digits after the point in both the decimals being multiplied and insert the decimal point in the answer. There are 2 decimal places in 3.27 and 1 in 0.2, so there will be 3 decimal places in the answer.	$2.37 \times 0.2 = 0.474$
3. Check the answer by rounding: $2 \times 0.2 = 0.4$, which is close to 0.474.	
c. 1. Multiply, ignoring the decimal places.	c. $\begin{array}{r} 4629 \\ \times\ 26 \\ \hline 27\,774 \\ +\,92\,580 \\ \hline 120\,354 \end{array}$
2. Count the number of digits after the point in both the decimals being multiplied. Insert the decimal point in that position in the answer. There are 4 decimal places in 0.4629 and 1 decimal place in 2.6, so there will be 5 decimal places in the answer.	$0.4629 \times 2.6 = 1.203\,54$

3.6.2 Squaring decimals

To square a decimal, multiply the number by itself. The following diagrams show how squaring decimal numbers can be represented visually.

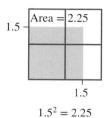

1.5² = 2.25

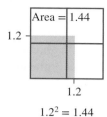

1.2² = 1.44

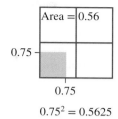

0.75² = 0.5625

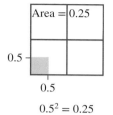

0.5² = 0.25

Squaring decimals

The number of decimal places in the square is twice the number of decimal places in the original number.

---

tlvd-3684

WORKED EXAMPLE 21 Squaring decimal numbers

Calculate the following.

a. 0.4^2

b. 1.4^2

THINK	WRITE
a. 1. Multiply the number by itself, ignoring the decimal places.	a. $4 \times 4 = 16$
2. Count the number of digits after the decimal point in both the decimals being multiplied and insert the decimal point in the answer. There will be 2 decimal places in the answer.	$0.4 \times 0.4 = 0.16$
3. Write the answer.	$0.4^2 = 0.16$
b. 1. Multiply the number by itself, ignoring the decimal places.	b. $14 \times 14 = 196$
2. Count the number of digits after the point in both the decimals being multiplied and insert the decimal point in the answer. There will be 2 decimal places in the answer.	$1.4 \times 1.4 = 1.96$
3. Write the answer.	$1.4^2 = 1.96$

3.6.3 Multiplying by multiples of 10

Multiplying by a multiple of 10 can be carried out by writing the multiple of 10 as the product of two factors. The factors used are the non-zero digits from the multiple of 10 and the corresponding power of 10.

For example, $8700 = 87 \times 100$, so multiplying by 8700 is the same as multiplying by 87 and then multiplying by 100.

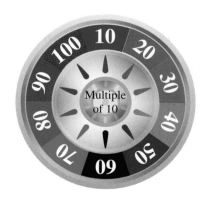

Multiplying numbers with decimals by powers of 10

To multiply a decimal number by a power of 10, move the decimal point to the right by the number of zeros.

For example:
- **to multiply by 10, move the decimal point 1 place to the right**
- **to multiply by 100, move the decimal point 2 places to the right**
- **to multiply by 1000, move the decimal point 3 places to the right.**

WORKED EXAMPLE 22 Multiplying by multiples of 10

Calculate the following.

a. 4.2×400

b. $0.024 \times 15\,000$

THINK	WRITE
a. 1. Multiplying by 400 is the same as first multiplying by 4 and then multiplying by 100. Calculate 4.2×4.	a. $\begin{array}{r} 4.2 \\ \times\ 4 \\ \hline 16.8 \end{array}$
2. Multiply the result by 100. Move the position of the decimal point 2 places to the right.	$16.8 \times 100 = 1680$
3. Write the answer.	$4.2 \times 400 = 1680$
b. 1. Multiplying by 15 000 is the same as first multiplying by 15 and then multiplying by 1000. Calculate 0.024×15.	b. $\begin{array}{r} 0.024 \\ \times\ 15 \\ \hline 120 \\ 240 \\ \hline 0.360 \end{array}$
2. Multiply the result by 1000. Move the position of the decimal point 3 places to the right.	$0.360 \times 1000 = 360$
3. Write the answer.	$0.024 \times 15\,000 = 360$

Recognising short cuts

When multiplying by some common decimals, such as 0.1 or 0.25, you can use short cuts to help you complete the calculation more efficiently.

For example, the fractional equivalent of 0.1 is $\dfrac{1}{10}$, so instead of multiplying by 0.1 you can multiply by $\dfrac{1}{10}$, which is the same as dividing by 10. Similarly, the fractional equivalent of 0.25 is $\dfrac{1}{4}$, so 0.25 of $60 is equivalent to $\dfrac{1}{4}$ of $60, which is equivalent to $60 ÷ 4.

 Resources

▶ **Video eLesson** Multiplication of decimals (eles-2311)

✦ **Interactivities** Multiplication of decimals (int-3985)

3.6 Exercise

Students, these questions are even better in jacPLUS

 Receive immediate feedback and access sample responses

 Access additional questions

 Track your results and progress

Find all this and MORE in jacPLUS ▶

1. Calculate the following.
   a. 2.5×3 **b.** 12.6×8 **c.** 10.4×5 **d.** 21.24×4

2. Calculate the following.
   a. 23.16×6 **b.** 64.87×8 **c.** 1.064×6 **d.** 0.245×3

3. **WE20a** Calculate the following.
   a. 1.8×0.4 **b.** 5.2×0.6 **c.** 0.8×0.4 **d.** 8.8×0.7

4. Calculate the following.
   a. 0.55×0.4 **b.** 0.56×0.3 **c.** 0.78×0.5 **d.** 0.58×0.7

5. Calculate the following.
   a. 0.002×0.04 **b.** 0.003×0.004 **c.** 0.7×0.09 **d.** 0.028×0.005

6. **WE20b** Calculate the following.
   a. 0.25×1.6 **b.** 0.42×2.4 **c.** 0.52×3.6 **d.** 0.05×4.5

7. **MC** When calculating 6.21×0.52, the number of decimal places in the answer is:
   A. 0   **B.** 1   **C.** 2   **D.** 3   **E.** 4

8. **MC** 0.3×0.3 equals:
   A. 0.009   **B.** 0.09   **C.** 0.9   **D.** 9   **E.** 90

9. **MC** 1.4×0.8 equals:
   A. 1.12   **B.** 8.2   **C.** 82   **D.** 11.2   **E.** 112

10. **MC** 0.0312×0.51 equals:
    A. 0.001 591 2   **B.** 0.015 912   **C.** 0.156 312   **D.** 0.159 12   **E.** 1.5912

11. **WE21** Calculate the following.
    a. 0.04^2 **b.** 1.2^2 **c.** 2.15^2

12. **WE22** Calculate the following.
    a. 0.134×6000 **b.** $0.565 \times 12\,000$ **c.** $9.2 \times 52\,000$

13. Calculate the following.
    a. 4.26×300 **b.** 6.48×500 **c.** 34.7×900

14. Calculate the following by moving the position of the decimal point.
    a. 6.48×10
    b. 13.896×100
    c. 0.2708×1000
    d. $217.148\,96 \times 1000$
    e. $0.820\,496\,783 \times 100\,000$

15. Evaluate the following by moving the position of the decimal point.
    a. $32.689\,043\,267 \times 100\,000$
    b. $0.984\,326\,641 \times 1\,000\,000$
    c. $0.000\,278\,498\,32 \times 1\,000\,000$
    d. 0.46×1000

16. For the following calculations:
    i. estimate the answer by rounding each number to the first digit
    ii. calculate the answers using written methods
    iii. check your answer using a calculator.
    a. 5.8×2200
    b. $12.1 \times 52\,410$
    c. 27.3×1542

17. Change the following amounts of money to cents. (*Hint:* There are 100 cents in one dollar.)
    a. $48
    b. $217
    c. $18
    d. $17.25
    e. $64.40

18. Allison, the school principal, is providing icy poles for 500 students. Calculate how much it will cost if each icy pole costs $1.25.

19. Calculate the area of the tennis court shown, given that area = length × width.

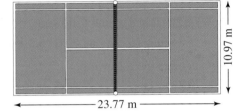

20. Jolie bought 45.5 litres of petrol at 95.9 cents per litre. She paid 4500 cents at the counter.
    a. Convert 4500 cents into dollars.
    b. Calculate how much change Jolie was given.

21. Miguel bought 0.85 kilograms of minced meat at a butcher's shop. The meat costs $9.50 per kilogram. Calculate how much Miguel paid for the meat. *Note:* Round to the nearest 5c.

22. James is using a recipe for chocolate chip muffins to make 1.5 times the amount in the recipe. If the recipe lists 0.25 litres of milk in the ingredients, determine the amount of milk James should use for his muffins.

3.7 Dividing decimals (including powers of 10)

LEARNING INTENTION

At the end of this subtopic you should be able to:
• divide a decimal number by a whole number or a decimal number
• divide a decimal number by a multiple of 10.

3.7.1 Dividing a decimal by a whole number

The method for dividing a decimal number by a whole number is the same as the method used for dividing whole numbers. A visual representation of $2.4896 \div 4$ is shown here.

	2.4896
	4

2.4896 ÷ 4 can be interpreted as 'how many times does 4 divide into 2.4896?' As shown, 4 divides into 2.4896 less than once.

WORKED EXAMPLE 23 Dividing a decimal by a whole number

Calculate the value of 2.4896 ÷ 4.

THINK	WRITE
1. Set out the question as you would for whole numbers.	$4\overline{)2.4896}$
2. Divide 4 into the first digit of 2.4896 (2 ÷ 4 = 0 remainder 2). Write 0 above 2, and write the remainder beside the next digit, as shown in black.	$4\overline{)2.^24896}$ with 0 above
3. The second digit in the number being divided is to the right of the decimal point, so write the decimal point in the answer directly above the decimal point in the question, as shown in pink.	$\begin{array}{r}0.\\4\overline{)2.^24896}\end{array}$
4. Divide 4 into the second digit, which includes the carried 2 (24 ÷ 4 = 6 remainder 0). Write 6 above 4, as shown in plum.	$\begin{array}{r}0.6\\4\overline{)2.^24896}\end{array}$
5. Divide 4 into the third digit (8 ÷ 4 = 2). Write 2 above 8, as shown in pink.	$\begin{array}{r}0.62\\4\overline{)2.^24896}\end{array}$
6. Divide 4 into the fourth digit (9 ÷ 4 = 2 remainder 1). Write 2 above 9, and write the remainder beside the next digit, as shown in purple.	$\begin{array}{r}0.622\\4\overline{)2.^24^{8}9^16}\end{array}$
7. Divide 4 into the fifth digit, which includes the carried 1 (16 ÷ 4 = 4). Write 4 above 6, as shown in orange.	$\begin{array}{r}0.6224\\4\overline{)2.^24^{8}9^16}\end{array}$
8. Write the answer.	2.4896 ÷ 4 = 0.6224

Sometimes, when you are dividing numbers, there will be a remainder.
For example, consider 15.3 ÷ 4:

$$\begin{array}{r}3.8\\4\overline{)15.^33^1}\end{array}\text{ remainder 1}$$

Instead of leaving a remainder, add zeros to the end of the decimal and keep dividing until there is no remainder.

$$\begin{array}{r}3.825\\4\overline{)15.^33^10^20}\end{array}$$

tlvd-3686

WORKED EXAMPLE 24 Dividing a decimal by a whole number

Calculate $18.58 \div 5$. Add zeros and keep dividing until there is no remainder.

THINK	WRITE
1. Set up the division. Write the decimal point in the answer directly above the decimal point in the question and divide as for short division, adding zeros as required.	$5\overline{\smash{\big)}18.{}^{3}58{}^{3}0}$ with quotient 3.716
2. Check the answer by rounding: $20 \div 5 = 4$, which is close to 3.716.	

3.7.2 Dividing a decimal number by a multiple of 10

When dividing by a multiple of 10, factorise the multiple to give a power of 10 and its other factor. Divide by the other factor first, and then by the power of 10.

For example, to divide by 500, divide first by 5 and then by 100.

> ### Dividing by a power of 10
>
> **To divide by a power of 10, move the decimal point to the left by the number of zeros.**
>
> **For example:**
> - **to divide by 10, move the decimal point 1 place to the left**
> - **to divide by 100, move the decimal point 2 places to the left**
> - **to divide by 1000, move the decimal point 3 places to the left, and so on.**

WORKED EXAMPLE 25 Dividing a decimal by a multiple of 10

Calculate the following.

a. $3.6 \div 30$

b. $13.6 \div 4000$

THINK	WRITE
a. 1. Dividing by 30 is the same as first dividing by 3 and then dividing by 10. First, divide by 3.	a. $3\overline{\smash{\big)}3.6}$ with quotient 1.2
2. To divide by 10, move the position of the decimal point 1 place to the left.	$1.2 \div 10 = 0.12$
3. Write the answer.	$3.6 \div 30 = 1.2$
b. 1. Dividing by 4000 is the same as dividing by 4 and then dividing by 1000. First, divide by 4.	b. $4\overline{\smash{\big)}13.6}$ with quotient 3.4
2. To divide by 1000, move the position of the decimal point 3 places to the left.	$3.4 \div 1000 = 0.0034$
3. Write the answer.	$13.6 \div 4000 = 0.0034$

3.7.3 Dividing a decimal number by another decimal number

A visual representation of $2.724 \div 0.4$ is shown below.

$2.724 \div 0.4$ can be interpreted as 'how many times does 0.4 divide into 2.724?'

$$2.724 \div 0.4 = 6.81$$

Dividing one decimal number by another

- **When dividing one decimal by another, multiply the decimal you are dividing by (divisor) by a power of 10 to make it a whole number.**

  Multiply the other decimal *by the same power of* 10, as shown in the following example.

  First multiply this number by 10 to make a whole number.

  Then multiply this number by 10 also.

  $$2.724 \div 0.4$$

- **This is the same as writing an equivalent fraction with a whole number as the denominator:**

$$\frac{2.724}{0.4} \times \frac{10}{10} = \frac{27.24}{4}$$

tlvd-3687

WORKED EXAMPLE 26 Dividing a decimal by a decimal

Calculate the following.

a. $32.468 \div 0.4$

b. $4.0284 \div 0.12$

THINK	WRITE
a. 1. Rewrite the question as a fraction.	a. $32.468 \div 0.4 = \dfrac{32.468}{0.4}$
2. Multiply both the numerator and the denominator by the appropriate power of 10.	$= \dfrac{32.468}{0.4} \times \dfrac{10}{10}$
	$= \dfrac{324.68}{4}$
3. Divide the decimal by the whole number.	$\begin{array}{r} 81.17 \\ 4\overline{)324.6^28} \end{array}$
4. Write the answer.	$32.468 \div 0.4 = 81.17$

b. 1. Rewrite the question as a fraction.

2. Multiply both the numerator and the denominator by the appropriate power of 10.

3. Divide the decimal by the whole number.

4. Write the answer.

b. $4.0284 \div 0.12$

$$= \frac{4.0284}{0.12} \times \frac{100}{100}$$

$$= \frac{402.84}{12}$$

$$12 \overline{\smash{)}40^4 2.^6 8^8 4} \quad \frac{33.57}{}$$

$4.0284 \div 0.12 = 33.57$

WORKED EXAMPLE 27 Dividing a decimal by a decimal

Calculate how many litres of petrol could be purchased for $74.88 if 1 litre costs $1.44.

THINK	WRITE
1. Write the question.	$74.88 \div 1.44$
2. Rewrite the question as a fraction.	$= \dfrac{74.88}{1.44}$
3. Multiply the numerator and denominator by the appropriate multiple of 10 (in this case 100). Alternatively, the decimal point could be moved twice to the right in both numbers so that the divisor is a whole number ($74.88 \div 1.44 = 7488 \div 144$).	$= \dfrac{74.88}{1.44} \times \dfrac{100}{100}$ $= \dfrac{7488}{144}$
4. Divide the decimal by the whole number. Alternatively, use a calculator.	$= 144 \overline{\smash{)}748^{28}8} \quad \frac{52}{}$ $74.88 \div 1.44 = 52$
5. Write the answer in a sentence.	Fifty-two litres of petrol could be purchased for $74.88.

---

on Resources

▶ **Video eLesson** Division of decimals (eles-1877)

✦ **Interactivities** Division of decimals by a multiple of 10 (int-3990)

Division of a decimal by another decimal (int-3991)

3.7 Exercise

1. **WE23** Calculate the following.
   a. $4.8 \div 8$
   b. $28.7 \div 7$
   c. $5.90 \div 5$
   d. $6.16 \div 4$
   e. $9.68 \div 4$

2. Calculate the following.
   a. $2.564 \div 2$
   b. $7.021 \div 7$
   c. $32.5608 \div 8$
   d. $27.4376 \div 4$
   e. $126.4704 \div 6$

3. Calculate the following.
   a. $49.92 \div 16$
   b. $35.706 \div 11$
   c. $17.108 \div 14$
   d. $77.052 \div 12$
   e. $144.2904 \div 12$

4. **WE24** Calculate the following. In each case, add zeros and keep dividing until there is no remainder.
   a. $4.7 \div 2$
   b. $9.5 \div 2$
   c. $9.3 \div 5$

5. Calculate the following. In each case, add zeros and keep dividing until there is no remainder.
   a. $14.7 \div 6$
   b. $8.5 \div 4$
   c. $55.6 \div 8$

6. Calculate the following by changing the position of the decimal point.
   a. $14.07 \div 10$
   b. $968.13 \div 100$
   c. $620.8 \div 1000$
   d. $3592.87 \div 1000$
   e. $2349.78 \div 100\,000$

7. Calculate the following by changing the position of the decimal point.
   a. $9.0769 \div 100\,000$
   b. $802\,405.6 \div 1\,000\,000$
   c. $152.70 \div 1\,000\,000$
   d. $0.7205 \div 10\,000$
   e. $0.0032 \div 1\,000\,000$

8. **WE25** Calculate the following.
   a. $13.25 \div 50$
   b. $23.7 \div 30$
   c. $424 \div 400$
   d. $178.5 \div 300$
   e. $5.22 \div 3000$

9. **WE26a** Calculate each of the following.
   a. $4.5 \div 0.5$
   b. $4.8 \div 0.8$
   c. $8.1 \div 0.9$
   d. $3.2 \div 0.8$
   e. $0.248 \div 0.8$

10. Calculate each of the following.
    a. $1.48 \div 0.8$
    b. $39.6 \div 0.6$
    c. $43.26 \div 0.6$
    d. $0.2556 \div 0.8$
    e. $0.6484 \div 0.4$

11. **WE26b** Calculate each of the following.
    a. $165.0572 \div 0.11$
    b. $0.510\,48 \div 0.12$
    c. $24.8685 \div 0.15$

12. Calculate each of the following.
    a. $142.888 \div 0.08$
    b. $0.028\,692 \div 0.06$
    c. $32.619 \div 0.02$

13. **MC** To calculate $9.84 \div 0.8$, it can be rewritten as:
    A. $9.84 \div 8$
    B. $0.984 \div 0.8$
    C. $98.4 \div 0.8$
    D. $98.4 \div 8$
    E. $984 \div 8$

14. **MC** To calculate $151.368 \div 1.32$, it can be rewritten as:
    A. $151.368 \div 132$
    B. $151.368 \div 13.2$
    C. $1513.68 \div 132$
    D. $15\,136.8 \div 132$
    E. $151\,368 \div 132$

15. **MC** $0.294 \div 0.7$ equals:

   A. 0.042   **B.** 0.42   **C.** 4.2   **D.** 42   **E.** 420

16. For the following calculations:

   i. estimate the answer by rounding each number to the first digit

   ii. calculate the answer using written methods

   iii. check your answer using a calculator.

      a. $5400 \div 18$ **b.** $1044 \div 4.5$ **c.** $288 \div 0.12$

17. Change the following to dollars ($) by dividing by 100.

   a. 435 cents   **b.** 170 cents   **c.** 4792 cents

   d. 58 cents   **e.** 90 cents   **f.** 7320 cents

18. Calculate $202\,532.525 \div 10^4$.

Round the answer to the nearest hundredth.

19. **WE27** The carat is a unit of measure used to weigh precious stones. A 2.9-carat diamond weighs 8.9494 grains. Calculate how many grains one carat is equal to.

20. Kimiko spent $8.95 on 15 chocolates. Determine the cost of each chocolate. Give your answer to the nearest 5 cents.

21. Calculate how many 1.25-litre bottles of Powerade could be poured into a 25-litre drink dispenser.

22. If you have $22.50 for train fares to school for the week, calculate how much you would spend on each of the 5 school days.

23. The area of an office floor is 85.8 square metres. Calculate how many people could fit in the office if each person takes up 1.2 square metres.

24. Calculate how many books can be stacked on top of each other on a shelf that is 30.8 centimetres high if each book is 1.1 centimetres thick.

3.8 Review

doc-38044

Hey students! Now that it's time to revise this topic, go online to:

Access the topic summary

Review your results

Watch teacher-led videos

Practise questions with immediate feedback

Find all this and MORE in jacPLUS

3.8 Exercise

Multiple choice

1. **MC** The number that has 7 in the hundredths position is:

   A. 47.935 B. 65.782 C. 93.827 D. 37.072 E. 73.027

2. **MC** The decimals 0.21, 0.12, 0.099, 0.301, 0.20 in ascending order are:

   A. 0.12, 0.099, 0.210, 0.20, 0.301 B. 0.21, 0.12, 0.099, 0.301, 0.20
   C. 0.099, 0.12, 0.20, 0.21, 0.301 D. 0.12, 0.20, 0.21, 0.099, 0.301
   E. 0.11, 0.21, 0.20, 0.31, 0.30

3. **MC** The decimal 0.36 in its simplest fraction form is:

   A. $\dfrac{36}{100}$ B. $\dfrac{18}{50}$ C. $\dfrac{18}{25}$ D. $\dfrac{1}{25}$ E. $\dfrac{9}{25}$

4. **MC** The decimal 4.48 as a mixed number in its simplest form is:

   A. $4\dfrac{12}{25}$ B. $4\dfrac{48}{100}$ C. $4\dfrac{13}{25}$ D. $4\dfrac{24}{50}$ E. $5\dfrac{4}{13}$

5. **MC** The decimal 2.036 59 rounded to 3 decimal places is:

   A. 2.036 B. 2.027 C. 2.037 D. 2.046 E. 2.064

6. **MC** The answer to $6.48 + 3.36$ is:

   A. 9.84 B. 9.74 C. 9.82 D. 9.48 E. 9.47

7. **MC** The answer to 6.4×6 is:

   A. 34.8 B. 34.7 C. 37.4 D. 38.6 E. 38.4

8. **MC** The answer to 23.0846×100 is:

   A. 230.846 B. 2.308 46 C. 2308.46 D. 23 084.6 E. 230 846

9. **MC** The answer to $7.2 \div 0.8$ is:

   A. 9 B. 5.76 C. 0.9 D. 9.2 E. 9.9

10. **MC** The answer to $2946.47 \div 1000$ is:

    A. 29.4647 B. 2.94 647 C. 2 946 470 D. 294.647 E. 294 647

Short answer

11. Calculate $24.76 \div 0.08$.

12. Jodie puts her loose change in a jar at the end of the week. Over the last 4 weeks, she has put in $10.50, $6.30, $7.70 and $3.40. Calculate how much money was put in the jar over the 4 weeks.

TOPIC 3 Decimals 113

13. Ryan drank 0.475 litres from a 1.25-litre bottle of Diet Coke. Calculate how much Diet Coke was left in the bottle.

14. Bradley purchased 0.55 kilograms of hot salami at $22.75 per kilogram. Calculate the cost of the hot salami to the nearest 5 cents.

15. Determine the decimal equivalent of each of the following fractions, correct to 4 decimal places:

a. $\dfrac{3}{13}$

b. $\dfrac{3}{7}$

Extended response

16. Nate purchased a new pair of runners. He got $15.05 change when he gave the salesperson $150.
Determine the cost of Nate's runners.

17. If postage costs $2.35 for each parcel, calculate the total cost of postage for the 200 parcels that were sent out.

18. Fredrick is making a sculpture out of copper wire. He has 6 m of copper wire and wants to cut the wire into 0.15-m lengths. Calculate how many 0.15-m lengths of wire he will have.

19. Kyle goes for a 7.23-km run. If it takes him 3.40 minutes on average to run 1 km, calculate how long his run takes.

20. Patricia goes grocery shopping at the supermarket. She buys 6 items with the following prices:
$1.24, $0.55, $3.78, $4.15, $0.99 and $2.44

a. Estimate the total cost by rounding each item's cost to the nearest unit.
b. Calculate the total cost.

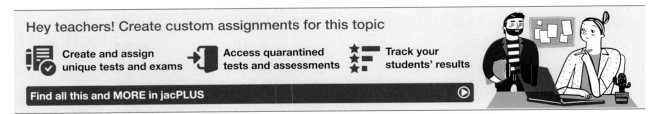

Answers

Topic 3 Decimals

3.2 Place value and comparing decimals

3.2 Exercise

1. a. $\dfrac{2}{10}$ b. $\dfrac{2}{100}$

   c. $\dfrac{2}{10}$ d. $\dfrac{2}{1000}$

2. a. $\dfrac{9}{10}$ b. $\dfrac{9}{10}$

   c. $\dfrac{9}{100}$ d. $\dfrac{9}{100}$

3. a. $\dfrac{6}{10}$ b. $2 + \dfrac{3}{10}$

   c. $6 + \dfrac{2}{10} + \dfrac{5}{100}$ d. $\dfrac{9}{10} + \dfrac{8}{100} + \dfrac{2}{1000}$

4. a. $\dfrac{2}{10} + \dfrac{1}{100} + \dfrac{2}{1000} + \dfrac{1}{10\,000}$

   b. $3 + \dfrac{8}{10} + \dfrac{7}{100} + \dfrac{4}{1000}$

   c. $\dfrac{6}{10} + \dfrac{7}{100} + \dfrac{3}{1000} + \dfrac{3}{10\,000}$

   d. $200 + \dfrac{9}{100}$

5. a. Two, 2

     Nine tenths, $\dfrac{9}{10}$

     Five hundredths, $\dfrac{5}{100}$

   b. Three tenths, $\dfrac{3}{10}$

     Seven hundredths, $\dfrac{7}{100}$

     Nine thousandths, $\dfrac{9}{1000}$

   c. Twelve, 12

     Zero tenths, $\dfrac{0}{10}$

     Zero hundredths, $\dfrac{0}{100}$

     Three thousandths, $\dfrac{3}{1000}$

   d. Three, 3

     Four tenths, $\dfrac{4}{10}$

     Zero hundredths, $\dfrac{0}{100}$

     Two thousandths, $\dfrac{2}{1000}$

     Seven ten thousandths, $\dfrac{7}{10\,000}$

6. a. 1 unit + 7 tenths + 5 hundredths

     $(1 \times 1) + \left(7 \times \dfrac{1}{10}\right) + \left(5 \times \dfrac{1}{100}\right)$

   b. 5 tens + 5 units + 0 tenths + 3 hundredths

     $(5 \times 10) + (5 \times 1) + \left(0 \times \dfrac{1}{10}\right) + \left(3 \times \dfrac{1}{100}\right)$

   c. 2 tens + 5 units + 0 tenths + 1 hundredths

     $(2 \times 10) + (5 \times 1) + \left(0 \times \dfrac{1}{10}\right) + \left(1 \times \dfrac{1}{100}\right)$

   d. 7 units + 9 tenths + 8 hundredths + 7 thousandths

     $(7 \times 1) + \left(9 \times \dfrac{1}{10}\right) + \left(8 \times \dfrac{1}{100}\right) + \left(7 \times \dfrac{1}{1000}\right)$

7. a. 2 tens + 3 units + 2 tenths + 7 hundredths

     $(2 \times 10) + (3 \times 1) + \left(2 \times \dfrac{1}{10}\right) + \left(7 \times \dfrac{1}{100}\right)$

   b. 3 tenths + 2 hundredths + 8 thousandths + 4 ten thousandths

     $\left(3 \times \dfrac{1}{10}\right) + \left(2 \times \dfrac{1}{100}\right) + \left(8 \times \dfrac{1}{1000}\right) + \left(4 \times \dfrac{1}{10\,000}\right)$

   c. 7 tenths + 0 hundredths + 0 thousandths + 1 ten thousandths

     $\left(7 \times \dfrac{1}{10}\right) + \left(0 \times \dfrac{1}{100}\right) + \left(0 \times \dfrac{1}{1000}\right) + \left(1 \times \dfrac{1}{10\,000}\right)$

   d. 5 units + 6 tenths + 5 hundredths

     $(5 \times 1) + \left(6 \times \dfrac{1}{10}\right) + \left(5 \times \dfrac{1}{100}\right)$

8. D

9. See the table at the bottom of the page.*

10. a. 0.35   b. 0.69   c. 2.110

11. a. 0.039 90   b. 2.559   c. 0.110 49

12. a. $3.2 > 2.9$ b. $8.6 < 8.9$ c. $0.64 < 0.67$
   d. $0.29 < 0.39$ e. $13.103 < 13.112$

13. a. $0.427 > 0.424$ b. $0.580 > 0.508$
   c. $0.0101 < 0.0120$ d. $0.048\,01 < 0.4801$
   e. $1.3830 > 1.3824$

14. a. 2

   b. i. Seventy-seven dollars and fifty-five cents

     ii. $77.55

---

*9.

		Tens	Units	.	Tenths	Hundredths	Thousandths
Example	37.684	3	7	.	6	8	4
a	0.205	0	0	.	2	0	5
b	1.06	0	1	.	0	6	0
c	74.108	7	4	.	1	0	8
d	0.108	0	0	.	1	0	8
e	50.080	5	0	.	0	8	0

15. a. $0.17, 0.21, 0.33, 0.39, 0.45$
    b. $0.314, 0.391, 0.413, 0.420, 0.502$
    c. $0.803, 0.807, 0.811, 0.821, 0.902$
16. a. $0.48, 0.39, 0.36, 0.31, 0.19$
    b. $0.99, 0.97, 0.95, 0.91, 0.90$
    c. $1.291, 1.288, 1.279, 1.273, 1.264$
17. E
18. D
19. E
20. a. False   b. True   c. False
    d. True   e. False   f. True
21. i. a. Tenths   b. Thousandths   c. Units
       d. Units   e. Tens
    ii. a. Yes   b. No   c. No
        d. Yes   e. Yes
22. a. Shelley
    b. Carolyn
    c. Mara
    d. 1st Shelley, 2nd Carolyn, 3rd Robyn
    e. 1st Carolyn, 2nd Shelley, 3rd Robyn
    f. 1st Mara, 2nd Jenika, 3rd Shelley
    g. Shelley

3.3 Converting decimals to fractions and fractions to decimals

3.3 Exercise

1. a. $\dfrac{1}{10}$ b. $\dfrac{1}{2}$ c. $\dfrac{37}{100}$
   d. $\dfrac{3}{5}$ e. $\dfrac{11}{25}$

2. a. $\dfrac{63}{100}$ b. $\dfrac{311}{500}$ c. $\dfrac{719}{1000}$
   d. $\dfrac{7}{10}$ e. $\dfrac{12}{25}$

3. a. $\dfrac{9}{25}$ b. $\dfrac{591}{625}$ c. $\dfrac{9209}{10\,000}$
   d. $\dfrac{4621}{10\,000}$ e. $\dfrac{7}{50}$

4. a. $1\dfrac{7}{10}$ b. $2\dfrac{3}{10}$ c. $8\dfrac{2}{5}$
   d. $1\dfrac{1}{5}$ e. $4\dfrac{1}{5}$

5. a. $6\dfrac{1}{2}$ b. $7\dfrac{27}{100}$ c. $19\dfrac{91}{500}$
   d. $6\dfrac{3}{20}$ e. $7\dfrac{1}{4}$

6. B
7. A
8. C
9. D
10. E
11. a. 0.75 b. 0.5 c. 0.6 d. 0.1

12. a. 0.75 b. 0.02 c. 0.24 d. 0.2
13. a. 0.375 b. 0.875 c. 0.0625
14. Don't agree – second jump is the biggest.
15. Maths: 0.8
    Science: 0.75
    English: 0.7
16. $\dfrac{30}{100}, \dfrac{8}{25}, \dfrac{6}{15}, \dfrac{5}{8}, \dfrac{3}{4}$

3.4 Rounding and repeating decimals

3.4 Exercise

1. a. 0.33 b. 0.87 c. 3.25
   d. 17.040 e. 3.1
2. a. 3.13 b. 110.7247 c. 67.37
   d. 0.01 e. 7.18
3. a. 0.5 b. 0.7 c. 12.2
   d. 32.4 e. 5.8
4. a. 100.0 b. 15 c. 99.9
   d. 93.99 e. 0.898
5. C
6. A
7. a. 13 b. 6 c. 6
   d. 88 e. 121
8. a. 32 b. 43 c. 3238
   d. 0 e. 1
9. a. $3.\dot{5}$ b. $0.\dot{3}$ c. $47.\dot{1}$ d. $0.3\dot{5}$
10. a. $0.1\dot{2}\dot{5}$ b. $7.4\dot{3}1\dot{2}$ c. $121.9\dot{2}\dot{1}$ d. $0.12\dot{7}$
11. A
12. A
13. $58.80
14. a. 0.3 b. 0.1 c. 0.42
15. a. 0.17 b. 0.182 c. 0.47
16. a. 4.5 b. $7.\dot{3}$ c. 0.25
17. a. 10 b. 70 c. 160
    d. 360 e. 380
18. a. 500 b. 500 c. 200
    d. 5800 e. 8700
19. a. 2000 b. $13\,000$ c. $22\,000$
    d. 9000 e. 2000
20. $0.295, 0.2\dot{9}\dot{5}, 0.29\dot{5}, 0.29\dot{5}$
21. $2.426\,\mathbf{78} \approx 2.427\,\text{m}$
    $2.426\,78$ is not realistic since it is to the hundredth of a mm, which is too small to measure with a tape measure.
22. $33\,°C$
23. a. i. 0.7 ii. 0.73 iii. 0.729
    b. The rules for rounding are the same no matter the number of decimal points.
24. $15

3.5 Adding and subtracting decimals

3.5 Exercise

1. a. 3.8   b. 3.58
   c. 11.821   d. 84.811

2. a. 8.95   b. 21.9060
   c. 2012.333   d. 159.7884

3. a. 5.8, estimate = 6   b. 8.9, estimate = 9
   c. 3.043, estimate = 3   d. 15.138, estimate = 15

4. a. 17.3271, estimate = 18   b. 20.4329, estimate = 21
   c. 11.69, estimate = 11   d. 105.22, estimate = 105

5. a. 0.52, estimate = 1   b. 8.428, estimate = 8
   c. 0.1839, estimate = 0   d. 2.4611, estimate = 2

6. a. 26.05, estimate = 26   b. 89.18, estimate = 89
   c. 167.8, estimate = 168   d. 2.733, estimate = 3

7. a. 3.98, estimate = 3   b. 12.86, estimate = 13
   c. 0.327, estimate = 1   d. 4.146, estimate = 4

8. a. 4.942, estimate = 5
   b. 3.289, estimate = 3
   c. 128.4358, estimate = 128
   d. 16.766, estimate = 17

9. B
10. A
11. C
12. B
13. B
14. $305.40
15. 33.56 km
16. 12.0 km
17. $6.60
18. $25.25
19. 50.48 kg
20. 1.72 m
21. 48.38 sec
22. a.

Team	Time
Team 1	153.0
Team 2	159.3
Team 3	148.9
Team 4	166.6
Team 5	155.9
Team 6	155.8

    b. Team 3
    c. Team 3 beats team 1 by 4.1 seconds.

23. There are multiple answers.
   One possibility is 32.03 g.
24. $7.70

3.6 Multiplying decimals (including powers of 10)

3.6 Exercise

1. a. 7.5   b. 100.8
   c. 52.0   d. 84.96

2. a. 138.96   b. 518.96
   c. 6.384   d. 0.735

3. a. 0.72   b. 3.12
   c. 0.32   d. 6.16

4. a. 0.220   b. 0.168
   c. 0.390   d. 0.406

5. a. 0.000 08   b. 0.000 012
   c. 0.063   d. 0.000 140

6. a. 0.400   b. 1.008
   c. 1.872   d. 0.225

7. E
8. B
9. A
10. B
11. a. 0.0016   b. 1.44   c. 4.6225
12. a. 804   b. 6780   c. 478 400
13. a. 1278   b. 3240   c. 31 230
14. a. 64.8   b. 1389.6   c. 270.8
   d. 217 148.96   e. 82 049.6783
15. a. 3 268 904.3267   b. 984 326.641
   c. 278.498 32   d. 460
16. a. i. 12 000   ii. 12 760   iii. 12 760
   b. i. 600 000   ii. 634 161   iii. 634 161
   c. i. 60 000   ii. 42 096.6   iii. 42 096.6
17. a. 4800 cents   b. 21 700 cents   c. 1800 cents
   d. 1725 cents   e. 6440 cents
18. $625
19. 260.7569
20. a. $45.00   b. 137 cents
21. $8.05
22. 0.375 litres

3.7 Dividing decimals (including powers of 10)

3.7 Exercise

1. a. 0.6   b. 4.1   c. 1.18
   d. 1.54   e. 2.42

2. a. 1.282   b. 1.003   c. 4.0701
   d. 6.8594   e. 21.0784

3. a. 3.12   b. 3.246   c. 1.222
   d. 6.421   e. 12.0242

4. a. 2.35   b. 4.75   c. 1.86

5. a. 2.45   b. 2.125   c. 6.95

6. a. 1.407   b. 9.6813   c. 0.6208
   d. 3.592 87   e. 0.023 497 8

7. a. 0.000 090 769   b. 0.802 405 6   c. 0.000 152 70
   d. 0.000 072 05   e. 0.000 000 003 2

8. a. 0.265   b. 0.79   c. 1.06
   d. 0.595   e. 0.001 74

9. a. 9   b. 6   c. 9
   d. 4   e. 0.31

10. a. 1.85   b. 66   c. 72.1
    d. 0.3195   e. 1.621

11. a. 1500.52   b. 4.254   c. 165.79

12. a. 1786.1   b. 0.4782   c. 1630.95

13. D

14. D

15. B

16. a. i. 250   ii. 300   iii. 300
    b. i. 200   ii. 232   iii. 232
    c. i. 3000   ii. 2400   iii. 2400

17. a. $4.35   b. $1.70   c. $47.92
    d. $0.58   e. $0.90   f. $73.20

18. 20.25

19. 3.086 grains

20. $0.60 to the nearest 5 cents

21. 20

22. $4.50

23. 71

24. 28

3.8 Review

3.8 Exercise

Multiple choice

1. D
2. C
3. E
4. A
5. C
6. A
7. E
8. C
9. A
10. B

Short answer

11. 309.5
12. 27.90
13. 0.775 litres
14. $12.50
15. a. 0.2308   b. 0.4286

Extended response

16. $134.95
17. $470
18. 40
19. 24.582 minutes or 24 minutes 34.92 seconds
20. a. $13   b. $13.15

4 Powers and roots

Fully worked solutions for this topic are available online.

4.1 Overview

4.1.1 Introduction

Whole numbers can be written in ways that can make calculations easier if you do not have your calculator handy. You have been using multiplication tables already without necessarily realising that they give you numbers that are multiples of each other. Knowing how to write numbers in different ways helps in understanding how numbers are connected.

Indices are the short way of writing repeated multiplications and are very useful in everyday life, because they allow us to write very large and very small numbers more easily. Indices can simplify calculations involving these very large or very small numbers. Astronomy is a branch of science in which very large distances are involved. Alpha Crucis, the brightest and closest star to us in the Southern Cross constellation, is approximately 3 million billion kilometres from Earth. That is a huge number — 3 followed by 15 zeros! Scientists and engineers in various disciplines need to be able to communicate and work with numbers of all sizes.

KEY CONCEPTS

This topic covers the following key concepts from the VCE Mathematics Study Design:
- application of integers, fractions and decimals, their properties and related operations.

Note: *Concepts shown in grey are covered in other topics.*

Source: VCE Mathematics Study Design (2023–2027) extracts © VCAA; reproduced by permission.

4.2 Index (power) notation

LEARNING INTENTION

At the end of this subtopic you should be able to:
- understand that index or exponent notation is a short way of writing a repeated multiplication
- write expressions in index notation
- simplify expressions using index notation
- use place value to write and evaluate numbers in expanded form with index notation.

4.2.1 Introduction to indices (powers)

An **index** (or **exponent** or **power**) is a short way of writing a repeated multiplication.

The **base** is the number that is being repeatedly multiplied by itself and the index (plural *indices*) is the number of times it is multiplied.

$$8^6 = 8 \times 8 \times 8 \times 8 \times 8 \times 8 = 262\ 144$$

Index form   Expanded form   Basic numeral

Numbers in index form are read using the value of both the base and the power.

8^6 is read as '8 to the power of 6'.

12^4 is read as '12 to the power of 4'.

WORKED EXAMPLE 1 Identifying the base and power

For the following expressions, state:
 i. **the number or pronumeral that is the base**
 ii. **the number or pronumeral that is the power.**
  a. 5^8 b. x^7 c. 5^n d. z^m

THINK

The base is the number or pronumeral that is repeatedly multiplied by itself, and the index is the number of times that it is multiplied.

WRITE

a. 5^8
  i. Base is 5.
  ii. Power is 8.

b. x^7
  i. Base is x.
  ii. Power is 7.

c. 5^n
  i. Base is 5.
  ii. Power is n.

d. z^m
  i. Base is z.
  ii. Power is m.

WORKED EXAMPLE 2 Writing an expression using index notation

Write the following expressions using index notation.
a. $5 \times 5 \times 5 \times 5 \times 5 \times 5 \times 5$ **b.** $3 \times 3 \times 3 \times 3 \times 7 \times 7$

THINK	WRITE
a. 1. Write the multiplication.	**a.** $5 \times 5 \times 5 \times 5 \times 5 \times 5 \times 5$
2. Write the number being repeatedly multiplied as the base, and the number of times it is written as the index (the number 5 is written 7 times).	$= 5^7$
b. 1. Write the multiplication.	**b.** $3 \times 3 \times 3 \times 3 \times 7 \times 7$
2. Write the number being multiplied as the base, and the number of times it is written as the index (the number 3 is written 4 times, and the number 7 is written 2 times).	$= 3^4 \times 7^2$

WORKED EXAMPLE 3 Simplifying expressions using index notation

Simplify each of the following expressions by first writing each expression as a repeated multiplication and then in index notation.
a. $5^3 \times 5^5$ **b.** $\left(3^4\right)^2$ **c.** $(3 \times 7)^2$

THINK	WRITE
a. 1. Write the expression.	**a.** $5^3 \times 5^5$
2. Write the expression using repeated multiplication (that is, in expanded form).	$= (5 \times 5 \times 5) \times (5 \times 5 \times 5 \times 5 \times 5)$ $= 5 \times 5 \times 5 \times 5 \times 5 \times 5 \times 5 \times 5$
3. Write the repeated multiplication using index notation. The number being repeatedly multiplied is 5 (base), and the number of times it is written is 8 (power).	$= 5^8$
b. 1. Write the expression.	**b.** $\left(3^4\right)^2$
2. Write the expression using repeated multiplication (that is, in expanded form).	$= 3^4 \times 3^4$ $= (3 \times 3 \times 3 \times 3) \times (3 \times 3 \times 3 \times 3)$ $= 3 \times 3 \times 3 \times 3 \times 3 \times 3 \times 3 \times 3$
3. Write the repeated multiplication using index notation. The number being repeatedly multiplied is 3 (base) and the number of times it is written is 8 (power).	$= 3^8$
c. 1. Write the expression.	**c.** $(3 \times 7)^2$
2. Write the expression using repeated multiplication (that is, in expanded form).	$= (3 \times 7) \times (3 \times 7)$ $= 3 \times 7 \times 3 \times 7$ $= 3 \times 3 \times 7 \times 7$
3. Write the repeated multiplication using index notation. The numbers being repeatedly multiplied are 3 and 7 (base) and the number of times they are written is 2 (power) in each case.	$= 3^2 \times 7^2$

4.2.2 Indices (powers) and place values

By using place value, you can write numbers in expanded form with index notation.

For example:

$$2700 = 2000 + 700$$
$$= 2 \times 10 \times 10 \times 10 + 7 \times 10 \times 10$$
$$= 2 \times 10^3 + 7 \times 10^2$$

WORKED EXAMPLE 4 Expressing numbers in expanded form using index notation

Write the following numbers in expanded form using index notation.
a. 59 176   **b. 108 009**

THINK	WRITE
a. 1. Write the number as the sum of each place value.	a. $59\,176 = 50\,000 + 9000 + 100 + 70 + 6$
2. Write each place value in multiples of 10.	$59\,176 = 5 \times 10 \times 10 \times 10 \times 10 + 9 \times 10 \times 10 \times 10 + 1 \times 10 \times 10 + 7 \times 10 + 6$
3. Write each place value in index notation.	$59\,176 = 5 \times 10^4 + 9 \times 10^3 + 1 \times 10^2 + 7 \times 10^1 + 6$
b. 1. Write the number as the sum of each place value.	b. $108\,009 = 100\,000 + 8000 + 9$
2. Write each place value in multiples of 10.	$108\,009 = 1 \times 10 \times 10 \times 10 \times 10 \times 10 + 8 \times 10 \times 10 \times 10 + 9$
3. Write each place value in index notation.	$108\,009 = 1 \times 10^5 + 8 \times 10^3 + 9$

tlvd-3659

WORKED EXAMPLE 5 Evaluating indices

Evaluate each of the following.
a. 3^4   **b. $5^2 \times 2^3$**   **c. $4^2 + 3^2 - 2^4$**

THINK	WRITE
a. 1. Write in expanded form.	a. $3^4 = 3 \times 3 \times 3 \times 3$
2. Multiply the terms.	$= 81$
3. Write the answer.	$3^4 = 81$
b. 1. Write both terms in expanded form.	b. $5^2 \times 2^3 = (5 \times 5) \times (2 \times 2 \times 2)$
2. Calculate the product of the numbers in the brackets.	$= 25 \times 8$
3. Multiply the terms.	$= 200$
4. Write the answer.	$5^2 \times 2^3 = 200$

c. 1. Write all terms in expanded form.

   2. Calculate the product of the numbers in the brackets.

   3. Remember the order of operations. Since the operations are addition and subtraction, work left to right.

   4. Write the answer.

c. $4^2 + 3^2 - 2^4 = (4 \times 4) + (3 \times 3) - (2 \times 2 \times 2 \times 2)$

$$= 16 + 9 - 16$$
$$= 25 - 16$$
$$= 9$$

$4^2 + 3^2 - 2^4 = 9$

 Resources

 Interactivity Index notation (int-3957)

4.2 Exercise

Students, these questions are even better in jacPLUS

 Receive immediate feedback and access sample responses

 Access additional questions

 Track your results and progress

Find all this and MORE in jacPLUS

1. **WE1** In each of the following expressions, enter the missing number into the box and state the values of the base and power.

   a. $6^{\square} = 36$ **b.** $4^{\square} = 4096$ **c.** $\square^3 = 27$ **d.** $\square^3 = 1000$

2. **WE2** Write each of the following expressions in index notation.

   a. $5 \times 5 \times 5 \times 5$ **b.** $7 \times 7 \times 7 \times 7 \times 7 \times 7$

   c. $2 \times 2 \times 2 \times 2 \times 2 \times 2 \times 2 \times 2 \times 2$ **d.** $12 \times 12 \times 12$

3. Write the following expressions in index notation.

   a. $2 \times 2 \times 2 \times 2 \times 2 \times 4 \times 4 \times 4$ **b.** $3 \times 3 \times 7$

   c. $5 \times 5 \times 3 \times 3 \times 3 \times 3$ **d.** $4 \times 4 \times 5 \times 5 \times 3 \times 3 \times 3$

4. Write the following using repeated multiplication (that is, in expanded form).

   a. 4^5 **b.** 13^3

5. **WE3** Simplify each of the following expressions by first writing each expression as a repeated multiplication and then using index notation.

   a. $5^6 \times 5^3$ **b.** $4^7 \times 4^2$ **c.** $\left(3^6\right)^2$

6. **WE4** Write the following numbers in expanded form using index notation.

   a. 300   **b.** 4500   **c.** 6705   **d.** 10 000

7. Simplify each of the following expressions by first writing each expression as a repeated multiplication and then using index notation.

   a. $\left(8^4\right)^3$ **b.** $(3 \times 13)^6$

8. **MC** The value of 4^4 is:

   A. 8   B. 16   C. 64   D. 256   E. 484

9. **WE5** Evaluate each of the following.

   a. 8^2 **b.** $5^2 \times 2^3$ **c.** $5^3 - 4^3 + 2^4$

10. Evaluate each of the following.

    a. $4^3 \times 2^3$ **b.** $3^4 \times 4^3$ **c.** $3^3 \times 9^3$

11. Evaluate each of following.

    a. $7^3 - 4^3$ **b.** $5^3 + 2^5 \times 9^2$ **c.** $2^7 - 4^5 \div 2^6$

12. **MC** Choose which of the following expressions has the greatest value.

    A. 2^8 B. 8^2 C. 3^4 D. 4^3 E. 9^2

13. Write < (less than) or > (greater than) between each pair of numbers to make a true statement.

    a. 1^3 3^1 **b.** 2^3 3^2 **c.** 5^3 3^5 **d.** 4^4 3^5

14. You have a choice of how your weekly allowance is increased for 15 weeks.
    Option 1: Start at 1 cent and double your allowance every week.
    Option 2: Start at $1 and increase your allowance by $1 every week.

    a. Explain which option you would choose and why.
    b. Explain whether powers can be used to help with these calculations.

15. A student sent an email to eight friends, who then each forwarded the email to eight different friends, who in turn forwarded the email to another eight friends each.
    Calculate how many emails in total were sent during this process.

16. You received a text message from your friend. After 5 minutes, you forward the text message to two of your other friends. After 5 more minutes, those two friends forward it to two more friends.

    If the text message is sent every 5 minutes in this way, calculate how many people have received it in 30 minutes.

17. Technetium-99m has a half-life of 6 hours. This means that every 6 hours it halves its mass.

    Calculate how long it will take for there to be $\dfrac{1}{16}$ of its original mass.

18. A knock-out tennis competition ends with players in the final. If the final is the seventh round of the competition, and half the players are knocked out in each round, determine how many players there were at the start of the competition.

4.3 Squares and square roots

LEARNING INTENTION

At the end of this subtopic you should be able to:
- determine values of squares and square roots
- estimate the value of the square root of other numbers by using the perfect squares that lie on either side of the number
- evaluate squares or square roots using BIDMAS.

4.3.1 Square numbers

The process of multiplying a number by itself is known as **squaring a number**. **Square numbers** or **perfect squares** are numbers that can be arranged in a square, as shown.

$1^2 = 1 \times 1$
$= 1$

$2^2 = 2 \times 2$
$= 4$

$3^2 = 3 \times 3$
$= 9$

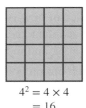

$4^2 = 4 \times 4$
$= 16$

A square number is the number we get after multiplying a whole number by itself. The first four square numbers are 1, 4, 9 and 16. All square numbers can be written in index notation; for example, $4^2 = 16$.

WORKED EXAMPLE 6 Determining square numbers

Determine the square numbers between 90 and 150.

THINK	WRITE
1. Use your knowledge of multiplication tables to determine the first square number after 90.	$10^2 = 10 \times 10 = 100$
2. Determine the square numbers that come after that one, but before 150.	$11^2 = 11 \times 11 = 121$ $12^2 = 12 \times 12 = 144$ $13^2 = 13 \times 13 = 169$ (too big)
3. Write the answer in a sentence.	The square numbers between 90 and 150 are 100, 121 and 144.

---

tlvd-3660

WORKED EXAMPLE 7 Determining approximate values of square numbers

Write the two whole square numbers between which 5.3^2 will lie.

THINK	WRITE
1. Write the whole numbers on either side of 5.3.	5.3 is between 5 and 6.
2. Consider the square of each whole number.	5.3^2 is between 5^2 and 6^2.
3. Simplify 5^2 and 6^2, then write the answer in a sentence.	So 5.3^2 is between 25 and 36.
4. Verify your answer with a calculator.	$5.3^2 = 28.09$, which lies between 25 and 36.

When a composite number is squared, the result is equal to the product of the squares of its factors. For example:

$$10^2 = (2 \times 5)^2$$
$$= 2^2 \times 5^2$$
$$= 4 \times 25$$
$$= 100$$

4.3.2 Square roots

Evaluating the **square root** of a number is the opposite of squaring the number; for example, since $4^2 = 16$, then $\sqrt{16} = 4$. The symbol for square root is called the radical symbol. It is written as $\sqrt{}$. Using the illustrations of the squares shown earlier, a square of area 16 square units must have a side length of 4 units. This may help you understand why $\sqrt{16} = 4$.

Square roots

When determining the square root of a number, you are determining the number that, when multiplied by itself, equals the number underneath the radical symbol.

For example, to determine $\sqrt{9}$, identify the number that when multiplied by itself gives 9.

Since $3 \times 3 = 9$, we can conclude that $\sqrt{9} = 3$.

To determine the square roots of larger numbers, it helps to break the number up as a product of two smaller square roots with which we are more familiar.

For example:

$$\sqrt{900} = \sqrt{9} \times \sqrt{100}$$
$$= 3 \times 10$$
$$= 30$$

WORKED EXAMPLE 8 Determining square roots

Evaluate the following square roots.

a. $\sqrt{64}$

b. $\sqrt{4900}$

THINK	WRITE
a. Determine the number that when multiplied by itself gives 64.	a. $\sqrt{64} = 8$ $(8 \times 8 = 64)$
b. 1. Write 4900 as the product of two smaller numbers for which you can calculate the square root.	b. $\sqrt{4900} = \sqrt{49} \times \sqrt{100}$
2. Take the square root of each of these numbers.	$= 7 \times 10$
3. Determine the product and write the answer.	$\sqrt{4900} = 70$

Only perfect squares have square roots that are whole numbers.

The value of the square root of other numbers can be estimated by using the perfect squares that lie on either side of the number.

WORKED EXAMPLE 9 Estimating square roots

Write the two numbers between which each of the following numbers lie.

a. $\sqrt{74}$

b. $\sqrt{342}$

THINK	WRITE
a. 1. Write the square numbers on either side of 74.	a. 74 is between 64 and 81.
2. Consider the square root of each number.	$\sqrt{74}$ is between $\sqrt{64}$ and $\sqrt{81}$.
3. Simplify $\sqrt{64}$ and $\sqrt{81}$.	So $\sqrt{74}$ is between 8 and 9.
4. Verify your answer with a calculator.	$\sqrt{74} \approx 8.6023$
b. 1. Write the square numbers on either side of 342.	b. 342 is between 324 $\left(18^2\right)$ and 361 $\left(19^2\right)$.
2. Consider the square root of each number.	$\sqrt{342}$ is between $\sqrt{324}$ and $\sqrt{361}$.
3. Simplify $\sqrt{324}$ and $\sqrt{361}$.	So $\sqrt{342}$ is between 18 and 19.
4. Verify your answer with a calculator.	$\sqrt{342} \approx 18.4932$

4.3.3 Order of operations

The square or square root is an index represented by the letter I (Indices) in BIDMAS, which determines the order in which operations are evaluated in a mathematical expression.

The order of operations

BIDMAS helps us to remember the correct order in which we should perform the various operations, working from left to right.

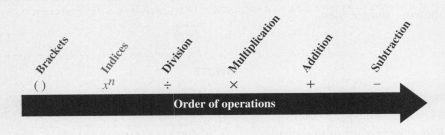

 Resources

 Interactivities Square numbers (int-3936)
Square roots (int-3937)

---

tlvd-3662

WORKED EXAMPLE 10 Evaluating square roots using the order of operations

Evaluate the following.

a. $\sqrt{(9+16)}$

b. $\sqrt{25}+7$

THINK	WRITE
a. 1. Simplify the expression inside the square root and write the answer.	a. $\sqrt{(9+16)} = \sqrt{25}$
2. Complete the calculation by evaluating the square root and write the answer.	$= 5$
b. 1. Evaluate the square root. *Note:* There is no expression inside the square root to evaluate first.	b. $\sqrt{25}+7 = 5+7$
2. Perform the addition to complete the calculation and write the answer.	$= 12$

4.3 Exercise

1. Evaluate the following and verify your answers with a calculator.

   a. 5^2
   b. 12^2
   c. 15^2
   d. 25^2

2. **WE6** State the square numbers between 50 and 130.

3. State the square numbers between 160 and 200.

4. **WE7** Write two whole square numbers between which each of the following will lie.

   a. 4.2^2
   b. 7.8^2
   c. 9.1^2
   d. 13.2^2

5. **WE8** Evaluate the following.

   a. $\sqrt{400}$
   b. $\sqrt{36}$
   c. $\sqrt{121}$
   d. $\sqrt{900}$

6. Evaluate the following and verify your answers with a calculator.

   a. $\sqrt{2500}$
   b. $\sqrt{14\,400}$
   c. $\sqrt{360\,000}$
   d. $\sqrt{160\,000}$

7. **WE9** Write the two whole numbers between which each of the following lies.

   a. $\sqrt{62}$
   b. $\sqrt{13}$
   c. $\sqrt{180}$
   d. $\sqrt{2}$

8. **WE10** Evaluate the following.

   a. $\sqrt{144+25}$
   b. $\sqrt{144}+25$

9. **MC** State for which of the following square roots we can calculate an exact answer.

   A. $\sqrt{10}$
   B. $\sqrt{25}$
   C. $\sqrt{50}$
   D. $\sqrt{75}$
   E. $\sqrt{82}$

10. **MC** State for which of the following square roots we cannot calculate the exact value.

   A. $\sqrt{160}$ **B.** $\sqrt{400}$ **C.** $\sqrt{900}$ **D.** $\sqrt{2500}$ **E.** $\sqrt{3600}$

11. Evaluate the following. Verify your answers with a calculator.

   a. $3^2 + \sqrt{16}$ **b.** $8^2 - \sqrt{49}$ **c.** $4^2 \times 3^2 \times \sqrt{36}$ **d.** $2^2 + 3^2 \times \sqrt{16}$

12. Evaluate the following. Verify your answers with a calculator.

   a. $3^2 - 2^2 \div \sqrt{4} + \sqrt{49}$ **b.** $\sqrt{9} \times 4^2 - \sqrt{144} \div 2^2$

13. **a.** Determine between which two whole numbers $\sqrt{130}$ lies.
   b. Use your answer to part **a** to estimate the value of $\sqrt{130}$ and check your answer with a calculator.

14. **a.** Evaluate 14^2.
   b. Evaluate $2^2 \times 7^2$.
   c. State if your answers to parts **a** and **b** are equal. Explain why or why not.

15. **a.** Evaluate $\sqrt{225}$.
   b. Evaluate $\sqrt{25} \times \sqrt{9}$.
   c. State if your answers to parts **a** and **b** are equal. Explain why or why not.

16. **a.** Evaluate $\sqrt{36} + \sqrt{64}$.
   b. Evaluate $\sqrt{36 + 64}$.
   c. State if your answers to parts **a** and **b** are equal. Explain why or why not.

17. **a.** Evaluate each of the following using a calculator.

   i. $25^2 - 24^2$ **ii.** $24^2 - 23^2$ **iii.** $23^2 - 22^2$ **iv.** $22^2 - 21^2$ **v.** $21^2 - 20^2$
   b. Comment on any pattern in your answers.
   c. Discuss whether this pattern occurs with other numbers that are squared. Try some examples.

18. Rewrite the following in ascending order.

   a. 8^2, $\sqrt{56}$, 62, $\sqrt{72}$, 8, 7, 3^2, $\sqrt{100}$ **b.** 100, $\sqrt{121}$, $\sqrt{10}$, 3×4, 9^2, $\sqrt{169}$, 2×7

4.4 Using powers and roots to solve practical problems

LEARNING INTENTION

At the end of this subtopic you should be able to:
- determine the nth root of a positive number
- use powers and roots to solve application problems.

4.4.1 Roots

The nth root of any positive number can be found. That is, $\sqrt[n]{b} = a$ or $a^n = b$.

For example, $\sqrt{25} = 5$ because $25 = 5 \times 5$.

Note: When the root is a square root, 2 is usually not written; for example, $\sqrt[2]{25}$ is simply written as $\sqrt{25}$. The cube root of 64 is written as

$\sqrt[3]{64} = 4$ because $64 = 4 \times 4 \times 4$.

WORKED EXAMPLE 11 Evaluating integer roots

Evaluate each of the following.

a. $\sqrt{16}$　　　　b. $\sqrt[3]{27}$　　　　c. $\sqrt[5]{32}$

THINK　　　　　　　　　　　　　　　　　　**WRITE**

a. Determine the number that when multiplied by itself gives 16: $16 = 4 \times 4$.

a. $\sqrt{16} = 4$

b. Determine the number that when multiplied by itself three times gives 27: $27 = 3 \times 3 \times 3$.

b. $\sqrt[3]{27} = 3$

c. Determine the number that when multiplied by itself five times gives 32: $32 = 2 \times 2 \times 2 \times 2 \times 2$.

c. $\sqrt[5]{32} = 2$

4.4.2 Non-integer roots

In Worked example 11, 16 is a perfect square, 27 is a perfect cube and 32 is a perfect fifth. Most numbers do not have integer (whole number) roots. Calculators can be used to determine an approximate value for the non-integer roots of numbers.

WORKED EXAMPLE 12 Evaluating non-integer roots

Evaluate each of the following. Give your answers correct to 3 decimal places.

a. $\sqrt{10}$　　　b. $\sqrt[3]{200}$　　　c. $\sqrt[4]{122.45}$　　　d. $\sqrt{1500}$

THINK　　　　　　　　　　　　　　　　　　**WRITE**

a. 1. As 10 is not a perfect square, locate the square root button on your calculator and enter $\sqrt{10}$.

a. $\sqrt{10}$

　　2. Write the answer correct to 3 decimal places.

　　$\sqrt{10} \approx 3.162$

　　3. Check your answer using the calculator.

　　$3.162 \times 3.162 = 9.998$

b. 1. As 200 is not a perfect cube, locate the cube root button on your calculator and enter $\sqrt[3]{200}$.

b. $\sqrt[3]{200}$

　　2. Write the answer correct to 3 decimal places.

　　$\sqrt[3]{200} \approx 5.848$

　　3. Check your answer using the calculator.

　　$5.848 \times 5.848 \times 5.848 = 199.996$

c. 1. As 122.45 is not a perfect fourth, locate the nth root button on your calculator and enter $\sqrt[4]{122.45}$.

c. $\sqrt[4]{122.45}$

　　2. Write the answer correct to 3 decimal places.

　　$\sqrt[4]{122.45} \approx 3.327$

　　3. Check your answer using the calculator.

　　$3.327 \times 3.327 \times 3.327 \times 3.327 = 122.521$

d. 1. As 1500 is not a perfect square, locate the square root button on your calculator and enter $\sqrt{1500}$.

d. $\sqrt{1500}$

　　2. Write the answer correct to 3 decimal places.

　　$\sqrt{1500} \approx 38.730$

　　3. Check your answer using the calculator.

　　$38.730 \times 38.730 = 1500.013$

4.4.3 Pythagoras' theorem

Pythagoras's theorem states that for a right-angled triangle, the square of the length of the hypotenuse is always equal to the sum of the squares of the lengths of the other two sides.

Calculating the length of the hypotenuse

Pythagoras' theorem

In any right-angled triangle, the square of the length of the hypotenuse is equal to the sum of the squares of the lengths of the two shorter sides. That is:

$$c^2 = a^2 + b^2$$

where a and b are the two shorter sides and c is the hypotenuse.

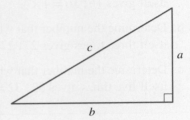

tlvd-3670

WORKED EXAMPLE 13 Calculating the length of the hypotenuse

Determine the length of the hypotenuse in the triangle shown. Give your answer correct to 2 decimal places.

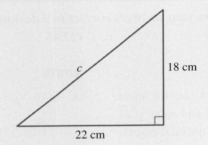

THINK	WRITE
1. This is a right-angled triangle with the lengths of two sides known. Use Pythagoras's theorem.	$c^2 = a^2 + b^2$
2. Substitute the values for the two shorter sides into the rule.	$c^2 = 22^2 + 18^2$ $= 484 + 324$ $= 808$
3. Calculate the value of c by determining the square root of 808. Round the answer correct to 2 decimal places.	$c^2 = 808$ $c = \sqrt{808}$ $= 28.43$
4. Write the answer. (Remember to add the correct unit.)	The length of the hypotenuse is 28.43 cm.

Calculating the length of a shorter side

Pythagoras' theorem can be rearranged to calculate the length of a shorter side in a right-angled triangle.

Pythagoras' theorem — hypotenuse and shorter sides

Calculating the hypotenuse:

$$c^2 = a^2 + b^2$$

Calculating the shorter side:

$$a^2 = c^2 - b^2$$

or

$$b^2 = c^2 - a^2$$

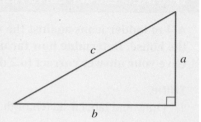

WORKED EXAMPLE 14 Calculating the length of a shorter side in a right-angled triangle

Determine the length of the side labelled x in the following right-angled triangle. Give your answer correct to 2 decimal places.

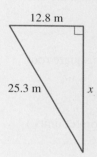

THINK	WRITE
1. This is a right-angled triangle with the lengths of two sides known. Use Pythagoras's theorem.	$c^2 = a^2 + b^2$
2. As it is the length of one of the shorter sides that needs to be determined, rearrange the rule to make a the subject.	$a^2 = c^2 - b^2$
3. Substitute the values for the hypotenuse, c, and the other side, b, into the rule.	$x^2 = 25.3^2 - 12.8^2$ $= 640.09 - 163.84$ $= 476.25$
4. Calculate the value of x by determining the square root of 476.25. Round the answer correct to 2 decimal places.	$x^2 = 476.25$ $x = \sqrt{476.25}$ $= 21.82$
5. Write the answer. (Remember to add the correct unit.)	The length of the side labelled x is 21.82 m.

Applying Pythagoras' theorem in a practical situation

Pythagoras' theorem is used in many practical situations, such as building and landscaping. When solving practical problems using Pythagoras' theorem, draw and label a diagram first.

WORKED EXAMPLE 15 Applying Pythagoras' theorem

A 4 m ladder leans against the wall of a house. The foot of the ladder is resting 1.8 m from the wall of the house. Determine how far up the wall of the house the ladder reaches.
Give your answer correct to 2 decimal places.

THINK	WRITE/DRAW
1. Draw and label a sketch of the situation.	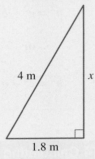
2. As it is the length of one of the shorter sides that needs to be determined, rearrange the rule to make a the subject.	$a^2 = c^2 - b^2$
3. Substitute the values for the hypotenuse, c, and the other side, b, into the rule.	$x^2 = 4^2 - 1.8^2$ $= 16 - 3.24$ $= 12.76$
4. Calculate the value of x by determining the square root of 12.76. Round the answer correct to 2 decimal places.	$x^2 = 12.76$ $x = \sqrt{12.76}$ $= 3.57$
5. Write the answer. (Remember to add the correct units.)	The ladder reaches 3.57 m up the wall.

4.4.4 Application of powers to problems involving growth and decay

Powers can be used to solve problems related to growth and decay, such as the growth of a population or the reduction in the number of an endangered species.

If the population of the town Reddan was 1500 people, and its population was doubling every 10 years, then after 50 years the population could be calculated as follows:

$$\text{Population after 50 years} = 1500 \times 2 \times 2 \times 2 \times 2 \times 2$$
$$= 1500 \times 2^5$$
$$= 48\,000$$

This is an example of **exponential growth**.

A rare and endangered plant species was found to have only 400 plants left in the wild.

Scientists calculated that the plants were reducing by $\frac{1}{20}$ every year. After 4 years the number of plants left in the wild could be calculated as follows:

$$\text{Number left after 4 years} = 400 \times \frac{19}{20} \times \frac{19}{20} \times \frac{19}{20} \times \frac{19}{20}$$
$$= 400 \times \left(\frac{19}{20}\right)^4$$
$$= 325.8025$$
$$\approx 326$$

This is an example of **exponential decay**.

Note: For exponential decay, subtract the decay factor from 1. In the case above, $1 - \left(\frac{1}{20}\right) = \frac{19}{20}$.

Exponential growth and decay rule

$$P = a \times r^n$$

where P is the population after n time periods,

a is the initial population and

r is the growth or decay factor.

---

tlvd-3672

WORKED EXAMPLE 16 Applying exponential growth and decay

Solve the following problems.
a. The population of a town is currently 3200 people, but due to the creation of new industries the population is tripling every 2 years. If this trend continues, calculate the size of the population in 8 years.
b. Another town has a population of 5500, but declining job opportunities have resulted in a decrease in population by one-fifth each year. If this trend continues, calculate the size of the population of this town in 6 years.

THINK	WRITE
a. 1. The population is tripling, so this is an exponential growth problem. Write the rule and identify the values of the variables.	a. $P = a \times r^n$ $a = 3200$ $r = 3$ $n = \dfrac{8}{2} = 4$
2. Substitute the values into the rule and calculate P.	$P = 3200 \times 3^4$ $= 259\,200$
3. State the answer.	In 8 years, the population will be 259 200.

▶

b. 1. The population is decreasing by one-fifth, so this is an exponential decay problem. Write the rule and identify the values of the variables.

b. $P = a \times r^n$
$a = 5500$
$r = 1 - \dfrac{1}{5} = \dfrac{4}{5}$
$n = 6$

2. Substitute the values into the rule and calculate P.

$P = 5500 \times \left(\dfrac{4}{5}\right)^6$
$= 1441.792$
≈ 1442

3. Write the answer.

In 6 years, the population will be 1442.

4.4 Exercise

Students, these questions are even better in jacPLUS

Receive immediate feedback and access sample responses

Access additional questions

Track your results and progress

Find all this and MORE in jacPLUS

1. **WE11** Evaluate each of the following.

   a. $\sqrt[2]{81}$ b. $\sqrt[3]{64}$ c. $\sqrt[3]{125}$ d. $\sqrt[4]{81}$ e. $\sqrt[4]{256}$ f. $\sqrt[5]{32}$

2. **WE12** Evaluate each of the following. Give your answers correct to 3 decimal places.

   a. $\sqrt{22}$ b. $\sqrt[4]{150}$ c. $\sqrt[3]{1.54}$ d. $\sqrt{2899}$

3. Evaluate each of the following. Where appropriate give your answer correct to 3 decimal places. (*Hint:* When using your calculator do not round until the last step.)

   a. $\left(\sqrt{3} + \sqrt{2}\right)^4$ b. $\left(\sqrt[3]{5}\right)^2$ c. $\dfrac{\sqrt{6} + \sqrt{12}}{\sqrt{2}}$

   d. $\left(\sqrt{2}\right)^6$ e. $8 - 2\sqrt{3} + \sqrt[3]{2} + 5 \times \sqrt{10}$

4. **WE13** Determine the length of the hypotenuse in the triangle shown. Give your answer correct to 2 decimal places.

8 cm c 20 cm

5. Calculate the length of the hypotenuse of the following triangles.

   a.
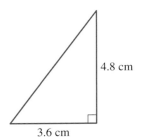
   4.8 cm   3.6 cm

   b.
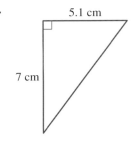
   5.1 cm   7 cm

6. Calculate the length of the hypotenuse of a right-angled triangle with perpendicular side lengths of 10 cm and 14 cm. Give your answer correct to 2 decimal places.

7. **WE14** Determine the length of the side labelled x in the following right-angled triangle. Give your answer correct to 2 decimal places.

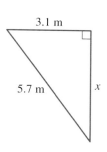

8. Calculate the length of the unknown side of each of the following triangles.

a.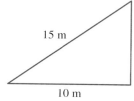

b.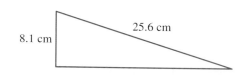

9. Calculate the length of a perpendicular side of a right-angled triangle where the other side lengths are 32 m and 18.4 m.

10. **WE15** A 5.5 m ladder leans against the wall of a house. The foot of the ladder is resting 3 m from the wall of the house. Calculate how far up the wall of the house the ladder reaches. Give your answer correct to 2 decimal places.

11. Calculate the length of the diagonal of a square of side length 5 cm.

12. **WE16** Solve the following questions.

   a. The population of a town is currently 6000 people, but due to the creation of new industries the population is doubling every 3 years. If this trend continues, calculate the size of the population in 12 years.

   b. Another town has a population of 15 000, but declining job opportunities have caused a decrease in population by one quarter each year. If this trend continues, calculate the size of the population of this town in 5 years.

13. A student tells their best friend something personal and asks that the friend only share it with three other close friends. The next day, the friend shares it with three others and tells them the same thing. If the pattern continues, calculate how many people will know the secret in 7 days.

14. The population of an endangered bird is under threat from the destruction of a native forest. The scientists researching the bird numbers note that there are currently approximately 2600 of these birds left in the wild. They estimate that the birds are decreasing in number by about one-eighth every year.
If this pattern continues, calculate how many of these birds will be left after 7 years.

15. Two Cape Barren geese are released into an area where there are no other geese. They produce two offspring every six months, and the offspring reproduce at the same rate.
    If there are no other factors, calculate how many Cape Barren geese will be in the area in 4 years' time.

16. You win a lottery and have a choice of two prizes. The first option is $1 000 000 immediately. The second option is $1 immediately and your money doubled every month for the next two years.
    Determine which is the better option financially and explain your answer using mathematics.

17. A vacant corner block of land is rectangular in shape. It is 50 m long and 20 m wide. A parent with a toddler in a pram keeps to the path and walks around the two sides of the block, but their other child runs diagonally across the block.
    Calculate how much shorter, in metres, is the diagonal path across the block.

18. A gift box in the shape of a cube has a volume of 1500 cm^3. Calculate the height of the box. (Give your answer correct to 2 decimal places.)

4.5 Review

4.5 Exercise

Multiple choice

1. **MC** The base of 6^5 is:

   A. 5   B. 36   C. 6   D. 7776   E. 65

2. **MC** The power of 4^3 is:

   A. 3   B. 16   C. 4   D. 64   E. 12

3. **MC** The evaluated value of 2^4 is:

   A. 2   B. 16   C. 4   D. 8   E. 12

4. **MC** 4.3^2 lies between the two whole square numbers:

   A. 9 and 16   B. 4 and 9   C. 16 and 25   D. 4 and 5   E. 5 and 10

5. **MC** The value of $\sqrt{121}$ is:

   A. 11   B. 10   C. 9   D.   E. 13

6. **MC** The value of $\sqrt[3]{1000}$ is:

   A. 3   B. 1000   C. 100   D.   E. 1

7. **MC** The index notation of $3 \times 3 \times 3 \times 3 \times 3$ is:

   A. 3   B. 3^5 C. 3^4 D.   E. 3^2

8. **MC** The value of $\sqrt[3]{1200}$ correct to 3 decimal places is:

   A. 10.627   B. 11.581   C. 10.63   D. 10.626   E. 10.620

9. **MC** The value of $3\sqrt{2} - 5 - 3\sqrt{3} + 4 \times \sqrt{6}$ correct to 2 decimal places is:

   A. 3.8444   B. 3.84   C. 3.8   D. 2.564   E. 3.85

10. **MC** The triangle shown has side lengths of 16 cm and 14 cm.
    The length of the third side is closest to:

    A. 15 cm
    B. 12 cm
    C. 20 cm
    D. 21 cm
    E. 22 cm

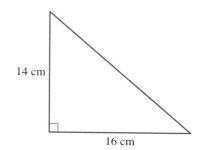

Short answer

11. Evaluate the following.

a. 6^2

b. 3^3

c. 2^5

12. Evaluate the following.

a. $\sqrt{49}$

b. $\sqrt{144}$

c. $\sqrt[3]{216}$

13. Write the following in index notation.

a. $7 \times 7 \times 7 \times 7 \times 7 \times 7$

b. $5 \times 5 \times 5 \times 5 \times 5 \times 5 \times 5 \times 6 \times 6 \times 6$

c. $3 \times 3 \times 5 \times 5 \times 5 \times 5 \times 5 \times 6 \times 6 \times 6$

14. Evaluate the following, giving your answers correct to 2 decimal places.

a. $\sqrt{3} + \sqrt{7} + 2\sqrt{3}$

b. $12\sqrt{5} + 8\sqrt{10} - 5\sqrt{5} - 7\sqrt{10}$

c. $2\sqrt{8} + 3\sqrt{3} - 2\sqrt{3} + \sqrt{2}$

15. Evaluate the following without using a calculator.

a. $\sqrt{25} - 3 + \sqrt{4}$

b. $3\sqrt{16} - 6 + 2\sqrt{9}$

c. $7\sqrt{16} - 5\sqrt{4} - 4\sqrt[3]{8}$

16. Calculate the length of the diagonal of a rectangle of side lengths 8 cm and 6 cm.

17. A ladder 8 m in length leans against a wall. If the ladder reaches 7 m up the wall, calculate how far from the bottom of the wall is the foot of the ladder. Give your answer correct to 2 decimal places.

18. The population of a small town is 2200. New housing developments are planned for the town, with the population expected to grow by 20% every 2 years for the foreseeable future. At this rate, calculate the size of the population in 10 years.

Extended response

19. a. Determine between which two whole numbers $\sqrt[3]{50}$ lies.

b. Use a calculator to determine the value of $\sqrt[3]{50}$.

20. Joel wants to landscape his garden by laying a rectangular area of instant turf. The dimensions of the rectangular area are $4\sqrt{12}$ by $3\sqrt{6}$. Calculate the area of the turf to 2 decimal places.

Hey teachers! Create custom assignments for this topic

Create and assign unique tests and exams

Access quarantined tests and assessments

Track your students' results

Find all this and MORE in jacPLUS

Answers

Topic 4 Powers and roots

4.2 Index (power) notation

4.2 Exercise

1. a. Missing: 2
      Base: 6
      Power: 2
   b. Missing: 6
      Base: 4
      Power: 6
   c. Missing: 3
      Base: 3
      Power: 3
   d. Missing: 10
      Base: 10
      Power: 3

2. a. 5^4 b. 7^6 c. 2^9 d. 12^3

3. a. $2^5 \times 4^3$ b. $3^2 \times 7$
   c. $5^2 \times 3^4$ d. $4^2 \times 5^2 \times 3^3$

4. a. $4 \times 4 \times 4 \times 4 \times 4$ b. $13 \times 13 \times 13$

5. a. 5^9 b. 4^9 c. 3^{12}

6. a. 3×10^2 b. $4 \times 10^3 + 5 \times 10^2$
   c. $6 \times 10^3 + 7 \times 10^2 + 5$ d. 10^4

7. a. 8^{12} b. $3^6 \times 13^6$

8. D

9. a. 64   b. 200   c. 77

10. a. 512   b. 5184   c. 19 683

11. a. 279   b. 2717   c. 112

12. A

13. a. $1^3 < 3^1$ b. $2^3 < 3^2$ c. $5^3 < 3^5$ d. $4^4 > 3^5$

14. a. Choose option 1 since it grows to $327.68, whereas option 2 only grows to $16.
    b. Doubling every week for 15 weeks is written as $\times 2^{15}$. Increasing by $1 per week is the same as adding $15 \times \$1$.

15. 512

16. 127

17. 24 hours

18. 254

4.3 Squares and square roots

4.3 Exercise

1. a. 25   b. 144   c. 225   d. 625

2. 64, 81, 100, 121

3. 169, 196

4. a. 16 and 25   b. 49 and 64
   c. 81 and 100   d. 169 and 196

5. a. 20   b. 6   c. 11   d. 30

6. a. 50   b. 120   c. 600   d. 400

7. a. 7 and 8; 7.87   b. 3 and 4; 3.61

     c. 13 and 14; 13.42   d. 1 and 2; 1.41

8. a. 13   b. 37

9. B

10. A

11. a. 13   b. 57   c. 864   d. 40

12. a. 14   b. 45

13. a. 11 and 12
    b. Estimate: 11.4; actual $= 11.4018$

14. a. 196
    b. 196
    c. The answers in parts a and b are the same. If two square roots are multiplied together, then the result is equal to the square roots of the base numbers multiplied together.

15. a. 15
    b. 15
    c. The answers in parts a and b are the same. If two square roots are multiplied together, then the result is equal to the square roots of the base numbers multiplied together.

16. a. 14
    b. 10
    c. The answers in parts a and b are different. The square root of two numbers that have been added together is not equal to adding the square roots of these two numbers together.

17. a. $25^2 - 24^2 = 49$
       $24^2 - 23^2 = 47$
       $23^2 - 22^2 = 45$
       $22^2 - 21^2 = 43$
       $21^2 - 20^2 = 41$
    b. The answer is the sum of the two bases.
    c. No

18. a. 7, $\sqrt{56}$, 8, $\sqrt{72}$, 3^2, $\sqrt{100}$, 62, 8^2
    b. $\sqrt{10}$, $\sqrt{121}$, 3×4, $\sqrt{169}$, 2×7, 9^2, 100

4.4 Using powers and roots to solve practical problems

4.4 Exercise

1. a. 9   b. 4   c. 5
   d. 3   e. 4   f. 2

2. a. 4.690   b. 3.500   c. 1.155
   d. 53.842

3. a. 97.990   b. 2.924   c. 4.182
   d. 8   e. 21.607

4. 12.81 cm

5. a. 6 cm   b. 8.66 cm

6. 17.20 cm

7. 4.78 m

8. a. 11.18 m   b. 24.28 cm

9. 26.18 m

10. 4.61 m

11. 7.07 cm

12. a. 96 000   **b.** 3560

13. In 7 days, 729 people will know the secret. (730 if the original person is included.)

14. 1021

15. 512

16. The second option.

17. 16.15 m

18. 11.45 cm

4.5 Review

4.5 Exercise

Multiple choice

  1. C

  2. A

  3. B

  4. C

  5. A

  6. D

  7. B

  8. A

  9. B

10. D

Short answer

11. a. 36   **b.** 27   **c.** 32

12. a. 7   **b.** 12   **c.** 6

13. a. 7^6 **b.** $5^7 \times 6^3$ **c.** $3^2 \times 5^5 \times 6^3$

14. a. 7.84   **b.** 18.81   **c.** 8.80

15. a. 4   **b.** 12   **c.** 10

16. 10 cm

17. 3.87 m

18. 5474

Extended response

19. a. 3 and 4

    b. 3.684

20. 101.82

5 Estimation and approximation

Fully worked solutions for this topic are available online.

5.1 Overview

5.1.1 Introduction

Estimation is an important skill that we use each day without necessarily realising we're doing so, whether we are comparing prices for different quantities of food at the supermarket or estimating how many spectators are at the MCG watching a sporting event.

When we estimate, we determine an answer that is close enough to the right answer. Different techniques can be used, such as rounding to the leading digit to make it easier to estimate. It is important to know the level of accuracy used in particular estimations and approximations, and why that level is used. This helps to first estimate the answer to a problem by making approximations and then verifying whether the answer from a calculator or computer is correct.

Calculators and computers can be used for accurate calculations, but they rely on the user to correctly input the data. Estimation is a way of checking that the data has been entered correctly. Estimations can be useful in many situations — for example, when we are working with calculators or calculating the total of a shopping bill. By mentally estimating an approximate answer, we are more likely to notice if we make a data-entry error.

To estimate the answer to a mathematical problem, round the numbers and calculate an approximate answer. Rounding is a simple form of estimation that is applied to quickly estimate a value.

KEY CONCEPTS

This topic covers the following key concepts from the VCE Mathematics Study Design:
- estimation, approximation and reasonableness of calculations and results.

Source: VCE Mathematics Study Design (2023–2027) extracts © VCAA; reproduced by permission.

5.2 Rounding and estimation

5.2.1 Rounding

An **estimate** is not the same as a guess, because it is based on information.

Numbers can be rounded to different degrees of accuracy.

For example, they can be rounded to 1, 2, 3 or more decimal places. The more decimal places a number has, the more accurate it is. However, more decimal places are not always necessary or relevant.

Rounding to the nearest 10

To round to the nearest 10, think about which multiple of 10 the number is closest to.

For example, 34 rounded to the nearest 10 is 30, because 34 is closer to 30 than 40.

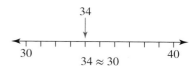

(*Note:* The ≈ symbol represents 'is approximately equal to'.)

Rounding to the nearest 100

To round to the nearest 100, think about which multiple of 100 the number is closest to.

For example, 375 rounded to the nearest 100 is 400, because 375 is closer to 400 than 300.

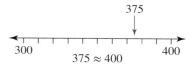

Rounding to the first digit

When rounding to the leading (first) digit, the second digit needs to be looked at.
- If the second digit is 0, 1, 2, 3 or 4, the first digit stays the same and all the following digits are replaced with zeros.
- If the second digit is 5, 6, 7, 8 or 9, the first digit is increased by 1 (rounded up) and all the following digits are replaced with zeros.

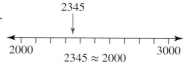

For example, if 2345 is rounded to the first digit, the result is 2000 because the second digit in the number is 3, so all digits following the first digit (in this instance 2) are replaced with 0.

---

tlvd-3663

WORKED EXAMPLE 1 Rounding integers

Round each of the following as directed.
a. 563 to the nearest 10   **b. 12 786 to the nearest 100**
c. 7523 to the leading digit

THINK	WRITE
a. 1. To round to the nearest 10, look at the units place value.	a. 563
2. Is the units place value less than 5? Since 3 is less than 5, we round down. Replace 3 with 0 and write the answer.	≈ 560

b. **1.** To round to the nearest 100, look at the tens place value.

2. Is the tens place value greater than or equal to 5? Since 8 is greater than or equal to 5, we round up. Increase 7 by 1 to 8 and replace the following digits with zeros.

3. Write the answer.

b. 12 786

12 786

↓↓↓

800

12 786 ≈ 12 800

c. **1.** When rounding to the leading digit, look at the second digit.

2. Since 5 is greater than or equal to 5, we round up. Increase the first digit (7) by 1 and replace the remaining digits with zeros.

3. Write the answer.

c. 7523

7523

↓↓↓↓

8000

7523 ≈ 8000

5.2.2 Estimation

Estimation is useful for checking that your answer is reasonable when performing calculations. This can help avoid simple mistakes, like typing an incorrect number into a calculator.

tlvd-3664

WORKED EXAMPLE 2 Estimating by rounding numbers to the leading (first) digit

Estimate 66 123 × 749 by rounding each number to the leading (first) digit and then completing the calculation.

THINK	WRITE
1. Round the first number (**66** 123) by focusing on the second digit. Since the second digit (6) is greater than or equal to 5, increase the leading digit (6) by 1 and replace the remaining digits with zeros.	$66\,123 \approx 70\,000$
2. Round the second number (**7**49) by focusing on the second digit. Since the second digit (4) is less than 5, leave the leading digit and replace the remaining digits with zeros.	$749 \approx 700$
3. Multiply the rounded numbers by multiplying the leading digits ($7 \times 7 = 49$) and then add the number of zeros ($4 + 2 = 6$).	$70\,000 \times 700$ $7 \times 7 = 49$ $70\,000 \times 700 = 49\,000\,000$
4. Write the estimated answer and compare to the actual answer.	$66\,123 \times 749 \approx 49\,000\,000$ This compares well with the actual answer of 49 526 127, found using a calculator.

 Resources

 **Interactivities** Rounding (int-3932)

Rounding to the first digit (int-3731)

Checking by estimating (int-3983)

5.2 Exercise

1. **WE1** Round each of the following as directed.
   a. 934 to the nearest 10
   b. 12 963 to the nearest 100
   c. 85 945 to the leading digit

2. Round the following.
   a. 347 to the nearest 10
   b. 86 557 to the nearest 100
   c. 65 321 to the leading digit

3. Round each of the following numbers to the nearest 10.
   a. 47
   b. 82
   c. 129
   d. 162
   e. 4836
   f. 7

4. Round each of the following numbers to the nearest 100.
   a. 43
   b. 87
   c. 177
   d. 3285
   e. 86 621
   f. 213 951

5. Round each of the following numbers to the nearest 1000.
   a. 512
   b. 3250
   c. 6300
   d. 13 487
   e. 435 721
   f. 728 433

6. Round each of the following numbers to the leading digit.
   a. 12
   b. 23
   c. 153
   d. 1388
   e. 16 845
   f. 492 385

7. **WE2** Estimate $74\,852 \times 489$ by rounding each number to its leading digit.

8. Estimate $87\,342 \div 449$ by rounding each number to its leading digit.

9. Estimate the answer to the following expressions by rounding each number to the leading digit and then completing the calculation.
   a. $482 + 867$
   b. $123 + 758$
   c. $1671 - 945$
   d. $2932 - 1455$

10. Estimate the answer to the following expressions by rounding each number to the leading digit and then completing the calculation.
    a. 88×543
    b. 57×2632
    c. $69\,523 \div 1333$
    d. $3600 \div 856$

11. In your own words, explain how to round using leading digit approximation.

12. Give three examples of situations where it is suitable to use an estimate or a rounded value instead of an exact value.

13. Emily purchased a number of items at the supermarket that cost $1.75, $5.99, $3.45, $5.65, $8.95, $2.35 and $7.45. She was worried she didn't have enough money, so she used leading digit approximation to do a quick check of how much her purchases would cost.
    Determine what approximation she came up with.

14. The crowds at the Melbourne Cricket Ground for each of the five days of the Boxing Day Cricket test were: 88 214, 64 934, 55 349, 47 567 and 38 431.

   a. Using leading digit approximation, determine the estimated total crowd over the five days of the test.
   b. Calculate the difference between the leading digit estimate and the actual crowd over the five days.

15. A car's GPS estimates a family's trip for Christmas lunch is going to take 45 minutes.

   a. Determine if the trip is likely to take exactly 45 minutes. Explain your answer.
   b. Explain how this time is estimated by the GPS.

16. A school student went to buy a new school uniform. They decided they needed a new blazer, a new winter dress and a new shirt costing $167, $89 and $55 respectively.

   a. Use the leading digit method to estimate how much the student spent, and calculate the difference between the total they spent and the leading digit estimate.
   b. Determine if the estimate was more or less than the actual price. Explain why this was the case.

5.3 Reasonableness and calculator skills

LEARNING INTENTION

At the end of this subtopic you should be able to:
- check for accuracy and reasonableness of calculations and results
- use technology effectively for accurate, reliable and efficient calculation
- use order of magnitude as powers of 10.

5.3.1 Reasonableness

When calculating mathematical answers, it is always important to understand the question so you have an idea of what a *reasonable* answer would be. When you complete arithmetic calculations, either manually or using your calculator, you can make mistakes. Checking the reasonableness of answers can indicate possible mistakes in your working.

tlvd-3665

WORKED EXAMPLE 3 Checking reasonableness of calculations

Cathy goes to the shop to purchase an outfit for dinner. She buys a $145 dress and a $180 jacket, and gets $30 off the combined price by purchasing the two together. From the shoe shop, she buys a pair of shoes for $120 and a pair of stockings for $15.
Calculate how much Cathy spent in total. Check your answer for reasonableness.

THINK	WRITE
1. Read the question carefully to understand what it is about.	The question asks you to add up Cathy's total shopping bill including a discount.

2. Have an idea of what sort of answer you expect.

Cathy purchased three items that were over $100 dollars, so the answer should be over $300.

3. Work out what needs to be added and subtracted.

$145 + $180 − $30 + $120 + $15

4. Calculate and write the answer.

= $430 spent

5. Does the answer seem reasonable?

This answer is above $300, so it seems reasonable. Since the price of two items end with 5 and the rest end with 0, the answer should end in 0.

5.3.2 Calculator skills

Even though your calculator will always give the correct answer, it relies on you entering the information correctly. To make sure you get the correct answer, it's always good to estimate it to check for reasonableness and accuracy.

You will notice that your calculator automatically follows the correct order of operations (BIDMAS).

WORKED EXAMPLE 4 Calculating values of expressions using a calculator

Calculate the value of the following expressions using your calculator.
a. $140 + 23 \times 8$
b. $7(28 + 43) - 18$
c. $283 - 13.78 \times 11.35$

THINK

a. 1. Estimate your answer using leading digit approximation.

2. Enter the exact equation into your calculator. Check the reasonableness of your estimate.

b. 1. Estimate your answer using leading digit approximation.

2. Enter the exact expression into your calculator. Check the reasonableness of your estimate.

c. 1. Estimate your answer using leading digit approximation.

2. Enter the exact expression into your calculator. Check the reasonableness of your estimate.

WRITE

a. $140 + 23 \times 8$
$\approx 100 + 20 \times 10$
≈ 300

$140 + 23 \times 8 = 324$

b. $7(28 + 43) - 18$
$\approx 7(30 + 40) - 20$
$\approx 7(70) - 20$
$\approx 490 - 20$
≈ 470

$7(28 + 43) - 18 = 479$

c. $283 - 13.78 \times 11.35$
$\approx 300 - 10 \times 10$
$\approx 300 - 100$
≈ 200

$283 - 13.78 \times 11.35 = 126.597$

When using your calculator, first estimate the result to compare with the answer from the calculator. Always check that what you type into the calculator matches the question.

WORKED EXAMPLE 5 Evaluating expressions using a calculator and applying rounding

tlvd-3666

Evaluate the following expressions using your calculator, rounding your answer to 4 decimal places.

a. $(2.56 + 3.83)^2 + 45.93$ b. $\sqrt{4.39 - 2.51} + (3.96)^2$ c. $\dfrac{4}{7} + \dfrac{5}{9} + 5$

THINK	WRITE
a. 1. Estimate your answer using leading digit approximation.	**a.** $(2.56 + 3.83)^2 + 45.93$ $\approx (3 + 4)^2 + 50$ $\approx 7^2 + 50$ $\approx 49 + 50$ ≈ 99
2. Enter into the calculator. Use the x^2 key to raise the brackets to the power of 2. Compare with your estimate to check for reasonableness.	$(2.56 + 3.83)^2 + 45.93$ $= 86.7621$
b. 1. Estimate your answer using leading digit approximation.	**b.** $\sqrt{4.39 - 2.51} + (3.96)^2$ $\approx \sqrt{4 - 3} + 4^2$ $\approx \sqrt{1} + 16$ $\approx 1 + 16$ ≈ 17
2. Enter into the calculator. Use the $\sqrt{}$ and x^2 keys. Make sure the square root covers both numbers. Compare with your estimate to check for reasonableness.	$\sqrt{4.39 - 2.51} + (3.96)^2$ $= 17.0527$
c. 1. Estimate your answer using leading digit approximation.	**c.** $\dfrac{4}{7} + \dfrac{5}{9} + 5$ $\approx 1 + 1 + 5$ ≈ 7
2. Enter into the calculator. Use the fraction key to enter each fraction. Compare with your estimate to check for reasonableness.	$\dfrac{4}{7} + \dfrac{5}{9} + 5$ $= 6.1270$

5.3.3 Scientific notation

Scientific notation is a special way of writing numbers. It is a useful way to write very large and very small numbers.

Number	Scientific notation
200	2×10^2
6000	6×10^3
120 000	1.2×10^5
0.000 000 5	5×10^{-7}

A number written in scientific notation has two parts:
- A number between 1 and 10
- A power of 10

For example,

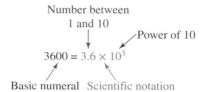

To write numbers in scientific notation, first write an appropriate number between 1 and 10 and then count the number of decimal places; positive to the left, negative to the right.

$$4500 = 4.5 \quad 0 \quad 0 \qquad \times 10^3$$
$$= 4.5 \times 10^3$$

$$0.000 006 7 = 0\ 0\ 0\ 0\ 0\ 6.7 \qquad \times 10^{-6}$$
$$= 6.7 \times 10^{-6}$$

Calculators often display numbers using scientific notation, so it is important to be able to read them. Scientists, engineers and people working in many other professions write numbers in scientific notation also.

tlvd-3667

WORKED EXAMPLE 6 Calculating values of expressions using scientific notation

Calculate the value of the following expressions using scientific notation.
a. $7.3 \times 10^4 + 5.8 \times 10^5$ **b.** $3.5 \times 10^8 \times 5.9 \times 10^6$

THINK

a. 1. Enter the expression into a calculator.

   2. Write the answer.

b. 1. Enter the expression into a calculator.

   2. Write the answer.

WRITE

a. $7.3 \times 10^4 + 5.8 \times 10^5$

   $= 6.53 \times 10^5$

b. $3.5 \times 10^8 \times 5.9 \times 10^6$

   $= 2.07 \times 10^{15}$

 Resources

🧩 **Interactivity** Scientific notation (int-6031)

5.3 Exercise

1. **WE3** Fred went to the sports department to get some clothes for the gym. He purchased a $120 tracksuit and a pair of $150 runners as a package and got $35 off the combined price. Fred also purchased a pair of running shorts for $30 and a singlet top for $25.
   Calculate how much he spent at the sports department. Check your answer for reasonableness.

2. Erika goes shopping for some party food. She buys 6 packets of chips at $2.50 each and 2 boxes of cola at $8.50 each; since she purchased 2 boxes of cola, she gets a $2 discount. She also buys 5 packets of biscuits at $1 each and 3 dips at $2 each.
   Calculate how much Erika spent in total. Check your answer for reasonableness.

3. **WE4** Calculate the value of each of the following expressions using your calculator.
   a. $269 + 12 \times 16$
   b. $9(78 + 61) - 45$
   c. $506.34 - 39.23 \times 17.04$

4. Calculate the value of each of the following expressions using your calculator.
   a. $497 - 13 \times 24 \div 6$
   b. $13(194 - 62) + 4 \times 8$
   c. $459.38 \div 18.5 \times 17.04 - (34.96 + 45.03)$

5. Evaluate the following expressions using your calculator.
   a. $3 + 6 \times 7$
   b. $47 + 8 \times 6$
   c. $285 + 21 \times 16$
   d. $2859 + 178 \times 79$

6. Evaluate the following expressions using your calculator.
   a. $12 - 5 \times 2$
   b. $68 - 4 \times 9$
   c. $385 - 16 \times 9$
   d. $1743 - 29 \times 45$

7. Evaluate the following expressions using your calculator.
   a. $4(6 + 24) - 58$
   b. $5(23.5 - 18.3) + 23$
   c. $2.56(89.43 - 45.23) - 92.45$
   d. $6(45.89 - 32.78) - 3(65.89 - 59.32)$

8. **WE5** Evaluate the following expressions using your calculator, rounding your answer to 4 decimal places.
   a. $(5.89 + 2.16)^2 + 67.99$
   b. $\sqrt{8.77 - 3.81} + (5.23)^2$
   c. $\dfrac{3}{5} + \dfrac{7}{8} + 10$

9. Evaluate the following expressions using your calculator, rounding your answer to 4 decimal places.
   a. $\left(\dfrac{7}{12}\right)^2 - \dfrac{3}{4}$
   b. $\sqrt{3.56 + 8.28} - \sqrt{5.29 - 3.14}$
   c. $\left(\dfrac{4}{5}\right)^2 + \sqrt{\dfrac{7}{12}}$

10. Evaluate the following expressions using your calculator, rounding your answers to 4 decimal places.
    a. $16.9 + 5.2^2 \div 4.3$
    b. $9.3^2 \div 4.5$
    c. $(3.7 + 5.9)^2 - 15.5$
    d. $\dfrac{(7.2)^2}{4.2}$

11. Evaluate the following expressions using your calculator, rounding your answers to 4 decimal places.

   a. $\sqrt{5.67 + 8.34}$

   b. $2.5 \times \sqrt{8.64} - 2.5$

   c. $12.8 + \sqrt{3.5 \times 5.8} \times 1.2$

   d. $\dfrac{\sqrt{4.7 - 3.6}}{5}$

12. Evaluate the following expressions to 2 decimal places using your calculator.

   a. $\dfrac{1}{2} + \dfrac{1}{4}$

   b. $\dfrac{3}{7} - \dfrac{1}{9}$

   c. $\dfrac{18}{13} - \dfrac{2}{5} \times \dfrac{2}{7}$

   d. $15 - \dfrac{7}{5} \times \dfrac{5}{3} + 7$

13. **WE6** Calculate the value of the following expressions in scientific notation.

   a. $7.9 \times 10^4 + 4.9 \times 10^4$

   b. $3.7 \times 10^6 \times 8.3 \times 10^9$

14. Calculate the value of the following expressions in scientific notation.

   a. $5.3 \times 10^6 - 8.1 \times 10^5$

   b. $2.7 \times 10^5 \times 5.6 \times 10^3$

15. Evaluate the following expressions using your calculator, leaving your answer in scientific notation.

   a. $1.45 \times 10^3 + 5.82 \times 10^3$

   b. $6.89 \times 10^6 - 4.71 \times 10^5$

   c. $3.1 \times 10^5 \times 8.49 \times 10^3$

   d. $4.21 \times 10^4 \div (8.32 \times 10^2)$

   e. $7.38 \times 10^5 \div (2.62 \times 10^8)$

   f. $3.82 \times 10^3 + (9.27 \times 10^5 \times 4.5 \times 10^2)$

16. Cathy is 7 years older than Marie, but Marie is twice the age of her younger brother Fergus. Determine Cathy's age if Fergus is 3 years old.

17. After a football party night, some empty soft drink cans were left and had to be collected for recycling. Of the cans found, 7 were half full, 6 cans were one-quarter full and the remaining 8 cans were completely empty. Use your calculator to find out how many equivalent full cans were wasted.

18. NASA landed the Curiosity Rover on Mars in 2012. The average distance from the Sun to Earth is 1.5×10^8 km and the average distance from the Sun to Mars is 2.25×10^8 km. Calculate the average distance between Earth and Mars when Earth is between Mars and the Sun.

19. Carbon is an atom that has a radius of 7.0×10^{-11} m. The circumference of a carbon atom can be calculated by evaluating $2 \times 3.14 \times 7.0 \times 10^{-11}$ m (that is $2\pi r$). Calculate the circumference of a carbon atom in scientific form.

20. Jupiter is the largest planet in our solar system with a radius of 7.1492×10^7 m. The volume $\left(V = \dfrac{4}{3}\pi r^3 \right)$ of Jupiter can be calculated by evaluating $\dfrac{4}{3} \times 3.14 \times \left(7.1492 \times 10^7\right)^3$ m³. Calculate the volume of Jupiter in scientific form.

5.4 Order of magnitude

LEARNING INTENTION

At the end of this subtopic you should be able to:
- understand that for a number written as a power of 10, the exponent (index) is the order of magnitude
- use orders of magnitude to compare different values and check that estimates are reasonable.

5.4.1 What is an order of magnitude?

An **order of magnitude** uses factors of 10 to give generalised estimates and relative scale to numbers.

To use orders of magnitude we need to be familiar with powers of 10.

Power of 10	Basic numeral	Order of magnitude
10^{-4}	0.0001	−4
10^{-3}	0.001	−3
10^{-2}	0.01	−2
10^{-1}	0.1	−1
10^{0}	1	0
10^{1}	10	1
10^{2}	100	2
10^{3}	1000	3
10^{4}	10 000	4

As you can see from the table, the exponents of the powers of 10 are equal to the orders of magnitude.

Order of magnitude

An increase of 1 order of magnitude means an increase in the basic numeral by a multiple of 10. Similarly, a decrease of 1 order of magnitude means a decrease in the basic numeral by a multiple of 10.

5.4.2 Using orders of magnitude

We use orders of magnitude to compare different values and to check that estimates we make are reasonable. For example, if we are given the mass of two objects as 1 kg and 10 kg, we can say that the difference in mass between the two objects is 1 order of magnitude, as one of the objects has a mass 10 times greater than the other.

WORKED EXAMPLE 7 Identifying the order of magnitude

Identify the order of magnitude that expresses the difference in distance between 3.5 km and 350 km.

THINK	WRITE
1. Calculate the difference in size between the two distances by dividing the larger distance by the smaller distance.	$\dfrac{350}{3.5} = 100$

2. Express this number as a power of 10.

$100 = 10^2$

3. The exponent of the power of 10 is equal to the order of magnitude. Write the answer.

The order of magnitude that expresses the difference in distance is 2.

5.4.3 Scientific notation and orders of magnitude

When working with orders of magnitude, it can be helpful to express numbers in scientific notation. If two numbers in scientific notation have the same coefficient, (that is, if the numbers in the first part of the scientific notation (between 1 and 10) are the same), then we can easily determine the order of magnitude by calculating the difference in value between the exponents of the powers of 10.

WORKED EXAMPLE 8 Determining the order of magnitude

Determine how many orders of magnitude the following distances differ.
Distance A: 2.6×10^{-3} km
Distance B: 2.6×10^2 km

THINK

1. Check that the coefficients of both numbers are the same in scientific notation.
2. Determine the order of magnitude difference between the numbers by subtracting the exponent of the smaller power of 10 (−3) from the exponent of the larger power of 10 (2).
3. Write the answer.

WRITE

In scientific notation, both numbers have a coefficient of 2.6.

$2 - (-3) = 5$

The distances differ by an order of magnitude of 5.

5.4.4 Units of measure

When using orders of magnitude to compare values it is important to factor in the units used. For example, if the weight of a fully grown male giraffe is 1.5×10^3 kilograms and a full bottle of milk weighs 1.5×10^3 grams, and the units were not considered, it would appear that the order of magnitude between the weight of the milk and the giraffe is 0. When converted into the same units (e.g. kilograms), the weight of the bottle of milk becomes 1.5×10^0 kg and the weight of the giraffe remains 1.5×10^3 kg, which is 3 orders of magnitude larger than the weight of the milk.

We can convert units of length and mass by using the following charts.

Converting length:

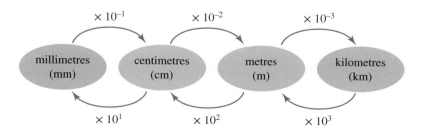

Converting mass:

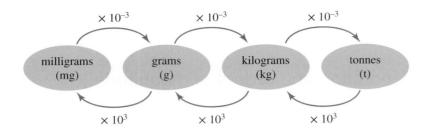

Note: You can see from the charts that multiplying by 10^{-1} is the same as dividing by 10^1, multiplying by 10^{-3} is the same as dividing by 10^3, etc.

Our conversion charts show that there is a difference of 1 order of magnitude between millimetres and centimetres, and 3 orders of magnitude between grams and kilograms.

WORKED EXAMPLE 9 Determining orders of magnitude to compare values with different units

The mass of a single raindrop is approximately 1×10^{-4} grams and the mass of an average apple is approximately 1×10^{-1} kilograms.
a. State by how many orders of magnitude these masses differ.
b. State how many times larger the mass of the apple is than that of the raindrop, expressed as a basic numeral.

THINK	WRITE
a. 1. In this situation it will be easier to work in grams. To convert 1×10^{-1} kilograms to grams, multiply by 10^3.	a. $1 \times 10^{-1} \times 10^3 = 1 \times 10^2$ g
2. Compare the orders of magnitude by subtracting the smaller exponent (-4) from the larger exponent (2).	Raindrop: 1×10^{-4} g Apple: 1×10^2 g $2 - -4 = 6$ The masses differ by 6 orders of magnitude.
b. 1. Each order of magnitude indicates a power of 10.	b. $10^6 = 1\ 000\ 000$
2. Write the answer.	The mass of the apple is 1 000 000 times larger than the mass of the raindrop.

5.4.5 Logarithmic scales

Earthquakes are measured by seismometers, which record the amplitude of the seismic waves of the earthquake. There are large discrepancies in the size of earthquakes, so rather than using a traditional scale to measure their amplitude, a **logarithmic scale** is used.

Logarithmic scale

A logarithmic scale represents numbers using a log (base 10) scale. This means that if we express all of the numbers in the form 10^a, the logarithmic scale will represent these numbers as a.

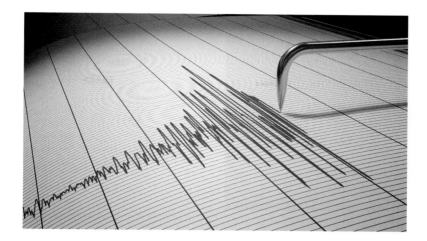

This means that for every increase of 1 in the magnitude in the scale, the amplitude or power of the earthquake is increasing by a multiple of 10. This allows us to plot earthquakes of differing sizes on the same scale.

The Richter scale was designed in 1934 by Charles Richter and is the most widely used method for measuring the magnitude of earthquakes.

Let's consider the following table of historical earthquake data.

World earthquake data			
Year	Location	Magnitude on Richter scale	Amplitude of earthquake
2021	Australia (Mansfield, Victoria)	5.9	$10^{5.9}$
2011	Japan (Tohoku)	9	10^9
2010	New Zealand (Christchurch)	7.1	$10^{7.1}$
1989	USA (San Francisco)	6.9	$10^{6.9}$
1960	Chile (Valdivia)	9.5	$10^{9.5}$

If we were to plot this data on a linear scale, the amplitude of the largest earthquake would be 3981 times bigger ($10^{9.5}$ compared to $10^{5.9}$) than the size of the smallest earthquake. This would create an almost unreadable graph. By using a logarithmic scale, the graph becomes easier for us to interpret.

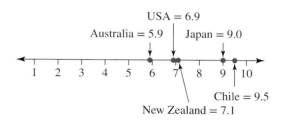

However, this scale doesn't highlight the real difference between the amplitudes of the earthquakes, which only becomes clear when these values are calculated.

The difference in amplitude between an earthquake of magnitude 1 and an earthquake of magnitude 2 on the Richter scale is 1 order of magnitude, or $10^1 = 10$ times.

Let's consider the difference between the 2021 Australian earthquake and the 2010 New Zealand earthquake. According to the Richter scale the difference in magnitude is 1.2, which means that the real difference in the amplitude of the two earthquakes is $10^{1.2}$. When evaluated, $10^{1.2} = 15.8$, which indicates that New Zealand earthquake was more than 15 times more powerful than the Australian earthquake.

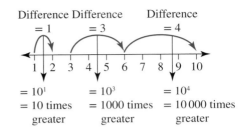

Difference = 1

Difference = 3

Difference = 4

$$= 10^1$$
$$= 10 \text{ times greater}$$

$$= 10^3$$
$$= 1000 \text{ times greater}$$

$$= 10^4$$
$$= 10\,000 \text{ times greater}$$

WORKED EXAMPLE 10 Using logarithmic scale to measure the amplitude of an earthquake

Using the information from the World Earthquake Data, compare the real amplitude of the Australian earthquake to that of the Japanese earthquake, which resulted in a tsunami that damaged the Fukushima power plant.

THINK	WRITE
1. First identify the order of magnitude difference between Japan's earthquake and Australia's earthquake.	$9 - 5.9$ $= 3.1$
2. Express the order of magnitude difference in real terms by displaying it as a power of 10.	$10^{3.1}$
3. Evaluate to express the difference in amplitude.	$= 1258.92$
4. Write the answer.	Japan experienced an earthquake that was nearly 1260 times larger than the earthquake in Australia.

5.4.6 Why use a logarithmic scale instead of a linear scale?

As you can see from Worked example 10, very large numbers are involved when dealing with magnitudes of earthquakes. It would be challenging to represent an increase in amplitude of 4000 times while also having a scale accurate enough to accommodate smaller changes. Using the logarithmic scale enables such diversity in numbers to be represented on the same plane with a functioning scale.

Logarithmic scales are also used in measuring pH levels. On the pH scale 7 is considered neutral, while values from 6 to 0 indicate an increase in acidity levels and values from 8 to 14 indicate an increase in alkalinity.

5.4 Exercise

1. **WE7** Identify the order of magnitude that expresses the difference between 0.3 metres and 3000 metres.

2. The Big Lobster in South Australia is a 4000 kg sculpture of a lobster. If a normal lobster has a mass of 4 kg, determine the order of magnitude by which the mass of the Big Lobster is greater than the mass of a normal lobster.

3. Convert the values shown to the units in the brackets, using scientific form.

   a. 1×10^3 m (km)
   b. 1×10^4 g (kg)
   c. 9×10^5 mm (cm)
   d. 5.4×10^2 t (kg)
   e. 1.2×10^{-5} kg (mg)
   f. 6.3×10^{12} mm (km)

4. State by how many orders of magnitude the following pairs of values differ.
   (*Note:* 1 kilotonne (kt) = 1000 tonnes (t).)

   a. 1.15×10^5 mm and 1.15×10^{-2} m
   b. 3.67×10^{-12} km and 3.67×10^7 cm
   c. 2.5×10^{17} km and 2.5×10^{26} mm
   d. 4.12×10^5 kt and 4.12×10^{15} t
   e. 5.4×10^{14} kg and 5.4×10^{20} mg
   f. 4.01×10^{-10} kt and 4.01×10^{10} mg

5. **WE8** The weight of a brown bear in the wild is 3.2×10^2 kg, while the weight of a cuddly teddy bear is 3.2×10^{-1} kg. State by how many orders of magnitude the weights of the bears differ.

6. A cheetah covers 100 m in 7.2 seconds. It takes a snail a time 3 orders of magnitude greater than the cheetah to cover this distance. Express the time it takes the snail to cover 100 m in seconds.

7. **MC** Select the smallest value.

   A. 1.4×10^{-5} km
   B. 1.4×10^{-6} m
   C. 1.4×10^{-3} cm
   D. 1.4×10^{-2} mm
   E. 1.4×10^4 cm

8. **MC** A virus has an approximate mass of 1×10^{-20} kg. Select the item that has a mass 10 000 000 000 times smaller than that of the virus.

   A. A hydrogen atom (mass $= 1 \times 10^{-27}$ kg)
   B. An electron (mass $= 1 \times 10^{-30}$ kg)
   C. A bacterium (mass $= 1 \times 10^{-15}$ kg)
   D. An ant (mass $= 1 \times 10^{-6}$ kg)
   E. A grain of fine sand (mass $= 1 \times 10^{-9}$ kg)

9. **WE9** The length of paddock A is 2×10^3 m, while the length of the adjoining paddock B is 2×10^5 km. State by how many orders of magnitude paddock B is longer than paddock A.

10. The mass of an amoeba is approximately 1×10^{-5} grams, while the mass of a one-year-old child is approximately 1×10^1 kilograms.
    a. State by how many orders of magnitude these masses differ.
    b. Express as a basic numeral the number of times lighter the amoeba is than the child.

11. In Tran's backyard the average height of a blade of grass is 6 cm. Tran has a tree that has grown to a height 2 orders of magnitude taller than the average blade of grass.
    a. Express the height of the grass and the tree in scientific notation.
    b. State the height of the tree.

12. **WE10** Many of the earthquakes experienced in Australia are between 3 and 5 in magnitude on the Richter scale. Compare the experience of a magnitude 3 earthquake to that of a magnitude 5 earthquake, expressing the difference in amplitude as a basic numeral.

13. A soft drink has a pH of 5, while lemon juice has a pH of 2.
    a. Identify the order of magnitude difference between the acidity of the soft drink and the lemon juice.
    b. State the number of times the juice is more acidic than the soft drink.

14. **MC** Water is considered neutral and has a pH of 7. State which of the following liquids is 10 000 times more acidic or alkaline than water.
    A. Soapy water, pH 12
    B. Detergent, pH 10
    C. Orange juice, pH 3
    D. Vinegar, pH 2
    E. Battery acid, pH 1

15. The largest ever recorded earthquake was in Chile in 1960 and had a magnitude of 9.5 on the Richter scale. In 2011, an earthquake of magnitude 9.0 occurred in Japan.
    a. Determine the difference in magnitude between the two earthquakes.
    b. Reflect this difference in real terms by calculating the difference in amplitude between the two earthquakes, giving your answer as a basic numeral correct to 2 decimal places.

16. Diluted sulfuric acid has a pH of 1, making it extremely acidic. If its acidity was reduced by 1000 times, state what pH it would register.

17. Andrew has a tennis ball with a mass of 5×10^1 grams. At practice his coach sets him a series of exercise using a 5×10^3 gram medicine ball.
    a. State the order of magnitude difference between the mass of the two balls.
    b. Determine the mass of each ball, written as a basic numeral.

18. The size of a hydrogen atom is 1×10^{-10} m and the size of a nucleus is 1×10^{-15} m.
    a. State by how many orders of magnitude the atom and nucleus differ in size.
    b. State how many times larger the atom is than the nucleus.

19. The volume of a 5000 mL container was reduced by 1 order of magnitude.
    a. Express this statement in scientific notation.
    b. Determine the new volume of the container.

20. The distance between Zoe's house and Gwendolyn's house is 25 km, which is 2 orders of magnitude greater than the distance from Zoe's house to her school.
    a. Express the distance from Zoe's house to her school in scientific notation.
    b. Determine the distance from Zoe's house to her school.

5.5 Approximation

5.5.1 Approximating calculations

Approximating or estimating an answer is useful when an accurate answer is not necessary. You can approximate the size of a crowd at a sporting event based on an estimate of the fraction of seats filled. This approach can be applied to a variety of situations, e.g. approximating the amount of water in a water tank.

When approximating size, it is important to use an appropriate unit, e.g. for the distance you drive you would use 24 km instead of 2 400 000 cm. Approximations can also be made by using past history to indicate what might happen in the future.

WORKED EXAMPLE 11 Approximating by carrying out relevant calculations

Approximate the size of the crowd on the first day of the cricket test at the Melbourne Cricket Ground, if it is estimated to be at $\dfrac{8}{10}$ of its capacity of 95 000.

THINK

1. Write down the information given in the question.

2. Calculate the fraction of the total capacity.

3. Write the answer.

WRITE

Estimate of crowd $= \dfrac{8}{10}$ of capacity

$$\dfrac{8}{10} \text{ of } 95\,000 = \dfrac{8}{10} \times 95\,000$$
$$= 0.8 \times 95\,000$$
$$= 76\,000$$

The approximate crowd is 76 000 people.

5.5.2 Approximation from graphs

When approximating from graphs:
- if the data is shown, read from the graph, noting the scale on the axes
- if the data is not shown, follow the trend (behaviour of the graph) to make an estimate of future or past values.

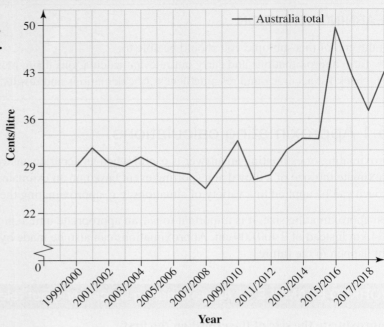

tlvd-3668

WORKED EXAMPLE 12 Approximating values from graphs

The following graph shows the farmgate price of milk in cents/litre. Approximate the price in 2019/2020.

THINK

1. Check what each axis represents.

2. Look for the trend that the graph has followed in previous years. Draw a line to represent this trend and extend it to the required period of 2019/2020.

3. Read the answer from the graph.

WRITE/DRAW

Vertical axis: cents/litre
Horizontal axis: years

The graph has been on an upward trend for the past 8 years.

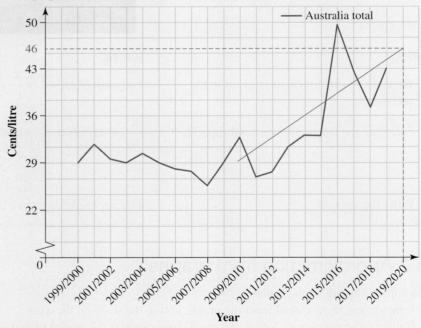

Approximately 46 cents/litre

Resources

⊞ **eLesson** Estimating and rounding (eles-0822)

5.5 Exercise

Students, these questions are even better in jacPLUS

1. **WE11** Approximate the number of people at a Bledisloe Cup match at Stadium Australia in Sydney if the crowd was estimated at $\dfrac{7}{10}$ of its capacity of 83 500.

2. When full, a pool holds 51 000 litres of water. Approximate the number of litres in the pool when it is estimated to be 90% full.

3. Choose the most appropriate unit from the options below for each of the following measurements.
centimetres (cm), metres (m), kilometres (km), millilitres (mL), litres (L), grams (g), kilograms (kg)

   a. The length of a football field
   b. The weight of a packet of cheese
   c. The volume of soft drink in a can
   d. The distance you travel to go on a holiday
   e. The size of an OLED TV
   f. The weight of a Rugby League player

4. **MC** Select the best approximation of the volume of milk in the container shown.
   A. 500 mL
   B. 750 mL
   C. 1 L
   D. 2 L
   E. 5 L

5. **MC** Select the best approximation of the weight of the tub of butter shown.

A. 500 kg

B. 125 g

C. 500 g

D. 750 g

E. 5 kg

6. **WE12** From the graph of the cost of tuition fees for one school, predict the approximate tuition fees in 2024.

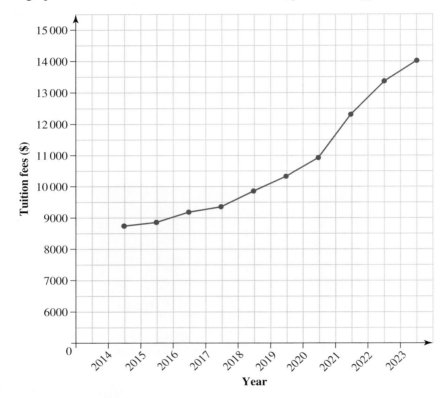

7. Use the following graph to approximate the volume of domestic air travel emissions in 2021.

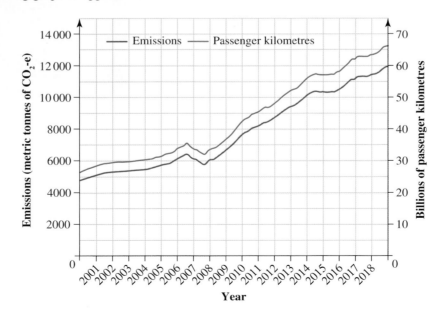

8. If the Sydney Olympics swimming venue was $\frac{9}{10}$ full and had a capacity of 17 000, determine the approximate crowd on this night.

9. The capacity at Sydney's Qudos Bank Arena, where the Sydney Kings play, is 18 200. Determine the approximate crowd at the game shown in the picture.

10. The capacity of Rod Laver Arena is 14 820. Approximate the crowd at this match from the Australian Open.

11. The picture shown is of Tiger Woods at the US Masters. State why it is difficult to estimate the size of the Masters crowd on this day by looking at the picture.

12. Explain in your own words the difference between an estimate and a guess.

13. Describe a way to approximate how many students are at your school. Investigate the actual number and compare this to your estimate. Explain how you could improve your initial estimate.

14. From the graph of the population of Australia since 1970, approximate how many people will live in Australia in 2030.

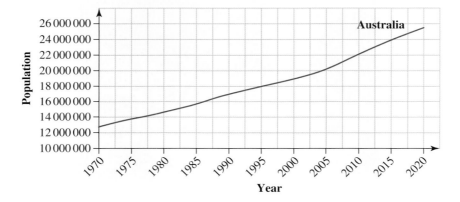

15. Use the graph of the value of the Australian dollar against the US dollar to approximate the value of the Australian dollar in January 2019.

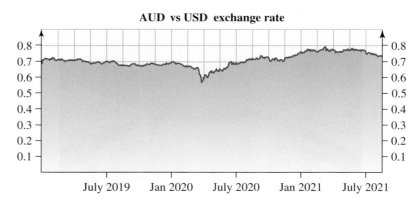

16. From the graph of Aboriginal and Torres Strait Islander population, predict the approximate population in:

    a. 2025

    b. 2030.

Historical population of Aboriginal and Torres Strait Islander Australians, 1986 to future

5.6 Review

5.6.1 Summary

doc-38046

Hey students! Now that it's time to revise this topic, go online to:

Access the topic summary

Review your results

Watch teacher-led videos

Practise questions with immediate feedback

Find all this and MORE in jacPLUS

5.6 Exercise

Multiple choice

1. **MC** The number 47 321 rounded using leading digit estimation is:

   A. 47 000   **B.** 4000   **C.** 5000   **D.** 7000   **E.** 50 000

2. **MC** If three purchases were made with the values of $7.34, $18.05 and $2.69, using leading digit estimation on the three purchases, the total purchase will approximately cost:

   A. $30   **B.** $28.08   **C.** $28   **D.** $28.10   **E.** $30.10

3. **MC** Using your calculator, the value of $\sqrt{6.78 + 19.83} + 3.91$ to 2 decimal places is closest to:

   A. 5.16   **B.** 9.07   **C.** 5.52   **D.** 1.25   **E.** $10.25

4. **MC** The volume of Earth can be calculated using the formula for the volume of a sphere. Since Earth has a radius of 6.38×10^6, its volume can be calculated by evaluating $\frac{4}{3} \times 3.14 \times \left(6.38 \times 10^6\right)^3$ m^3.

   The volume of Earth to 2 decimal places and in scientific notation is closest to:

   A. 1.70×10^{14} m^3
   B. 3.26×10^{21} m^3
   C. 1.09×10^{14} m^3
   D. 1.09×10^{21} m^3
   E. 1.70×10^{21} m^3

5. **MC** State by how many orders of magnitude the distances 3×10^{-4} and 3×10^1 differ.

   A. 2   **B.** −4   **C.** 100   **D.** −2   **E.** 5

6. **MC** The difference in magnitude between two earthquakes measuring 5.9 and 8.2 on the Richter scale is:

   A. 199.5   **B.** 1.1   **C.** 2.3   **D.** 9.2   **E.** 20.0

7. **MC** 786 rounded to the nearest 10 is:

   A. 760   **B.** 770   **C.** 780   **D.** 790   **E.** 800

8. **MC** Estimate the answer to $49 \times 971 + 711$ by rounding each number using the leading digit first.

   A. 50 700   **B.** 50 000   **C.** 45 700   **D.** 40 700   **E.** 35 700

9. **MC** The graph shows the highest year of school completed by Aboriginal and Torres Strait Islander persons aged 20 years and over.

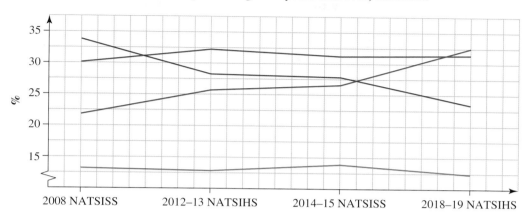

Highest year of school completed by Aboriginal and Torres Strait Islander persons aged 20 years and over, Australia

— Year 12 or equivalent — Year 11 or equivalent — Year 10 or equivalent — Year 9 or below

Using the graph, the best estimate of the percentage of people who complete Year 12 or equivalent in 2022–23 is:

A. 31%   B. 37%   C. 20%   D. 17%   E. 25%

10. **MC** The most appropriate unit to use when measuring the distance from your school to the MCG is:

A. m   B. cm   C. mm   D. km   E. L

Short answer

11. The distance from Earth to Sun is approximately 1×10^{11} m, whereas the distance to the nearest star is 1×10^{16} m.

   a. State by how many orders of magnitude these distances differ.
   b. Convert the units to km.

12. Round the following to the nearest 100.

   a. 5832
   c. 780
   b. 341
   d. 1977

13. Use leading digit approximation to estimate the following.

   a. 49×821
   c. $\dfrac{7563}{676}$
   b. $1396 + 183$
   d. $17 \times 873 + 47$

14. Use you calculator to evaluate the following to 2 decimal places.

   a. $\sqrt{3(5.94 - 1.48)}$
   b. $(5.74)^2 - \dfrac{5}{8}$
   c. $\dfrac{6}{9} + \dfrac{2}{5} \times \dfrac{3}{7}$
   d. $\sqrt{(3.25)^2 + 1.5}$

15. Round the following fractions to 2 decimal places.

   a. $\dfrac{1}{8}$
   b. $\dfrac{37}{50}$
   c. $\dfrac{6}{25}$
   d. $\dfrac{2}{3}$

Extended response

16. Copy and complete the table shown.
    - In the column headed 'Estimate', round each number to the leading digit.
    - In the column headed 'Estimated answer', calculate the answer.
    - In the column headed 'Prediction', guess whether the actual answer will be higher or lower than your estimate.
    - Use a calculator to work out the actual answer and record it in the column headed 'Calculation' to determine whether it is higher or lower than your estimate.

		Estimate	Estimated answer	Prediction	Calculation
Example	$4000 \div 200$	$4000 \div 200$	20	Lower	16.784 553 so lower
a.	$487 + 962$				
b.	$33\,041 + 82\,629$				
c.	$184\,029 + 723\,419$				
d.	$93\,261 - 37\,381$				
e.	$321 - 194$				
f.	$468\,011 - 171\,962$				
g.	36×198				
h.	$623 \times 12\,671$				
i.	$29\,486 \times 39$				
j.	$31\,690 \div 963$				
k.	$63\,003 \div 2590$				
l.	$69\,241 \div 1297$				

17. A student is selling tickets for their school's theatre production of *South Pacific*. So far they have sold 439 tickets for Thursday night's performance, 529 for Friday's and 587 for Saturday's.
The tickets cost $9.80 for adults and $4.90 for students.

    a. Round the figures to the first digit to estimate the number of tickets the student has sold so far.
    b. If approximately half the tickets sold were adult tickets and the other half were student tickets, estimate how much money has been received so far by rounding the cost of the tickets to the first digit.

18. Approximate each of the following:

    a. A stadium estimated to be $\dfrac{3}{4}$ full with a capacity of 50 000 people
    b. A water tank estimated to be 80% full with a capacity of 7500 litres

19. Bob the builder intends to build a new decking out the back of his house so he can entertain friends. He wants to get a rough idea of how much it is going to cost him in materials so he can see if he can afford to go ahead with the job. The hardware store gave him prices on the materials he needs as shown.

Material	Price
Timber	$789
Ready-mix bags	$32.50
Nails	$67.25
Decking stain	$77.95
General equipment	$65.45

a. Estimate the rough price of materials if rounding each price to the leading digit.
b. Estimate the rough price if rounding each price to the nearest dollar.
c. State which of the two rounded prices is greater. Explain why this is so.

20. The surface area of a sphere is calculated by multiplying 12.56 by the radius squared.

a. Calculate the surface area of the following planets with the given radii, correct to 2 decimal places.

Radius of Mercury $= 2.44 \times 10^6$ m
Radius of Mars $= 3.40 \times 10^6$ m
Radius of Earth $= 6.38 \times 10^6$ m
Radius of Jupiter $= 7.15 \times 10^7$ m
Radius of Neptune $= 2.48 \times 10^7$ m

b. State which planet has the greatest surface area.
c. State which planet has the smallest surface area.

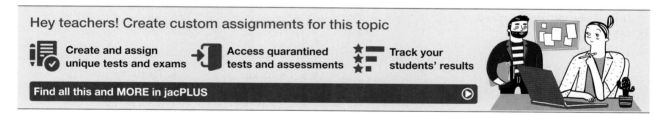

Answers

Topic 5 Estimation and approximation

5.2 Rounding and estimation

5.2 Exercise

1. a. 930   b. 13 000   c. 90 000
2. a. 350   b. 86 600   c. 70 000
3. a. 50   b. 80   c. 130
   d. 160   e. 4840   f. 10
4. a. 0   b. 100   c. 200
   d. 3300   e. 86 600   f. 214 000
5. a. 1000   b. 3000   c. 6000
   d. 13 000   e. 436 000   f. 728 000
6. a. 10   b. 20   c. 200
   d. 1000   e. 20 000   f. 500 000
7. 35 000 000
8. 225
9. a. 1400   b. 900   c. 1100   d. 2000
10. a. 45 000   b. 180 000   c. 70
    d. $\dfrac{40}{9}$
11. Since it is leading digit approximation, you look at the second digit. If it is 5 or greater, then you increase the leading digit by 1 and replace the rest of the digits with zeros. If the second digit is less than 5, then replace it and all of the other digits with zeros.
12. Adding the cost of groceries when shopping. Calculating the cost of petrol needed for a trip. Determining the size of a crowd.
13. $35
14. a. 300 000
    b. Actual crowd = 294 495
       Estimated crowd ≈ 300 000
       Difference = 5505
15. a. It could be correct, but the GPS cannot predict traffic or red lights on the way, so it is only an estimate.
    b. By calculating the distance of the trip and the speed limits on the roads to come up with an estimate for time
16. a. Estimated = $350
       Actual = $311
       Difference = $39
    b. The estimate was more than the actual price since the items purchased were rounded up.

5.3 Reasonableness and calculator skills

5.3 Exercise

1. $290
   The answer is reasonable.
2. $41
   Answer reasonable since roughly chips ≈ $10,

Coke ≈ $20, biscuits ≈ $5 and dips ≈ $5.
This totals $40, which reduces to $38 after the discount.
This value is close to $41, so it is reasonable.

3. a. 461   b. 1206   c. −162.1392
4. a. 445   b. 1748   c. 343.136 227
5. a. 45   b. 95   c. 621
   d. 16 921
6. a. 2   b. 32   c. 241
   d. 438
7. a. 62   b. 49   c. 20.702
   d. 58.95
8. a. 132.7925   b. 29.5800   c. 11.4750
9. a. −0.4097   b. 1.9746   c. 1.4038
10. a. 23.1884   b. 19.2200   c. 76.6600
    d. 12.3429
11. a. 3.7430   b. 4.8485   c. 18.2067
    d. 0.2098
12. a. 0.75   b. 0.32   c. 1.27   d. 19.67
13. a. 1.28×10^5 b. 3.071×10^{16}
14. a. 4.49×10^6 b. 1.512×10^9
15. a. 7.27×10^3 b. 6.419×10^6 c. 2.6319×10^9
    d. 5.06×10^1 e. 2.8168×10^{-3} f. 4.1715×10^8
16. Cathy is 13 years old.
17. 5 cans
18. 7.5×10^7 km
19. 4.396×10^{-10} m
20. 1.53×10^{24} m^3

5.4 Order of magnitude

5.4 Exercise

1. 4
2. 3
3. a. 1×10^0 km   b. 1×10^1 kg
   c. 9×10^4 cm   d. 5.4×10^5 kg
   e. 1.2×10^1 mg   f. 6.3×10^6 km
4. a. 4   b. 14   c. 3
   d. 7   e. 0   f. 8
5. 3
6. 7200 seconds
7. B
8. B
9. 5
10. a. 9   b. 1 000 000 000
11. a. Grass: 6×10^0 cm; Tree: 6×10^2 cm
    b. 600 cm or 6 m
12. An earthquake of magnitude 5 is 100 times stronger than an earthquake of magnitude 3.
13. a. 3   b. 1000
14. C
15. a. 0.5   b. 3.16
16. 4

17. a. 2

  b. Tennis ball: 50 g; Medicine ball: 5000 g

18. a. 5   **b.** 100 000

19. a. $5 \times 10^3 \times 10^{-1}$ **b.** 500 mL

20. a. 2.5×10^{-1} km   **b.** 0.25 km or 250 m

5.5 Approximation

5.5 Exercise

1. 58 450

2. 45 900 L

3. **a.** m   **b.** g   **c.** mL
   d. km   **e.** cm   **f.** kg

4. D

5. C

6. $14 500

7. 14 000 metric tonnes of CO_2-e

8. 15 300

9. Around 17 000

10. Around 11 115

11. Even though the crowd behind Tiger Woods looks large, we don't know the capacity of the golf course. Even if we did, the picture doesn't show us the crowd at the other holes. There might not be many people watching the other groups because they are not as popular as Tiger Woods.

12. A guess doesn't really have any thought or logic behind it, but an estimate does.

13. Estimate the size of each year level and add them up. This could be improved by finding how many home-room or tutor groups there are at each year level, estimating the number of students in each, and adding them all up.

14. Estimate: 28 000 000

15. Estimate: $1.04

16. **a.** Estimate: 750 000
    b. Estimate: 820 000

5.6 Review

5.6 Exercise

Multiple choice

1. E
2. A
3. B
4. D
5. A
6. C
7. D
8. A
9. B
10. D

Short answer

11. **a.** 5
    b. 1×10^{11} m $= 1 \times 10^8$ km
       1×10^{16} m $= 1 \times 10^{13}$ km

12. a. 5800   **b.** 300   **c.** 800   **d.** 2000

13. a. 40 000   **b.** 1200   **c.** $\dfrac{80}{7}$ **d.** 18 050

14. a. 3.66   **b.** 32.32   **c.** 0.84   **d.** 4.75

15. a. 0.13   **b.** 0.74   **c.** 0.24   **d.** 0.67

Extended response

16.

	Estimate	Estimated answer	Actual answer
a.	500 + 1000	1500	1449
b.	33 000 + 80 000	110 000	115 670
c.	200 000 + 700 000	900 000	907 448
d.	90 000 − 40 000	50 000	55 880
e.	300 − 200	100	127
f.	500 000 − 200 000	300 000	296 049 s
g.	40 × 200	8000	7128
h.	600 × 10 000	6 000 000	7 894 033 s
i.	30 000 × 40	1 200 000	1 149 954
j.	30 000 ÷ 1000	30	32.9076
k.	60 000 ÷ 3000	20	24.325 483
l.	70 000 ÷ 1000	70	53.3855

17. a. 1500 tickets   **b.** $11 250

18. a. 37 500 people

  b. Between 5000 and 7000 litres (6000 L)

19. a. Around $1050

  b. $1032

  c. The price in part **a** is higher since four of the five values were rounded up using first digit rounding.

20. a. Mercury: 7.48×10^{13} m^2
   Mars: 1.45×10^{14} m^2
   Earth: 5.11×10^{14} m^2
   Jupiter: 6.42×10^{16} m^2
   Neptune: 7.72×10^{15} m^2

  b. Jupiter has the greatest surface area of 6.42×10^{16} m^2.

  c. Mercury has the smallest surface area of 7.48×10^{13} m^2.

6 Ratios and proportions

LEARNING SEQUENCE

Fully worked solutions for this topic are available online.

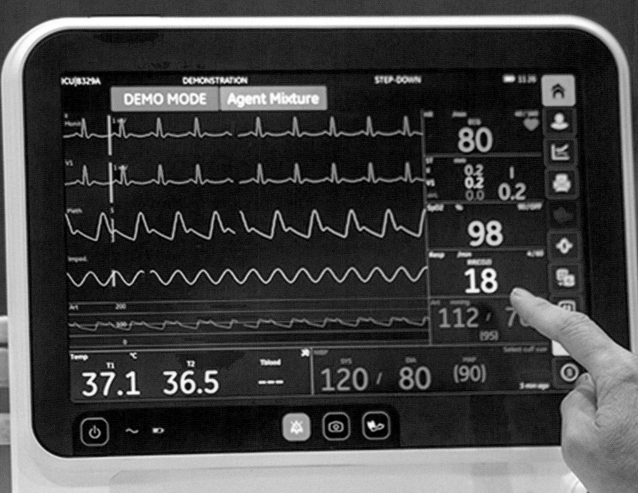

6.1 Overview

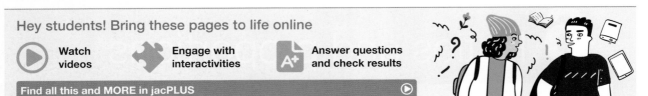
6.1.1 Introduction

Ratios and proportions are often used to compare different values or quantities. They tell us how much there is of one thing compared to another.

Ratios are used in our daily lives, both at home and at work. The use of ratios is required in areas such as building and construction, cooking, engineering, hairdressing, landscaping, carpentry and pharmaceuticals, just to name a few. Understanding ratios is important as they are used in money transactions, perspective drawings, enlarging or reducing measurements in building and construction, and dividing items equally within a group.

The golden ratio, approximately 1.618, is a proportion that is found in geometry, art and architecture and has been made famous in the illustrations of Leonardo da Vinci (1452–1519). In the construction industry, concrete is made in different proportions of gravel, sand, cement and water depending on the particular application. Artists, architects and designers use the concept of proportion in many spheres of their work.

KEY CONCEPTS

This topic covers the following key concepts from the VCE Mathematics Study Design:
• use of ratios, proportions, percentages and rates to solve problems.

Note: Concepts shown in grey are covered in other topics.

Source: VCE Mathematics Study Design (2023–2027) extracts © VCAA; reproduced by permission.

6.2 Introduction to ratios

LEARNING INTENTION

At the end of this subtopic you should be able to:
- compare quantities using ratios
- convert between equivalent ratios
- simplify ratios.

6.2.1 What is a ratio?

Ratios are used to compare two or more quantities of the same kind (e.g. two scoops of chocolate ice cream and one scoop of vanilla ice cream). The numbers are separated by a colon and they are always whole numbers.

The basic notation for ratios is:

$$a : b$$

where a and b are two whole numbers. This is read as 'a to b'.

In the following diagram, the ratio of blue squares to pink squares is $2 : 4$ (the quantities are separated by a colon). This is read as '2 to 4'.

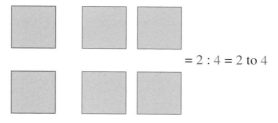

$$= 2 : 4 = 2 \text{ to } 4$$

When writing a ratio, the order in which it is written is very important.

The picture shows two scoops of chocolate ice cream and one scoop of vanilla ice cream. The ratio of chocolate ice cream to vanilla ice cream is 2 scoops to 1 scoop or $2 : 1$ (two to one), while the ratio of vanilla ice cream to chocolate ice cream is $1 : 2$ (one to two).

There are also ratios with more than two values. For example, the ratio of flour to sugar to water in a recipe could be flour : sugar : water $= 3 : 1 : 4$. In other words, for every 3 parts of flour, the recipe requires 1 part of sugar and 4 parts of water.

Determine the ratio of black to coloured layers of licorice in the picture.

THINK	WRITE
1. Determine the order in which to write the ratio.	Black layers : coloured layers
2. Write the ratio in simplest form.	2 : 3

6.2.2 Ratios as fractions

Ratios can be written in fraction form: $a : b = \dfrac{a}{b}$. Fractions should always be written in simplest form.

$$= 2 : 4 \Leftrightarrow \frac{2}{4} = \frac{1}{2}$$

($\Leftrightarrow$ means *is equivalent to*.)

Note: This does not mean that the quantity on the left of the ratio is one half of the total.

Determine the ratio of red-bellied macaw parrots to orange-bellied macaw parrots in the picture. Express the ratio in fraction form.

THINK	WRITE
1. Count the number of red-bellied macaws.	2 red-bellied macaw parrots
2. Count the number of orange-bellied macaws.	5 orange-bellied macaw parrots
3. Express the ratio in simplest form and in order.	Red-bellied macaws : orange-bellied macaws $= 2 : 5$
	$= \dfrac{2}{5}$

6.2.3 Equivalent ratios

Equivalent ratios are equal ratios. To determine equivalent ratios, multiply or divide both sides of the ratio by the same number. (This is a similar process to that of obtaining equivalent fractions).

Consider the diagram shown. A cordial mixture can be made by adding one part of cordial concentrate to four parts of water.

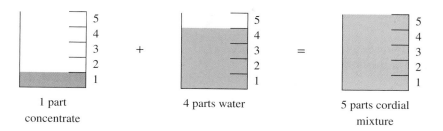

1 part concentrate   +   4 parts water   =   5 parts cordial mixture

The ratio of concentrate to water is 1 to 4 and is written as $1:4$.

By applying the knowledge of 'equivalent ratios', the amount of cordial can be doubled. This can be achieved by keeping the ratio of concentrate to water the same by doubling the amounts of both concentrate and water.

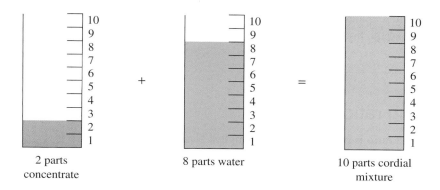

2 parts concentrate   +   8 parts water   =   10 parts cordial mixture

The relationship between the amount of concentrate to water is now $2:8$, as shown in the above diagram. The ratios $2:8$ and $1:4$ are equivalent. In both ratios, there is 1 part of concentrate for every 4 parts of water.

Some examples of equivalent fractions include:

Some equivalent ratios to 2 : 5	Some equivalent ratios to 100 : 40
$2 \times 2 : 5 \times 2 = 4 : 10$	$100 \div 2 : 40 \div 2 = 50 : 20$
$2 \times 3 : 5 \times 3 = 6 : 15$	$100 \div 20 : 40 \div 20 = 5 : 2$
$2 \times 10 : 5 \times 10 = 20 : 50$	$100 \div 10 : 40 \div 10 = 10 : 4$

Equivalent fractions are calculated in the same way as equivalent ratios. Both the numerator and denominator are multiplied or divided by the same number to give an equivalent fraction.

$$\frac{2}{5} \times \frac{3}{3} = \frac{6}{15} \quad \text{and} \quad \frac{100}{40} \div \frac{5}{5} = \frac{20}{8}$$

WORKED EXAMPLE 3 Equivalent ratios

tlvd-3757

Use equivalent ratios to determine the values of the pronumerals in the following.
a. $3 : 10 = 24 : a$
b. $45 : 81 : 27 = b : c : 3$

THINK	WRITE
a. 1. The left-hand sides of both ratios are given. To determine the factor, divide 24 by 3. The result is 8. Hence, to determine a, multiply the right-hand side of the first ratio by 8.	a. $3 : 10 = 24 : a$ $\overset{\times 8}{\overgroup{3 : 10 = 24}} : a$ $\times 8$ $3 \times 8 = 24$ $10 \times 8 = a$
2. Write the answer.	$a = 80$
b. 1. The right-hand sides of both ratios are given. To determine the factor, divide 27 by 3. The result is 9. Hence, to determine b and c, divide the left and middle side of the first ratio by 9.	b. $45 : 81 : 27 = b : c : 3$ $45 \div 9 : 81 \div 9 : 27 \div 9 = 5 : 9 : 3$
2. Write the answer.	$b = 5$ and $c = 9$

6.2.4 Simplifying ratios

There are times when the ratio is not written in its simplest form. A ratio is simplified by dividing all numbers in the ratio by their highest common factor (HCF). In the picture shown, there are 16 chocolates: 4 white chocolates and 12 milk chocolates. The ratio between the white and the milk chocolates is 4 to 12.

$$\text{white chocolates} : \text{milk chocolates} = 4 : 12$$

Notice that this ratio can be simplified by dividing both values by 4, because 4 is the highest common factor of both 4 and 12. This ratio then becomes:

$$\text{white chocolates} : \text{milk chocolates} = 1 : 3$$

In other words, this means that for every white chocolate, there are 3 milk chocolates.

We can say that $4 : 12 = 1 : 3$.

Expressed in fraction form, the ratio becomes:

$$\frac{4}{12} = \frac{1}{3}$$

Unless otherwise stated, ratios should always be written in their simplest form.

tlvd-3758

WORKED EXAMPLE 4 Expressing ratios in their simplest form

a. **Examine the image and express the ratio of antique chairs to tables in simplest form.**

b. **Express the ratio in fraction form.**

THINK	WRITE

a. 1. Count the number of the first part of the ratio (chairs).

a. 10 chairs

2. Count the number of the second part of the ratio (tables).

4 tables

3. Write the ratio.

Chairs : tables $= 10 : 4$

4. Both 10 and 4 can be divided exactly by 2. So, the highest common factor of 10 and 4 is 2.
Write the ratio in its simplest form by dividing both sides by their highest common factor.

$= \dfrac{10}{2} : \dfrac{4}{2}$

$= 5 : 2$

b. Write the ratio in fraction form.

b. $5 : 2 = \dfrac{5}{2}$

6.2.5 Ratios and decimal numbers

What do we do with ratios written in the form 2 : 3.5? We said before that ratios must be expressed as whole numbers.

Step 1: We can convert 3.5 into a whole number by multiplying it by 10.

$$3.5 \times 10 = 35$$

Step 2: If we multiply the right-hand side of the ratio by 10, we have to multiply the left-hand side of the ratio by 10 also, in order to keep the ratio unchanged.

$$2 \times 10 : 3.5 \times 10 = 20 : 35$$

Step 3: Ratios should be expressed in simplest form, so divide both sides of the ratio by the highest common factor of 20 and 35, which is 5.

$$20 \div 5 : 35 \div 5 = 4 : 7$$

Decimal numbers with 2 decimal places will need to be multiplied by 10^2 or 100.

Decimal numbers with 3 decimal places will need to be multiplied by 10^3 or 1000.

The number of decimal places will give the power of 10 required.

WORKED EXAMPLE 5 Simplifying ratios involving decimals

tlvd-3759

Express the ratio 4.9 : 0.21 in its simplest form using whole numbers.

THINK	WRITE
1. Multiply both sides of the ratio by 100, because the largest number of decimal places is 2 on the right-hand side of the ratio.	$4.9 \times 100 : 0.21 \times 100 = 490 : 21$
2. Divide both sides of the ratio by the highest common factor of the two numbers (HCF = 7). The highest common factor of both 490 and 21 is 7.	$490 \div 7 : 21 \div 7 = 70 : 3$
3. Write the ratio in its simplest form.	$4.9 : 0.21 = 70 : 3$

6.2 Exercise

1. **WE1** Determine the ratio of white flowers (chamomile) to purple flowers (bluebell) in the picture shown.

2. The abacus shown has white, green, yellow, red and blue beads. State the following as ratios:

   a. Blue beads to yellow beads
   b. Red beads to green beads
   c. White beads to red beads

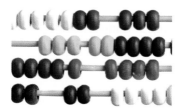

3. **WE2** Convert the following ratios to fractions.

   a. 3 : 5   b. 4 : 28   c. 6 : 11   d. 5 : 2

4. Using the letters in the word MATHEMATICS, express the following as ratios.

   a. Vowels to consonants   b. Consonants to vowels
   c. Letters E to letters A   d. Letters M to letters T

5. There are 12 shapes in this image.

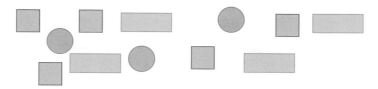

Express the following as ratios:
   a. Circles to quadrilaterals
   b. Squares to rectangles
   c. Rectangles to circles
   d. Squares to circles

6. Express the following ratios in fraction form.
   a. $2:3$
   b. $9:8$
   c. $15:49$
   d. $20:17$

7. **WE3** Determine the value of the pronumerals in the following equivalent ratios.
   a. $121:66=a:6$
   b. $3:2=b:14$

8. Determine the value of the pronumerals in the following equivalent ratios.
   a. $2:7:8=10:m:n$
   b. $81:a:54=9:5:b$

9. Determine the value of the pronumerals in the following equivalent ratios.
   a. $7:4=x:16$
   b. $a:9:b=1:3:7$
   c. $5:b=15:30$
   d. $18:12=3:m$

10. **MC** Determine which of the following ratios is not equivalent to $12:8$.
    A. $6:4$
    B. $24:16$
    C. $60:40$
    D. $11:7$
    E. $9:6$

11. **MC** Determine which of the following fractions is equivalent to $\dfrac{39}{65}$.

    A. $\dfrac{78}{120}$
    B. $\dfrac{13}{21}$
    C. $\dfrac{6}{10}$
    D. $\dfrac{13}{195}$
    E. $\dfrac{6}{11}$

12. **WE4** Determine the ratio of orange balls to black and white balls in the picture shown. Express your answer in its simplest fraction form.

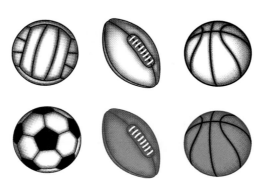

13. Convert the ratio $56:42$ into a fraction in its simplest form.

14. Simplify the following ratios.
    a. $\dfrac{120}{36}$
    b. $1000:200$
    c. $\dfrac{33}{2200}$
    d. $58:116$
    e. $36:24:60$
    f. $315:180:360$

15. **WE5** Express the given ratios in simplest form using whole numbers.

   a. 2.5 : 1.5
   b. 3.6 : 4.2
   c. 0.11 : 1.1
   d. 1.8 : 0.32
   e. 1.6 : 0.24
   f. 1.15 : 13.8

16. Express the following decimal ratios in simplest form using whole numbers.

   a. 0.3 : 1.2
   b. 7.5 : 0.25
   c. 0.64 : 0.256
   d. 1.2 : 3 : 0.42

17. In a class there are 28 students. If the number of girls in this class is 16, determine:

   a. the number of boys in the class
   b. the ratio of boys to girls in simplest form.

18. a. State the ratio of length to width for the rectangle shown.

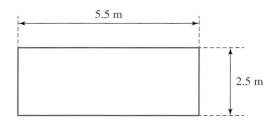

5.5 m

2.5 m

   b. State the ratio of length to width to height for the rectangular prism shown.

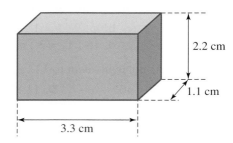

2.2 cm

1.1 cm

3.3 cm

19. Using a calculator, a spreadsheet or otherwise, express the following ratios as fractions in simplest form.

   a. 18 : 66
   b. 20 : 25
   c. 30 : 21
   d. 56 : 63
   e. 10 : 100
   f. 728 : 176

20. Using a calculator, a spreadsheet or otherwise, determine ten equivalent fractions to the following fractions.

   a. $\dfrac{3}{5}$
   b. $\dfrac{1080}{840}$

6.3 Ratios of two quantities

LEARNING INTENTION

At the end of this subtopic you should be able to:
   • determine ratios of ingredients in recipes and other mixtures
   • determine ratios in art, building, landscaping and other common applications.

6.3.1 Weight ratios

Ratios are used to compare the quantities of ingredients when making a mixture. For example, the mortar necessary for building a brick wall is usually prepared using the cement-to-sand ratio of 1 : 3. Pharmacists are required to use ratios in the preparation of medicines. Chefs use ratios in every meal they cook.

Food is a very important part of everyone's life. A recipe involves ratios. If a recipe says 3 cups of flour and 1 cup of sugar, it means that the recipe requires 3 cups of flour for every 1 cup of sugar, so the ratio flour : sugar = 3 : 1.

What if the recipe requires 300 grams of flour and 100 grams of sugar? We can calculate the ratio of flour to sugar by simplifying the quantities given.

$$\text{flour : sugar} = 300\,\text{g} : 100\,\text{g}$$

The highest common factor of 300 and 100 is 100. To simplify the ratio, divide both sides of the ratio by 100.

$$\text{flour : sugar} = 3 : 1$$

Note:
1. Usually, units are not written in ratios.
2. To calculate ratios of quantities, always write all quantities in the same unit.

Metric conversions of weight

Metric units of weight are converted in the following way:

$$1\,\text{kg} = 1000\,\text{g}$$
$$1\,\text{tonne} = 1000\,\text{kg}$$

WORKED EXAMPLE 6 Determining weight ratios

Peas and corn is a common combination of frozen vegetables. Determine the ratio of 500 g peas to 400 g corn.

THINK	WRITE
1. Ensure that the two quantities are written in the same unit (both are in grams).	500 g peas 400 g corn
2. Write the quantities as a ratio without units.	Peas : corn = 500 : 400
3. Simplify the ratio by dividing both sides by their highest common factor. As 100 is the highest common factor of 500 and 400, divide both sides of the ratio by 100.	= 500 ÷ 100 : 400 ÷ 100 = 5 : 4
4. Write the ratio in its simplest form.	Peas : corn = 5 : 4

6.3.2 Length ratios

Leonardo da Vinci was a famous painter, architect, sculptor, mathematician, musician, engineer, inventor, anatomist and writer. He painted the famous Mona Lisa, whose face has the golden ratio. The golden ratio distributes height and width according to the ratio 34 : 21. For Mona Lisa, this is:

$$\text{height of face} : \text{width of face} = 34 : 21$$

Ratios of lengths are calculated in the same way as the ratios for weights.

Metric conversions of length

Recall the relationship between the metric units of length. The following diagram can be used for quick reference.

$$\times 1000 \quad \times 100 \quad \times 10$$

km   m   cm   mm

$$\div 1000 \quad \div 100 \quad \div 10$$

WORKED EXAMPLE 7 Calculating length ratios

The window in the picture has a width of 1000 mm and a height of 1.75 m. Calculate the ratio of width to height.

THINK	WRITE
1. Make sure to convert both lengths into cm. The most appropriate unit is cm because converting 1.75 m into cm will change the decimal number into a whole number.	1000 mm = 100 cm 1.75 m = 175 cm
2. Write the lengths as a ratio without units.	Width : height = 100 : 175
3. Simplify the ratio by dividing both sides by their highest common factor (HCF = 25).	$= \dfrac{100}{25} : \dfrac{175}{25}$
As 25 is the highest common factor of 100 and 175, divide both sides of the ratio by 25.	$= 4 : 7$
4. Write the ratio in its simplest form.	Width : height = 4 : 7

6.3.3 Area ratios

Areas are enclosed surfaces. Many municipalities have strict building and landscaping guidelines. Landscape architects use ratios to calculate area of land to area of building and area of pavement to area of grass.

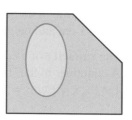

The diagram shown is a sketch of a paved backyard (blue) with an area of 50 m² and a garden (green) with an area of 16 m².

The ratio of the paved area to the garden area is:

$$\text{paved area : garden area} = 50 \text{ m}^2 : 16 \text{ m}^2$$
$$= 50 : 16$$
$$= 25 : 8$$

Metric conversions of area

Recall the relationship between the metric units of area. The diagram shown can be used for quick reference.

$$\times 1000^2 \quad \times 100^2 \quad \times 10^2$$

$$km^2 \quad m^2 \quad cm^2 \quad mm^2$$

$$\div 1000^2 \quad \div 100^2 \quad \div 10^2$$

---

tlvd-3761

WORKED EXAMPLE 8 Determining area ratios

Kazem has a bedroom with an area of 200 000 cm². The whole house has an area of 160 m².

Determine the ratio of the area of the bedroom to the area of the whole house.

THINK	WRITE
1. Make sure to convert both areas into m², because converting 200 000 cm² into m² will simplify this value.	Area of bedroom = 200 000 cm² $= 20 \text{ m}^2$ Area of house = 160 m²
2. Write the areas as a ratio without units.	Area of bedroom : area of house = 20 : 160
3. Simplify the ratio by dividing both sides by their highest common factor of 20 (HCF = 20).	$= \dfrac{20}{20} : \dfrac{160}{20}$
4. Write the ratio in its simplest form.	Area of bedroom : area of house = 1 : 8

6.3.4 Volume and capacity ratios

The **volume** of an object is the space taken up by the object. A cube of length 1 cm, width 1 cm and height 1 cm takes up a space of $1 \times 1 \times 1 = 1 \text{ cm}^3$.

The **capacity** of a container is the maximum volume it can hold. A container in the shape of a cube of length 1 cm, width 1 cm and height 1 cm can hold $1 \times 1 \times 1 = 1 \text{ cm}^3 = 1 \text{ mL}$ of liquid.

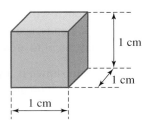

Metric conversions of volume and capacity

Recall the relationship between the metric units of volume.

The diagram shown can be used for quick reference.

$$\times 1000^3 \quad \times 100^3 \quad \times 10^3$$

$$\text{km}^3 \qquad \text{m}^3 \qquad \text{cm}^3 \qquad \text{mm}^3$$

$$\div 1000^3 \quad \div 100^3 \quad \div 10^3$$

Volume units are converted to capacity units using the following conversions.

$$1 \text{ cm}^3 = 1 \text{ mL}$$

$$1000 \text{ cm}^3 = \frac{1}{1000} \text{ m}^3 = 1 \text{ L}$$

$$1 \text{ L} = 1000 \text{ mL}$$

---

tlvd-3762

WORKED EXAMPLE 9 Determining volume ratios

Salila and Devi have just installed two rainwater tanks: one for household use and one for watering the garden.
The household tank has a volume of $10\,000\,000 \text{ cm}^3$, while the tank for watering the garden has a volume of 45 m^3.
a. Determine the ratio of the volume of the household water tank to the volume of the tank for watering the garden.
b. Determine the ratio of the capacities of the two water tanks.

THINK	WRITE
a. 1. Convert both quantities to the same unit.	a. Volume of household tank $= 10\,000\,000 \text{ cm}^3$
	$= 10 \text{ m}^3$
The most appropriate unit is m^3 because converting $10\,000\,000 \text{ cm}^3$ into m^3 will simplify this value.	Volume of garden tank $= 45 \text{ m}^3$
2. Write the volumes as a ratio without units.	Household tank : garden tank $= 10 : 45$

3. Simplify the ratio by dividing both sides by their highest common factor.

As 5 is the highest common factor of 10 and 45, divide both sides of the ratio by 5 (HCF = 5).

$$= \frac{10}{5} : \frac{45}{5}$$
$$= 2 : 9$$

4. Write the ratio in its simplest form.

Household tank : garden tank = 2 : 9

b. 1. Convert both quantities to the same unit. The most appropriate unit is L because converting $10\,000\,000\ cm^3$ into L will simplify this value.

b. Capacity of household tank $= 10\,000\,000\ cm^3$
$$= 10\,000\ L$$
Capacity of garden tank $= 45\ m^3$
$$= 45\,000\ L$$

2. Write the volumes as a ratio without units.

Capacity of household tank : capacity of garden tank
$$= 10\,000 : 45\,000$$

3. Simplify the ratio by dividing both sides by their highest common factor.

$$= \frac{10\,000}{5000} : \frac{45\,000}{5000}$$
$$= 2 : 9$$

As 5000 is the highest common factor of 10 000 and 45 000, divide both sides of the ratio by 5000 (HCF = 5000).

4. Write the ratio in its simplest form.

Capacity of household tank : capacity of garden tank
$$= 2 : 9$$

6.3.5 Time ratios

Time is a physical quantity we use for many of our daily activities. Have you ever considered the ratio of the time you spend on homework compared to the time you spend on social media?

If you spend 30 minutes completing homework and 10 minutes on social media, the ratio is:

$$\text{time for homework} : \text{time on social media} = 30 : 10$$
$$= 3 : 1$$

This means that for every three minutes you spend completing homework, you spend one minute using social media.

Conversions of time

Recall the relationship between the metric units of time. The diagram shown can be used for quick reference.

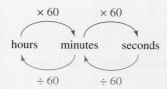

1 hour = 60 minutes
1 minute = 60 seconds
1 hour = 60 minutes (3600 seconds)

Emori is baking lemon cupcakes and chocolate cupcakes. It takes him $\frac{3}{4}$ of an hour to make the lemon cupcakes and 36 minutes to make the chocolate cupcakes.

Determine the ratio of the time taken to make the chocolate cupcakes to the time taken to make the lemon cupcakes.

THINK	WRITE
1. Convert both quantities to the same unit. The most appropriate unit is minutes because converting $\frac{3}{4}$ of an hour into minutes will change the fraction into a whole number.	$\frac{3}{4}$ of an hour $= \frac{3}{4} \times 60$ minutes $= 45$ minutes
2. Write the times as a ratio without units.	Chocolate cupcake time : lemon cupcake time $= 36 : 45$
3. Simplify the ratio by dividing both sides by their highest common factor. As 9 is the highest common factor of 36 and 45, divide both sides of the ratio by 9 (HCF $= 9$).	$= \frac{36}{9} : \frac{45}{9}$ $= 4 : 5$
4. Write the answer.	Chocolate cupcake time : lemon cupcake time $= 4 : 5$

6.3.6 Money ratios

A budget planner is a way of planning where money earned is going to be spent. Budget planners are used by people, businesses and governments.

A family budgets $50 per week for electricity. If the family's total weekly wage is $2100, what is the ratio of money spent on electricity per week to the total weekly wage?

$$\text{Money on electricity} \; : \; \text{wage} = 50 : 2100 \; (\text{HCF} = 50)$$
$$= 1 : 42$$
$$= \frac{1}{42}$$

Conversions of money

Recall the relationship between dollars ($) and cents.

$$\$1 = 100 \text{ cents}$$
$$1 \text{ cent} = \$0.01$$
$$= \$\frac{1}{100}$$

WORKED EXAMPLE 11 Calculating money ratios

Jackie earns $240 per month from her part-time job. She decides to spend $40 per month to buy her lunch from the school canteen. If her monthly phone bill is $20, calculate the ratio of the lunch money to her phone bill and to her income per month.

THINK

1. Ensure that all the amounts are written in the same unit.

2. Write the amounts as a ratio without units.

3. Simplify the ratio by dividing all 3 parts by their highest common factor.
   As 20 is the highest common factor of 20, 40 and 240, divide all 3 parts of the ratio by 20 (HCF = 20).

4. Write the answer.

WRITE

Monthly wage = $240
Monthly lunch money = $40
Monthly phone bill = $20

Lunch money : phone bill : income = 40 : 20 : 240

$$= \frac{40}{20} : \frac{20}{20} : \frac{240}{20}$$

$$= 2 : 1 : 12$$

Lunch money : phone bill : income = 2 : 1 : 12

6.3 Exercise

Students, these questions are even better in jacPLUS

 Receive immediate feedback and access sample responses

 Access additional questions

 Track your results and progress

Find all this and MORE in jacPLUS

1. **WE6** Determine the ratio of 240 000 g sand to 80 kg cement.

2. A recipe for a carrot cake has the ratio 0.5 kg flour, 100 g sugar and 250 g grated carrots. Determine the ratio of sugar : flour : grated carrots in its simplest form.

3. A puff pastry recipe requires 225 g flour, 30 g lard and 0.210 kg butter. Determine the ratio of:
   a. flour to lard
   b. butter to lard
   c. flour to butter
   d. flour to lard to butter.

4. Shane makes dry ready mix concrete using 2.750 kg of stone and gravel, 2500 g of sand, and 1 kg of cement and water.
   Calculate the ratios of:

   a. cement and water to stone and gravel
   b. sand to cement and water
   c. cement and water to sand to stone and gravel
   d. cement and water to the total quantity of materials.

5. **WE7** The wheelchair ramp shown in the diagram has a length of 21 m and a height of 150 cm. Determine the ratio of height to length.

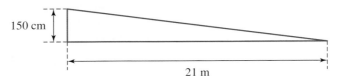

6. A rectangular swimming pool has length 50 m, width 2500 cm and depth 1000 mm. Calculate the ratio length : width : depth in its simplest form.

7. Determine the ratio of length to width of a football field 162 m long and 144 m wide.

8. **WE8** The area of the small triangle, ABC, is 80 000 cm² and the area of the large triangle, DEF, is 40 m². Write the ratio of the area of the small triangle to the area of the large triangle in its simplest form.

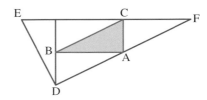

9. Earth's surface area is approximately 510 100 000 km². Water covers 361 100 000 km². Calculate the ratio of the surface area covered with water to the total surface area of Earth.

10. The landscape in the diagram represents the design of a garden with two ponds of areas 48 m² and 640 000 cm² respectively. The sand area is 320 000 cm² and the area of the table is 4 m².
For the areas given, determine the ratio of:

a. the area of the large pond to the area of the small pond
b. the area of the table to the area of the sand
c. the area of the sand to the area of the large pond
d. the area of the small pond to the area of the table.

11. **MC** Determine which of the following conversions is incorrect.

A. $3 \text{ m} = 0.003 \text{ km}$
B. $8.7 \text{ m}^2 = 870\,000 \text{ cm}^2$
C. $1.5 \text{ km} = 1\,500\,000 \text{ mm}$
D. $19.6 \text{ L} = 19\,600 \text{ cm}^3$
E. $3.25 \text{ hours} = 195 \text{ minutes}$

12. **WE9** The capacity of the glass in the picture is 250 mL and the capacity of the carafe is 1.5 L. Determine the ratio of the capacity of the carafe to the capacity of the glass.

13. The two jugs shown hold 0.300 L of oil and 375 mL of water. Determine the ratio volume of water to volume of oil required for this recipe.

14. The volumes of the orange juice boxes in the picture shown are 1.250 L and 750 mL respectively. Write the ratio of the large volume to the smaller volume in simplest fraction form.

15. Write the following ratios in simplest fraction form.
    a. $70 \text{ cm}^3 : 20 \text{ cm}^3$
    b. $560\,000 \text{ cm}^3 : 0.6 \text{ m}^3$
    c. $15 \text{ cm}^3 : 45\,000 \text{ mm}^3$
    d. $308\,000\,000 \text{ cm}^3 : 385 \text{ m}^3$

16. **WE10** The time it takes Izzy to get to school by bus is 22 minutes while the time it takes her by train is 990 seconds. Determine the ratio of the time taken by bus to the time taken by train.

17. If a person spends 8.5 hours sleeping and 15.5 hours being awake, calculate the ratio of sleeping time to time awake.

18. **MC** The ratio of 1 h 25 min to 55 min in its simplest form is:
    A. $\dfrac{11}{17}$
    B. $\dfrac{85}{55}$
    C. $\dfrac{5}{11}$
    D. $\dfrac{17}{11}$
    E. $\dfrac{170}{110}$

19. The capacities of the three bottles shown are 150 mL, 200 mL and 0.175 L respectively. Determine the following ratios in simplest form:
    a. The capacity of the smallest bottle to the capacity of the largest bottle
    b. The capacity of the largest bottle to the capacity of the middle bottle
    c. The capacity of the middle bottle to the capacity of the smallest bottle
    d. The capacity of the smallest bottle to the capacity of the middle bottle to the capacity of the largest bottle

20. **WE11** Su Yi is a sales consultant and has a budget of $154 for a mobile phone, $198 for a computer and $66 for lunch. Determine the ratio between the budgets for:
    a. mobile phone to computer to lunch
    b. mobile phone to computer
    c. computer to lunch
    d. mobile phone to lunch.

21. Every week Sam spends $27.00 for petrol, $5.40 for his favourite magazine and $3.60 for his prepaid calls. Calculate the ratio:
    a. petrol money to the price of the magazine
    b. the price of the magazine to the money spent on the prepaid calls
    c. the money spent on the prepaid calls to petrol money
    d. petrol money to the price of the magazine and to the money spent on the prepaid calls.

22. a. State the ratio volume of tennis ball to volume of basketball.

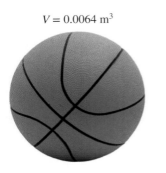

$V = 0.0064 \text{ m}^3$

$V = 160 \text{ cm}^3$

b. State the ratio of the capacity of the lunch box to the capacity of the drink bottle.

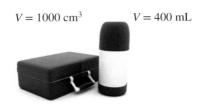

$V = 1000 \text{ cm}^3$ $V = 400 \text{ mL}$

23. Using a calculator, a spreadsheet or otherwise, write the following ratios as fractions in simplest form.

    a. 360 : 450   **b.** 455 : 195   **c.** 2412 : 4288

    d. 5858 : 32 219   **e.** 53 856 : 20 196

24. Using a calculator, a spreadsheet or otherwise, determine 10 equivalent ratios to each of the following ratios.

    a. 19 : 7   **b.** 720 : 480

6.4 Divide into ratios

LEARNING INTENTION

At the end of this subtopic you should be able to:
- divide a number of items into a given ratio
- divide quantities such as weight, length, volume and time into given ratios.

6.4.1 Divide a number of items into a ratio

There are situations when a quantity needs to be divided into a given ratio.

The total of 12 pencils in the figure can be divided into the ratio 2 : 3 : 1. This means dividing the 12 pencils into 3 groups where the first group has twice as many pencils as the third group, and the second group has 3 times as many pencils as the third group.

To determine the number of pencils in each group, we need to calculate the total number of parts of the given ratio.

Add up all parts to determine the total number of parts:

$$\text{Total pencils} = 2 + 3 + 1 = 6 \text{ parts}$$

To calculate the number of pencils per part, divide the total number of pencils by the total number of parts required:

$$\text{Total pencils} = \frac{12}{6} = 2 \text{ pencils per part.}$$

The ratio is 2 : 3 : 1. To determine the number of pencils in each group, multiply the parts by the number of pencils per part (in this case by 2).

$$2 \times 2 : 3 \times 2 : 1 \times 2 = 4 : 6 : 2$$

Check your answer by adding up all the quantities in the last answer:

$$4 + 6 + 2 = 12$$

WORKED EXAMPLE 12 Dividing into a given ratio

tlvd-3764

Divide the jelly beans in the figure shown into a ratio of 3 : 4 : 1.

THINK

1. Calculate the total number of parts of the ratio.

2. Determine the number of jelly beans per part by dividing the total number of jelly beans by the total number of parts.

3. Multiply each number in the ratio by the number of jelly beans per part.

4. Check the answer by adding up all quantities in the ratio. This number should be equal to the total number of jelly beans.

5. Write the answer.

WRITE

$3 + 4 + 1 = 8$

The total number of jelly beans is 16.

$$\frac{\text{total number of jelly beans}}{\text{total number of parts}} = \frac{16}{8}$$
$$= 2 \text{ jelly beans per part}$$

$3 : 4 : 1 = 3 \times 2 : 4 \times 2 : 1 \times 2$
$ = 6 : 8 : 2$

$6 + 8 + 2 = 16$

16 jelly beans divided into the ratio 3 : 4 : 1 is 6, 8 and 2.

6.4.2 Dividing weights into ratios

Chemists work with mixtures every day. Sometimes they need to divide a mixture into given ratios.

WORKED EXAMPLE 13 Dividing a total weight into a given ratio

A mixture of three chemicals has a weight of 150 g.
The three chemicals must be combined in the ratio
orange : red : brown = 3 : 2 : 1.
Determine how many grams of each chemical are required.

THINK	WRITE
1. Calculate the total number of parts of the ratio.	$3 + 2 + 1 = 6$ parts
2. Determine the weight per part by dividing the total weight by the total number of parts.	The total weight of the mixture is 150 g. $\dfrac{\text{weight of the mixture}}{\text{total number of parts}} = \dfrac{150}{6}$ $= 25$ g per part
3. Multiply each number in the ratio by the weight per part.	$3 : 2 : 1 = 3 \times 25 : 2 \times 25 : 1 \times 25$ $= 75 : 50 : 25$
4. Check the answer by adding up all quantities in the ratio. This number should be equal to the total weight of the mixture.	$75 + 50 + 25 = 150$
5. Write the answer.	The 150 g of chemicals should be divided into the following amounts: Orange 75 g Red 50 g Brown 25 g

---

6.4.3 Dividing lengths into ratios

Carpenters use ratios to cut planks of timber, plastic pipes or metal rods to specified lengths. A 2-m-long piece of timber is being cut in a ratio 1 : 3. There are a total of four parts in this ratio, with each part being 0.5 m. This makes the two pieces of timber 0.5 m and 1.5 m respectively.

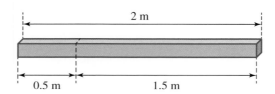

tlvd-3765

WORKED EXAMPLE 14 Dividing lengths into given ratios

The lamp pole in the diagram is 3.2 m tall. The lamp is placed at a distance 1 : 4 from the top of the pole. Determine how high up the pole the lamp post is placed.

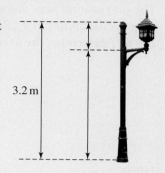

3.2 m

THINK

1. Calculate the total number of parts of the ratio.

2. Determine the length per part by dividing the total height by the total number of parts.

3. Multiply each number in the ratio by the length per part.

4. Check the answer by adding up all lengths in the ratio. This number should be equal to the total length.

5. Write the answer.

WRITE

$1 + 4 = 5$

The total height is 3.2 m.

$$\frac{\text{height of the pole}}{\text{total number of parts}} = \frac{3.2}{5}$$

$$= 0.64 \text{ m per part}$$

$1 : 4 = 1 \times 0.64 : 4 \times 0.64$
$\quad\quad = 0.64 : 2.56$

$0.64 + 2.56 = 3.2$ m

The lamp post must be placed 2.56 m from the ground.

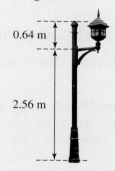

0.64 m

2.56 m

6.4.4 Dividing areas into ratios

Surveyors divide areas into ratios. Architects use these ratios to design accurate plans.

If the area of a property is 800 m^2 and must be divided into the ratio house : backyard $= 5 : 3$, then we can calculate the area for the house and the area for the backyard.

The total number of parts is $5 + 3 = 8$.

The number of m^2 per part is $\dfrac{\text{total area}}{\text{total number of parts}} = \dfrac{800}{8} = 100$ m^2 per part

The corresponding areas are $5 \times 100 = 500$ m^2 for the house, and $3 \times 100 = 300$ m^2 for the backyard.

WORKED EXAMPLE 15 Dividing areas into given ratios

The playground area shown must be divided into a ratio of grass to sandpit of 7 : 2. If the area of the playground is 270 m², determine the area required for the sandpit.

THINK	WRITE
1. Calculate the total number of parts of the ratio.	$7 + 2 = 9$
2. Determine the area per part by dividing the total area by the total number of parts.	The total area is 270 m². $\dfrac{\text{area of playground}}{\text{total number of parts}} = \dfrac{270}{9}$ $= 30$ m² per part
3. Multiply each number in the ratio by the area per part.	Grass : sandpit $= 7 : 2$ $= 7 \times 30 : 2 \times 30$ $= 210 : 60$
4. Check the answer by adding up all areas in the ratio. This sum should be equal to the total area.	$210 + 60 = 270$ m²
5. Write the answer.	The area of the sandpit is 60 m².

6.4.5 Dividing volumes and capacities into ratios

Recipes use mass for solid ingredients and volume or capacity for liquid ingredients. Volume and capacity are measured in different units.

Consider a 2-L cordial drink containing 500 mL cordial and 1.5 L water in the ratio cordial : water = 1 : 3. The cordial drink can be made in a 2000 cm³ jug where the volume of water fills 1500 cm³ of the jug and the cordial fills 500 cm³ of the jug. The ratio for the volumes filled by cordial and water in the jug is also 1 : 3.

WORKED EXAMPLE 16 Dividing volumes into given ratios

The two containers shown have a total volume of 343 750 cm³.
The volumes of the two containers are in a ratio of 4 : 7.
Calculate the volumes of the two containers.

THINK	WRITE
1. Calculate the total number of parts of the ratio.	$4 + 7 = 11$
2. Determine the volume per part by dividing the total volume by the total number of parts.	The total area is $343\,750$ cm^3. $$\frac{\text{total volume}}{\text{total number of parts}} = \frac{343\,750}{11}$$ $$= 31\,250 \text{ cm}^3 \text{ per part}$$
3. Multiply each number in the ratio by the volume per part.	Small container : large container $= 4 : 7$ $= 4 \times 31\,250 : 7 \times 31\,250$ $= 125\,000 : 218\,750$
4. Check the answer by adding up the volumes in the ratio. This sum should be equal to the total volume.	$125\,000 + 218\,750 = 343\,750$ cm^3
5. Write the answer.	The volume of the small box is $125\,000$ cm^3. The volume of the large box is $218\,750$ cm^3.

6.4.6 Dividing time into ratios

There are times when we travel by various modes of transport. Suppose that you catch a bus and then a train to visit a friend. The whole trip takes 40 minutes. If the ratio of the time travelling by bus to the time travelling by train is 3 : 1, what will be the travel time on the bus?

The total number of parts of the ratio is $3 + 1 = 4$.

The number of minutes per part is $\dfrac{40}{4} = 10$ minutes per part.

Bus time : train time $= 3 \times 10 : 1 \times 10$
$= 30 : 10$

This means that travel time on the bus is 30 minutes and the travel time on the train is 10 minutes.

WORKED EXAMPLE 17 Dividing time into given ratios

There are three stages involved in the manufacturing of shoes:
Stage 1: cutting out all of the parts
Stage 2: stitching them together
Stage 3: attaching the sole
The average time taken to produce a basic pair of shoes is 1 hour.
The times for the three stages of cutting, stitching and attaching are in the ratio 4 : 5 : 1.
Determine how long it would take to cut out all the pieces for one pair of shoes.

THINK	WRITE
1. Calculate the total number of parts of the ratio.	$4 + 5 + 1 = 10$
2. Determine the time per part by dividing the total time by the total number of parts. As we are going to divide 1 hour by 10 parts, it is easier to work with minutes than hours.	The total time is 1 hour = 60 minutes $$\frac{\text{total time}}{\text{total number of parts}} = \frac{60}{10}$$ $$= 6 \text{ minutes per part}$$
3. Multiply each number in the ratio by the number of minutes per part.	$4 : 5 : 1 = 4 \times 6 : 5 \times 6 : 1 \times 6$ $= 24 : 30 : 6$
4. Check the answer by adding up the times in the ratio. This sum should be equal to the total time.	$24 + 30 + 6 = 60 \text{ minutes}$ $= 1 \text{ hour}$
5. Write the answer.	The time taken to cut out all the pieces for a pair of shoes is 24 minutes.

6.4 Exercise

1. **WE12** The bouquet in the picture is made of 30 tulips. Flora wants to make two smaller bouquets in the ratio 2 : 3.
   Determine how many tulips there are in each smaller bouquet.

2. The school canteen has 250 mandarins. The mandarins must be divided in the ratio 5 : 3 : 2. Determine how many mandarins there are in each group.

3. Determine the number of children in a group of 1320 people if the ratio of children to adults is:
   a. 3 : 1   b. 5 : 1   c. 3 : 2   d. 7 : 4.

4. Sophia bought 5.2 kg of fruits and vegetables in the fruits to vegetables ratio of 3 : 2.
   a. Determine how many kilograms of vegetables Sophia bought.
   b. Determine how many kilograms of fruit she bought.

5. Carbon steel contains amounts of manganese, silicon and copper in the ratio 8 : 3 : 2.
   a. Calculate the quantity of manganese required to produce carbon steel if the total amount of manganese, silicon and copper is 936 kg.
   b. Calculate the number of kilograms of silicon in this quantity.

6. **WE13** A chef has a recipe for a homemade soup that requires 500 g of tomatoes and capsicum in the ratio of 3 : 2. Calculate the quantity of tomatoes and capsicum that the chef should buy to make this soup.

7. The ratio of sand to cement to make mortar is 3 : 1. A bricklayer wants to make 17 kg of mortar. Calculate the quantities of sand and cement required.

8. **WE14** Monika has 4.5 m of fabric to make a top and a skirt. She wants to divide the fabric in the ratio 2 : 3. Determine the lengths of the two pieces of fabric.

9. A 1.4-m steel beam is cut into three lengths in the ratio 1 : 2 : 4. Determine the length of each of the cut beams. Give the answers in mm.

10. A picture frame has a perimeter of 123 cm. Calculate the size of the width of the picture frame if the ratio of width to length is 1 : 3.

11. **MC** The ratio of length to width to height of the container shown is 4 : 2 : 3. If the sum of the three dimensions is 6.75 m, the height of the container is:

A. 2.25 m
B. 150 cm
C. 1.50 m
D. 2250 cm
E. 15 m

12. **WE15** The area of a window is 1.5 m^2. The ratio of the areas of the glass used is 1 : 2 : 3. Calculate the area of each part of glass.

13. The total area of a tennis court is approximately 260 m^2 for doubles matches. The ratio of the court area for singles matches to the extension area added on for doubles is approximately 13 : 3. Calculate the area of the court used for singles matches.

14. **MC** Three areas add up to 26 000 000 m^2. If they are in the ratio of 3 : 2 : 8, the smallest area is:

A. 4 km^2 B. 16 000 m^2 C. 6 km^2 D. 4000 m^2 E. 160 000 cm^2

15. The total surface area of Victoria and Western Australia is approximately 2 760 000 km^2. If the ratio of Victoria's surface area to Western Australia's surface area is 1 : 11, calculate the approximate surface area of Victoria.

16. **WE16** The ratio of the volume of the container with the green lid to the volume of the container with the red lid is 5 : 6. If the total volume of the two containers is 506 cm³, determine the individual volumes of the two containers.

17. Hot air balloons are each fitted with three propane fuel tanks, with a total volume of 171 L. Their individual volumes are in the ratio 2 : 3 : 4. Calculate the individual volumes of the fuel tanks.

18. If the total capacity of three rainwater tanks is 42 500 L, determine the capacity of each tank if their capacities are in the ratio of 2 : 5 : 10.

19. **WE17** Claire has 25 minutes to tidy up her bedroom and have breakfast before she goes to school. If the ratio of the tidy-up time and breakfast time is 3 : 2, calculate how many minutes she will have to tidy up her room.

20. Rich has 1 hour and 45 minutes to complete his Science, English and Maths homework. If the time he needs to spend on the three subjects is in the ratio 1 : 2 : 4 respectively, determine how much time he has to spend on his Maths homework.

21. Bayside College has a ratio of class time to lunch to recess of 10 : 2 : 1. If the total time for a school day at Bayside College is 6 hours and 30 minutes, calculate

    a. the class time per day
    b. the time for lunch
    c. recess time.

22. Determine how much a family will spend on their mortgage, car loan and health insurance if they allow a budget of $1225 per month and the ratio of mortgage to car loan to health insurance is 10 : 3 : 1.

23. A company spends $3 875 000 on wages and company cars in a ratio of 30 : 1. Calculate how much money the company spends on the company cars.

24. Three friends won a prize of $1520. Calculate how much money each of them wins if they contributed to the price of the winning ticket in the ratio 2 : 5 : 9.

25. Using a calculator, a spreadsheet or otherwise, divide the quantities below in the ratios given.

    a. $420 in the ratio 7 : 3
    b. 6 hours in the ratio 1 : 5 : 6
    c. 18 293 m in the ratio 5 : 2 : 4
    d. 15 000 m² in the ratio 10 : 3 : 2
    e. 960 L in the ratio 7 : 13
    f. 7752 in the ratio 6 : 2 : 9

26. Using a calculator, a spreadsheet or otherwise, determine the quantities obtained when $5040 is divided in the following ratios:

    a. $\dfrac{1}{2}$
    b. $\dfrac{2}{3}$
    c. $\dfrac{1}{5}$
    d. $\dfrac{2}{7}$
    e. $\dfrac{3}{4}$
    f. 2 : 3 : 4

6.5 Scales, diagrams and maps

LEARNING INTENTION

At the end of this subtopic you should be able to:
- use equivalent ratios to determine scale factors on maps and models
- solve problems using a scale factor.

6.5.1 Introduction to scales

Scales are used to draw maps and represent objects, such as cars and boats. Scales are commonly used when it is not feasible to construct full-size models.

A **scale** is a ratio of the length on a drawing or map to the actual distance or length of an object. It is the length on drawing : actual length. Scales are usually written with no units. If a scale is given in two different units, the larger unit is usually converted into the smaller unit.

A **scale factor** is the ratio of two corresponding lengths in two similar shapes. A scale factor of $\frac{1}{2}$ or a scale (ratio) of 1 : 2 means that 1 unit on the drawing represents 2 units in actual size. The unit can be mm, cm, m or km.

tlvd-3767

WORKED EXAMPLE 18 Calculating the scale of a drawing

Calculate the scale of a drawing where 2 cm on the diagram represents 1 km in reality.

THINK	WRITE
1. Convert the larger unit (km) to the smaller unit (cm).	$1 \text{ km} = 100\,000 \text{ cm}$
2. Now that both values are in the same unit, write the scale of the drawing with no units.	$2 : 100\,000$
3. Simplify the scale by dividing both sides of the scale by the highest common factor.	$\frac{2}{2} : \frac{100\,000}{2}$
Divide both sides of the scale by 2, as 2 is the highest common factor of 2 and 100 000.	$= 1 : 50\,000$
4. Write the answer.	$1 : 50\,000$

6.5.2 Calculating dimensions

The scale factor must always be included when a diagram is drawn. It is used to calculate the dimensions needed. To calculate the actual dimensions, measure the dimensions on the diagram and then divide them by the scale factor.

Calculate the length and the width of the bed shown if the scale of the drawing is 1 : 100. Use a ruler to measure the length and the width of the bed.

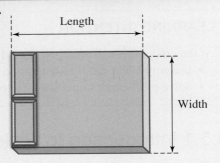

THINK

1. Use a ruler to measure the length and the width of the diagram.

2. Write the scale of the drawing.

3. Divide both dimensions by the scale factor.

4. Write the answer.

WRITE

Length = 2 cm
Width = 1.5 cm

1 : 100
This means that every 1 cm on the drawing represents 100 cm of the real bed.

Length of the bed $= 2 \div \dfrac{1}{100}$
$= 2 \times 100$
$= 200$ cm

Width of the bed $= 1.5 \div \dfrac{1}{100}$
$= 1.5 \times 100$
$= 150$ cm

The length is 200 cm and the width is 150 cm.

6.5.3 Scale drawing

To be able to draw to scale, we need to know the dimensions on the diagram. To calculate these dimensions, we multiply the actual dimension by the scale factor.

tlvd-3768

A drawing of a car uses a scale of 1 : 50. Calculate the diameter of a car wheel on the diagram if its real diameter is 60 cm.

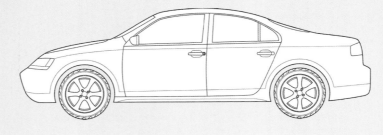

THINK	WRITE
1. Calculate the length on the diagram.	Length on the diagram = real measurement × scale factor $$= 60 \times \frac{1}{50}$$ $$= \frac{60}{50}$$ $$= 1.2 \text{ cm}$$
2. Write the answer.	An actual diameter of 60 cm will be 1.2 cm on a diagram of scale 1 : 50.

6.5.4 Maps and scales

Maps are always drawn at a much smaller scale. All maps have the scale written or drawn on the map.

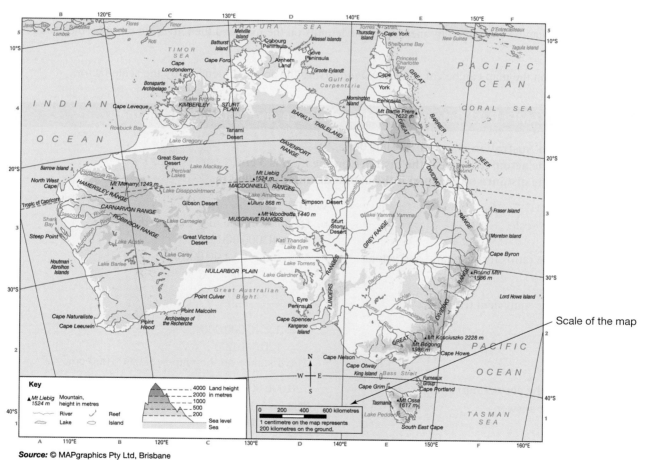

Scale of the map

Source: © MAPgraphics Pty Ltd, Brisbane

WORKED EXAMPLE 21 Determining the scale of a map as a ratio

Determine the scale of the map as a ratio using the information in the diagram.

```
┌──┬───┬──┐
0   200   400 km
```

THINK	WRITE
1. Write the scale as a ratio.	1 cm : 200 km
2. Convert the larger unit into the smaller unit.	The larger unit is km. This has to be converted into cm.
	200 km $= 20\,000\,000$ cm
	1 cm : 200 km $= 1$ cm : $20\,000\,000$ cm
3. Write the answer.	The scale is 1 : $20\,000\,000$.

6.5.5 Maps and distances

Both actual distances and distances on the map can be calculated if the scale of the map is known.

Actual distances can be calculated by measuring the lengths on the map and then *dividing* these by the scale factor.

The dimensions on the map can be calculated by *multiplying* the actual dimension by the scale factor.

tlvd-3770

WORKED EXAMPLE 22 Calculating actual distance and distance on the map

The scale of an Australian map is 1 : 40 000 000.
a. Calculate the actual distance if the distance on the map is 3 cm.
b. Calculate the distance on the map if the actual distance is 2500 km.

THINK	WRITE
a. 1. Write the scale.	a. Distance on the map : actual distance
	$= 1 : 40\,000\,000$
2. Set up the ratios for map ratio and actual ratio.	Map ratio is 1 : $40\,000\,000$.
	Actual ratio is $3 : x$.
3. Construct equivalent fractions.	$\dfrac{x}{3} = \dfrac{40\,000\,000}{1}$
	$x = 3 \times 40\,000\,000$
	$\quad = 120\,000\,000$
	The distance on the map is $120\,000\,000$ cm.
4. Convert the answer into km.	$120\,000\,000$ cm $= 1200$ km
5. Write the answer.	The actual distance represented by 3 cm on the map is 1200 km.
b. 1. Write the scale.	b. Distance on the map : actual distance
	$= 1 : 40\,000\,000$
2. The actual distance is 2500 km. Convert this to cm.	2500 km $= 250\,000\,000$ cm

3. Construct equivalent fractions.

$$1 : 40\,000\,000$$
$$x : 250\,000\,000$$

$$\frac{x}{250\,000\,000} = \frac{1}{40\,000\,000}$$

$$x = \frac{250\,000\,000}{40\,000\,000}$$

$$= 6.25$$

4. Write the answer.

The distance on the map is 6.25 cm if the actual distance is 2500 km.

6.5 Exercise

Students, these questions are even better in jacPLUS

 Receive immediate feedback and access sample responses

 Access additional questions

 Track your results and progress

Find all this and MORE in jacPLUS

1. **WE18** Calculate the scale of a drawing where 100 mm on the diagram represents 2 m in reality.

2. If the scale of a diagram is 5 cm : 100 km, write the scale using the same unit.

3. Calculate the scale of a drawing where:
   a. 4 cm on the diagram represents 5 km in reality
   b. 20 mm on the diagram represents 100 m in reality
   c. 5 cm on the diagram represents 10 km in reality
   d. 3 cm on the diagram represents 600 m in reality.

4. **WE19** Estimate the actual diameter of the tabletop shown if the scale of the diagram is 1 : 50 and the diameter of the table in the diagram is 4 cm.

5. If the scale of the diagram shown is 1 : 50 and the dimensions of the Aboriginal flag in the diagram are length = 3.6 cm and width = 1.8 cm, calculate the actual dimensions, in metres, of the Aboriginal flag design shown.

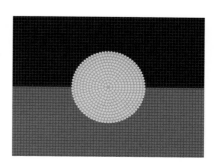

6. Calculate the actual dimensions of the nut and bolt shown if the scale of the drawing is 1 : 2 and the dimensions on the diagram are as follows:

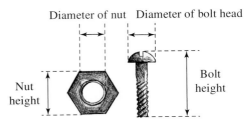

Diameter of nut   Diameter of bolt head

Nut height

Bolt height

- Diameter of the nut is 5 mm.
- Diameter of bolt head is 7 mm.
- Nut height is 10 mm.
- Bolt height is 14 mm.

7. **WE20** The floor plan of a kitchen is drawn to a scale of 1 : 100. Determine the width of the kitchen on the plan if its real width is 5 m.

8. The real height of a building is 147 m. Calculate the height of the building on a diagram drawn at a scale of 1 : 3000.

9. The floor plan shown is planned to be drawn at a scale of 1 : 200. The actual dimensions of the house are shown on the diagram.

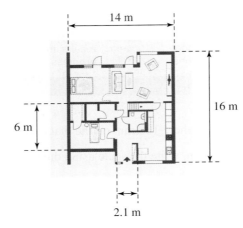

14 m

16 m

6 m

2.1 m

The floor plan is not yet drawn to scale. Calculate the lengths of the dimensions shown if the floor plan was drawn to scale.

10. Calculate the actual lengths for the following lengths measured on a diagram with the given scale.
    a. 12 mm; scale 1 : 1000
    b. 3.8 cm; scale 1 : 150
    c. 27 mm; scale 1 : 200
    d. 11.6 cm; scale 1 : 7000

11. **WE21** Determine the scale of the map as a ratio using the information in the diagram.

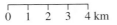

0 1 2 3 4 km

12. Determine the scale of the map as a ratio using the information in the diagram.

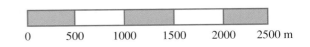

0   500   1000   1500   2000   2500 m

13. **MC** The scale of the map is:

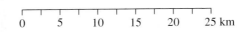

0   5   10   15   20   25 km

A. 1 : 2 500 000
B. 1 : 100 000
C. 1 : 500 000
D. 1 : 5 000 000
E. 1 : 10 000

14. For each of the map scales shown, state the scale and determine the actual distance for the map distance given.

   a. 5.1 cm

   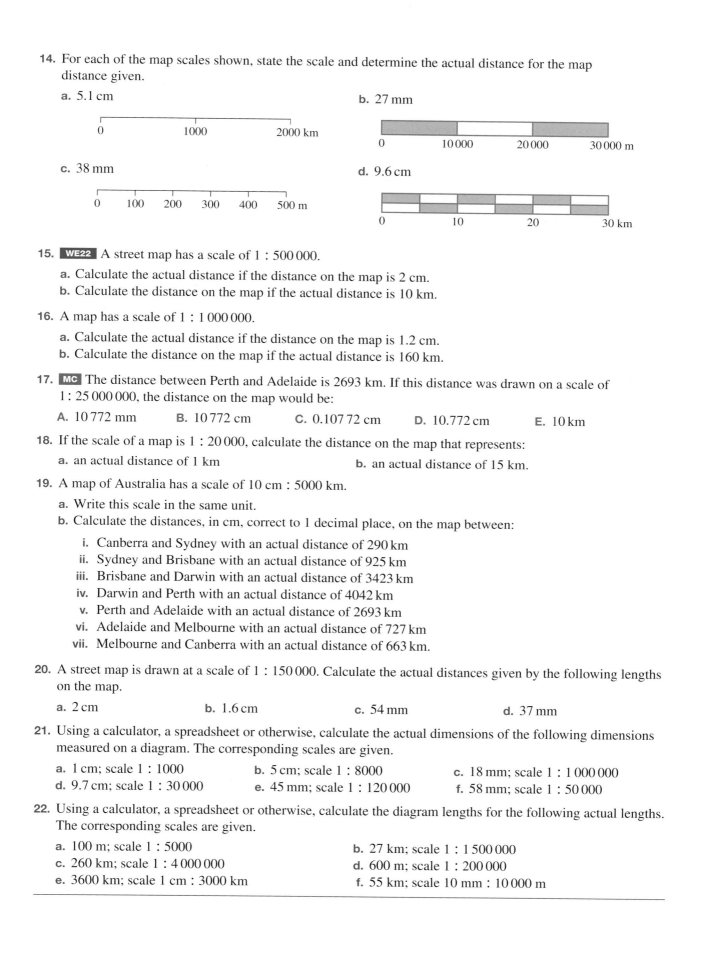

   b. 27 mm

   c. 38 mm

   d. 9.6 cm

15. **WE22** A street map has a scale of 1 : 500 000.

   a. Calculate the actual distance if the distance on the map is 2 cm.
   b. Calculate the distance on the map if the actual distance is 10 km.

16. A map has a scale of 1 : 1 000 000.

   a. Calculate the actual distance if the distance on the map is 1.2 cm.
   b. Calculate the distance on the map if the actual distance is 160 km.

17. **MC** The distance between Perth and Adelaide is 2693 km. If this distance was drawn on a scale of 1 : 25 000 000, the distance on the map would be:

   A. 10 772 mm   **B.** 10 772 cm   **C.** 0.107 72 cm   **D.** 10.772 cm   **E.** 10 km

18. If the scale of a map is 1 : 20 000, calculate the distance on the map that represents:

   a. an actual distance of 1 km
   b. an actual distance of 15 km.

19. A map of Australia has a scale of 10 cm : 5000 km.

   a. Write this scale in the same unit.
   b. Calculate the distances, in cm, correct to 1 decimal place, on the map between:

   i. Canberra and Sydney with an actual distance of 290 km
   ii. Sydney and Brisbane with an actual distance of 925 km
   iii. Brisbane and Darwin with an actual distance of 3423 km
   iv. Darwin and Perth with an actual distance of 4042 km
   v. Perth and Adelaide with an actual distance of 2693 km
   vi. Adelaide and Melbourne with an actual distance of 727 km
   vii. Melbourne and Canberra with an actual distance of 663 km.

20. A street map is drawn at a scale of 1 : 150 000. Calculate the actual distances given by the following lengths on the map.

   a. 2 cm   **b.** 1.6 cm   **c.** 54 mm   **d.** 37 mm

21. Using a calculator, a spreadsheet or otherwise, calculate the actual dimensions of the following dimensions measured on a diagram. The corresponding scales are given.

   a. 1 cm; scale 1 : 1000   **b.** 5 cm; scale 1 : 8000   **c.** 18 mm; scale 1 : 1 000 000
   d. 9.7 cm; scale 1 : 30 000   **e.** 45 mm; scale 1 : 120 000   **f.** 58 mm; scale 1 : 50 000

22. Using a calculator, a spreadsheet or otherwise, calculate the diagram lengths for the following actual lengths. The corresponding scales are given.

   a. 100 m; scale 1 : 5000   **b.** 27 km; scale 1 : 1 500 000
   c. 260 km; scale 1 : 4 000 000   **d.** 600 m; scale 1 : 200 000
   e. 3600 km; scale 1 cm : 3000 km   **f.** 55 km; scale 10 mm : 10 000 m

6.6 Proportion

6.6.1 Proportion

A **proportion** is a statement of equality of two ratios. For example, $12 : 18 = 2 : 3$. When objects are in proportion, they have the same fractional relationship between the size of their parts and the size of the whole.

For example, the rectangles below are in proportion to each other.

The following observations can be made:

When you compare the two-coloured parts of each rectangle, there is 1 blue square for every 2 pink squares; the number of blue squares is $\frac{1}{2}$ of the number of pink squares.

When you compare the parts with the whole rectangle, the blue squares form $\frac{1}{3}$ of the whole rectangle, and the pink squares form the other $\frac{2}{3}$.

If $a : b = c : d$ or $\frac{a}{b} = \frac{c}{d}$, then using cross-multiplication, as shown below:

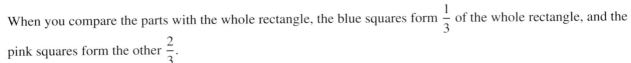

gives:

$$a \times d = c \times b$$

If the cross-products ($a \times d$ and $c \times b$) are equal, then the ratios form a proportion and therefore the two ratios are equivalent.

---

tlvd-3771

▶ WORKED EXAMPLE 23 Determining whether two ratios are in proportion

Use the cross-multiplication method to determine whether the following two ratios are in proportion.

$$6 : 9; \quad 24 : 36$$

THINK	WRITE
1. Write the ratios in fraction form.	$\dfrac{6}{9} = \dfrac{24}{36}$
2. Perform cross-multiplication.	
	$6 \times 36 = 216; \; 24 \times 9 = 216$

3. Check that the two products are equal. If they are, the two ratios are in proportion.

$216 = 216$

4. Write your answer.

Therefore, the ratios are in proportion.

WORKED EXAMPLE 24 Determining the value of a pronumeral in a proportion

Determine the value of a in the proportion $\dfrac{a}{3} = \dfrac{6}{9}$.

THINK	WRITE
1. Write the proportion statement.	$\dfrac{a}{3} = \dfrac{6}{9}$
2. Cross-multiply and equate the products.	$\dfrac{a}{3} \diagup\!\!\!\!\diagdown \dfrac{6}{9}$ $a \times 9 = 6 \times 3$
3. Solve for a by dividing both sides of the equation by 9.	$9a = 18$ $\dfrac{9a}{9} = \dfrac{18}{9}$ $a = 2$
4. Write your answer.	The value of a is 2.

WORKED EXAMPLE 25 Determining the value of an unknown in a proportion

The ratio of children to adults on the bus was 4 : 3. If there were 28 children, calculate the number of adults.

THINK	WRITE
1. Let the number of adults be a and write a proportion statement. (Since the first number in the ratio represents children, place the number of children, 28, as the numerator.)	$\dfrac{4}{3} = \dfrac{28}{a}$
2. Cross-multiply and equate the products.	$\dfrac{4}{3} \diagup\!\!\!\!\diagdown \dfrac{28}{a}$ $4 \times a = 28 \times 3$
3. Solve for a by dividing both sides by 4.	$4a = 84$ $\dfrac{4a}{4} = \dfrac{84}{4}$ $a = 21$
4. Write the answer.	There were 21 adults on the bus.

Resources ───

 Interactivities Proportion (int-3735)

6.6 Exercise

Students, these questions are even better in jacPLUS

Receive immediate feedback and access sample responses

Access additional questions

Track your results and progress

Find all this and MORE in jacPLUS ⊙

1. **WE23** Use the cross-multiplication method to determine whether the following pairs of ratios are in proportion.
   a. $2 : 3$; $8 : 12$
   b. $4 : 7$; $8 : 14$
   c. $5 : 7$; $10 : 14$
   d. $5 : 8$; $10 : 16$

2. Use the cross-multiplication method to determine whether the following pairs of ratios are in proportion.
   a. $\dfrac{7}{9}$; $\dfrac{21}{25}$
   b. $\dfrac{3}{8}$; $\dfrac{12}{32}$
   c. $\dfrac{14}{16}$; $\dfrac{5}{9}$
   d. $\dfrac{11}{12}$; $\dfrac{7}{8}$

3. **WE 24** Determine the value of a in each of the following proportions.
   a. $\dfrac{a}{2} = \dfrac{4}{8}$
   b. $\dfrac{a}{6} = \dfrac{8}{12}$
   c. $\dfrac{a}{9} = \dfrac{2}{3}$
   d. $\dfrac{3}{a} = \dfrac{9}{12}$
   e. $\dfrac{7}{a} = \dfrac{14}{48}$
   f. $\dfrac{10}{a} = \dfrac{3}{15}$

4. Determine the value of a in each of the following proportions.
   a. $\dfrac{3}{7} = \dfrac{a}{28}$
   b. $\dfrac{12}{10} = \dfrac{a}{5}$
   c. $\dfrac{8}{12} = \dfrac{a}{9}$
   d. $\dfrac{35}{7} = \dfrac{5}{a}$
   e. $\dfrac{24}{16} = \dfrac{6}{a}$
   f. $\dfrac{30}{45} = \dfrac{2}{a}$

5. **WE25** Solve each of the following, using a proportion statement and the cross-multiplication method.
   a. The ratio of children to adults on a bus is $3 : 4$. If there are 12 adults, calculate how many children there are on the bus.
   b. In a room, the ratio of length to width is $5 : 4$. If the width is 8 m, calculate the length.
   c. The team's win–loss ratio is $7 : 5$. Calculate how many wins it has had if it has had 15 losses.
   d. The ratio of flour to milk in a mixture is $7 : 2$. If 14 cups of flour are used, calculate how much milk is required.
   e. In a supermarket, the ratio of 600 mL cartons of milk to 1-litre cartons of milk is $4 : 5$. If there are sixty 600 mL cartons, calculate how many 1-litre cartons there are.
   f. In a crowd of mobile-phone users, the ratio of men to women is $7 : 8$. Calculate the number of women if there are 2870 men.

6. Although we know that only whole numbers are used in ratios, sometimes in a proportion statement the answer can be a fraction or a mixed number. Consider the following proportion:

$$\frac{a}{6} = \frac{7}{4}$$
$$a \times 4 = 7 \times 6$$
$$4a = 42$$
$$a = 10.5$$

Calculate the value of a in each of the following proportion statements. Write your answer correct to 1 decimal place.

a. $\dfrac{a}{7} = \dfrac{8}{5}$

b. $\dfrac{a}{6} = \dfrac{4}{5}$

c. $\dfrac{a}{3} = \dfrac{7}{10}$

d. $\dfrac{a}{9} = \dfrac{9}{10}$

e. $\dfrac{5}{a} = \dfrac{7}{10}$

f. $\dfrac{8}{a} = \dfrac{6}{7}$

7. Write a proportion statement for each situation and then solve the problem. If necessary, write your answer correct to 1 decimal place.

a. A rice recipe uses the ratio of 1 cup of rice to 3 cups of water.
Calculate how many cups of rice can be cooked in 5 cups of water.

b. Another recipe states that 2 cups of rice are required to serve 6 people. If you have invited 11 people, determine how many cups of rice you will need.

c. In a chemical compound, there should be 15 g of chemical A to every 4 g of chemical B. If a compound contains 50 g of chemical A, calculate how many grams of chemical B it should contain.

d. A saline solution contains 2 parts of salt to 17 parts of water. Calculate how much water should be added to 5 parts of salt.

e. To mix concrete, 2 buckets of sand are needed for every 3 buckets of blue metal. For a big job, calculate how much blue metal will be needed for 15 buckets of sand.

8. **MC** a. If $\dfrac{p}{q} = \dfrac{l}{m}$, then:

A. $p \times q = l \times m$

B. $p \times l = q \times m$

C. $p \times m = l \times q$

D. $\dfrac{p}{m} = \dfrac{l}{q}$

E. none of these is true.

MC b. If $\dfrac{x}{3} = \dfrac{y}{6}$, then:

A. $x = 2$ and $y = 4$

B. $x = 1$ and $y = 2$

C. $x = 3$ and $y = 6$

D. $x = 6$ and $y = 12$

E. all of these are true.

9. **MC** a. If $\dfrac{23}{34} = \dfrac{x}{19}$, then, correct to the nearest whole number, x equals:

A. 13   B. 12   C. 34   D. 28   E. 17

MC b. The directions on a cordial bottle suggest mixing 25 mL of cordial with 250 mL of water. Determine how much cordial should be mixed with 5.5 L of water.

A. 0.55 mL

B. 5.5 mL

C. 55 mL

D. 550 mL

E. 5500 mL

10. In a family, 3 children receive their allowances in the ratio of their ages, which are 16 years, 14 years and 10 years. If the oldest child receives $32, determine how much the other two children receive.

11. In jewellery, gold is often combined with other metals. 'Pink gold' is a mixture of pure gold, copper and silver in the ratio of 15 : 4 : 1.
'White gold' is a mixture of pure gold and platinum in the ratio of 3 : 1.

   a. Calculate what fraction of both pink and white gold jewellery is pure gold.
   Pure gold is 24 carats and is not mixed with other metals. For most jewellery, however, 18-carat gold is used.
   b. Using your answer to part a, show why jewellery gold is labelled 18 carats.
   c. If the copper and silver in an 18-carat bracelet weigh a combined 2 grams, calculate the weight of gold in the bracelet.
   d. If the price of gold is $35 per gram, calculate the cost of the gold in the bracelet from part c.

12. Two jars each contain 8 green lollies. In one jar, the ratio of green lollies to red lollies is 1 : 2; in the other, it is 2 : 1. If the contents of the two jars combine, determine the new ratio.

13. The nutritional panel for a certain cereal is shown. Sugars are one form of carbohydrate. Calculate the ratio of sugars to total carbohydrates in the cereal.

	Quantity per 30-g serving	Percentage daily intake per 30-g serving	Quantity per 30-g serving with $\frac{1}{2}$ cup skim milk
Energy	480 kJ	5.5%	670 kJ
Protein	6.6 g	13.1%	11.2 g
Fat — total — saturated	0.2 g 0.1 g	0.3% 0.1%	0.3 g 0.2 g
Carbohydrate — total — sugars	20.8 g 9.6 g	6.7% 10.7%	27.3 g 16.1 g

14. The solution to this question contains some gaps. Rewrite the solution, replacing the empty boxes with the appropriate numbers.

On a map for which the scale is 1 : 20 000, determine what distance in cm represents 2.4 km on the ground.

$$1 : 20\,000 = x\,\text{cm} : 2.4\,\text{km}$$

$$1 : 20\,000 = x\,\text{cm} : 2.4 \times 1000 \times 100\,\text{cm}$$

$$1 : 20\,000 = x : 240\,000$$

$$\frac{1}{20\,000} = \frac{x}{\square}$$

$$\frac{1}{20\,000} \times \square = \frac{x}{240\,000} \times \square$$

$$x = \square$$

So, $\square$ cm on the map represents 2.4 km on the ground.

15. Simplify the following three-part ratios.

a. 50 : 20 : 15

b. 0.4 : 1.8 : 2.2

16. A concreter needs to make 1.0 m³ of concrete for the base of a garden shed. To make concrete mix, 1 part cement, 2 parts sand and 4 parts gravel are needed.

Calculate how many cubic metres of each component should be used.

6.7 Review

6.7.1 Summary

doc-38047

6.7 Exercise

Multiple choice

1. **MC** The ratio of blue pawns to red pawns, in simplest form, in the picture shown is:

   A. $6:9$　　　　B. $3:2$　　　　C. $2:3$

   D. $9:6$　　　　E. $1:2$

2. **MC** The ratio $385:154$ in simplest fraction form is:

   A. $\dfrac{5}{2}$　　　　B. $\dfrac{25}{14}$　　　　C. $7:5$

   D. $\dfrac{2}{5}$　　　　E. $\dfrac{85}{54}$

3. **MC** The ratio of 364 m^2 to 455 m^2 to 819 m^2, in simplest form, is:

   A. $28:35:63$　　　　B. $4:5:9$　　　　C. $63:35:28$

   D. $52:65:117$　　　　E. $64:55:19$

4. **MC** The ratio between the volume of the sphere and the volume of the cube in the diagram shown, in simplest fractional form, is:

   A. $\dfrac{45}{100}$　　　　B. $9:20$　　　　C. $45:100$

   D. $\dfrac{9}{20}$　　　　E. $450:10$

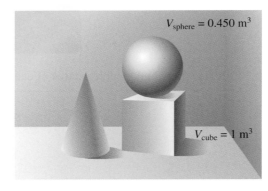

$V_{\text{sphere}} = 0.450 \text{ m}^3$

$V_{\text{cube}} = 1 \text{ m}^3$

5. **MC** If an area of 2730 cm^2 is divided into a ratio $5:3:7$, then the corresponding areas are:

   A. 546 cm^2, 390 cm^2 and 910 cm^2

   B. 455 cm^2, 273 cm^2 and 637 cm^2

   C. $546 \text{ cm}^2 : 1274 \text{ cm}^2$ and 910 cm^2

   D. 910 cm^2, 546 cm^2 and 1274 cm^2

   E. $500 \text{ cm}^2 : 30 \text{ cm}^2 : 7 \text{ cm}^2$

6. **MC** If a road map is drawn to a scale of $1:900\,000$, then 3.5 cm on the map represents a road length of:

   A. 31.5 cm　　　　B. 3.9 km　　　　C. 3.5 km　　　　D. 13.5 km　　　　E. 31.5 km

7. **MC** A netball court is 30.5 m long and 15.25 m wide. If the dimensions of the court in the diagram shown are 6.1 cm and 3.05 cm, then the scale factor of the diagram is:

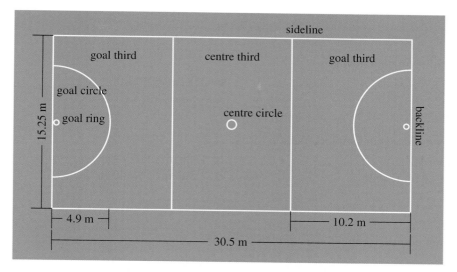

A. $\dfrac{1}{5}$ B. $\dfrac{15.25}{3.05}$ C. $\dfrac{1}{500}$ D. $\dfrac{6.1}{30.5}$ E. $\dfrac{5}{1}$

8. **MC** The ratio $4\dfrac{1}{3} : 1\dfrac{2}{3}$ in simplest form is:

A. $13 : 5$ B. $3 : 13$ C. $4 : 1$ D. $5 : 13$ E. $1 : 4$

9. **MC** Three areas add up to $56\,000\ \text{m}^2$. If they are in a ratio of $5 : 3 : 8$, the smallest area is:

A. $4\ \text{km}^2$ B. $1500\ \text{m}^2$ C. $3500\ \text{km}^2$ D. $14\,000\ \text{m}^2$ E. $10\,500\ \text{m}^2$

10. **MC** The ratio of length to width to height of a storage container is $3 : 1 : 5$. If the sum of the three dimensions is 6.75 m, the length of the container is:

A. 2.25 m   B. 150 cm   C. 1.50 m   D. 2250 cm   E. 15 cm

Short answer

11. Two bus stops, A and B, are 45 km apart. Two new bus stops, C and D, are going to be placed in a ratio $2 : 3 : 4$ between A and B. Determine how far apart bus stops C and D are.

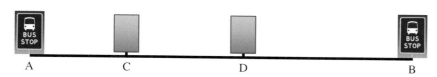

12. While camping, the Blake family use powdered milk. They mix the powder with water in the ratio of $1 : 24$. Determine how much of each ingredient they would need to make up:

a. 600 mL   b. 1.2 L.

13. Yoke made 3 L of a refreshing drink using orange juice, apple juice and carrot juice in the ratio $5 : 8 : 2$.

a. Calculate how much orange juice she used.
b. Calculate how much apple juice she used.
c. Calculate how much carrot juice she used.
d. Determine the ratio of carrot juice to apple juice.

14. Calculate the value of n in each of the following proportions.

    a. $\dfrac{n}{3} = \dfrac{20}{15}$ **b.** $\dfrac{n}{28} = \dfrac{5}{7}$

    c. $\dfrac{2}{3} = \dfrac{8}{n}$ **d.** $\dfrac{4}{5} = \dfrac{12}{n}$

15. Calculate the following unknowns.

    a. $160 : x = 5 : 7$ **b.** $258 : a = 3 : 10$

    c. $b : 39.6 = 2 : 9$ **d.** $5 : 4 = 21.5 : x$

    e. $8 : 1 = 134.9 : a$

16. State the following ratios in simplest fraction form.

    a. $945 \text{ m}^2 : 1365 \text{ m}^2$ **b.** $1386 \text{ m} : 2.079 \text{ km}$

    c. $10 \text{ h } 12 \text{ min} : 61 \text{ h } 12 \text{ min}$ **d.** $2.592 \text{ m}^3 : 3.240 \text{ m}^3$

    e. $13.68 \text{ L} : 12.24 \text{ L}$

Extended response

17. Three swimming pools have the dimensions shown in the table.

Pool	Length (m)	Width (m)	Depth (m)
Large	12.6	3.50	1.20
Medium	8.4	3.50	1.20
Small	4.2	3.50	1.20

    a. Determine the ratio between the lengths of the three pools.
    b. Calculate the volumes of the three swimming pools.
    c. Calculate the ratio between the volumes of the three swimming pools.
    d. Convert the volume measurements $\left(\text{m}^3\right)$ into capacity measurements (L).
    e. If water is poured into the swimming pools at a rate of $5 \text{ L}/\text{min}$, calculate how long it would take, to the nearest hour, to fill in the three swimming pools.

18. For a main course at a local restaurant, guests can select from a chicken, fish or vegetarian dish. On Friday night the kitchen served 72 chicken plates, 56 fish plates and 48 vegetarian plates.

    a. Express the number of dishes served as a ratio in the simplest form.
    b. On a Saturday night the restaurant can cater for 250 people. If the restaurant was full, determine how many people would be expected to order a non-vegetarian dish.
    c. The Elmir family of five and the Cann family of three dine together. The total bill for the table was $268.

        i. Calculate the cost of dinner per head.
        ii. If the bill is split according to family size, calculate the proportion of the bill that the Elmir family will pay.

19. A recipe for shortbread cookies requires 500 g flour, 250 g unsalted butter, 125 g of sugar, 150 g walnuts and 200 g chocolate chips.

    a. Determine the total weight of ingredients.
    b. Determine the ratio of flour to sugar.
    c. Determine the ratio of chocolate chips to flour.
    d. Determine the ratios of all the ingredients in the total mixture.
    e. Calculate the weight of each ingredient if the total quantity of ingredients is 2.205 kg.
    f. Calculate the total cost for the original recipe, correct to 2 decimal places, if 1 kg flour cost $1.50, 250 g unsalted butter $2.10, 1 kg sugar $1.25, 1 kg walnuts $9.50 and 200 g chocolate chips $3.25.

20. The floor plan shown is drawn to a scale of 1 : 200. The dimensions written represent the actual dimensions in millimetres (i.e. 4000 = 4 m).

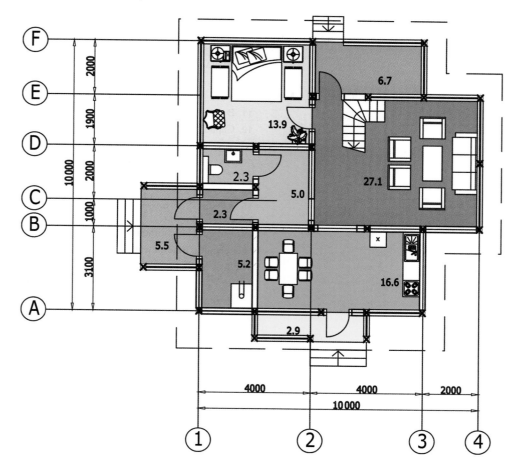

a. Calculate the actual dimensions (length and width) of the bedroom.
b. Calculate the area of the bedroom.
c. Calculate the dimensions from part a on the diagram.
d. Calculate the actual dimensions (length and width) of the lounge room.
e. Calculate the area of the lounge room.
f. Calculate the dimensions from part d on the diagram.

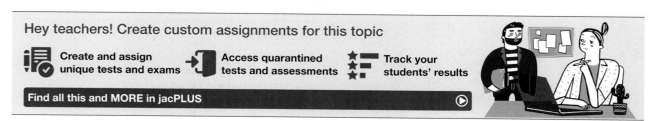

Answers

Topic 6 Ratios and proportions

6.2 Introduction to ratios

6.2 Exercise

1. $5 : 4$

2. a. $8 : 5$ b. $5 : 9$ c. $2 : 1$

3. a. $\dfrac{3}{5}$ b. $\dfrac{1}{7}$ c. $\dfrac{6}{11}$ d. $\dfrac{5}{2}$

4. a. $4 : 7$ b. $7 : 4$ c. $1 : 2$ d. $1 : 1$

5. a. $1 : 3$ b. $5 : 4$ c. $4 : 3$ d. $5 : 3$

6. a. $\dfrac{2}{3}$ b. $\dfrac{9}{8}$ c. $\dfrac{15}{49}$ d. $\dfrac{20}{17}$

7. a. $a = 11$ b. $b = 21$

8. a. $m = 35, n = 40$ b. $a = 45, b = 6$

9. a. $x = 28$ b. $a = 3, b = 21$
   c. $b = 10$ d. $m = 2$

10. D

11. C

12. Orange balls : black and white balls $= \dfrac{1}{2}$

13. $\dfrac{4}{3}$

14. a. $\dfrac{10}{3}$ b. $5 : 1$ c. $\dfrac{3}{200}$
    d. $1 : 2$ e. $3 : 2 : 5$ f. $7 : 4 : 8$

15. a. $5 : 3$ b. $6 : 7$ c. $1 : 10$
    d. $45 : 8$ e. $20 : 3$ f. $1 : 12$

16. a. $1 : 4$ b. $30 : 1$ c. $5 : 2$
    d. $20 : 50 : 7$

17. a. 12 boys   b. $3 : 4$

18. a. $11 : 5$ b. $3 : 1 : 2$

19. a. $\dfrac{3}{11}$ b. $\dfrac{4}{5}$ c. $\dfrac{10}{7}$
    d. $\dfrac{8}{9}$ e. $\dfrac{1}{10}$ f. $\dfrac{91}{22}$

20. a. Some equivalent fractions could be:
    $$\dfrac{6}{10}, \dfrac{9}{15}, \dfrac{12}{20}, \dfrac{15}{25}, \dfrac{18}{30}, \dfrac{27}{45}, \dfrac{30}{50}, \dfrac{33}{55}, \dfrac{300}{500}, \dfrac{3000}{5000}$$
    b. Some equivalent fractions could be:
    $$\dfrac{540}{420}, \dfrac{270}{210}, \dfrac{180}{140}, \dfrac{135}{105}, \dfrac{108}{84}, \dfrac{90}{70}, \dfrac{54}{42}, \dfrac{45}{35}, \dfrac{18}{14}, \dfrac{9}{7}$$

6.3 Ratios of two quantities

6.3 Exercise

1. $3 : 1$

2. $2 : 10 : 5$

3. a. Flour : lard $= 15 : 2$
   b. Butter : lard $= 7 : 1$
   c. Flour : butter $= 15 : 14$
   d. Flour : lard : butter $= 15 : 2 : 14$

4. a. $4 : 11$ b. $5 : 2$
   c. $4 : 10 : 11$ d. $4 : 25$

5. $1 : 14$

6. $50 : 25 : 1$

7. $9 : 8$

8. $\text{Area}_{\triangle ABC} : \text{Area}_{\triangle DEF} = 1 : 5$

9. $\text{Area}_{\text{water}} : \text{Area}_{\text{total}} = 3611 : 5101$

10. a. $4 : 3$ b. $1 : 8$ c. $1 : 2$ d. $12 : 1$

11. B

12. $\text{Capacity}_{\text{carafe}} : \text{Capacity}_{\text{glass}} = 6 : 1$

13. $V_{\text{water}} : V_{\text{oil}} = 5 : 4$

14. $\dfrac{5}{3}$

15. a. $\dfrac{7}{2}$ b. $\dfrac{14}{15}$ c. $\dfrac{1}{3}$ d. $\dfrac{4}{5}$

16. Time by bus : time by train $= 4 : 3$

17. Sleeping time : awake time $= 17 : 31$

18. D

19. a. $3 : 4$ b. $8 : 7$
    c. $7 : 6$ d. $6 : 7 : 8$

20. a. $7 : 9 : 3$ b. $7 : 9$
    c. $3 : 1$ d. $7 : 3$

21. a. $5 : 1$ b. $3 : 2$
    c. $2 : 15$ d. $15 : 3 : 2$

22. a. $1 : 40$ b. $5 : 2$

23. a. $\dfrac{4}{5}$ b. $\dfrac{7}{3}$ c. $\dfrac{9}{16}$
    d. $\dfrac{2}{11}$ e. $\dfrac{8}{3}$

24. a. $38 : 14, 57 : 21, 76 : 28, 95 : 35, 114 : 42, 133 : 49,$
    $152 : 56, 171 : 63, 190 : 70$
    Other answers are possible.
    b. $3 : 2, 6 : 4, 12 : 8, 72 : 48, 720 : 480, 1872 : 1248,$
    $5040 : 3360, 18\,018 : 12\,012, 32\,760 : 21\,840,$
    $55\,440 : 36\,960$
    Other answers are possible.

6.4 Divide into ratios

6.4 Exercise

1. 12 tulips and 18 tulips

2. 125 mandarins, 75 mandarins and 50 mandarins

3. a. 990 children and 330 adults
   b. 1100 children and 220 adults
   c. 792 children and 528 adults
   d. 840 children and 480 adults

4. a. $2.08 \, \text{kg}$ b. $3.12 \, \text{kg}$

5. a. $576 \, \text{kg}$ b. $216 \, \text{kg}$

6. 300 g tomatoes, 200 g capsicum

7. 12.75 kg sand, 4.25 kg cement

8. 1.8 m and 2.7 m

9. 200 mm, 400 mm, 800 mm

10. $15.375 \, \text{cm}$

11. A

12. $0.25\,\text{m}^2$, $0.5\,\text{m}^2$, $0.75\,\text{m}^2$

13. $211.25\,\text{m}^2$

14. D

15. $230\,000\,\text{km}^2$

16. $230\,\text{cm}^3$, $276\,\text{cm}^3$

17. 38 L, 57 L, 76 L

18. 5000 L, 12 500 L, 25 000 L

19. 15 minutes

20. 60 minutes

21. a. 5 hours   b. 1 hour   c. 30 minutes

22. $875, $262.50, $87.50

23. $125 000

24. $190, $475, $855

25. a. $294, $126
    b. 30 minutes, 2.5 hours, 3 hours
    c. 8315 m, 3326 m, 6652 m
    d. $10\,000\,\text{m}^2$, $3000\,\text{m}^2$, $2000\,\text{m}^2$
    e. 336 L, 624 L
    f. 2736, 912, 4104

26. a. $1680, $3360   b. $2016, $3024
    c. $840, $4200   d. $1120, $3920
    e. $2160, $2880   f. $1120, $1680, $2240

6.5 Scales, diagrams and maps

6.5 Exercise

1. 1 : 20

2. 1 : 2 000 000

3. a. 1 : 125 000   b. 1 : 5000   c. 1 : 200 000
   d. 1 : 20 000

4. 200 cm

5. Length = 1.8 m, width = 0.9 m

6. Nut height = 2 cm, diameter of the nut = 1 cm,
   bolt height = 2.8 cm, diameter of bolt head = 1.4 cm

7. 5 cm

8. 4.9 cm

9. 7 cm, 8 cm, 3 cm, 1.05 cm

10. a. 12 m   b. 5.7 m   c. 5.4 m   d. 812 m

11. 1 : 200 000

12. 1 : 50 000

13. C

14. a. Scale 5 : 200 000 000, 2040 km
    b. Scale 1 : 500 000, 13 500 m
    c. Scale 1 : 10 000, 380 m
    d. Scale 1 : 500 000, 48 km

15. a. 10 km   b. 2 cm

16. a. 12 km   b. 16 cm

17. D

18. a. 5 cm   b. 75 cm

19. a. 1 : 50 000 000
    b. i. 0.6 cm
       ii. 1.9 cm
       iii. 6.8 cm
       iv. 8.1 cm
       v. 5.4 cm
       vi. 1.5 cm

20. a. 3 km   b. 2.4 km   c. 8.1 km   d. 5.55 km

21. a. 10 m   b. 400 m   c. 18 km
    d. 2910 m   e. 5.4 km   f. 2.9 km

22. a. 2 cm   b. 18 mm   c. 65 mm
    d. 3 mm   e. 1.2 cm   f. 55 mm

6.6 Proportion

6.6 Exercise

1. a. Yes   b. Yes   c. Yes   d. Yes

2. a. No   b. Yes   c. No   d. No

3. a. 1   b. 4   c. 6
   d. 4   e. 24   f. 50

4. a. 12   b. 6   c. 6
   d. 1   e. 4   f. 3

5. a. 9 children   b. Length = 10 m
   c. Wins = 21   d. Milk = 4 cups
   e. 1-litre cartons = 75   f. Women = 3280

6. a. 11.2   b. 4.8   c. 2.1
   d. 8.1   e. 7.1   f. 9.3

7. a. 1.7   b. 3.7   c. 13.3
   d. 42.5   e. 22.5

8. a. C   b. E

9. a. A   b. D

10. 14-year-old = $28
    10-year-old = $20

11. a. Pink gold = $\dfrac{3}{4}$; white gold = $\dfrac{3}{4}$

    b. Since $\dfrac{18}{24} = \dfrac{3}{4}$

    c. 6 grams

    d. $210

12. 4 : 5

13. 9.6 : 20.8

14.
$$1 : 20\,000 = x\,\text{cm} : 2.4\,\text{km}$$
$$1 : 20\,000 = x\,\text{cm} : 2.4 \times 1000 \times 100\,\text{cm}$$
$$1 : 20\,000 = x : 240\,000$$

$$\frac{1}{20\,000} = \frac{x}{240\,000}$$

$$\frac{1}{20\,000} \times 240\,000 = \frac{x}{240\,000} \times 240\,000$$

$$x = 12$$

So 12 cm on the map represents 2.4 km.

15. a. 3 : 4 : 3   b. 2 : 9 : 11

16. Cement = $\dfrac{1}{7} \approx 0.14\,\text{m}^3$

    Sand = $\dfrac{2}{7} \approx 0.29\,\text{m}^3$

    Gravel = $\dfrac{4}{7} \approx 0.57\,\text{m}^3$

6.7 Review

6.7 Exercise

Multiple choice

1. C
2. A
3. B
4. D
5. D
6. E
7. C
8. A
9. E
10. A

Short answer

11. 15 km
12. a. 24 mL of powder and 576 mL of water
    b. 48 mL of powder and 1152 mL of water
13. a. 1 L   b. 1.6 L   c. 400 mL   d. 1 : 4
14. a. 4   b. 20   c. 12   d. 15
15. a. 224   b. 860   c. 8.8
    d. 17.2   e. 16.86
16. a. $\dfrac{9}{13}$ b. $\dfrac{2}{3}$ c. $\dfrac{1}{6}$
    d. $\dfrac{4}{5}$ e. $\dfrac{19}{17}$

Extended response

17. a. 1 : 2 : 3
    b. 52.92 m^3, 35.28 m^3 and 17.64 m^3
    c. 1 : 2 : 3
    d. 52 920 L, 35 280 L and 17 640 L
    e. 7 days 8 hours, 4 days 22 hours and 2 days 11 hours
18. a. 9 : 7 : 6
    b. 182
    c. i. $33.50
       ii. 62.5%
19. a. 1.225 kg
    b. 4 : 1
    c. 2 : 5
    d. 20 : 10 : 5 : 6 : 8
    e. 900 g flour, 450 g unsalted butter, 225 g sugar,
       270 g walnuts and 360 g chocolate chips
    f. $7.68
20. a. Length = 4 m and width = 3.9 m
    b. 15.6 m^2
    c. Length = 20 mm and width = 19.5 mm
    d. Length = 6 m and width = 4.9 m
    e. 29.4 m^2
    f. Length = 30 mm and width = 24.5 mm

7 Rates

Fully worked solutions for this topic are available online.

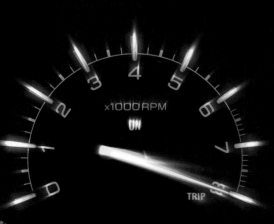

7.1 Overview

7.1.1 Introduction

Rates allow us to compare quantities expressed in different units. This is important to know when it comes to supermarket shopping, so you can ensure you are getting the best value for money. For example, is it cheaper to buy a large box of Corn Flakes (725 g for $4.50) or multiple packets (220 g for $1.95)? Using rates helps you determine which option is the best value. Other common rates are those where some quantity changes with respect to time.

For example, how far will a racing car travel on 60 litres of fuel when travelling at an average speed of 275 km/h? Or how long will it take to fill a 5000-litre water tank if water is pumped in at a rate of 15 litres per second? From these examples you can see that rates are used regularly in daily activities.

KEY CONCEPTS

This topic covers the following key concepts from the VCE Mathematics Study Design:
- use of ratios, proportions, percentages and rates to solve problems.

Note: Concepts shown in grey are covered in other topics.

Source: VCE Mathematics Study Design (2023–2027) extracts © VCAA; reproduced by permission.

7.2 Identifying rates

7.2.1 Rates

A **rate** is a fraction that compares the size of one quantity to that of another quantity. Rates have units. Unlike ratios, which compare the same quantities measured in the same unit, a rate compares two different quantities measured in different units. Examples of rates are cost per kilogram or kilometres per hour.

The unit for rates contains the word per, which is written as /. For example, a car travelling at a speed of 60 km/h is covering a distance of 60 km for every one hour of time.

$$60 \text{ km/h} = \frac{60 \text{ km}}{1 \text{ hour}}$$

Calculating rates

To calculate a rate, divide the first quantity by the second quantity.

$$\text{rate} = \frac{\text{one quantity}}{\text{another quantity}}$$

Common rates involving time

$$\text{speed} = \frac{\text{distance}}{\text{time}}$$

$$\text{flow rate} = \frac{\text{litres}}{\text{minute}}$$

$$\text{heart rate} = \frac{\text{number of heartbeats}}{\text{minute}}$$

---

tlvd-3773

WORKED EXAMPLE 1 Calculating the rate

A car travels 80 kilometres in 2 hours. Calculate the rate at which the car is travelling using units of:
a. kilometres per hour   **b. metres per second (to 1 decimal place).**

THINK	WRITE
a. 1. Identify the units in the rate. The rate 'kilometres per hour' indicates that the number of kilometres should be divided by the number of hours taken.	**a.** Rate $= \dfrac{\text{distance (km)}}{\text{time (h)}}$
2. Write the rate using the correct units.	$= \dfrac{80\,\text{km}}{2\,\text{h}}$
3. Simplify the rate by dividing the numbers.	$= \dfrac{40\,\text{km}}{1\,\text{h}}$
4. Write the answer.	80 kilometres in 2 hours is equivalent to a rate of 40 km/h.
b. 1. Identify the units in the rate. The rate 'metres per second' indicates that the number of metres should be divided by the number of seconds taken.	**b.** Rate $= \dfrac{\text{distance (m)}}{\text{time (s)}}$
2. Convert the distance to metres by multiplying by 1000. Convert time to seconds by multiplying by 60 to change to minutes and 60 again to change to seconds.	$80\,\text{km} = 80 \times 1000$ $\quad\quad\quad = 80\,000\,\text{m}$ $2\,\text{h} = 2 \times 60 \times 60$ $\quad\quad = 7200\,\text{s}$
3. Write the rate using the correct units.	Rate $= \dfrac{80\,000\,\text{m}}{7200\,\text{s}}$
4. Simplify the rate.	$= 11.1\,\text{m/s}$
5. Write the answer.	80 kilometres in 2 hours is equivalent to a rate of 11.1 m/s.

When comparing the difference between two quantities, you need to subtract the smaller quantity from the larger quantity.

tlvd-3774

WORKED EXAMPLE 2 Calculating rates between two quantities

Using the growth chart shown, determine the rate at which the child has grown between the ages of 10 and 13. Give your answer in centimetres per year.

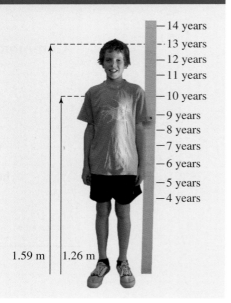

THINK	WRITE
1. Amount of growth = final height − initial height. To determine how much he has grown, calculate the difference in heights.	Amount of growth = 1.59 m − 1.26 m $\qquad\qquad\qquad = 0.33$ m
2. The final rate is in centimetres per year, so convert from metres to centimetres by multiplying by 100.	$= 33$ cm
3. He has grown 33 cm between the ages of 10 and 13 — i.e. in 3 years.	Rate of growth $= \dfrac{\text{amount of growth (cm)}}{\text{time (years)}}$ $\qquad\qquad\quad = \dfrac{33}{3}$ $\qquad\qquad\quad = 11$ cm/year
4. Write the answer.	The child has grown at a rate of 11 cm per year or 11 cm/year.

7.2.2 Common rates

Some common rates that don't involve time include gradient, concentration and density.

Gradient is a measure of steepness. This can be calculated by:

$$\textbf{Gradient} = \frac{\textbf{rise}}{\textbf{run}}$$

Concentration is a measure of how much material is dissolved in a liquid. This can be calculated by:

$$\textbf{Concentration} = \frac{\textbf{amount of solution}}{\textbf{amount of solvent}}$$

Density is a measure of how much mass the material has for a certain volume. This can be calculated by:

$$\textbf{Density} = \frac{\textbf{mass}}{\textbf{volume}}$$

WORKED EXAMPLE 3 Calculating common rates

A material has a mass of 4.5 kilograms and a volume of 0.75 cubic metres. Calculate the density of the material in grams per cubic metre.

THINK	WRITE
1. Density is the mass (g) divided by the volume (m^3).	Density $= \dfrac{\text{mass (g)}}{\text{volume } (m^3)}$
2. Convert the mass to grams by multiplying by 1000.	$4.5 \text{ kg} = 4.5 \times 1000$ g $\qquad\quad\; = 4500$ g
3. Write the rate using the correct units.	Density $= \dfrac{4500 \text{ g}}{0.75 \text{ m}^3}$
4. Write the answer.	$= 6000 \text{ g/m}^3$

7.2 Exercise

1. **WE1** A cyclist travels 120 km in 3 hours. Calculate the rate at which the cyclist is travelling using units of:

   a. kilometres per hour   **b.** metres per second.

2. An athlete's heart beats 250 beats in 5 minutes. Calculate the athlete's heart rate in beats per minute.

3. Rewrite the fractions below as rates.

   a. $\dfrac{\text{mass}}{\text{volume}}$

   b. $\dfrac{\text{distance}}{\text{time}}$

   c. $\dfrac{\text{rise}}{\text{run}}$

   d. $\dfrac{\text{amount of solution}}{\text{amount of solvent}}$

4. Express each of the following as a rate using the units given.

   a. A truck travelling 560 km in 7 hours (km/h)
   b. A 5.4-m length of timber costing $21.06 ($/m)
   c. A 9-litre bucket taking 45 seconds to fill (L/s)
   d. A hiker walking 9.6 km in 70 minutes (km/min)

5. A student walks 1.6 km to school in 22 minutes. Calculate the speed at which she walks to school in m/s.

6. In the 2000 Sydney Olympics, Cathy Freeman won gold in the 400 m race. Her time was 49.11 seconds.
   Calculate her average speed in km/h.

7. The Bathurst 1000 is a 1000-km car race. In 2010, it was won by Craig Lowndes and Mark Skaife in 6 hours, 12 minutes and 51.4153 seconds.
   Calculate their average speed in km/h.

8. **WE2** Noah has grown from 1.12 m at 5 years old to 1.41 m at 7 years old. Calculate Noah's growth rate in centimetres per year.

9. Calculate the growth rate in centimetres per year of a child who has grown 0.32 m in the past 4 years.

10. A child's height at 5 years of age is 85 cm. On reaching 12 years of age, the child's height is 128 cm. Calculate the average rate of growth (in cm/year) over the 7 years.

11. A school had 300 students in 2020 and 450 students in 2022. Calculate the average rate of growth in students per year.

12. **WE3** A material has a mass of 12.5 kilograms and a volume of 0.25 cubic metres. Calculate the density of the material in grams per cubic metres.

13. A 2-litre drink container was filled with a mix of 200 millilitres of cordial and 1.8 litres of water. Calculate the concentration of the cordial mix in millilitres per litre.

14. Three sugar cubes, each with a mass of 1 gram, are placed into a 300-millilitre mug of coffee. Calculate the concentration of sugar in grams per millilitre (g/mL).

15. Beaches are sometimes unfit for swimming if heavy rain washes pollution into the water. A beach is declared unsafe for swimming if the concentration of bacteria is more than 5000 organisms per litre. A sample of 20 millilitres was tested and found to contain 55 organisms.
Calculate the concentration in the sample (in organisms/litre) and state whether or not the beach should be closed.

16. A kettle is filled with tap water at 16 °C and then turned on. It boils (at 100 °C) after 2.75 minutes. Calculate the average rate of increase in temperature in °C/s.

17. Simon weighs 88 kg and has a volume of 0.11 m^3, whereas Ollie weighs 82 kg and has a volume of 0.08 m^3.
    a. Calculate and simplify Simon's density.
    b. Calculate and simplify Ollie's density.
    c. Determine who is more dense and by how much.

18. Uluru is in the middle of Australia and is 348 m high and 3.6 km long. Calculate the average gradient (to 1 decimal place) from the bottom to the top of Uluru, assuming the top is in the middle.

7.3 Conversion of rates

LEARNING INTENTION

At the end of this subtopic you should be able to:
- identify and express a rate in simplest form
- convert a rate from one set of units to another
- express a rate in the most appropriate units.

7.3.1 Expressing a rate in simplest form

A rate is a measure of the change in one quantity with respect to another.

> ### Rate of change
>
> The Rate of change of Y with respect to $X = \dfrac{\text{change in Y}}{\text{change in X}}$

WORKED EXAMPLE 4 Expressing a rate in simplest from

A car travels 320 km in 4 hours. Express this rate in km/h.

THINK	WRITE
1. Write out the rate formula in terms of the two units.	$\text{rate} = \dfrac{\text{change in Y}}{\text{change in X}}$ $= \dfrac{\text{change in km}}{\text{change in hours}}$
2. Substitute the known quantities into the formula.	$\text{Rate} = \dfrac{320 \text{ km}}{4 \text{ h}}$
3. Simplify the rate.	$= 80 \text{ km/h}$

7.3.2 Equivalent rates

Rates can be converted from one unit to another; for example, speed in km/h can be expressed in m/s.

To change the unit of a rate, follow these steps.
1. Convert the numerator of the rate to the new unit.
2. Convert the denominator of the rate to the new unit.
3. Divide the new numerator by the new denominator.

Converting units

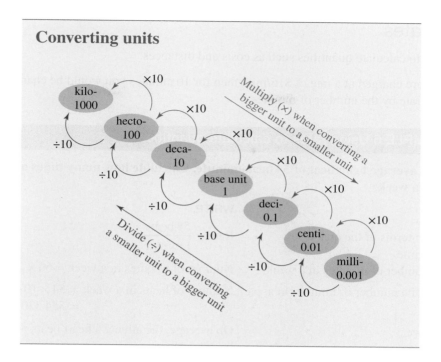

tlvd-3776

WORKED EXAMPLE 5 Converting rates

Covert the following rates, as shown, to 2 decimal places.
a. 60 km/h to m/s
b. 25 L/h to mL/min

THINK	WRITE
a. 1. Convert the numerator from km to m by multiplying by 1000.	**a.** $60 \text{ km} = 60 \times 1000$ $= 60\,000 \text{ m}$
2. Convert the denominator from hours to seconds by multiplying first by 60 to change to minutes, and then by 60 again to change to seconds.	$1 \text{ h} = 1 \times 60 \times 60$ $= 3600 \text{ sec}$
3. Divide the numerator by the denominator.	$\text{Rate} = \dfrac{60\,000}{3600}$ $= 16.6667 \text{ m/s}$
4. Write the new rate to 2 decimal places.	$60 \text{ km/h} = 16.67 \text{ m/s}$
b. 1. Convert the numerator from L to mL by multiplying by 1000.	**b.** $25 \text{ L} = 25 \times 1000$ $= 25\,000 \text{ mL}$
2. Convert the denominator from hours to minutes by multiplying by 60.	$1 \text{ h} = 1 \times 60$ $= 60 \text{ min}$
3. Divide the numerator by the denominator.	$\text{Rate} = \dfrac{25\,000}{60}$ $= 416.6667 \text{ mL/min}$
4. Write the new rate to 2 decimal places.	$25 \text{ L/h} = 416.67 \text{ mL/min}$

7.3.3 Using rates

Rates are often used to calculate quantities such as costs and distances.

For example, if you are charged at a rate of $15/min, then for 10 minutes you would be charged $150, since you need to multiply the rate by the number of minutes.

tlvd-3777

WORKED EXAMPLE 6 Using rates to calculate quantities

An athlete has an average heartbeat of 54 beats/minute. Calculate how many times on average their heart will beat in a week.

THINK	WRITE
1. Write the rate in terms of the quantities: heartbeats and time.	$\text{Rate} = \dfrac{54 \text{ beats}}{1 \text{ minute}}$
2. Calculate the number of minutes in a week.	Number of minutes in a week $= 60 \times 24 \times 7 = 10\,080$
3. Multiply 54 by the number of minutes in a week.	Number of beats in a week $= 54 \times 10\,080$ $= 544\,320$
4. Write your answer.	On average, the athlete's heart beats 544 320 beats/week.

7.3 Exercise

Students, these questions are even better in jacPLUS

 Receive immediate feedback and access sample responses

 Access additional questions

 Track your results and progress

Find all this and MORE in jacPLUS

1. **WE4** A cyclist travels 102 km in 3 hours. Express this rate in km/h.

2. A person purchases a 2.5-kg bag of chocolates for $37.50. Express this rate in $/kg.

3. Given the following information, express the rate in the units shown in the brackets.
   a. A car travels 240 km in 4 hours (km/h).
   b. A hiker covers 1400 m in 7 minutes (m/min).
   c. 9 litres of water runs out of a tap in 30 seconds (L/s).
   d. Grass has grown 4 mm in 2 days (mm/day).

4. Given the following information, express the rate in the units shown in the brackets.
   a. A runner travels 800 m in 4 minutes (km/min).
   b. Water runs into a water tank at 800 mL in 2 minutes (L/min).
   c. A phone call costs 64 cents for 2 minutes ($/min).
   d. Apples cost $13.50 for 3 kg (cents/gram).

5. **WE5** Convert the following rates as shown to 2 decimal places.
   a. 100 km/h to m/s
   b. 40 L/h to mL/min

6. Convert the following rates as shown to 2 decimal places.

   a. $5.50/m to cents/cm

   b. 10 m/s to km/h

7. Convert the following rates as shown.

   a. 70 km/h to m/h

   b. 40 kg/min to g/min

   c. $120/min to $/h

   d. 27 g/mL to g/L

8. Convert the following rates as shown, giving your answer to 2 decimal places.

   a. 80 km/h to m/s

   b. 25 m/s to km/h

   c. 50 g/ml to kg/L

   d. 360 mL/min to L/h

9. **WE6** If a person has an average heartbeat of 72 beats/minute, calculate how many times their heart would beat on average in a week.

10. If a phone call to the USA costs 10c/min, calculate how much a 1.5-hour call to the USA would cost.

11. If an electricity company charges 26.5 cents/kWh, calculate how much they would charge you if you used 750 kWh.

12. An athlete has an average heart rate of 63 beats/minute. Calculate how many times their heart would beat on average in one year (assume 365 days in a year).

13. Christi wants to hire a jukebox for her party and has received a quote of $45 per hour. If Christi has a party that goes from 8.30 pm to 1 am, calculate how much she will have to pay for the jukebox.

14. On one trip, a driver took an average of 45 minutes to travel 60 km.

   a. Calculate their average speed in km/h.

   b. If the driver travelled from 11.30 am to 3.45 pm continuously, calculate how far they have travelled.

15. A gum tree grows at a rate of 80 cm/year. Calculate how long it will take to grow from the ground to a height of 10 m.

16. Tiger cubs put on approximately 100 g of weight per day. If Simba the tiger cub weighed 950 g at birth, calculate how long it would take him to weigh 2 kg. Give your answer in days and hours.

7.4 Comparison of rates

7.4.1 Comparing rates

Rates are commonly used to make comparisons. For example, useful comparisons for consumers could be comparing the price per 100 grams of different sizes of packaged cheese in the supermarket, or the fuel economy of different models of cars in litres per 100 kilometres.

When you are comparing two rates, it is important to have both rates in the same unit.

Price per unit

$$\text{Price per unit} = \frac{\text{price}}{\text{number of units}}$$

Remember to represent the same units when you are making comparisons.

tlvd-3778

WORKED EXAMPLE 7 Comparing rates

Johan wants to purchase a new car and fuel economy is important when considering which car to buy. The first car he looked at used 34.56 litres for 320 kilometres and the second car used 29.97 litres for 270 kilometres. Calculate the fuel consumption rate for each car in L/100 km and determine which car has the better fuel economy.

THINK	WRITE
1. The first car used 34.56 L/320 km.	First car rate = 34.56 L/320 km
2. Convert the rate to L/100 km by dividing the numerator and denominator by 3.2.	First car rate $= \dfrac{34.56 \div 3.2}{320 \div 3.2}$ $= \dfrac{10.8}{100}$ $= 10.8 \text{ L}/100 \text{ km}$
3. The second car used 29.97 L/270 km.	Second car rate = 29.97 L/270 km

4. Convert the rate to L/100 km by dividing the numerator and denominator by 2.7.

$$\text{Second car rate} = \frac{29.97 \div 2.7}{270 \div 2.7}$$

$$= \frac{11.1}{100}$$

$$= 11.1 \text{ L/100 km}$$

5. Now that the rates have the same units, a direct comparison can be made.

The first car has better fuel economy than the second car.

7.4 Exercise

1. Two cars used fuel as shown.
   - Car A used 52.89 litres to travel 430 kilometres.
   - Car B used 40.2 litres to travel 335 kilometres.

   Determine which car has the better fuel economy by comparing the rates of L/100 km.

2. **WE7** Car A uses 43 L of petrol travelling 525 km. Car B uses 37 L of petrol travelling 420 km. Show mathematically which car is more economical.

3. Josh wants to be an Ironman, so he eats Ironman food for breakfast. Nutri-Grain comes in a 290-g pack for $4.87 or an 805-g pack for $8.67.
   Compare the price per 100 g of cereal for each pack, and hence calculate how much he will save per 100 g if he purchases the 805-g pack.

4. A solution has a concentration of 45 g/L. The amount of solvent is doubled to dilute the solution. Calculate the concentration of the diluted solution. Explain your reasoning.

5. Tea bags in a supermarket can be bought for $1.45 per pack of 10 or for $3.85 per pack of 25. Determine which of the two is the cheaper way of buying tea bags.

6. Coffee can be bought in 250-g jars for for $9.50 or in 100-g jars for $4.10. Determine which is the cheaper way of buying the coffee and how much the saving would be.

7. In cricket, a batter's strike rate indicates how fast the batter is scoring their runs by comparing the number of runs scored per 100 balls.

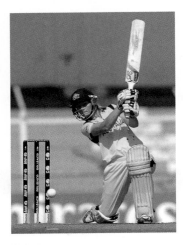

Ellyse Perry scored 72 runs in 56 balls and Meg Lanning scored 42 runs in 31 balls.
Correct to 2 decimal places:

a. calculate the strike rate of Ellyse Perry
b. calculate the strike rate of Meg Lanning
c. determine who has the better strike rate and by how much.

8. Eating a balanced diet is important for your overall health. One area that should be looked at is minimising the amount of total fats you eat. From the two labels below, determine which product is the healthier according to the amount of total fats.

Product A	Product B

Product A

Nutrition Facts
Serving Size 1 Cake (43g)
Servings Per Container 5

Amount Per Serving

Calories 200 Calories from Fat 90

	% Daily Value*
Total Fat 10g	15%
Saturated Fat 5g	25%
Trans Fat 0g	
Cholesterol 0mg	0%
Sodium 100mg	4%
Total Carbohydrate 26g	9%
Dietary Fiber 0g	0%
Sugars 19g	
Protein 1g	

Vitamin A 0%	•	Vitamin C 0%
Calcium 0%	•	Iron 2%

* Percent Daily Values are based on a 2,000 calorie diet. Your daily values may be higher or lower depending on your calorie needs:

	Calories:	2,000	2,500
Total Fat	Less than	65g	80g
Sat. Fat	Less than	20g	25g
Cholesterol	Less than	300mg	300mg
Sodium	Less than	2,400mg	2,400mg
Total Carbohydrate		300g	375g
Dietary Fiber		25g	30g

Product B

Nutrition Facts
Serving Size 1/4 Cup (30g)
Servings Per Container About 38

Amount Per Serving

Calories 200 Calories from Fat 150

	% Daily Value*
Total Fat 17g	26%
Saturated Fat 2.5g	13%
Trans Fat 0g	
Cholesterol 0mg	0%
Sodium 120mg	5%
Total Carbohydrate 7g	2%
Dietary Fiber 2g	8%
Sugars 1g	
Protein 5g	

Vitamin A 0%	•	Vitamin C 0%
Calcium 4%	•	Iron 8%

*Percent Daily Values are based on a 2,000 calorie diet.

9. Two athletes had their heart rates tested as shown.
   • Athlete A's heart beats 255 beats in 5 minutes.
   • Athlete B's heart beats 432 beats in 9 minutes.
      a. Calculate Athlete A's heart rate in beats/minute.
      b. Calculate Athlete B's heart rate in beats/minute.
      c. Determine who has the lower heart rate and by how many beats/minute.
      d. Determine how many times each athlete's heart would beat in one day.

10. a. The Earth is a sphere with a mass of 6.0×10^{24} kg and a radius of 6.4×10^6 m.

      i. Use the formula $V = \frac{4}{3}\pi r^3$ to calculate the volume of the Earth.

      ii. Hence calculate the density of the Earth.

    b. Jupiter has a mass of 1.9×10^{27} kg and a radius of 7.2×10^7 m.

      i. Calculate the volume of Jupiter.
      ii. Calculate the density of Jupiter.

    c. Different substances have their own individual density. State whether your results suggest that the Earth and Jupiter are made of the same substance. Explain.

11. Supermarkets use rates to make it easier for the consumer to compare prices of different brands and different-sized packages. To make the comparison, they give you the price of each product per 100 grams, known as unit pricing. Two different brands have the following pricing:
    • Brand A comes in a 2.5-kg pack for $3.10.
    • Brand B comes in a 1-kg pack for $1.05.
    Calculate the unit price (the price per 100 g) of each of the brands and determine which brand is best value.

12. The Stawell Gift is a 120-m handicap footrace. Runners who start from scratch run the full 120 m; for other runners, their handicap is how far in front of scratch they start.
    Joshua Ross has won the race twice. In 2003, with a handicap of 7 m, his time was 11.92 seconds. In 2005, from scratch, he won in 12.36 seconds.
    Determine in which race he ran faster.

7.5 Rates and costs

LEARNING INTENTION

At the end of this subtopic you should be able to:
   • compare quotes from tradespeople
   • calculate and compare the cost of living: food, transport and clothing.

7.5.1 Costs for trades

As consumers, we all need to be able to compare costs. Using rates helps us compare costs because we are comparing the same quantities. If we need work done around the home, we often ask for two or three quotes from different tradespeople to compare costs.

Most tradespeople charge a call-out fee and then a rate per hour. These costs may vary from one tradesperson to another.

Some tradespeople have fixed rates for certain jobs. An electrician, for example, may charge a $60 call-out fee (fixed cost) and then $75 per hour. If the job requires 3 hours to be completed, the final cost will be:

$$\begin{aligned} \text{Cost} &= 60 + 75 \times 3 \\ &= 60 + 225 \\ &= \$285 \end{aligned}$$

Calculating cost for trades

Cost = fixed fee + charge per hour × number of hours

WORKED EXAMPLE 8 Calculating trade rates

tlvd-3779

Adam and Jess are both electricians. Adam charges a $70 call-out fee and $80 per hour, while Jess charges a $50 call-out fee and $85 per hour.
Determine which electrician will have the cheaper quote for a:
a. 2-hour job
b. 5-hour job.

THINK	WRITE
a. 1. Write the formula.	**a.** Cost = fixed cost + charge per hour × hours
2. Substitute the known values for each electrician and simplify.	Adam's cost = $70 + 80 \times 2$ $= \$230$ Jess's cost = $50 + 85 \times 2$ $= \$220$
3. Compare the two costs.	Adam charges $10 more for this two-hour job than Jess does, so Jess is cheaper.
b. 1. Write the formula.	**b.** Cost = fixed cost + charge per hour × hours
2. Substitute the known values for each electrician and simplify.	Adam's cost = $70 + 80 \times 5$ $= \$470$ Jess's cost = $50 + 85 \times 5$ $= \$475$
3. Compare the two costs and write the answer.	Jess charges $5 more for this five-hour job than Adam does, so Adam is cheaper.

7.5.2 Living costs: food

To calculate how much we spend per week on food, we simply add up all the money we spend on food products over a week. However, food products like sugar and cooking oil are not all going to be consumed in that week. The money spent per week on these products could be estimated.

If we use 1 litre of cooking oil that costs $6 over a period of 4 weeks, then the price estimate is $1.50 per week. This is regardless of whether we use more cooking oil in one week than another.

Note: Throughout this section we will use an average of 30 days per month.

WORKED EXAMPLE 9 Calculating cost-of-living rates

Georgia drinks 1 L of milk per day, eats one loaf of bread per week and uses one 500-g container of butter per month. Calculate Georgia's daily, weekly and monthly spending for the three types of food if 1 L of milk costs $1, one loaf of bread costs $4.20 and a 500-g container of butter costs $3.15.

THINK

1. Draw a table with foods listed in the first vertical column and time periods listed on the first horizontal row.

WRITE/DRAW

		Time period	
		Day ($)	Week ($)	Month ($)
Type of food	Milk			
	Bread			
	Butter			
	Total			

2. Fill in the table with the information given.

We know that Georgia drinks 1 L of milk per day at a cost of $1 per litre. This means that the milk costs her $1 per day.

We also know that she eats a loaf of bread per week. This means that she spends $4.20 per week on bread. She also spends $3.15 on butter per month.

		Time period	
		Day ($)	Week ($)	Month ($)
Type of food	Milk	1		
	Bread		4.20	
	Butter			3.15
	Total			

3. Calculate the missing entries.

Price of milk per week = $1 × 7 days
= $7
Price of milk per month = $1 × 30 days
= $30
Price of bread per day = $4.20 ÷ 7 days
= $0.60
Price of bread per month = $0.60 × 30 days
= $18
Price of butter per day = $3.15 ÷ 30 days
= $0.105
Price of butter per week = $0.105 7 days
= $0.735

4. Fill in the rest of the table.

		Time period		
		Day ($)	**Week ($)**	**Month ($)**
Type of food	**Milk**	1	7	30
	Bread	0.60	4.20	18
	Butter	0.105	0.735	3.15
	Total	1.705	11.935	51.15

5. Write the answer.

Georgia spends $1.71 on food per day, $11.94 on food per week and $51.15 on food per month.

7.5.3 Living costs: transport

Means of transport to and from school are personal cars, bicycles or public transport. There are people who go to work or school using one or more forms of transport.

The costs for transport when using a personal car include petrol, maintenance and insurance.

Note: Throughout this section we will use an average of 30 days per month.

tlvd-3780

WORKED EXAMPLE 10 Calculating transport costs

Tim uses his personal car as a means of transport. He spends $720 per year on car insurance, $64.40 per week on petrol and $42 per month on maintenance. Tabulate this data and calculate Tim's daily, weekly, monthly and yearly spending for running the car.

THINK

1. Construct a table with the types of service listed in the first vertical column and time periods listed on the first horizontal row.

WRITE/DRAW

		Time period			
		Day ($)	**Week ($)**	**Month ($)**	**Year ($)**
Type of service	**Petrol**				
	Insurance				
	Maintenance				
	Total				

2. Fill in the table with the information given.

We know that Tim spends $720 per year on insurance, $64.40 on petrol per week and $42 on maintenance per month.

		Time period			
		Day ($)	**Week ($)**	**Month ($)**	**Year ($)**
Type of service	**Petrol**		64.40		
	Insurance				720
	Maintenance			42	
	Total				

3. Calculate the missing entries.

Price of petrol per day $= \$64.40 \div 7$ days
$$= \$9.20$$
Price of petrol per month $= \$9.20 \times 30$ days
$$= \$276$$
Price of petrol per year $= \$276 \times 12$ months
$$= \$3312$$
Price of insurance per month $= \$720 \div 12$ months
$$= \$60$$
Price of insurance per day $= \$60 \div 30$ days
$$= \$2$$
Price of insurance per week $= \$2 \times 7$ days
$$= \$14$$
Price of maintenance per day $= \$42 \div 30$ days
$$= \$1.40$$
Price of maintenance per week $= \$1.4 \times 7$ days
$$= \$9.80$$
Price of maintenance per year $= \$42 \times 12$ months
$$= \$504$$

4. Fill in the rest of the table.

<table>
<tr><th rowspan="2"></th><th rowspan="2"></th><th colspan="4">Time period</th></tr>
<tr><th>Day ($)</th><th>Week ($)</th><th>Month ($)</th><th>Year ($)</th></tr>
<tr><th rowspan="4">Type of service</th><th>Petrol</th><td>9.20</td><td>64.40</td><td>276</td><td>3312</td></tr>
<tr><th>Insurance</th><td>2</td><td>14</td><td>60</td><td>720</td></tr>
<tr><th>Maintenance</th><td>1.40</td><td>9.80</td><td>42</td><td>504</td></tr>
<tr><th>Total</th><td>12.60</td><td>88.20</td><td>378</td><td>4536</td></tr>
</table>

5. Write the answer as a sentence. Tim's car costs him $12.60 per day, $88.20 per week, $378 per month and $4536 per year.

7.5.4 Living costs: clothing

There are people who like to buy clothes on a regular basis to keep up with trends. Other people are not concerned about trends and buy only what they need.

How much do you spend on clothing?

Vince is creating a budget for clothing. He made four categories: small items, trousers, shirts and jumpers. He added up the money he spent on these items and recorded the amounts in the two-way table shown. Complete the table.

		Time period		
		Day ($)	Week ($)	Month ($)
Category	Small items		12.60	
	Trousers	5.00		
	Shirts		35.70	
	Jumpers			93
	Total			

THINK

1. Calculate the missing entries.

WRITE/DRAW

Cost of small items per day = $12.60 ÷ 7 days
= $1.80

Cost of small items per month = $1.80 × 30 days
= $54

Cost of trousers per week = $5 × 7 days
= $35

Cost of trousers per month = $5 × 30 days
= $150

Cost of shirts per day = $35.70 ÷ 7 days
= $5.10

Cost of shirts per month = $5.10 × 30 days
= $153

Cost of jumpers per day = $93 ÷ 30 days
= $3.10

Cost of jumpers per week = $3.10 × 7 days
= $21.70

2. Complete the table.

		Time period		
		Day ($)	Week ($)	Month ($)
Category	Small items	1.80	12.60	54
	Trousers	5	35	150
	Shirts	5.10	35.70	153
	Jumpers	3.10	21.70	93
	Total	15	105	450

7.5.5 Living costs on spreadsheets

Excel spreadsheets are a practical and useful tool when calculating budgets. For the purpose of this exercise we are going to consider the same situation as in Worked example 10.

WORKED EXAMPLE 12 Calculating living costs using spreadsheets

Tim uses his personal car as a means of transport. He spends $720 per year on car insurance, $64.40 per week on petrol and $42 per month on maintenance. Use a spreadsheet to tabulate this data and calculate Tim's daily, weekly, monthly and yearly spending for running the car.

THINK

WRITE/DRAW

1. Construct a table in a spreadsheet with the type of service in the first vertical column, column A, and time periods on the horizontal row, row 2.

	A	B	C	D	E
1		Time period			
2	Type of service	Day	Week	Month	Year
3	Petrol		4.30	30.10	129.00
4	Maintenance		2.60	18.20	78.00
5	Insurance		0.30	2.10	9.00
6			7.20	50.40	216.00

2. Fill in the table with the information given.

	A	B	C	D	E
1		Time period			
2	Type of service	Day	Week	Month	Year
3	Petrol		64.4		
4	Maintenance			42	
5	Insurance				720
6					

3. Set up the calculations required. Remember to start each formula with the = sign. Click in cell B3, type =C3/7 and press ENTER.

	A	B	C	D	E
1		Time period			
2	Type of service	Day	Week	Month	Year
3	Petrol	=C3/7	64.4		
4	Maintenance			42	
5	Insurance				720
6					

4. Click in cell D3, type =B3*30 and press ENTER.

	A	B	C	D	E
1		Time period			
2	Type of service	Day	Week	Month	Year
3	Petrol	9.2	64.4	=B3*30	
4	Maintenance			42	
5	Insurance				720
6					

5. Click in cell E3, type =D3*12 and press ENTER.

	A	B	C	D	E
1		Time period			
2	Type of service	Day	Week	Month	Year
3	Petrol	9.2	64.4	276	=D3*12
4	Maintenance			42	
5	Insurance				720
6					

6. In the bottom right corner of cell E3 you will notice a small circle. Drag it down with your mouse to copy the formula to the cell below.

	A	B	C	D	E
1		Time period			
2	Type of service	Day	Week	Month	Year
3	Petrol	9.2	64.4	276	3312
4	Maintenance			42	504
5	Insurance				720
6					

7. Click in cell B4, type =D4/30 and press ENTER.

	A	B	C	D	E
1		Time period			
2	Type of service	Day	Week	Month	Year
3	Petrol	9.2	64.4	276	3312
4	Maintenance	=D4/30		42	504
5	Insurance				720
6					

8. Drag down the small circle in the bottom right corner of cell B4 to copy the formula to the cell below.

	A	B	C	D	E
1		Time period			
2	Type of service	Day	Week	Month	Year
3	Petrol	9.2	64.4	276	3312
4	Maintenance	1.4		42	504
5	Insurance	0			720
6					

9. You will notice a '0' in cell B5. This is because at the moment there is no value in cell D5.
Click in cell D5, type =E5/12 and press ENTER.

	A	B	C	D	E
1		Time period			
2	Type of service	Day	Week	Month	Year
3	Petrol	9.2	64.4	276	3312
4	Maintenance	1.4		42	504
5	Insurance	0		=E5/12	720

10. Click in cell C4, type =B4*7 and press ENTER.

	A	B	C	D	E
1			Time period		
2	Type of service	Day	Week	Month	Year
3	Petrol	9.2	64.4	276	3312
4	Maintenance	1.4	=B4*7	42	504
5	Insurance	2		60	720
6					

11. Drag down the small circle in the bottom right corner of cell C4 to copy the formula to the cell below.

	A	B	C	D	E
1			Time period		
2	Type of service	Day	Week	Month	Year
3	Petrol	9.2	64.4	276	3312
4	Maintenance	1.4	9.8	42	504
5	Insurance	2	14	60	720
6					

12. Now, we are going to add a Total row by typing 'Total' in cell A6. Click in cell B6, type =SUM (B3:B5) and press ENTER. This command will add up the values in cells B3, B4 and B5.

	A	B	C	D	E
1			Time period		
2	Type of service	Day	Week	Month	Year
3	Petrol	9.2	64.4	276	3312
4	Maintenance	1.4	9.8	42	504
5	Insurance	2	14	60	720
6	Total	12.6			

13. This time drag the small circle in the bottom right corner of cell B6 to the right to copy the formula in cells C6, D6 and E6.

	A	B	C	D	E
1			Time period		
2	Type of service	Day	Week	Month	Year
3	Petrol	9.2	64.4	276	3312
4	Maintenance	1.4	9.8	42	504
5	Insurance	2	14	60	720
6	Total	12.6	88.2	378	4536

7.5 Exercise

Students, these questions are even better in jacPLUS

 Receive immediate feedback and access sample responses

 Access additional questions

 Track your results and progress

Find all this and MORE in jacPLUS ▶

1. Compare the two quotes given by two computer technicians, Joanna and Dimitri, for a four-hour job. Joanna charges a $65 fixed fee and $80 per hour. Dimitri charges a $20 fixed fee and $100 per hour.

2. Hamish and Hannah are both tilers. Hamish charges $200 for surface preparation and $100 per square metre, while Hannah charges $170 for surface preparation and $120 per square metre. Compare their charges for a job that covers an area of $10\,m^2$.

3. Calculate the following costs for a tradesperson.
   a. A fixed fee of $35 and $50 per hour for 3 hours
   b. A fixed fee of $45 and $65 per hour for 2 hours
   c. A fixed fee of $20 and $30 per half an hour for 1.5 hours
   d. A fixed fee of $80 and $29.50 per 15 minutes for 30 minutes
   e. A fixed fee of $70 and $37.50 per half an hour for 2.5 hours
   f. A fixed fee of $100 and $50.25 per 15 minutes for 45 minutes

4. **WE8** Lynne needs a plumber to fix the drainage of her house. Lynne obtained three different quotes:
   Quote 1: Fixed fee of $80 and $100 per hour
   Quote 2: Fixed fee of $25 and $120 per hour
   Quote 3: Fixed fee of $70 and $110 per hour

   a. Determine which quote is the best if the job requires 2 hours.
   b. Determine which quote is the best if the job requires 7 hours.

5. a. Calculate the fuel consumption (in $L/100\,km$) for:
      i. a sedan using 35 L of petrol for a trip of 252 km
      ii. a wagon using 62 L of petrol for a trip of 428 km
      iii. a 4WD using 41 L of petrol for a trip of 259 km.

   b. Determine which of the three cars is the most economical.

6. A lawyer charges a $250 application fee and $150 per hour. Calculate how much money you would spend for a case that requires 10 hours of the lawyer's time.

7. Calculate and compare the fuel consumption rates for a sedan travelling in city conditions for 37 km on 5 L and a 4WD travelling on the highway for 216 km on 29 L.

8. **WE9** Ant spends $129 a month on fruit, $18.20 a week on vegetables and $0.30 per day on eggs.
   a. Calculate Ant's daily, weekly and monthly spending for the three types of food and construct a table to display these costs.
   b. **WE12** Construct an Excel spreadsheet for part a.

9. A family spend $2.80 a week on potatoes, $45 a month on tomatoes and $1.60 per day on cheese. Calculate their daily, weekly and monthly spending for the three types of food and construct a table to display these costs.

10. **WE10** Halina spends $4.10 per day on petrol, $18.20 a week on insurance and $39 per month on maintenance for her car. Calculate Halina's daily, weekly, monthly and yearly spending for the three types of services and construct a table to display these costs.

11. Andy spends $17.50 per week on petrol, $72 a month on insurance and $612 per year on maintenance for his car. Calculate Andy's daily, weekly, monthly and yearly spending for the three types of services and construct a table to display these costs.

12. Elaine spends $15 a day on food, Tatiana spends $126 a week on food and Anya spends $420 per month on food.
    a. Calculate how much money Anya spends on food per day.
    b. Calculate how much money Tatiana spends on food per day.
    c. Determine who, out of the three friends, spends the most on food per month. Explain your answer.

13. **WE11** Carol has noticed that she spends too much money buying tops and wants to cut down on her spending. Carol separated the tops into three categories: singlets, casual and formal. She then added up the money spent on these items and recorded them in the table shown. Fill in the rest of the table.

		Time period		
		Day ($)	Week ($)	Month ($)
Category	Singlets	3.42		
	Casual			198.30
	Formal		52.50	
	Total			

14. Nadhea made a budget for clothing. She made four categories: small items, bottoms (skirts and trousers), dresses and tops. She then added up the money spent on these items and recorded the amounts in the table shown.

		Time period		
		Day ($)	Week ($)	Month ($)
Category	Small items	2.65		
	Bottoms			325.50
	Dresses		22.40	
	Tops	6.20		
	Total			

a. Fill in the rest of the table.
b. Construct an Excel spreadsheet for part a.

15. Paris uses both her personal car and public transport. Paris spends $630 per year on car insurance, $9.10 per week on petrol and $21 per month on maintenance. She also spends $5.80 per day on public transport.
Tabulate this data and calculate Paris' daily, weekly, monthly and yearly spending for running the car and catching public transport. (Assume Paris catches public transport every day of the year.)

16. If the total cost for running a car for 12 days is $27, determine this cost per day, per week, per month and per year.

17. Using a calculator, a spreadsheet or otherwise, calculate the weekly and monthly spending on food, clothes and transport for the following daily costs:

a. $10 on food, $6.70 on clothes and $5.20 on transport
b. $12.15 on food, $9.65 on clothes and $10.80 on transport
c. $5.21 on food, $7.26 on clothes and $2.95 on transport
d. $8.93 on food, $12.54 on clothes and $3.72 on transport

18. Using a calculator, a spreadsheet or otherwise, calculate the daily and monthly spending on food, clothes and transport for the following yearly costs:

a. $3720 on food, $2700 on clothes and $2100 on transport
b. $5610 on food, $3200 on clothes and $4016 on transport
c. $7200 on food, $3410 on clothes and $1800 on transport
d. $6960 on food, $4230 on clothes and $3500 on transport

7.6 Review

7.6.1 Summary

doc-38048

7.6 Exercise

Multiple choice

1. **MC** If a runner travels 3 km in 10 minutes and 20 seconds, the runner's average speed in m/s is:
   A. 5 m/s   B. 5.24 m/s   C. 4.84 m/s   D. 4.64 m/s   E. 4 m/s

2. **MC** If an escalator rises 3.75 m for a horizontal run of 5.25 m, then the escalator's gradient is:
   A. 1.4   B. 1.21   C. 0.91   D. 0.71   E. 1.61

3. **MC** The speed limit around schools is 40 km/h. This rate when converted to m/s is:
   A. 144 m/s   B. 120 m/s   C. 10.1 m/s   D. 12.5 m/s   E. 11.1 m/s

4. **MC** On average, Noah kicks 2.25 goals per game of soccer. In a 16-game season the number of goals Noah would kick is:
   A. 32   B. 36   C. 30
   D. 38   E. 16

5. **MC** If car A travels at 60 km/h and car B travels at 16 m/s, select the correct statement from the following.
   A. They travel at the same speed.
   B. Car A travels 0.67 m/s slower than car A.
   C. Car A travels 44 km/h faster than car B.
   D. Car B travels 0.67 m/s slower than car A.
   E. Car B travels 0.44 m/s slower than car A.

6. **MC** At the gym Lisa and Simon were doing a skipping exercise. Lisa completed 130 skips in one minute, whereas Simon skipped at a rate of 40 skips in 20 seconds. For 3 minutes of skipping, select the correct statement.
   A. Simon completed 20 more skips.
   B. Lisa completed 20 more skips.
   C. Lisa completed 30 more skips.
   D. Simon completed 30 more skips.
   E. They completed the same number of skips.

7. **MC** If the cost of food per week for a family is $231 and the cost of petrol per month is $159.60, the corresponding costs for food and petrol per day, respectively, are:
   A. $7.7 and $5.32.   B. $5.32 and $32.   C. $7.7 and $22.8.   D. $33 and $5.32.   E. $3.30 and $53.20.

8. **MC** A rate of 86.4 L/h is equivalent to:
   A. 24 mL/sec   B. 24 L/sec   C. 1.44 mL/min   D. 24 mL/min   E. $2.44 mL/min

9. **MC** The Shinkansen train in Japan travels at a speed of approximately 500 km/h. The distance from Broome to Melbourne is approximately 4950 km.

Calculate how long, to the nearest minute, it would take the Shinkansen train to cover this distance.

A. 1 hour 54 minutes

B. 0.1 hours

C. 10 hours

D. 9 hours 56 minutes and 48 seconds

E. 9 hours 54 minutes

10. **MC** A sedan, a wagon and a 4WD are travelling the same distance of 159 km. The sedan uses 20 L, the wagon uses 22 L and the 4WD uses 25 L of fuel for this trip. The fuel consumptions for the 4WD, wagon and sedan are respectively:

A. 13.8 L/km, 15.7 L/100km and 12.6 L/100 km.

B. 15.7 L/100km, 12.6 L/km and 13.8 L/100 km.

C. 15.7 L/100km, 13.8 L/km and 12.6 L/100 km.

D. 13.8 L/km, 15.7 L/100km and 12.5 L/100 km.

E. 1.8 L/km, 5.1 L/100km and 2.5 L/100 km.

Short answer

11. Convert the following rates to the rates shown in the brackets.

a. 17 m/s (km/h)

c. 0.0125c/sec ($/min)

b. 24.2 L/min (L/h)

d. 2.5 kg/mL (g/L)

12. A dragster covers a 400-m track in 8.86 sec.

a. Calculate its average speed in m/s.

b. If it could maintain this average speed for 1 km, determine how long it would take to cover the 1 km.

13. Oliver's hair grows at a rate of 125 mm/month, whereas Emilio's grows at 27.25 mm/week. If their hair is initially 5 cm long, determine how long their hair will be after 1 year (assume 52 weeks in a year).

14. Isabella worked 4 weeks at the rate of $693.75 per week before her wage was increased to $724.80 per week.

a. Calculate how much Isabella was paid for the 4 weeks.

b. Calculate how much Isabella will earn at the new rate for the next 4 weeks.

c. Calculate how much more money Isabella is receiving per week.

d. Determine the difference between Isabella's earnings for the first four weeks and the following four weeks.

Extended response

15. It takes me 2 hours to mow my lawn. My daughter takes 2.5 hours to mow the same lawn. If we work together using two lawnmowers, calculate how long it will take to mow the lawn.

Give your answer in hours, minutes and seconds.

16. To renovate your house, you may require a number of different tradespeople. Given the charges of each tradesperson and the hours required to complete their section of the work, calculate the total cost of the renovation.
    - Carpenter at $60 per hour, for 40 hours
    - Electrician at $80 per hour for 14.5 hours
    - Tiler at $55 per hour for 9 hours and 15 minutes
    - Plasterer at $23 per half-hour for 8 hours and 20 minutes
    - Plumber with a call-out charge of $85 and $90 per hour for 6.5 hours

17. Following the Australian tax rates shown, calculate the amount of tax to be paid on the following taxable incomes.

Taxable income	Tax on this income
0–$18 200	Nil
$18 201–$45 000	19c for each $1 over $18200
$45 001–$120 000	$5092 plus 32.5c for each $1 over $45000
$120 001–$180 000	$29467 plus 37c for each $1 over $120000
$180 001 and over	$51667 plus 45c for each $1 over $180000

    a. $21 000   b. $15 600   c. $50 000   d. $100 000   e. $210 000

18. Consider the nutrition information shown in the label.

NUTRITION INFORMATION		
SERVINGS PER PACKAGE: 3	SERVING SIZE: 150 g	
	QUANTITY PER SERVING	QUANTITY PER 100 g
Energy	608 KJ	405 KJ
Protein	4.2 g	2.8 g
Fat, total	7.4 g	4.9g
– Saturated	4.5 g	3.0 g
Carbohydrate	18.6 g	12.4 g
– Sugars	18.6 g	12.4 g
Sodium	90 mg	60 mg

    a. If you ate 400 g, determine how much energy that would include.
    b. If you ate 550 g, determine how much saturated fat you have consumed.
    c. If you had one serving, determine how much protein you have consumed.
    d. If you had three serves, determine how much sugar you have consumed.
    e. If you ate 350 g, determine what percentage of carbohydrates you have consumed.
    f. If you ate 250 g, determine what percentage of total fat you have consumed.

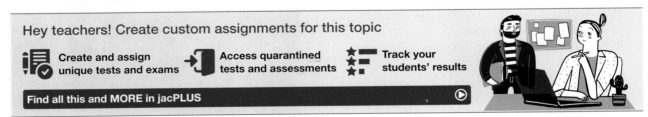

Hey teachers! Create custom assignments for this topic

Create and assign unique tests and exams   Access quarantined tests and assessments   Track your students' results

Find all this and MORE in jacPLUS

Answers

Topic 7 Rates

7.2 Identifying rates

7.2 Exercise

1. a. 40 km/h   b. 11.11 m/s
2. 50 beats/minute
3. a. $\dfrac{\text{mass}}{\text{volume}} = \text{density}$

   b. $\dfrac{\text{distance}}{\text{time}} = \text{speed}$

   c. $\dfrac{\text{rise}}{\text{run}} = \text{gradient}$

   d. $\dfrac{\text{amount of solution}}{\text{amount of solvent}} = \text{concentration}$
4. a. 80 km/h   b. \$3.90/m
   c. 0.2 L/s   d. 0.14 km/min
5. 1.21 m/s
6. 29.32 km/h
7. 160.92 km/h
8. 14.5 cm/year
9. 8 cm/year
10. 6.14 cm/year
11. 75 students/year
12. 50 000 g/m^3
13. 111.11 mL/L
14. 0.01 g/mL
15. 2750 organisms/L; the beach is safe for swimming.
16. 0.51 °C/s
17. a. 800 kg/m^3
    b. 1025 kg/m^3
    c. Ollie is 225 kg/m^3 denser than Simon.
18. 0.2

7.3 Conversion of rates

7.3 Exercise

1. 34 km/h
2. \$15/kg
3. a. 60 km/h   b. 200 m/min
   c. 0.3 L/s   d. 2 mm/day
4. a. 0.2 km/min   b. 0.4 L/min
   c. \$0.32/min   d. 0.45 cents/g
5. a. 27.78 m/s   b. 666.67 mL/min
6. a. 5.50 cents/cm   b. 36.00 km/h
7. a. 70 000 m/h   b. 40 000 g/min
   c. \$7200/h   d. 27 000 g/L
8. a. 22.22 m/s   b. 90.00 km/h
   c. 50.00 kg/L   d. 21.60 L/h
9. 725 760 beats
10. \$9.00
11. \$198.75
12. 33 112 800 beats

13. \$202.50
14. a. 80 km/h   b. 340 km
15. 12.5 years
16. 10 days and 12 hours

7.4 Comparison of rates

7.4 Exercise

1. Car A rate = 12.3 L/100 km
   Car B rate = 12 L/100 km
   Therefore car B has better fuel economy.
2. Car A = 8.19 L/100 km
   Car B = 8.81 L/100 km
   Therefore car A is more economical.
3. 290 g: rate = \$1.68/100 g
   805 g: rate = \$1.08/100 g
   805 g is cheaper by \$1.68 − \$1.08 = \$0.60 per 100 g.
4. 22.5 g/L
   Since the solvent is what the solution is dissolved in, doubling the solvent halves the concentration.
5. Pack of 10: \$0.145 each
   Pack of 25: \$0.154 each
   Technically speaking, the pack of 10 is cheaper, but when rounded off to the nearest cent, each tea bag has the same price.
6. 250 g: \$3.80/100 g
   100 g: \$4.10/100 g
   It is cheaper to buy the 250 gram jar by 30 cents per 100 grams.
7. a. 128.57 runs per 100 balls
   b. 135.48 runs per 100 balls
   c. Meg Lanning has the greater strike rate by 6.91 runs per 100 balls.
8. Product A has 15% fat and product B has 26% fat, so product A is healthier according to total fat.
9. a. 51 beats/min
   b. 48 beats/min
   c. Athlete B has the lower heart rate by 3 beats/minute.
   d. Athlete A: 73 440
      Athlete B: 69 120
10. a. i. 1.10×10^{21} m^3 ii. 5464.15 kg/m^3
    b. i. 1.56×10^{24} m^3 ii. 1215.26 kg/m^3
    c. No, because they have different densities.
11. Brand B is better because it is 10.5 cents per 100 grams compared to 12.4 cents per 100 grams for brand A.
12. Speed 2003: 9.48 m/s
    Speed 2005: 9.71 m/s
    Joshua Ross ran faster in winning the Stawell Gift in 2005.

7.5 Rates and costs

7.5 Exercise

1. Joanna = \$385; Dimitri = \$420
2. Hamish = \$1200; Hannah = \$1370
3. a. \$185   b. \$175   c. \$110
   d. \$139   e. \$257.50   f. \$250.75

4. a. Quote 2   **b.** Quote 1

5. a. i. 13.89 L/100 km   **ii.** 14.49 L/100 km
   iii. 15.83 L/100 km

  b. The sedan

6. $1750

7. Sedan — 13.51 L/100 km
  4WD — 13.43 L/100 km
The fuel consumption rate is better when travelling on a highway compared to city driving.

8. a.

		Time period		
		Day ($)	Week ($)	Month ($)
Type of food	Fruit	4.30	30.10	129.00
	Vegetables	2.60	18.20	78.00
	Eggs	0.30	2.10	9.00
	Total	7.20	50.40	216.00

b.

	A	B	C	D
1		Time period		
2	Type of food	Day	Week	Month
3	Fruit	4.30	30.10	129.00
4	Vegetables	2.60	18.20	78.00
5	Eggs	0.30	2.10	9.00
6	Total	7.20	50.40	216.00

9.

		Time period		
		Day ($)	Week ($)	Month ($)
Type of food	Potatoes	0.40	2.80	12.00
	Tomatoes	1.50	10.50	45.00
	Cheese	1.60	11.20	48.00
	Total	3.50	24.50	105.00

10.

		Time period			
		Day ($)	Week ($)	Month ($)	Year ($)
Type of service	Petrol	4.10	28.70	123.00	1476.00
	Insurance	2.60	18.20	78.00	936.00
	Maintenance	1.30	9.10	39.00	468.00
	Total	8.00	56.00	240.00	2880.00

11.

		Time period			
		Day ($)	Week ($)	Month ($)	Year ($)
Type of service	Petrol	2.50	17.50	75.00	900.00
	Insurance	2.40	16.80	72.00	864.00
	Maintenance	1.70	11.90	51.00	612.00
	Total	6.60	46.20	198.00	2376.00

12. a. $14   **b.** $18   **c.** Tatiana

13.

		Time period		
		Day ($)	Week ($)	Month ($)
Category	Singlets	3.42	23.94	102.60
	Casual	6.61	46.27	198.30
	Formal	7.50	52.50	225.00
	Total	17.53	122.71	525.90

14. a.

		Time period		
		Day ($)	Week ($)	Month ($)
Category	Small items	2.65	18.55	79.50
	Bottoms	10.85	75.95	325.50
	Dresses	3.20	22.40	96.00
	Tops	6.20	43.40	186.00
	Total	22.90	160.30	687.00

b.

	A	B	C	D
1		Time period		
2	Category	Day	Week	Month
3	Small items	2.65	18.55	79.50
4	Bottoms	10.85	75.95	325.50
5	Dresses	3.20	22.40	96.00
6	Tops	6.20	43.40	186.00
7	Total	22.90	160.30	687.00

15.

		Time period			
		Day ($)	Week ($)	Month ($)	Year ($)
Type of service	Insurance	1.75	12.25	52.50	630
	Fuel	1.30	9.10	39	468
	Maintenance	0.70	4.90	21	252
	Public transport	5.80	40.60	174	2088
	Total	$9.55	$66.85	$286.50	$3438.00

16. Per day: $2.25
Per week: $15.75
Per month: $67.50
Per year: $810

17. a. Weekly: $153.30   **b.** Weekly: $228.20
    Monthly: $657   Monthly: $978
c. Weekly: $107.94   **d.** Weekly: $176.33
    Monthly: $462.60   Monthly: $755.70

18. a. Daily: $23.67   **b.** Daily: $35.63
    Monthly: $710   Monthly: $1068.83
c. Daily: $34.47   **d.** Daily: $40.81
    Monthly: $1034.17   Monthly: $1224.17

7.6 Exercise

Multiple choice

1. C
2. D
3. E
4. B
5. D
6. C
7. D
8. A
9. E
10. C

Short answer

11. a. 61.2 km/h   b. 1452 L/h
    c. $0.0075/ min   d. 2 500 000 g/L

12. a. 45.15 m/s   b. 22.15 s

13. Olivia: 155 cm
    Emily: 146.7 cm

14. a. $2775   b. $2899.20
    c. $31.05   d. $124.20

Extended response

15. 1 hour 6 minutes 40 seconds

16.

Trade		Calculation	Total
	Carpenter	$60 \times 40 = \$2400$	$2400
	Electrician	$80 \times 14.5 = \$1160$	$1160
	Tiler	$55 \times 9\dfrac{15}{60} = \$508.75$	$508.75
	Plasterer	$(\$23 \times 2) \times 8\dfrac{20}{60} = \$383.33$	$383.33
	Plumber	$85 + \$90 \times 6.5 = \$670$	$670
	Total		$5122.08

17. a. $532   b. $0   c. $6717
    d. $22 967   e. $65 167

18. a. 1620 kJ   b. 16.5 g   c. 4.2 g
    d. 55.8 g   e. 12.4%   f. 4.9%

8 Percentages

Fully worked solutions for this topic are available online.

8.1 Overview

8.1.1 Introduction

Percentages are used to describe many different kinds of information and are regularly used in day-to-day activities. They are so common that they have their own symbol, %. A *per cent* is a hundredth, so using percentages is an alternative to using decimals and fractions. Percentages are a convenient way to describe how much of something you have and how meaningful information is. You are most likely familiar with percentages as a way of recording a test result, but they are used in many different areas and are therefore very important to understand.

Some examples include items advertised at a Black Friday sale at 25% discount, the percentage charged on a loan or a credit card, the percentage tax you need to pay or even the percentage chance of rain on a particular day.

KEY CONCEPTS

This topic covers the following key concepts from the VCE Mathematics Study Design:
- use of ratios, proportions, percentages and rates to solve problems.

Note: Concepts shown in grey are covered in other topics.

Source: VCE Mathematics Study Design (2023–2027) extracts © VCAA; reproduced by permission.

8.2 Calculating percentages

LEARNING INTENTION

At the end of this subtopic you should be able to:
- convert percentages into fractions and decimals.

8.2.1 Converting percentages to fractions

The term **per cent** means 'per hundred' or 'out of 100'. The symbol for per cent is %. For example, 13% means 13 out of 100.

Since all percentages are out of 100, they can be converted to a fraction. To convert a percentage to a fraction, divide by 100 or multiply by $\dfrac{1}{100}$ and simplify; for example:

$$35\% = \frac{^{7}\cancel{35}}{_{20}\cancel{100}} = \frac{7}{20}$$

$$\frac{25}{4}\% = \frac{25}{4 \times 100} = \frac{^{1}\cancel{25}}{^{16}\cancel{400}} = \frac{1}{16}$$

$$12\frac{4}{5}\% = \frac{64}{5}\% = \frac{64}{5 \times 100} = \frac{^{16}\cancel{64}}{^{125}\cancel{500}} = \frac{16}{125}$$

$$42.5\% = \frac{42.5}{100} = \frac{42.5 \times 10}{100 \times 10} = \frac{^{17}\cancel{425}}{^{40}\cancel{1000}} = \frac{17}{40}$$

Expressing percentages as fractions

To convert a percentage to a fraction, divide by 100.

---

tlvd-3800

WORKED EXAMPLE 1 Converting percentages to fractions

Write the following percentages as fractions in their simplest form.

a. 45% b. $\dfrac{5}{8}\%$ c. $25\dfrac{1}{4}\%$ d. 0.75%

THINK	WRITE
a. 1. Divide the integer value by 100.	a. $45\% = \dfrac{45}{100}$
2. Simplify by dividing the numerator and denominator by the highest common factor (in this case 5).	$= \dfrac{^{9}\cancel{45}}{_{20}\cancel{100}}$ $= \dfrac{9}{20}$

b. 1. Convert the fraction percentage to a fraction by multiplying by $\frac{1}{100}$.

b. $\frac{5}{8}\% = \frac{5}{8} \times \frac{1}{100}$

2. Simplify by dividing the numerator and denominator by the highest common factor (in this case 5).

$$= \frac{\overset{1}{\cancel{5}}}{\underset{160}{\cancel{800}}} = \frac{1}{160}$$

c. 1. Convert the mixed number to an improper fraction.

c. $25\frac{1}{4}\% = \frac{4 \times 25 + 1}{4}\% = \frac{101}{4}\%$

2. Convert the improper fraction percentage to a fraction by multiplying by $\frac{1}{100}$.

$\frac{101}{4}\% = \frac{101}{4} \times \frac{1}{100}$

3. Simplify if possible; if not, write the fractional answer.

$$= \frac{101}{400}$$

d. 1. Divide the decimal value by 100.

d. $0.75\% = \frac{0.75}{100}$

2. Convert the decimal value to an integer by multiplying by an appropriate power of 10. Also multiply the denominator by this power of 10. (Here 0.75 has 2 decimal places, so to make it an integer, multiply by 10^2 or 100.)

$$\frac{0.75}{100} = \frac{0.75 \times 100}{100 \times 100}$$
$$= \frac{75}{10\,000}$$

3. Simplify by dividing the numerator and denominator by the highest common factor (in this case 25).

$$\frac{\overset{3}{\cancel{75}}}{\underset{400}{\cancel{10\,000}}} = \frac{3}{400}$$

8.2.2 Converting percentages to decimals

In order to convert percentages to decimals, divide by 100.

When dividing by 100, the decimal point moves 2 places to the left.

For example:

$$60\% = \frac{60}{100}$$
$$= \frac{60.0}{100}$$
$$= 0.60$$

Expressing percentages as decimals

To convert a percentage to a decimal, divide by 100.

tlvd-3801

Write the following percentages as decimals.

a. **67%**

b. **34.8%**

THINK

WRITE

a. 1. Convert the percentage to a fraction by dividing by 100.

a. $67\% = \dfrac{67}{100}$

2. Divide the numerator by 100 by moving the decimal point 2 places to the left. Remember to include a zero in front of the decimal point.

$\dfrac{67}{100} = 67 \div 100$

$= 0.67$

b. 1. Convert the percentage to a fraction by dividing by 100.

b. $34.8\% = \dfrac{34.8}{100}$

2. Divide the numerator by 100 by moving the decimal point 2 places to the left. Remember to include a zero in front of the decimal point.

$\dfrac{34.8}{100} = 34.8 \div 100$

$= 0.348$

8.2.3 Converting a fraction or a decimal to a percentage

To convert a decimal or fraction to a percentage, multiply by 100%.

When multiplying by 100, move the decimal point 2 places to the right.

For example:

$$0.60 = 0.\overset{\frown}{60} \times 100\%$$
$$= 60\%$$

Expressing fractions and decimals as percentages

To convert a fraction or decimal to a percentage, multiply by 100%.

Convert the following to percentages.

a. **0.51**

b. $\dfrac{2}{5}$

THINK

WRITE

a. 1. To convert to a percentage, multiply by 100.

2. Move the decimal point 2 places to the right and write the answer.

a. $0.51 = 0.\overset{\frown}{51} \times 100\%$

$= 51\%$

b. 1. To convert to a percentage, multiply by 100.

b. $\dfrac{2}{5} \times 100\% = \dfrac{\overset{40}{\cancel{200}}}{\cancel{5}}\%$

$= \dfrac{40}{1}\%$

2. Simplify the fraction by dividing the numerator and denominator by the highest common factor (in this case 5).

$= 40\%$

8.2.4 Determining the percentage of a given amount

Quantities are often expressed as a percentage of a given amount. To determine the percentage of an amount using fractions, follow these steps:

Step 1: Convert the percentage into a fraction.

Step 2: Multiply by the amount.

For example, the percentage of left-handed tennis players is around 10%. So, for every 100 tennis players, approximately 10 of them will be left-handed.

$$10\% \text{ of } 100 \text{ tennis players} = \dfrac{10}{100} \times 100$$
$$= 0.10 \times 100$$
$$= 10 \text{ tennis players}$$

tlvd-3802

WORKED EXAMPLE 4 Determining the percentage of a given amount

Calculate the following.

a. 35% of 120

b. 76% of 478 kg

THINK	WRITE
a. 1. Write the problem.	**a.** 35% of 120
2. Express the percentage as a fraction and multiply by the amount.	$= \dfrac{35}{100} \times 120$
3. Write the answer.	$= 42$
b. 1. Write the problem as a decimal.	**b.** 76% of 478 kg
2. Express the percentage as a fraction and multiply by the amount.	$= \dfrac{76}{100} \times 478$
3. Write the answer. Remember the result has the same unit.	$= 363.28$ kg

---

tlvd-3803

WORKED EXAMPLE 5 Determining the percentage of a given amount

60% of Year 11 students said Mathematics is their favourite subject. If there are 165 students in Year 11, calculate to the nearest person how many students claimed that Mathematics was their favourite subject.

THINK	WRITE
1. Write the problem as a mathematical expression.	60% of 165
2. Write the percentage as a fraction and multiply by the amount.	$= \dfrac{60}{100} \times 165$
3. Simplify.	$= 0.60 \times 165$ $= 99$
4. Write the answer.	99 Year 11 students said their favourite subject was Mathematics.

8.2.5 A shortcut for determining 10% of a quantity

To determine 10% of an amount, divide by 10 or move the decimal point 1 place to the left; for example:

$$10\% \text{ of } \$23.00 = \frac{10}{100} \times 23.00 = \frac{23.00}{10} = \$2.30$$

$$10\% \text{ of } 45.6 \text{ kg} = \frac{10}{100} \times 45.6 = \frac{45.6}{10} = 4.56 \text{ kg}$$

This shortcut can be adapted to other percentages, as listed below:
- 5% — first determine 10% of the amount, then halve this amount.
- 20% — first calculate 10% of the amount, then double this amount.
- 15% — first calculate 10%, then 5% of the amount, then add the totals together.
- 25% — first calculate 10% of the amount and double it, then calculate 5%, then add the totals together (or calculate 50% and then halve it).

WORKED EXAMPLE 6 Determining a percentage of a quantity

Determine the following:
a. 10% of \$84   **b. 5% of 160 kg**   **c. 15% of \$162**

THINK	WRITE
a. To calculate 10% of 84, move the decimal point 1 place to the left.	a. 10% of \$84 = \$8.40
b. 1. First calculate 10% of 160. To do this, move the decimal point 1 place to the left.	b. 10% of 160 kg = 16 kg

2. To calculate 5%, divide the 10% value by 2.

$$\frac{16}{2} = 8$$

3. Write the answer.

5% of 160 kg = 8 kg

c. 1. First calculate 10% of 162. To do this, move the decimal point 1 place to the left.

c. 10% of \$162 = \$16.20

2. To calculate 5%, divide the 10% value by 2.

$$5\% \text{ of } 162 = \frac{16.20}{2}$$
$$= 8.10$$

3. Determine 15% by adding the 10% and 5% values together.

$16.20 + 8.10 = 24.30$

4. Write the answer.

15% of \$162 = \$24.30

 Resources

 Interactivities Fractions as percentages (int-3994)
Percentages as fractions (int-3992)
Percentages as decimals (int-3993)
Decimals as percentages (int-3995)
Percentage of an amount using decimals (int-3997)
Common percentages and shortcuts (int-3999)

8.2 Exercise

Students, these questions are even better in jacPLUS

Receive immediate feedback and access sample responses

Access additional questions

Track your results and progress

Find all this and MORE in jacPLUS

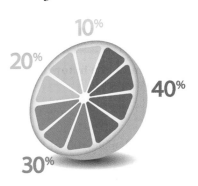

1. **WE1** Write the following percentages as fractions in their simplest form.

   a. 48% b. 26.8% c. $12\frac{2}{5}\%$ d. $\frac{4}{5}\%$

2. Write the following percentages as fractions in their simplest form.

   a. 92% b. 74.125% c. $66\frac{2}{3}\%$

3. Write the following percentages as fractions.

   a. 88% b. 25% c. 0.92%

   d. 35.5% e. $30\frac{2}{5}\%$ f. $72\frac{3}{4}\%$

4. **WE2** Write the following percentages as decimals.

   a. 73% b. 94.3%

5. Write the following percentages as decimals.
   a. 2.496% b. 0.62%

6. Write the following percentages as decimals.
   a. 43% b. 39% c. 80% d. 47.25% e. 24.05% f. 0.83%

7. **WE3** Write the following as a percentage.
   a. 0.21 b. $\dfrac{4}{5}$

8. Write the following as a percentage.
   a. 0.652 b. $\dfrac{8}{25}$ c. 0.55 d. 0.83

9. Write the following as a percentage.
   a. $\dfrac{1}{2}$ b. $\dfrac{7}{8}$ c. $2\dfrac{1}{5}$ d. 1.65

10. **WE4** Calculate each of the following.
    a. 45% of 160 b. 28% of 288 kg c. 63% of 250 d. 21.5% of $134

11. Determine the following to the nearest whole number.
    a. 34% of 260 b. 55% of 594 kg c. 12.5% of 250 m
    d. 45% of 482 e. 60.25% of 1250 g f. 37% of 2585

12. **WE5** If a student spent 70% of their part-time weekly wage of $85 on their mobile phone bill, calculate how much they were charged.

13. If a dress was marked down by 20% and it originally cost $120, calculate how much the dress has been marked down by.

14. **WE6** Determine the following:
    a. 10% of $56 b. 5% of 250 kg c. 15% of 370 d. 20% of 685

15. If 35% of the 160 Year 11 students surveyed prefer to watch a movie in their free time, use mental arithmetic to calculate how many Year 11 students prefer watching movies. Check your answer using a calculator.

16. In 2021 Australia had a population of around 25 million people, approximately 3.3% of whom were Aboriginal and Torres Strait Islander people. Using these figures, calculate how many Aboriginal and Torres Strait Islander people there were in Australia in 2021.

17. Bella's weekly wage is $1100, and 45% of this is spent on rent. Calculate the amount Bella spends on rent each week.

18. After a good 54 mm rainfall overnight, Jacko's 7500-litre water tank was 87% full. Calculate how much water (in litres) is in Jacko's water tank following the rainfall.

19. Last year a loaf of bread cost $3.10. Over the last 12 months, the price increased by 4.5%.
    a. Calculate how much, to the nearest cent, the price of bread has increased over the past 12 months.
    b. Calculate the current price of a loaf of bread.
    c. If the price of bread is to increase by another 4.5% over the next year, calculate the increase in price over the next 12 months, to the nearest cent.
    d. Calculate what a loaf of bread will cost 12 months from now.

20. Ryan has had his eyes on his dream pair of shoes. To his surprise, they were advertised with a 25% discount. If the full price of the shoes is $175, calculate the price Ryan paid for them.

21. The Goods and Services Tax (GST) requires that 10% be added to the price of certain goods and services. Calculate the total cost of each of the following items after GST has been added. (Prices are given as pre-tax prices.)

a.

$1250

b.

$145

c.

$370

22. If a surf shop had a 20% discount sale on for the weekend, calculate what price you would pay for each of the following items.

a. i. Board shorts with the original price of $65.00
   ii. T-shirt with the original price of $24.50
   iii. Hoodie with the original price of $89.90
b. Calculate how much you would have saved by purchasing these three items on sale compared to paying the full price.

23. Catherine wants to find the best price for a new iPad mini. She found an ad saying that if you purchase the iPad online you can get 15% off the price. The original price was advertised at $585.
Calculate how much Catherine would save if she made the purchase online.

24. If John spends 35% of his $1200 weekly wage on rent and Jane spends 32% of her $1050 weekly wage on rent:

a. calculate who spends more money per week on rent
b. calculate the difference in John's and Jane's weekly rent.

8.3 Express as a percentage

LEARNING INTENTION

At the end of this subtopic you should be able to:
• express an amount as a percentage of another amount.

8.3.1 Expressing one amount as a percentage of another

To express an amount as a percentage of another, write the two numbers as a fraction, with the first number on the numerator and the second number on the denominator. Then convert this fraction to a percentage by multiplying by 100%.

Expressing one amount as a percentage of another

$$\frac{\text{amount}}{\text{total}} \times \frac{100}{1}\%$$

When expressing one amount as a percentage of another, make sure that both amounts are in the same unit.

tlvd-3804

WORKED EXAMPLE 7 Expressing one amount as a percentage of another

Express the following to the nearest percentage.
a. 37 as a percentage of 50   **b. 18 as a percentage of 20**   **c. 23 as a percentage of 30**

THINK	WRITE
a. 1. Write the amount as a fraction of the total.	**a.** $\dfrac{\text{amount}}{\text{total}} = \dfrac{37}{50}$
2. Multiply the fraction by 100% and express as a percentage.	$\dfrac{37}{50} \times 100\% = 74\%$
b. 1. Write the amount as a fraction of the total.	**b.** $\dfrac{\text{amount}}{\text{total}} = \dfrac{18}{20}$
2. Multiply the fraction by 100% and express as a percentage.	$\dfrac{18}{20} \times 100\% = 90\%$
c. 1. Write the amount as a fraction of the total.	**c.** $\dfrac{\text{amount}}{\text{total}} = \dfrac{23}{30}$
2. Multiply the fraction by 100% and express as a percentage.	$\dfrac{23}{30} \times 100\% = 76.666\%$
3. Round to the nearest percentage.	77%

8.3.2 Expressing one amount as a percentage of another for different units

When expressing one amount as a percentage of another, ensure that both amounts are expressed in the same unit.

WORKED EXAMPLE 8 Expressing one amount as a percentage of another for different units (1)

Write 92 cents as a percentage of $5, correct to 1 decimal place.

THINK	WRITE
1. Since the values are in different units, convert the larger unit into the smaller one.	$\$5 = 500$ cents
2. Write the first amount as a fraction of the second.	$\dfrac{92}{500}$

3. Multiply the fraction by 100%.

$$\frac{92}{500} \times 100\%$$

4. Calculate the percentage and round to 1 decimal place as required.

$$= 18.4\%$$

WORKED EXAMPLE 9 Expressing one amount as a percentage of another for different units (2)

If Rebecca Allen scored 12 baskets out of her 15 shots from the free throw line, calculate the percentage of free throws that she got in.

THINK

1. Write the amount as a fraction of the total.

2. Multiply the fraction by 100% and express as a percentage.

WRITE

$$\frac{\text{amount}}{\text{total}} = \frac{12}{15}$$

$$\frac{12}{15} \times 100\% = 80\%$$

8.3.3 Percentage discounts

To determine a percentage discount, determine the fraction $\dfrac{\text{amount saved}}{\text{original price}}$ and multiply by 100.

Determining a percentage discount

$$\text{Percentage discount} = \frac{\text{amount saved}}{\text{original price}} \times 100$$

---

tlvd-3806

WORKED EXAMPLE 10 Determining a percentage discount

A T-shirt was on sale, reduced from $89.90 to $75.45. Calculate what percentage was taken off the original price. Give your answer to 1 decimal place.

THINK

1. Calculate the amount saved.

2. Divide the amount saved by the original price.

WRITE

Amount saved $= \$89.90 - \$75.45$
$= \$14.45$

$$\frac{\text{amount saved}}{\text{original amount}} = \frac{14.45}{89.90}$$

---

3. Multiply the fraction by 100% to get the percentage.	$\dfrac{14.45}{89.90} \times 100\%$
4. Calculate the value of the expression.	$= 16.0734\%$
5. Round to 1 decimal place.	16.1% discount on the original price

 Resources

 Interactivites One amount as a percentage of another (int-3998)
Discount (int-3744)

8.3 Exercise

Students, these questions are even better in jacPLUS

 Receive immediate feedback and access sample responses

 Access additional questions

 Track your results and progress

Find all this and MORE in jacPLUS

1. **WE7** Express each of the following to the nearest percentage.
   a. 64 as a percentage of 100
   b. 13 as a percentage of 20
   c. 9 as a percentage of 30

2. Express each of the following to the nearest percentage.
   a. 37 as a percentage of 50
   b. 21 as a percentage of 40
   c. 15 as a percentage of 23

3. Express the following to the nearest percentage.
   a. 36 as a percentage of 82
   b. 45 as a percentage of 120
   c. 12 as a percentage of 47
   d. 9 as a percentage of 15
   e. 15 as a percentage of 44
   f. 67 as a percentage of 175

4. Express the following as a percentage, giving your answers to 1 decimal place.
   a. 23.5 of 69
   b. 59.3 of 80
   c. 45.75 of 65
   d. 23.82 of 33
   e. 0.85 of 5
   f. 1.59 of 2.2

5. **WE8** Write 53c as a percentage of $3, to 1 decimal place.

6. Write 1500 m as a percentage of 5 km.

7. Express the following as a percentage, giving your answers to 1 decimal place.
   a. 68c of $2
   b. 31c of $5
   c. 67 g of 2 kg
   d. 0.54 g of 1 kg
   e. 546 m of 2 km
   f. 477 m of 3 km
   g. 230 mm of 400 cm
   h. 36 min of 3 hours

8. **WE9** If Buddy Franklin kicks 6 goals from his 13 shots at goal, calculate the percentage of goals he kicked. Give your answer correct to 2 decimal places.

9. If Kate Moloney has 8 of her team's 32 intercepts in a game, calculate the percentage of the team's intercepts that Kate had.

10. If a student received 48 out of 55 for a Mathematics test, calculate the percentage, correct to 2 decimal places, that the student got on the test.

11. Corn Flakes have 7.8 g of protein per 0.1 kg. Calculate the percentage of protein in Corn Flakes.

12. **WE10** If a pair of boots were reduced from $155 to $110, calculate the percentage discount offered on the boots. Give your answer to 1 decimal place.

13. Olivia saved $35 on the books she purchased, which were originally priced at $187.
    a. Calculate how much was paid for the books.
    b. Calculate what percentage, correct to 1 decimal place, Olivia saved from the original price.

14. A store claims to be taking 40% off the prices of all items, but your friend is not so sure. Calculate the percentage discount on each of the items shown and determine if the store has been completely truthful!

8.4 Percentage decrease and increase

LEARNING INTENTION

At the end of this subtopic you should be able to:
- calculate discounted prices
- calculate percentage increase
- calculate repeated percentage change.

8.4.1 Percentage decrease

A percentage decrease is often used to discount goods that are on sale. A **discount** is a reduction in price from the original marked price.

When the discount is expressed as a percentage, to determine the amount of the discount, we need to multiply the original price by the discount percentage expressed as a decimal or $\dfrac{\text{percentage change}}{100} \times$ original price.

Percentage decrease

$$\text{Discount} = \frac{\text{percentage change}}{100} \times \text{original price}$$

For example, a 10% discount on an item marked at $150 gives a discount of:

$$\frac{10}{100} \times \$150 = 0.1 \times \$150$$
$$= \$15$$

Calculating the discounted price

The discounted price can be calculated in two ways.

Method 1

- Calculate the discount by multiplying the original price by the percentage expressed as a decimal.
- Subtract the discount from the original price.

Method 2

- Discounted price $= \left(1 - \dfrac{\text{percentage discount}}{100}\right) \times$ original amount

WORKED EXAMPLE 11 Calculating the discounted price

tlvd-3807

A store has a 15%-off-everything sale. Molly purchases an item that was originally priced at $160.
a. Calculate the price Molly paid for the item.
b. Calculate how much Molly saved from the original price.

THINK	WRITE
a. **Method 1:**	a. Discount $= 0.15 \times \$160$
1. Discount is 15% of $160, so multiply 0.15 (15% as a decimal) by $160.	$= \$24$
2. Determine the discounted price by subtracting the discount from the original price.	Discounted price $= \$160 - \$24$ $= \$136$
3. Write the discounted price.	Molly paid $136.
Method 2:	
4. Calculate the discounted price.	Discounted price $= \left(1 - \dfrac{15}{100}\right) \times 160$ $= (1 - 0.15) \times 160$ $= 0.85 \times 160$ $= \$136$
b. **Method 1:**	b. Saving $= 0.15 \times \$160$
1. Calculate the saving as a percentage of the original price.	$= \$24$
2. State the amount saved.	Molly saved $24.
Method 2:	
3. Calculate the difference between the original price and the discounted price (found in part a).	Saving $= \$160 - \$136$ $= \$24$
4. Write the amount saved.	Molly saved $24.

8.4.2 Percentage increase

Just as some items are on sale or reduced, some items or services increase in price. Examples of increases in price can be the cost of electricity or a new model of car.

We calculate the percentage increase in price in a similar way to the percentage decrease; however, instead of subtracting the discounted value, we add the extra percentage value.

WORKED EXAMPLE 12 Calculating percentage increase

Ashton paid $450 for his last electricity bill, and since then the charges have increased by 5%.
a. Assuming he used the same amount of electricity, determine how much extra he would expect to pay on his next bill.
b. Calculate his total bill.

THINK	WRITE
a. 1. To calculate the increase amount, multiply the percentage increase by the previous cost.	a. 5% of $\$450 = 0.05 \times \$450$ $= \$22.50$
2. Write the amount of the extra charge.	The extra charge is $22.50.

b. 1. Calculate the total cost of the bill by adding the extra charge to the original cost.

b. Total bill = $450 + $22.50

2. Write the total cost of the bill.

Total bill = $472.50

8.4.3 Repeated percentage change

In many situations, the percentage change differs over the same time period. For example, the population of a small town increases by 10% in one year and only by 4% the next year. If the initial population of the town is known to be 15 500, then percentage multipliers can be used to calculate the population of the town after the 2 years.

> **Percentage change over a time period**
>
> The multiplying factor is $1 + \dfrac{r}{100}$, where r is the percentage increase,
>
> and $1 - \dfrac{r}{100}$, where r is the percentage decease.

An increase of 10% means the population has increased by a factor of $1 + \dfrac{10}{100} = 1.10$ and an increase of 4% means the population has increased by a factor of $1 + \dfrac{4}{100} = 1.04$.

Therefore, the population after 2 years is:

$$15\,500 \times 1.10 \times 1.04 = 17\,732$$

Overall percentage change

To calculate the overall percentage change in the situation described previously, multiply the two percentage factors, $1.10 \times 1.04 = 1.144$. This is 0.144 greater than 1. Convert 0.144 to a percentage by multiplying by 100. That gives a percentage change of $0.144 \times 100 = 14.4\%$.

tlvd-3808

WORKED EXAMPLE 13 Calculating repeated percentage change

Calculate the overall change in the following quantities.
a. A small tree 1.2 metres tall grew by 22% in the first year and 16% in the second year. Determine its height at the end of the second year correct to 2 decimal places.
b. An investment of $12 000 grew by 6% in the first year and decreased by 1.5% in the second year. Calculate the value of the investment at the end of the second year.

THINK

WRITE

a. 1. The multiplying factors are found using the two percentages, 22% and 16%, and $1 + \dfrac{r}{100}$.

a. First percentage change $= 1 + \dfrac{22}{100}$
$= 1.22$

Second percentage change $= 1 + \dfrac{16}{100}$
$= 1.16$

2. Multiply the initial height of the tree by 1.22 and 1.16.

$1.2 \times 1.22 \times 1.16 = 1.698$

3. Write the answer.

The height of the tree after two years is 1.70 m correct to 2 decimal places.

b. 1. The multiplying factors are found using 6% and $1 + \dfrac{r}{100}$ for the increase and 1.5% and $1 - \dfrac{r}{100}$ for the decrease.

b. First percentage change $= 1 + \dfrac{6}{100}$
$= 1.06$

Second percentage change $= 1 - \dfrac{1.5}{100}$
$= 0.985$

2. Multiply the initial investment by 1.06 and 0.985.

$12\,000 \times 1.06 \times 0.985 = 12\,529.20$

3. Write the answer.

The value of the investment after two years is $12 529.20.

---

on Resources

 Interactivity Percentage increase and decrease (int-3742)

8.4 Exercise

Students, these questions are even better in jacPLUS

 Receive immediate feedback and access sample responses

Access additional questions

Track your results and progress

Find all this and MORE in jacPLUS

1. **WE11** A store has a 25%-off-everything sale. Bradley purchases an item that was originally priced at $595.
   a. Calculate how much Bradley paid for the item.
   b. Calculate how much he saved from the original price.

2. Nicole purchases a new pair of shoes that were marked down by 20%. Their original price was $139.95.
   a. Calculate how much Nicole paid for the new shoes.
   b. Calculate how much Nicole saved.

3. A new suit is priced at $550. Calculate how much you would pay if you received the following percentage discount:
   a. 5%
   b. 10%
   c. 15%
   d. 20%
   e. 30%
   f. 50%

4. A pool holds 55 000 L of water. During a hot summer week, the pool lost 5.5% of its water due to evaporation.
   a. Calculate how much water was lost during the week.
   b. Calculate how much water the pool held at the end of the week.

5. **WE12** This year it costs $70 for a ticket to watch the Australian Open at the Rod Laver Arena. Tickets to the Australian Open are due to increase by 10% next year.

a. Determine the amount the tickets will increase by.
b. Determine how much you would expect to pay for a ticket to the Australian Open next year.

6. A television costs $2500 this week, but next week there will be a price increase of 7.5%. Calculate how much you would expect to pay for the television next week.

7. The Australian government introduced the Goods and Services Tax (GST) in 2000. It added 10% to the price of a variety of goods and services. Calculate the price after GST has been added to the following pre-tax prices.

   a. $189 car service
   b. $1650 dining table
   c. $152.50 pair of runners
   d. $167.85 pair of sunglasses

8. When you purchase a new car you also need to pay government duty on the car. You must pay a duty of 3% of the value of the car for cars priced up to $59 133 and 5% of the value for cars priced over $59 133.

   a. Calculate the duty on each of the following car values.

      i. $15 500
      ii. $8000
      iii. $21 750
      iv. $45 950
      v. $65 000
      vi. $78 750

   b. Calculate the cost of each of the cars including the government duty.

9. A sports shop is having an end-of-year sale on all stock. A discount of 20% is applied to clothing and a discount of 30% is applied to all sports equipment. Calculate the discount you will receive on the following items.

   a. A sports jumper initially marked at $175.
   b. A basketball ring marked at $299.
   c. A tennis outfit priced at $135.
   d. A table tennis table initially priced at $495.

10. A clothing store has a rack with a 50% off sign on it. As a weekend special, for an hour the shop owner takes a further 50% off all items in the store. Explain whether that means that you can get clothes for free and give reasons for your answer.

11. A new car was purchased for $35 000. It depreciates from its current value each year by the percentage shown in the table.

    a. Complete the table.

		% depreciation	Depreciation ($)	Value ($)
End year	First	10%	$3500	$31 500
	Second	7.5%	$2362.50	
	Third	6%		
	Fourth	5%		
	Fifth	4%		

    b. If the car continued to lose 2% each year in the sixth, seventh, eighth and ninth year, calculate how much the car would be worth at the end of the ninth year.

12. The following items are discounted as shown.

$260 with $33\frac{1}{3}$ % discount

$380 with 25% discount

$450 with 20% discount

$600 with 15% discount

a. Determine which item has the largest discount.
b. Determine which of the purchases give the same discount.
c. State the difference between the largest and smallest discount.
d. State what percentage discount you would need if you only have $420 to buy the bike.

13. **WE13** Calculate the overall change for each of the following situations.

a. A new IT company of 2400 employees grew its workforce by 3% in the first year and 4.5% in the second year. Calculate how many employees the company had at the end of the second year.

b. Roya had invested $250 000 in her superannuation fund. The fund grew at 6% in the first year but had negative growth of 3.5% in the second year.
Determine how much was in the superannuation fund at the end of the second year.

14. Chris purchased a new car for $42 000. After 1 year, the value of the car had decreased by 8%. In the following year, the value of the car decreased by a further 5.2%.

a. Calculate the value of the car at the end of the two years.

b. Calculate the overall percentage change of the car at the end of the two years. Give your answer correct to the nearest per cent.

15. An amount of $55 000 is invested for 7 years in an account that grows by 6.5% every year. Calculate how much the investment is worth at the end of the 7 years.

16. Priscilla takes her savings of $17 000 and invests it in a fixed term deposit that pays 5% each year on the money that is in the account. At the end of the fixed term, Priscilla's savings are worth $20 663.60. Using a trial and error method, determine the length in years of the fixed term.

17. Alex has been saving his money for a new tennis racquet. Alex has been monitoring the website that sells top brands and notices a Yonex Ezone racquet for $329.95.

The next week he notices there is a 20% off sale on all racquets on this website. After talking to some friends, Alex decides to wait until the end of financial year sale in case it is marked down further. To Alex's delight, another 15% is taken off the sale price and he then purchases the racquet.

Calculate how much Alex paid for the racquet.

18. Heidi decided to purchase a bottle of perfume for her mother on Mother's Day. Heidi was able to get her mother's favourite perfume at 20% off the original price of $115.50.

Calculate the price Heidi paid for the perfume.

19. Store A had to increase their prices by 10% to cover expenses, whereas store B was having a 15%-off sale over the entire store. Sarah wanted to purchase a soundbar and assumed it would be better to get it from store B since it was having a sale. The original price of the soundbar at store A was $185, whereas at store B it was originally priced at $240.

   a. State if Sarah's assumption was correct. Prove this mathematically.
   b. Calculate the difference in the prices of the soundbar.

20. A classmate was completing a discount problem where they needed to calculate a 25% discount on $79. They misread the question and calculated a 20% discount to get $63.20. They then realised their mistake and took a further 5% from $63.20. Explain if this is the same as taking 25% off $79.

   Use calculations to support your answer.

8.5 Simple interest

LEARNING INTENTION

At the end of this subtopic you should be able to:
- calculate simple interest
- calculate interest rates.

8.5.1 Principal, interest and value

When you put money into a financial institution such as a bank or credit union, the amount of money you start with is called the **principal**.

People who place money in a bank or financial institution (**investors**) receive a payment called **interest** from the financial institution in return for investing their money in the financial institution.

The amount of interest is determined by the **interest rate**. An interest rate is a percentage for a given time period, usually a year. For example, a bank might offer 5.8% per year interest on its savings accounts. This is also written as 5.8% per annum, or 5.8% p.a.

Simple interest calculations

Simple interest is the interest paid on the principal amount of an investment.

The amount of interest paid each time period is based on the principal, so the amount of interest is constant. For example, $500 placed in an account that earns 10% simple interest per year earns $50 each year, as shown in the following table.

Time period (years)	Amount of money at the start of the year	Amount of interest after one year
1	$500	$50
2	$550	$50
3	$600	$50
4	$650	$50

The total interest earned is $4 \times 50 = \$200$, so the value of the investment after 4 years is $\$500 + 200 = \$700$.

Simple interest is calculated using the formula below.

Simple interest

$$\text{Simple interest} = \frac{\text{principal} \times \text{rate} \times \text{time}}{100}$$

$$I = \frac{Prn}{100}$$

Where:
- **principal is the money invested or borrowed, P**
- **rate is the interest as a percentage per time period, r**
- **time is the length of time of the investment or loan, n.**

If r is given as a percentage per year, then the time n must be given in years. If r is given as a percentage per month, then n must be given in months.

The total amount of money is known as the value of the investment.

Total value of the investment

$$\text{Value} = \text{principal } (P) + \text{interest } (I)$$

---

tlvd-3809

WORKED EXAMPLE 14 Calculating simple interest

A real estate developer offers investors a chance to invest in the company's latest development. If $10 000 is invested, the developer will pay 11.5% simple interest per year for 5 years. Calculate the value of the investment.

THINK	WRITE
1. Identify the known quantities.	$P = 10\,000$ $r = 11.5\%$ $n = 5$

2. Calculate the amount of interest (I) using the formula $I = \dfrac{Prn}{100}$.

$$\text{Interest} = \frac{10\,000 \times 11.5 \times 5}{100}$$
$$= \$5750$$

3. The value of the investment after 5 years is the sum of the interest ($5750) and the principal ($10 000).

$$\text{Value} = 10\,000 + 5750$$
$$= 15\,750$$

4. Write the answer as a sentence.

The value of the investment after 5 years is $15 750.

In some cases, the time period for which the money is invested is not a multiple of the time quantity that the interest rate is quoted for — for example, if the interest rate is given as a per-year rate and the money is invested for 3 months. In these instances it is necessary to calculate the equivalent interest rate for the time units for which the money is invested.

tlvd-3810

WORKED EXAMPLE 15 Calculating simple interest if the time period is not a multiple of time

A bank offers interest on its savings account at 4% per year. If a Year 11 student opens an account with $600 and leaves the money there for 4 months, calculate how much interest they earned.

THINK

1. The interest rate period is per year and the amount of time the money is invested for is in months.
Calculate an equivalent monthly interest rate. Since there are 12 months in a year, divide the annual interest rate by 12 to get a monthly interest rate.

2. Identify the known quantities. Note that we are now dealing in months, rather than years.

3. Calculate the amount of interest (I) using the formula $I = \dfrac{Prn}{100}$.

4. Write the answer as a sentence.

WRITE

$$\frac{4}{12} = \frac{1}{3}$$

4% per year is $\dfrac{1}{3}$% per month.

$P = 600$

$r = \dfrac{1}{3}$% per month

$n = 4$ months

$$\text{Interest} = \frac{600 \times \frac{1}{3} \times 4}{100}$$
$$= 8$$

The total interest earned for 4 months is $8.

8.5.2 Calculating the interest rate

If the interest (I), principal (P) and time period (n) are known, it is possible to calculate the interest rate using the simple interest formula.

WORKED EXAMPLE 16 Calculating the interest rate

tlvd-3811

A Year 11 student is paid $65.40 in interest for an original investment of $800 for 2 years. Calculate the annual interest rate.

THINK	WRITE
1. Identify the known quantities: • interest (I) • principal (P) • time period (n)	$I = \$65.40$ $P = \$800$ $n = 2$ years
2. Substitute the values into the formula $I = \dfrac{P \times r \times n}{100}$.	$65.40 = \dfrac{800 \times r \times 2}{100}$
3. Solve the equation for r to calculate the annual interest rate by: • multiplying both sides by 100 • dividing both sides by 1600.	$65.40 = \dfrac{1600 \times r}{100}$ $65.40 \times 100 = 1600 \times r$ $\dfrac{6540}{1600} = r$ $r = 4.09$
4. Write the answer.	The annual interest rate is 4.09%.

Determining P if I, r and n are known

$$I = \frac{Prn}{100}$$

$$P = \frac{100I}{rn}$$

Determining n if I, P and r are known

To determine the time: $\qquad n = \dfrac{100I}{Pr}$

To determine the interest rate: $\qquad r = \dfrac{100I}{Pn}$

To determine the principal: $\qquad P = \dfrac{100I}{rn}$

 Resources

 Interactivity Simple interest (int-6074)

8.5 Exercise

1. In your own words, explain the difference in meaning in the following pairs of terms:
   a. *Principal* and *value of the investment*
   b. *Amount of interest* and *interest rate*

2. **WE14** A film producer offers investors the chance to invest in their latest movie. If $20 000 is invested, the producer will pay 22.3% simple interest per year for 2 years. Determine the value of the investment at the end of the 2 years.

3. **MC** If an investment of $400 pays 8% simple interest per year, the value of the investment at the end of 3 years is:
   A. $32   B. $96   C. $432   D. $496   E. $196

4. Calculate the simple interest paid on the following investments.
   a. $500 at 6.7% per year for 2 years   b. $500 at 6.7% per year for 4 years
   c. $1000 at 6.7% per year for 4 years

5. Calculate the amount of interest paid on a $1000 investment at 5% for 5 years.

6. **WE15** A bank offers interest on its savings account of 6% p.a.
   If a Year 11 student opens an account with $750 and leaves the money there for 5 months, calculate how much interest they earned.

7. **MC** If the annual interest rate is 8%, then the monthly interest rate is closest to:
   A. 0.8%   B. 0.77%   C. 0.67%   D. 0.6%   E. 0.7%

8. Calculate the interest paid on the following investments.
   a. $500 invested at 8% per annum for 1 month   b. $500 invested at 8% per annum for 3 months
   c. $500 invested at 8% per annum for 6 months

9. Calculate the value of the following investments.
   a. $1000 invested at 10% p.a. for 10 years   b. $1000 invested at 12% p.a. for 10 months
   c. $1000 invested at 6% p.a. for 3 years

10. A bank offers investors an annual interest rate of 9% if they buy a term deposit. If a customer has $5600 and leaves the money in the term deposit for 2.5 years, calculate the value of the investment at the end of the 2.5 years.

11. **WE16** A Year 11 student is paid $79.50 in interest for an original investment of $500 for 3 years. Calculate the annual interest rate.

12. Bank A offers an interest rate of 7.8% on investments, while Bank B offers an interest rate of 7.4% in the first year and 7.9% in subsequent years. If a customer has $20 000 to invest for 3 years, determine which is the better investment.

13. **MC** If the total interest earned on a $6000 investment is $600 after 4 years, then the annual interest rate is:
    A. 10%   B. 7.5%   C. 5%   D. 2.5%   E. 1.5%

14. Determine the annual interest rate on the following investments.
    a. Interest = $750, principal = $6000, time period = 4 years
    b. Interest = $924, principal = $5500, time period = 3 years
    c. Interest = $322, principal = $7000, time period = 3 months

The *following statement relates to questions* 15 *and* 16.

A loan is an investment in reverse; you *borrow* money from a bank and are *charged* interest. The value of a loan becomes its total cost.

15. Your parents decide to borrow money to improve their boat but cannot agree which loan has better value. They would like to borrow $2550. Your mother goes to Bank A and finds that they will lend the money at $11\frac{1}{3}$% simple interest per year for 3 years. Your father finds that Bank B will lend the $2550 at 1% per month simple interest.

    a. Determine which bank offers the best rate over the three years.
    b. Provide reasons for your answer to part a.

16. A worker wishes to borrow $10 000 from a bank, which charges 11.5% interest per year. If the loan is over 2 years:
    a. calculate the total interest paid
    b. calculate the total cost of the loan.

    Most loans require a monthly payment. The monthly payment of a simple interest loan is calculated by dividing the total cost of the loan by the number of payments made during the term of the loan.

    c. Determine the monthly payment for the loan in part b.

17. An online bank has a special offer on term deposits as shown below.

Term deposit special offer:

| **4.50** % p.a.* for 5 months | **4.35** % p.a.* for 6 months | **4.50** % p.a.* for 24 months |

*Interest is calculated daily and paid at maturity.

The fine print at the bottom of the web page contains the following information:
Rates shown above are the nominal interest rates for a term deposit of $10 000. Interest is payable at maturity and if you choose a 24-month deposit, you can request interest to be transferred to you every month, once a quarter, at six months or at maturity.

    a. For each of the offers, use technology to determine the total amount of the term deposit at the end of the term.
    b. The formula for compound interest is $A = (1 + i)^n \times P$, where A is the total amount, i is the interest rate, n is the number of periods and P is the principal. Use this formula to calculate the final amount in a 24-month term deposit that was opened with $10 000, if the interest is calculated on the new total when the interest is added to the account:

       i. annually
       ii. at the end of every month
       iii. at the end of every quarter
       iv. at the end of each six-month period.
    c. State which of the interest payment options from part b you would choose and why.

8.6 Review

8.6.1 Summary

doc-38049

Hey students! Now that it's time to revise this topic, go online to:

 Access the topic summary

 Review your results

 Watch teacher-led videos

Practise questions with immediate feedback

Find all this and MORE in jacPLUS

8.6 Exercise

Multiple choice

1. **MC** The percentage 45% expressed as a simplified fraction is:

   A. $\dfrac{1}{4}$ B. $\dfrac{9}{20}$ C. $\dfrac{3}{20}$ D. $\dfrac{9}{25}$ E. $\dfrac{4}{5}$

2. **MC** 64% of 280 is closest to:

   A. 210 B. 79 C. 101 D. 117 E. 179

3. **MC** 270 m of 1.5 km as a percentage is closest to:

   A. 18% B. 24% C. 14% D. 27% E. 31%

4. **MC** The price of a new $16 000 car increased by 7.5%. The amount by which the car increased in price is closest to:

   A. $750 B. $950 C. $1200 D. $1000 E. $210

5. **MC** The annual salary of a sales assistant increased from $45 000 to $48 600. The percentage salary increase is:

   A. 4% B. 5% C. 6% D. 7% E. 8%

6. **MC** Air contains 21% oxygen, 0.9% argon, 0.1% trace gases and the remainder is nitrogen. The percentage of nitrogen in the air is:

   A. 100% B. 50% C. 35% D. 78% E. 88%

7. **MC** A tree was planted when it was 2.3 m tall. It grew by 30% in the first year and 12% in the second year after planting. At the end of the second year, the height of the tree is closest to:

   A. 3.35 m B. 2.66 m C. 4.56 m D. 3.95 m E. 1.95 m

8. **MC** If a house that cost $200 000 to build is sold for $10 000 less, the percentage loss is:

   A. 95% B. 5% C. 90% D. 10% E. 15%

9. **MC** The sale price of the car is:

   A. $25 000 B. $27 000 C. $28 000
   D. $28 500 E. $29 000

$30 000

5% discount

10. **MC** If an investment of $400 pays 8% simple interest per year, the value of the investment at the end of 3 years is:

    A. $32 B. $96 C. $432 D. $496 E. $196

Short answer

11. A person earns total interest of $600 on an investment of $6000 after 4 years. Calculate the annual interest rate.

12. The winner of a football tipping competition wins 65% of the prize pool, second place wins 20% and third place receives what is left over. The prize pool is $860. Calculate the amount:

    a. first prize wins   **b.** second prize wins   **c.** third prize wins.

13. On a weekend sale, the price of all laptops was reduced by 30%.

    For a laptop valued at $1500, calculate:
    a. the amount of money saved
    b. the sale price of the laptop.

14. Express the following as percentages.

    a. 756 m of 2.7 km   **b.** 45 g of 0.85 kg
    c. 15c of $2.90   **d.** 45 s of 1.5 min

15. A person invests $100 each month, earning simple interest of 1% per month for 4 months.

    a. Calculate the interest on the first investment, which earns interest for 4 months.
    b. Calculate the interest on the second, third and fourth investments, which earn interest for 3 months, 2 months and 1 month respectively.
    c. Calculate the total value of the investment at the end of the 4 months.

Extended response

16. An expensive lamp has a regular price of $200. At that price, the shop expects to sell 4 lamps per month.
    - If offered at a discount of 10%, the shop will sell 8 lamps per month.
    - If offered at a discount of 20%, the shop will sell 16 lamps per month.
    - If offered at a discount of 30%, the shop will sell 20 lamps per month.
    - These lamps cost the shop $100 each.

    a. Determine the total value of the sales in each of the 3 discount cases above.
    b. State which discount results in the highest total sales.

17. Anastasia is holding a birthday party to which she has invited 25 friends and 15 family members.

    a. Express the number of friends as a percentage of the total number of people invited.
    b. Express the number of family members as a percentage of the total number of people invited.
    c. For catering purposes, 25% of people are vegetarians and 75% of people like dessert. Determine how many people are vegetarians and how many like dessert.
    d. On the day, not everybody turns up. Anastasia couldn't remember exactly how many turned up, but she knows that they had to reduce the amount of food by 20%. Determine how many people turned up.

18. Answer the following questions.

   a. If you increase $100 by 30% and then decrease the new amount by 30%, state whether you will end up with more than $100, less than $100 or exactly $100.
   b. Explain your answer to part **a**, using mathematics to support your answer.
   c. If you decrease $100 by 30% and then increase the new amount by 30%, state whether you will end up with more than $100, less than $100 or exactly $100.
   d. Explain your answer to part **c**, using mathematics to support your answer.
   e. Determine the percentage change needed in parts **a** and **c** above to get back to the original $100 from the new amount.
   f. Determine if the percentage change in part **a** is greater than or less than the percentage change in part **c** and explain your answer.

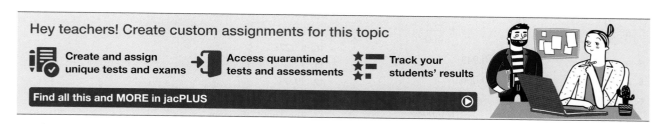

Answers

Topic 8 Percentages

8.2 Calculating percentages

8.2 Exercise

1. a. $\frac{12}{25}$ b. $\frac{67}{250}$ c. $\frac{31}{250}$ d. $\frac{1}{125}$

2. a. $\frac{23}{25}$ b. $\frac{593}{800}$ c. $\frac{2}{3}$

3. a. $\frac{22}{25}$ b. $\frac{1}{4}$ c. $\frac{23}{2500}$

   d. $\frac{71}{200}$ e. $\frac{38}{125}$ f. $\frac{291}{400}$

4. a. 0.73   b. 0.943

5. a. 0.024 96   b. 0.0062

6. a. 0.43   b. 0.39   c. 0.80
   d. 0.4725   e. 0.2405   f. 0.0083

7. a. 21%   b. 80%

8. a. 65.2%   b. 32%
   c. 55%   d. 83%

9. a. 50%   b. 87.5%
   c. 220%   d. 165%

10. a. 72   b. 80.64 kg
    c. 157.5   d. $28.81

11. a. 88   b. 327 kg   c. 31 m
    d. 217   e. 753 g   f. 956

12. $59.50

13. $24

14. a. $5.60   b. 12.5 kg
    c. 55.5   d. 137

15. 56

16. 825 000

17. $495

18. 6525 litres

19. a. 14 cents   b. $3.24
    c. 15 cents   d. $3.39

20. $131.25

21. a. $1375   b. $159.50   c. $407

22. a. i. $52.00   ii. $19.60   iii. $71.92
    b. $35.88

23. $87.75

24. a. John spends $420.
       Jane spends $336.
       John spends more than Jane.
    b. $84

8.3 Express as a percentage

8.3 Exercise

1. a. 64%   b. 65%   c. 30%

2. a. 74%   b. 53%   c. 65%

3. a. 44%   b. 38%
   c. 26%   d. 60%
   e. 34%   f. 38%

4. a. 34.1%   b. 74.1%
   c. 70.4%   d. 72.2%
   e. 17.0%   f. 72.3%

5. 17.7%

6. 30%

7. a. 34.0%   b. 6.2%
   c. 3.4%   d. 0.1%
   e. 27.3%   f. 15.9%
   g. 5.8%   h. 20.0%

8. 46.15%

9. 25%

10. 87.27%

11. 7.8%

12. 29.0%

13. a. $152   b. 18.7%

14. Top 1: 40% discount
    Shoes 1: 40% discount
    Top 2: 40% discount
    Shoes 2: 35.5% discount

8.4 Percentage decrease and increase

8.4 Exercise

1. a. $446.25   b. $148.75

2. a. $111.96   b. $27.99

3. a. $522.50   b. $495
   c. $467.50   d. $440
   e. $385   f. $275

4. a. 3025 L   b. 51 975 L

5. a. $7   b. $77

6. $2687.50

7. a. $207.90   b. $1815
   c. $167.75   d. $184.64

8. a. i. $465   ii. $240
      iii. $652.50   iv. $1378.50
      v. $3250   vi. $3937.50
   b. i. $15 965   ii. $8240
      iii. $22 402.50   iv. $47 328.50
      v. $68 250   vi. $82 687.50

9. a. $35   b. $89.70
   c. $27   d. $148.50

10. No, you cannot get clothes for free.
    You get 50% off the already reduced price. This means the
    price is first halved, then another 50% off means it is halved
    again, so you are paying one quarter of the original price.
    For example, if something originally costs $160:
    $0.5 \times \$160 = \$80$
    $0.5 \times \$80 = \$40$
    $40 is one quarter of $160.

11. a. See the table at the bottom of this page.*
    b. $23 039.83
12. a. Camera: $86.67
    iPhone: $95
    Speakers: $90
    Bike: $90
    The iPhone is reduced by $95, which is the largest reduction.
    b. Both the speakers and bike are reduced by $90.
    c. $8.33
    d. 30%
13. a. 2583   b. $255 725
14. a. $36 630.72   b. 13%
15. $85 469.26
16. 4 years
17. $224.37
18. $92.40
19. a. Store A: $203.50
    Store B: $204
    No, Sarah was not correct; store A was cheaper.
    b. 50 cents
20. Method 1: $60.04
    Method 2: $59.25
    No, it is not the same, since after the 20% discount, the 5% discount is on the reduced value of $63.20, so this method gives a smaller discount.

8.5 Simple interest

8.5 Exercise

1. a. The principal is the amount of money you start with and the value of the investment is the combination of the principal and interest.
   b. Interest is a payment from the financial institution and interest rate is the percentage that determines the amount of interest.
2. $28 920
3. D
4. a. $67   b. $134   c. $268
5. $250
6. $18.75
7. C
8. a. $3.33   b. $10   c. $20
9. a. $2000   b. $1100   c. $1180
10. $6860

11. 5.3%
12. Bank A
13. D
14. a. 3.125%   b. 5.6%   c. 18.4%
15. a. Bank A
    b. Total interest for Bank A: $867
       Total interest for Bank B: $918
16. a. $2300   b. $12 300   c. $512.50
17. a. 5 months: $10 187.50
       6 months: $10 217.50
       2 years: $10 900
    b. i. $10 920.25   ii. $10 939.90
       iii. $10 936.25   iv. $10 930.83
    c. Receiving interest payments at the end of each month is the best option. This is because the interest received at the end of each month is reinvested to earn 'interest on interest' more quickly than the other option.

8.6 Review

8.6 Exercise
Multiple choice
1. B
2. E
3. A
4. C
5. E
6. D
7. A
8. B
9. D
10. D

Short answer
11. 25%
12. a. $559   b. $172   c. $129
13. a. $450   b. $1050
14. a. 28%   b. 5.29%   c. 5.17%   d. 50%
15. a. $4   b. $3, $2, $1   c. $410

Extended response
16. a. $1440, $2560, $2800   b. 30%
17. a. 62.5%   b. 37.5%
    c. Vegetarians = 10   d. 32
       Dessert = 30

*11a

		% depreciation	Depreciation ($)	Value ($)
End year	First	10%	$3500	$31 500
	Second	7.5%	$2362.5	$31 500 − 2362.5 = $29 137.50
	Third	6%	$0.06 × 29 137.5 = $1748.25	$27 389.25
	Fourth	5%	$1369.46	$26 019.79
	Fifth	4%	$1040.79	$24 979

18. **a.** Less than $100

   b. 30% of $100 is $30.
   $100 + $30 = $130
   30% of $130 is $39.
   $130 − $39 = $91
   This shows that the final result is less than $100.

   c. Less than $100

   d. 30% of $100 is $30.
   $100 − $30 = $70
   30% of $70 is $21.
   $70 + $21 = $91
   This shows that the final result is less than $100.

   e. 23%, 43%

   f. The percentage change in part **a** is less than the percentage change in part **c**. This is because percentage change values are larger when a value is increased to a set point (part **c**) than when it is decreased to a set point (part **a**).

9 Equations

Fully worked solutions for this topic are available online.

9.1 Overview

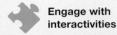

9.1.1 Introduction

Equations are used to describe and model everyday situations using variables and mathematical symbols. Most likely you have been solving equations without even realising it. For example, if you pay for an item that costs $4.25 and you hand a $5.00 note to the shop assistant, you will expect to receive 75 cents in change. This answer is obtained using an equation. Solving equations means determining the numbers you didn't originally know.

Understanding how to solve equations is one of the most useful skills that you could apply in many situations. For example, to work out which mobile phone plan gives the best value, you need to solve equations. If you are travelling overseas to a country that uses a different currency, you can use the exchange rate and equations to work out the cost of an item in Australian dollars. The computer chips used in washing machines, dryers, laptops, cars and other items are based on mathematical equations and algorithms. Equations are also put to use in traffic control systems, space programs, aircraft, medicine and in many other areas.

Meteorologists use equations with many variables to predict the weather for days into the future. Equations are also used in mathematical modelling that helps us determine and predict trends such as investments, house prices or the spread of COVID-19.

KEY CONCEPTS

This topic covers the following key concepts from the VCE Mathematics Study Design:
- construction, use and interpretation of formulas and symbolic expressions to describe relationships between variables and to model and represent generalisations and patterns
- manipulation of symbolic expressions and solution of equations.

Source: VCE Mathematics Study Design (2023–2027) extracts © VCAA; reproduced by permission.

9.2 Number patterns and pronumerals

LEARNING INTENTION

At the end of this subtopic you should be able to:
- determine the patterns and rules for sequences
- use pronumerals (variables) to model and represent patterns
- construct expressions with pronumerals (variables) from given information.

9.2.1 Number patterns

Algebra is a branch of mathematics that deals with number patterns and rules.

The number of sticks required to complete the triangles shown follows a pattern.

 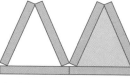 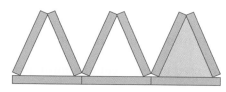

| 3 sticks | 3 sticks + 3 sticks = 6 sticks | 6 sticks + 3 sticks = 9 sticks |

The number of sticks required to make each shape follows the pattern 3, 6, 9, ... ; we can see that the **rule** is 'add 3 sticks for each additional triangle'.

A pattern can often be described in more than one way. Another rule to describe this pattern is 'the number of sticks is equal to 3 times the number of triangles'.

WORKED EXAMPLE 1 Determining the next three numbers and the rule for a sequence

Determine the next three numbers in the sequence 5, 9, 13, 17, ... and state the rule.

THINK

1. Examine the numbers carefully and look for a pattern. The sequence starts with 5 and increases by 4 at first. Check that this works for all the numbers in the sequence; $9 + 4 = 13$ and $13 + 4 = 17$.

2. Write the next three numbers by adding 4 to the previous number each time. State the rule.

WRITE

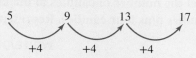

The next three numbers are 21, 25, 29. The rule is 'start with 5 and add 4 each time'.

WORKED EXAMPLE 2 Determining the next five numbers and the rule for a sequence

Determine the next five numbers in the sequence 1000, 500, 250, ... and state the rule.

THINK

1. Examine the numbers carefully and look for a rule.
   - The sequence starts with 1000 and decreases by 500 at first ($1000 - 500 = 500$), but the next number is 250, which doesn't fit this pattern, so subtraction doesn't work.

WRITE

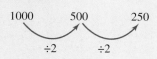

- The rule could be division, because the numbers are decreasing. Check $1000 \div 2 = 500$ and $500 \div 2 = 250$; this works. (Note that the rule could also be $\times \frac{1}{2}$.)

2. Write the next five numbers by dividing the previous number by 2 each time. State the rule.

The next five numbers are: 125, 62.5, 31.25, 15.625, 7.8125. The rule is 'start with 1000 and divide by 2 each time'.

9.2.2 Pronumerals (Variables)

A **pronumeral** or **variable** is a letter or symbol that is used to represent a number. For example, p could represent the number of pies in a shop. Any letter can be used as a variable.

Using pronumerals or variables

Variables are used to represent unknown quantities.

For example, without knowing the number of fish in a pond, the number of fish in the pond could be represented by the variable f.

In the pattern shown by the three sticks in subsection 9.2.1, if t represents the number of triangles, then a rule for the number of sticks needed for the pattern could be written as: number of sticks $= 3 \times t$.

WORKED EXAMPLE 3 Representing information using variables

We don't know exactly how many candies there are in a jar, but we know each jar has the same number of candies. We will refer to this unknown number as c, where c is the variable.
a. Draw a jar and label the number of candies in the jar as c.
b. There is a jar of candies plus four candies. Represent this information using pictures and symbols.

THINK	WRITE/DRAW
a. The number of candies in the jar is unknown.	a. Let c be the number of candies in the jar.

c

b. Draw a jar of candies and four candies. Use c in an algebraic expression to represent this information.

b. $c + 4$

9.2.3 Terms and expressions

Terms can be numbers or variables, or a combination of numbers and variables multiplied together.

For example, $m, 5, 2a$ and $7mn$ are all terms.

The **coefficient** is the number in front of the variable(s). For example, in the term $3t$, the coefficient is 3.

Terms such as $a \times a$ are expressed as a^2. Similarly, $b \times b \times b$ is expressed as b^3.

Constant terms are numbers without variables.

Expressions are mathematical sentences made up of terms separated by + or − signs.

For example, $3a + 5$, $9 - 2m$ and $a^2 + 2ab + b^2$ are all expressions.

The definitions for this section are summarised in the following diagram.

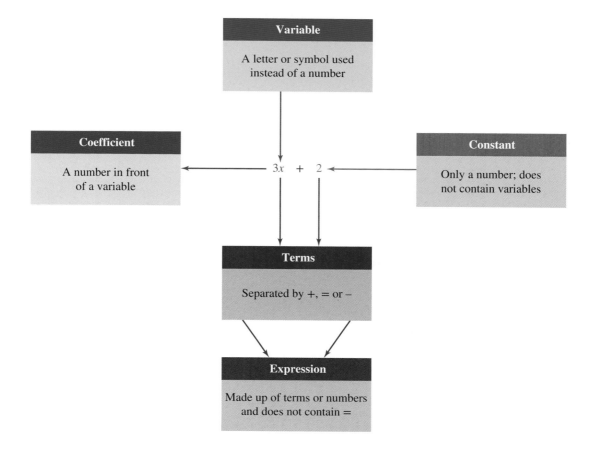

Remember

In terms that contain numbers and variables multiplied together, the number is placed at the front and the multiplication signs are removed.

For example, $3 \times x \times y$ would be simplified to $3xy$.

Variables in a term are written in alphabetical order.

For example, $4 \times c \times b \times a$ is written as $4abc$.

---

tlvd-4683

WORKED EXAMPLE 4 Writing expressions for given information

Write an expression for:
a. the total number of pencils in the pencil case shown
b. the total number of candies left if eight are eaten from the jar.

THINK	WRITE
a. 1. The number of pencils in the pencil case is unknown. Choose a variable to represent this amount.	**a.** Let $p =$ the number of pencils in the pencil case.

THINK	WRITE
2. The total number of pencils includes all the pencils in the pencil case (p) and (+) the five pencils outside the pencil case. Write an expression to represent this.	$p + 5$
b. 1. The number of candies in the jar is unknown. Choose a variable to represent this amount.	**b.** Let $c =$ the number of candies in the jar.

THINK	WRITE
2. The total number of candies includes all the candies in the jar (c) minus ($-$) the eight candies that were eaten. Write an expression to represent this.	$c - 8$

WORKED EXAMPLE 5 Simplifying when terms or variables are multiplied

If one jar of candies has c candies, then two jars of candies will have double the number of candies — that is, $2 \times c = 2c$. In algebra, $2c$ means two lots of c, $3c$ means three lots of c, and so on.

a. There are two jars of candies and two loose candies. Each jar has the same number of candies. Write an expression for the total number of candies using symbols.

b. There are four jars of candies, three packets of mini chocolate eggs and three loose eggs. Each jar has the same number of candies and each packet of mini chocolate eggs has the same number of mini chocolate eggs. Write an expression for the total number of candies and mini chocolate eggs using symbols.

THINK

a. There are two jars of candies (c) plus (+) the two that are loose.

b. There are three jars of candies (c) and three packets of mini chocolate eggs (e) plus (+) the three loose candies.

WRITE

a. Let c be the number of candies in the jar.
$$2 \times c + 2 = 2c + 2$$

b. Let c be the number of candies in each jar and p be the number of mini chocolate eggs in each packet.
$$4 \times c + 3 \times e + 3 = 4c + 3e + 3$$

9.2 Exercise

1. Describe a strategy for determining the rule for a number pattern.

2. **WE1, 2** For each of the following number patterns, determine the next three numbers and state the rule.

   a. 6, 8, 10, 12, ...

   b. 19, 18, 17, 16, ...

   c. 25, 50, 75, 100, ...

   d. 80, 40, 20, 10, ...

   e. $\frac{1}{2}, 1\frac{1}{2}, 2\frac{1}{2}, 3\frac{1}{2}, ...$

   f. $\frac{1}{3}, \frac{1}{6}, \frac{1}{12}, \frac{1}{24}, ...$

3. Write a rule for each of the following patterns.

a.

b.

c.

4. a. Define the meaning of *variable*.
   b. Explain why it is important to define the variable in an expression.

5. **WE3** If b represents the total number of apples in a fruit bowl, write a short paragraph to explain the following table.

Day	Number of apples in the fruit bowl
Monday	b
Tuesday	$b - 4$
Wednesday	$b - 6$
Thursday	$b + 8$
Friday	$b + 5$

6. Answer True or False for each of the statements below.

a. $3x$ is a term.
b. $3mn$ is a term.
c. $g = 23 - t$ is an expression.
d. $g = 5t - 6$ is an equation.
e. $3x = 9$ is an expression.
f. The expression $g + 2t$ has two terms.

7. **WE4** The variable t represents the number of televisions in your house. Write an expression to represent the number of televisions in each of the following locations.

a. Your school has 35 times as many televisions as you have in your house.
b. Your cousin's house has one television fewer than your house.
c. Your friend has five more televisions at their house than you have at your house.
d. Your neighbour has half as many televisions as you have at your house.

8. **MC** In each of the following, M is used to represent an unknown whole number. Choose the answer that matches the description given in each case.

a. Six more than the number

A. $6 + M$ B. $6M$ C. $M - 6$ D. $M \div 6$ E. $6 - M$

b. A fifth of the number

A. $5M$ B. $M - 5$ C. $\dfrac{M}{5}$ D. $M + \dfrac{1}{5}$ E. $\dfrac{1}{5} - M$

c. The number just before that number

A. $M + 1$ B. $M - 1$ C. M D. $M + 2$ E. $M - 2$

d. Eight more than the product of that number and 10

A. $10(M + 8)$ B. $10M + 8$ C. $8M + 10$ D. $8(M + 10)$ E. $88M$

e. Five more than three times the number

A. $5M + 3$ B. $3M - 5$ C. $3M + 5$ D. $5M - 3$ E. $3(M - 5)$

9. Using n to represent the unknown number, write an expression for each of the following.

a. Add 3 to the number.
b. Take 5 from the number.
c. Double the number and then add 6.
d. Multiply the number by 7 and then add 52.
e. Multiply the number by 3 and then subtract 15.

10. Write the following expressions in ascending order.

a. $m + 3$, $m - 2$, $m + 1$, $m - 8$
b. $n - 15$, $n + 25$, $n - 10.3$

11. Write your own descriptions for what the following expressions could represent.

   a. $d + 2$, d, $d - 1$
   b. t, $2t$, $t + 1$
   c. y, $3y$, $10y$

12. Write expressions for the following rules.

   a. The number of students left in the class if X students leave for the canteen out of a total group of T students.
   b. The amount of money earned by selling B cakes, where each cake is sold for $4.00.
   c. The total number of sweets if there are G bags of sweets with 45 sweets in each bag.
   d. The cost of one concert ticket if 5 tickets cost T.

13. There are n mice in a cage. If the number of mice doubles each month, write an expression for the number of mice:

   a. 1 month from now
   b. 2 months from now
   c. 3 months from now.

14. Write an expression for the total number of coins in each of the following. The symbol  represents a full moneybox and ⬭ represents one coin. Use c to represent the number of coins in a full moneybox.

   a.
   b.
   c.
   d.

15. **WE5** There are a apples at your house. At your friend's house there are six more apples than there are at your house. Your friend uses eight apples to make an apple pie.
   Use this information to write an expression describing the number of apples at your friend's house:

   a. before making the apple pie
   b. after making the apple pie.

9.3 Building expressions

LEARNING INTENTION

At the end of this subtopic you should be able to:
- draw a flow chart to describe the steps in an expression.

9.3.1 Flow charts

A **flow chart** is a diagram of the separate steps in a process.

For example, instructions for ordering a pizza are shown in the diagram.

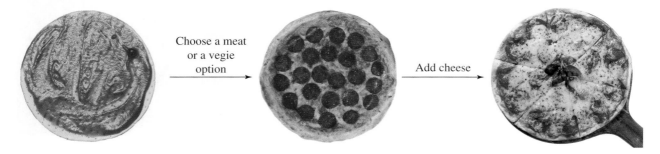

Choose a meat or a vegie option → Add cheese

The instructions for each step are written above the arrows that join the boxes.

In the flow charts, the **starting number** is placed in the first box. Move through the flow chart by carrying out each operation on the number in the previous box. The answer is the number in the last box.

$$9 \xrightarrow{+5} 14 \xrightarrow{\times 2} 28 \xrightarrow{\div 7} 4$$

Starting number   Answer

In the flow chart above, the starting number is 9. Carrying out the operations in the order displayed results in the answer of 4 — the number in the last box.

Flow charts can be used to build **algebraic expressions**. A variable represents the starting number. Each operation is then performed on the *whole expression* in the previous box. The flow chart below shows how the expression $2(m + 5) + 7$ is built.

$$m \xrightarrow{+5} m + 5 \xrightarrow{\times 2} 2(m + 5) \xrightarrow{+7} 2(m + 5) + 7$$

Add 5 to m to give $m + 5$.

Multiply $m + 5$ by 2 to give $2(m + 5)$.

Add 7 to $2(m + 5)$ to give $2(m + 5) + 7$.

tlvd-4684

Draw a flow chart to show how the expression $3\left(\dfrac{n-5}{8}\right)$ is built.

THINK

1. The starting number is represented by the variable n. Subtract 5 to give $n-5$, as shown in blue.

2. Divide the expression $n-5$ by 8 to give $\dfrac{n-5}{8}$, as shown in red.

3. Multiply the expression $\dfrac{n-5}{8}$ by 3 to give $3\left(\dfrac{n-5}{8}\right)$, as shown in green.

WRITE

$$n \xrightarrow{-5} n-5$$

$$n \xrightarrow{-5} n-5 \xrightarrow{\div 8} \dfrac{n-5}{8}$$

$$n \xrightarrow{-5} n-5 \xrightarrow{\div 8} \dfrac{n-5}{8} \xrightarrow{\times 3} 3\left(\dfrac{n-5}{8}\right)$$

9.3.2 Substituting numerical values

Expressions can be evaluated by substituting numerical values for the variables, if the values of the variables are known. Remember the order of operations (BIDMAS) when evaluating expressions.

Build a flow chart to represent the expression $\dfrac{4g-8}{3}+7$, and then use substitution to evaluate the expression when $g=11$.

THINK

1. Start the flow chart with g in the first box. To build the expression $4g$, we need to multiply g by 4.

2. To build the expression $4g-8$, subtract 8 from $4g$.

3. To build the expression $\dfrac{4g-8}{3}$, divide $4g-8$ by 3.

4. To build the expression $\dfrac{4g-8}{3}+7$, add 7 to $\dfrac{4g-8}{3}$.

5. Substitute $g=11$ into the flow chart.

6. Write the answer.

WRITE

$$g \xrightarrow{\times 4} 4g$$

$$g \xrightarrow{\times 4} 4g \xrightarrow{-8} 4g-8$$

$$g \xrightarrow{\times 4} 4g \xrightarrow{-8} 4g-8 \xrightarrow{\div 3} \dfrac{4g-8}{3}$$

$$g \xrightarrow{\times 4} 4g \xrightarrow{-8} 4g-8 \xrightarrow{\div 3} \dfrac{4g-8}{3} \xrightarrow{+7} \dfrac{4g-8}{3}+7$$

$$11 \xrightarrow{\times 4} 44 \xrightarrow{-8} 36 \xrightarrow{\div 3} 12 \xrightarrow{+7} 19$$

$\dfrac{4g-8}{3}+7$ is equal to 19 when $g=11$.

1. Explain why flow charts are helpful when building algebraic expressions.

2. Build an expression by following the instructions on each of the following flow charts.

   a.

   n $\times 2$ ☐ $+ 1$ ☐

   b.

   k $\div 2$ ☐ $+ 1$ ☐ $\times 3$ ☐

   c.

   m $\times 2$ ☐ $- 7$ ☐ $\div 5$ ☐

3. Build an expression by following the instructions on each of the following flow charts.

   a.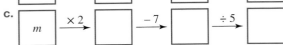

   c $\times -1$ ☐ $+ 5$ ☐ $\div 8$ ☐

   b.

   a $- 12$ ☐ $\times 3$ ☐ $\div 6$ ☐

   c.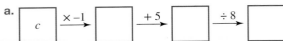

   s $\div 2$ ☐ $+ 1$ ☐ $\times 7$ ☐

4. Use a flow chart to build the following information into an expression.
   a. Start with y. Add 15 and divide by 10.
   b. Start with r. Divide by 4, subtract 1 and multiply by 2.
   c. Start with f. Divide by 8, subtract 1 and multiply by 8.
   d. Start with e. Multiply by 11, subtract 100 and divide by 3.

5. When building each of the following expressions, determine the first operation performed on the variable.

   a. $\dfrac{2w + 8}{7}$
   b. $\dfrac{3(4d + 16)}{27}$
   c. $-v + 7$

6. When building each of the following expressions, determine the first operation performed on the variable.

   a. $19\left(2p + \dfrac{7}{11}\right)$
   b. $y^2 + 7$
   c. $\dfrac{k^3 + 7}{4}$

7. Determine the last operation performed on the variable in each of the expressions in questions **5** and **6**.

8. **WE6** Construct a flow chart for each of the following expressions to show how the expressions were built.

   a. $2(x + 8)$
   b. $-3(n - 9)$
   c. $\dfrac{p}{5} - 3$
   d. $\dfrac{t - 3}{5}$

9. Construct a flow chart for each of the following expressions to show how the expressions were built.

   a. $-r + 18$
   b. $\dfrac{3(s - 15)}{7}$
   c. $4\left(\dfrac{2w}{7} - 29\right)$
   d. $\dfrac{1}{2}\left(\dfrac{5y}{7} + 3\right)$

10. **WE7** Evaluate each of the expressions in question **8** for each of the following values.

   a. $x = 2$ **b.** $n = 10$ **c.** $p = 15$ **d.** $t = 23$

11. Evaluate each of the expressions in question **9** for each of the following values.

   a. $r = 5$ **b.** $s = 22$ **c.** $w = 14$ **d.** $y = 3$

12. Fill in the missing operations on the following flow charts.

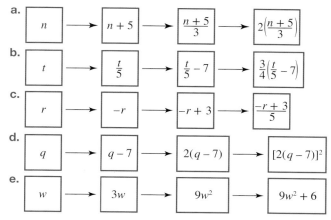

   a. $n \longrightarrow n + 5 \longrightarrow \dfrac{n+5}{3} \longrightarrow 2\left(\dfrac{n+5}{3}\right)$

   b. $t \longrightarrow \dfrac{t}{5} \longrightarrow \dfrac{t}{5} - 7 \longrightarrow \dfrac{3}{4}\left(\dfrac{t}{5} - 7\right)$

   c. $r \longrightarrow -r \longrightarrow -r + 3 \longrightarrow \dfrac{-r+3}{5}$

   d. $q \longrightarrow q - 7 \longrightarrow 2(q-7) \longrightarrow [2(q-7)]^2$

   e. $w \longrightarrow 3w \longrightarrow 9w^2 \longrightarrow 9w^2 + 6$

13. Translate the following problems into flow charts using letters and numbers.

   a. A classmate had some money for lunch and was then given $6. They then gave half of their money to a friend, and afterwards lost $2 down the drain.

   b. Asha is collecting bottle caps. In a week, she doubles her collection, but then gives five to you. Later, her uncle gives her another 10, but then she has to give half of her collection to her younger brother.

   c. You have a number of balloons for a party. When you were blowing them up, five of them burst. You try to return them to the shop, but instead you buy double the number you already have. When you blow these up, only three burst.

14. **a.** Write your own short story involving numbers and build an expression using a flow chart.

   b. Swap your flow chart with a classmate's and see if you can think of a story for your classmate's flow chart.

15. The formula for calculating the perimeter (P) of a rectangle of length l and width w is $P = 2l + 2w$. Use this formula to calculate the perimeter of the rectangular singles tennis court shown.

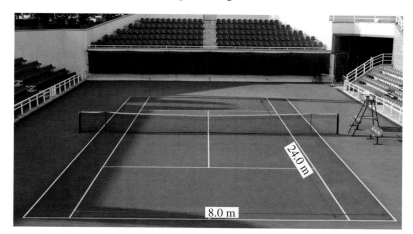

24.0 m

8.0 m

9.4 Backtracking to solve equations

9.4.1 Backtracking

Flow charts used to build expressions can be modified to represent equations.

The flow chart shown represents the equation $3x + 7 = 52$. The left-hand side of the equation, $3x + 7$, is built through the flow chart. The right-hand side of the equation, 52, is placed underneath the last box.

Backtracking through a flow shown can be used to solve equations. This involves working backwards through the flow chart, starting with the last box and working against the arrows using **inverse operations**.

The following is a list of operations and their inverses (opposites).

> ## Inverse operations
>
> $-$ **is the inverse operation of** $+$
>
> $+$ **is the inverse operation of** $-$
>
> $\div$ **is the inverse operation of** $\times$
>
> $\times$ **is the inverse operation of** $\div$

Consider the flow chart and the operations that were applied to obtain the answer 52 (shown with purple arrows). Backtracking can be used to solve the equation by applying inverse operations when working backwards through the flow chart (shown in pink).

The number under the variable is the solution to the equation.

$$3x + 7 = 52$$
$$3x = 45$$
$$x = 15$$

> ## Checking whether the solution is correct
>
> **To check whether a solution to an equation is correct, work out the left-hand side (LHS) of the equation separately from the right-hand side (RHS). If the solution is correct, the LHS will equal the RHS.**

tlvd-4685

WORKED EXAMPLE 8 Using a flow chart and backtracking to solve an equation

Use a flow chart and backtracking to solve the equation $\dfrac{x}{6} - 5 = 3$.

THINK

1. Build a flow chart to represent the expression on the LHS of the equation. Start with x, divide by 6 and then subtract 5. Write the number 3 under the last box.

2. Backtrack through the flow chart to determine the value of x. Start with 3 and apply the inverse operation at each step; the inverse of -5 is $+5$ ($3 + 5 = 8$), so write 8 under the middle box. The inverse of $\div 6$ is $\times 6$ ($8 \times 6 = 48$), so write 48 under the first box.

3. The solution is the number underneath the variable in the first box.

4. Check your answer by substituting 48 into the equation.

WRITE

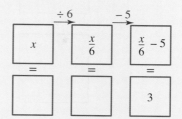

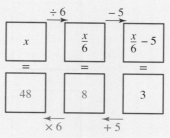

$x = 48$

$$\dfrac{x}{6} - 5 = 3$$
$$\dfrac{48}{6} - 5 = 3$$
$$8 - 5 = 3$$
$$3 = 3$$

As the LHS is equal to the RHS, the answer is correct.

To solve problems presented in words, turn the problem into an equation, solve the equation and use the solution to answer the problem.

WORKED EXAMPLE 9 Using a flow chart and backtracking to solve a worded problem

In a book sale, the price of a particular book was reduced by $11. If you purchase two copies of the book and a $4 magazine during the sale, it will cost you $20.
a. Write an equation to represent the situation.
b. Use backtracking to calculate the original price of the book.

Book sale

THINK

a. 1. Read the question carefully. Identify the unknown variables.

2. Two copies of the book at the reduced price and a $4 magazine cost $20. Represent this as an equation and simplify.

WRITE

a. Let $r =$ original price of the book.
The discounted price of the book is $r - 11$.

$2(r - 11) + 4 = 20$

b. 1. Represent the equation as a flow chart.

b.

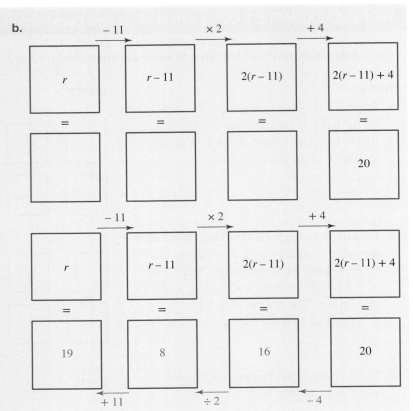

2. Backtrack to solve for r.

3. Check the solution. If the books were originally $19 each, then on sale they were $8 each. Two books at $8 each plus a $4 magazine cost $20, as required.

4. Write the answer in words.

The original price of the book was $19.00.

9.4 Exercise

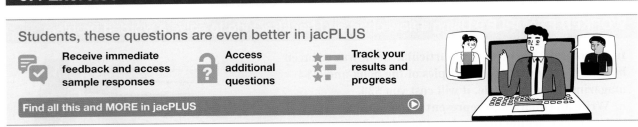

Students, these questions are even better in jacPLUS

Receive immediate feedback and access sample responses

Access additional questions

Track your results and progress

Find all this and MORE in jacPLUS

1. Explain how to solve an equation using backtracking.

2. Copy and complete the following sentences.
   a. To undo a step where you have added, you _____ .
   b. To undo a step where you have subtracted, you _____ .
   c. To undo a step where you have multiplied, you _____ .
   d. To undo a step where you have divided, you _____ .

3. For each of the following flow charts:

   i. write the equation that is represented by the flow chart

   ii. use backtracking to solve the equation.

   a.
   $$x \xrightarrow{\times 2} 2x \xrightarrow{+7} 2x+7$$
   $$-29$$

   b.
   $$x \xrightarrow{+7} x+7 \xrightarrow{\times 2} 2(x+7)$$
   $$32$$

   c.
   $$k \xrightarrow{\div 3} \frac{k}{3} \xrightarrow{+7} \frac{k}{3}+7 \xrightarrow{\times 3} 3\left(\frac{k}{3}+7\right)$$
   $$-93$$

4. For each of the following flow charts:

   i. write the equation that is represented by the flow chart

   ii. use backtracking to solve the equation.

   a.
   $$e \xrightarrow{\times 2} 2e \xrightarrow{\div 5} \frac{2e}{5} \xrightarrow{+11} \frac{2e}{5}+11$$
   $$17$$

   b.
   $$a \xrightarrow{+7} a+7 \xrightarrow{\div 8} \frac{a+7}{8} \xrightarrow{+9} \frac{a+7}{8}+9$$
   $$-4$$

   c.
   $$w \xrightarrow{\times -2} -2w \xrightarrow{+8} -2w+8 \xrightarrow{\div 3} \frac{-2w+8}{3}$$
   $$-10$$

5. **WE8** Use a flow chart and backtracking to solve each of the following equations.

   a. $3x - 8 = -29$
   b. $5x - 8 = 22$
   c. $2m - 7 = 11$
   d. $8(x + 6) = 24$
   e. $-6(x + 5) = 72$
   f. $4(h - 1.5) = 30$

6. Use a flow chart and backtracking to solve each of the following equations.

   a. $\dfrac{k}{3} + 1 = 10$
   b. $\dfrac{2s + 3}{5} = 7$
   c. $\dfrac{5r}{7} + 6 = 8$
   d. $7\left(\dfrac{x + 5}{6} + 1\right) = -56$

7. **WE9** In each of the following:

   i. write an equation to represent the situation

   ii. use backtracking to answer the question.

   a. The sum of an unknown number and 42 is 21. Determine the unknown number.

   b. A paddock is twice as long as it is wide. If the perimeter of the paddock is 120 m, determine the width of the paddock.

8. In each of the following:

   i. write an equation to represent the situation

   ii. use backtracking to answer the question.

   a. The three angles in a triangle add to 180°. If the second angle is twice the size of the first angle and the third angle is three times the size of the first angle, determine the size of each of the angles.

   b. The average of the values x, $x + 2$ and $3x + 5$ is 10. Calculate the values of the numbers.
   (*Hint:* To calculate the average, add all the values and divide the result by the number of values.)

9. Bread was on special at the supermarket. Your friend bought two loaves and received $5.50 change from $10. Use a flow chart and backtracking to calculate the cost of one loaf of bread.

10. Two friends add their ages in years and discover that together they are 25 years old. If one of the two friends is three years older than the other, determine the age of the younger friend.

11. Today, while training for a half marathon, I ran 1 km more than double the distance I ran yesterday. Calculate how far I ran on each day.
(*Hint:* Use the photo shown.)

I've run a total of 19 km today and yesterday.

12. Your friend tried to complete the following backtracking problems. Explain how he can get the correct answers.

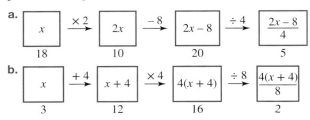

a.

| x | $\xrightarrow{\times 2}$ | $2x$ | $\xrightarrow{-8}$ | $2x - 8$ | $\xrightarrow{\div 4}$ | $\dfrac{2x-8}{4}$ |
| 18 | | 10 | | 20 | | 5 |

b.

| x | $\xrightarrow{+4}$ | $x + 4$ | $\xrightarrow{\times 4}$ | $4(x + 4)$ | $\xrightarrow{\div 8}$ | $\dfrac{4(x+4)}{8}$ |
| 3 | | 12 | | 16 | | 2 |

9.5 Balancing equations

LEARNING INTENTION

At the end of this subtopic you should be able to:
- solve equations by keeping equations balanced
- write equivalent equations.

9.5.1 Keeping equations balanced

Equations are mathematical statements that show two equal expressions.

A **balanced** equation is an equation whose left-hand side and right-hand side are equal.

For example, $6 \times 2 = 12$ is a balanced equation.

If either side of the equation has a greater value than the other side, the equation is **unbalanced**.

For example, the following equation is unbalanced as the left-hand side is not equal to the right-hand side.

6 + 6 = 12   6 × 2 = 12

18 + 3 = 21
8 × 3 = 24

$$18 + 3 = 8 \times 3$$
$$21 \neq 24$$
$$\text{LHS} \neq \text{RHS}$$

For example, the scale in this diagram can be described by the equation $x + 2 = 6$.

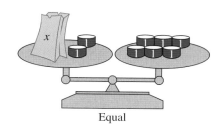

Making changes to both sides of the scale lets you work out how many weights are in the bag (the value of the variable).

If you remove 2 weights from each side of the scale, you can see that the bag with x weights in it weighs the same as 4 weights. That means that $x = 4$.

tlvd-4686

WORKED EXAMPLE 10 Solving an equation by keeping equations balanced

For the following pan balance scale:

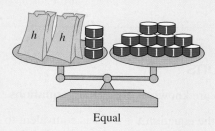

Equal

a. **write the equation represented by the scale**
b. **calculate the value of the variable.**

THINK

a. 1. Examine the balance scale. On the LHS there are 2 bags with h weights in each, and 3 weights outside the bags. On the RHS there are 11 weights.

2. Write this as an equation.

b. 1. Remove 3 weights from both sides so that the balance is maintained (-3). This leaves 2 bags on the LHS and 8 weights on the RHS.

WRITE

a.

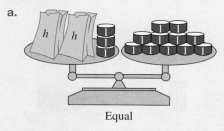

Equal

$2h + 3 = 11$

b.

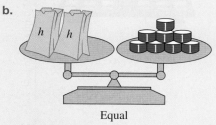

Equal
$$2h + 3 - 3 = 11 - 3$$
$$2h = 8$$

2. To work out the value of h (the value of 1 bag), divide both sides by 2 and then simplify.

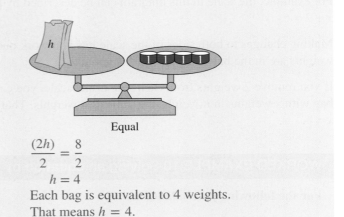

Equal

$$\frac{(2h)}{2} = \frac{8}{2}$$

$$h = 4$$

Each bag is equivalent to 4 weights.
That means $h = 4$.

9.5.2 Equivalent equations

Equations that have the same solution are known as **equivalent equations**.

For example, the diagram shows that the equation $x + 3 = 7$ is equivalent to $x + 5 = 9$, $x = 4$ and $2x + 6 = 14$, if the same operation is done to both sides of the equation.

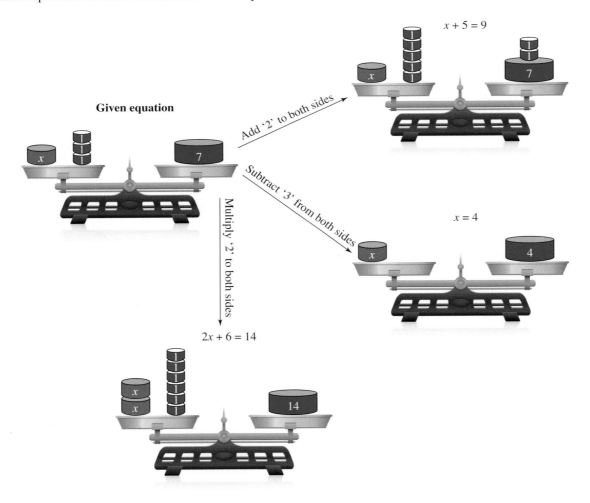

WORKED EXAMPLE 11 Writing an equivalent equation

Write equivalent equations to $4m + 7 = 19$ by:
a. subtracting 7 from both sides
b. multiplying both sides of the equation by 2.

THINK

a. 1. Write the equation and subtract 7 from both sides.

2. Simplify both sides.

b. 1. Write the equation.

2. Multiply both sides of the equation by 2.

3. Expand the brackets and simplify.

4. Write the answer.

WRITE

a.
$$4m + 7 = 19$$
$$4m + 7 - 7 = 19 - 7$$
$$4m = 12$$

b.
$$4m + 7 = 19$$
$$2(4m + 7) = 19 \times 2$$
$$8m + 14 = 38$$

$8m + 14 = 38$ is an equivalent equation to $4m + 7 = 19$.

9.5 Exercise

Students, these questions are even better in jacPLUS

Receive immediate feedback and access sample responses

Access additional questions

Track your results and progress

Find all this and MORE in jacPLUS

1. Explain what you must do to keep a pan balance scale balanced if you want to make a change to one side.

2. **WE10** For each of the following pan balance scales:
   i. write the equation represented by the scale
   ii. calculate the value of the variable.

a.
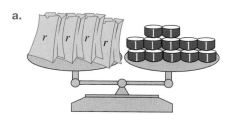
Equal

b.
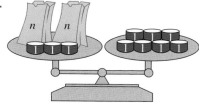
Equal

c.
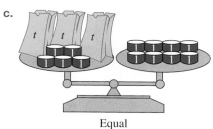
Equal

d.
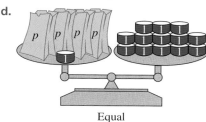
Equal

3. For each of the following equations:
   i. draw a pan balance scale to represent the equation
   ii. calculate the value of the variable.

   a. $q + 5 = 8$
   b. $8d = 24$
   c. $2m + 3 = 7$
   d. $3p + 8 = 11$
   e. $4 + 7x = 32$
   f. $13 = 3 + 5x$

4. Starting with the equation $y = 3$, make equivalent equations by:
   a. multiplying both sides by 10
   b. subtracting 5 from both sides
   c. dividing both sides by 2
   d. adding 28 to both sides.

5. Starting with the equation $a = 5$, make equivalent equations by:
   a. multiplying both sides by -2
   b. multiplying both sides by $\dfrac{4}{5}$
   c. taking $\dfrac{2}{3}$ from both sides
   d. multiplying both sides by 0.3.

6. Write at least three other equations that are equivalent to $x = 1$.

7. **WE11** Make an equivalent equation for each of the equations listed below by performing the operation in brackets on both sides.
   a. $m + 8 = 9$ $(+8)$
   b. $n - 3 = 5$ (-2)
   c. $2p + 3 = 9$ $(+3)$
   d. $3m = 12$ $(\times 4)$

8. Make an equivalent equation for each of the equations listed below by performing the operation given in brackets to both sides.
   a. $12n = 36$ $(\div 6)$
   b. $\dfrac{p}{3} = 5$ $(\times 6)$
   c. $3p + 7 = 2$ $(\times 4)$
   d. $5m - 3 = -1$ $(\div 3)$

9. Write at least three other equations that are equivalent to $4x + 2 = 18$.

10. Explain how you can know if two equations are equivalent.

11. Some simple equations are hard to represent on a pan balance scale.
    a. State what you would need to do to both sides of the equation so that the variable is on its own on the left hand side of the equation.
       i. $x - 4 = 10$
       ii. $x - 6 = 3$
       iii. $\dfrac{w}{3} = 7$
       iv. $\dfrac{e}{2} = 9$
       v. $\dfrac{q}{4} - 2 = 10$
       vi. $\dfrac{q - 2}{4} = 10$
    b. Explain why it is difficult to represent the problems in part **a** on a pan balance scale.
    c. Explain the difference between equations **v** and **vi** in part **a**.
    d. Solve the equations in part **a**.

12. Represent the following problems on pan balance scales and solve for the unknown quantity.
    a. Five bags of sugar weigh three kilos. Determine the weight of one bag of sugar.
    b. Four chocolate bars and an ice cream cost $9.90. If the ice cream costs $2.70, determine the cost of a chocolate bar.
    c. A shopkeeper weighed three apples at 920 g. They thought this was heavy and realised somebody had left a 500-g weight on their balance scale. Calculate the average weight of one apple.

9.6 Solving equations

9.6.1 Solving equations using inverse operations

Both backtracking and the balance method of solving equations use inverse operations to determine the value of a variable. For example, the equation $2q + 3 = 11$ can be solved as follows.

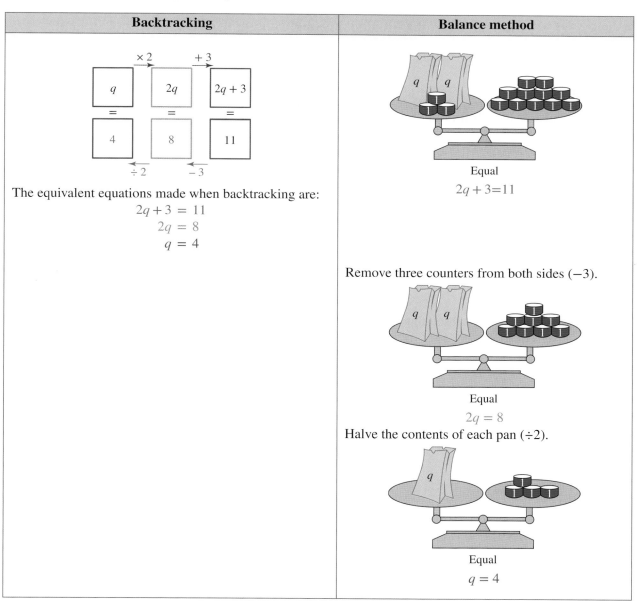

Backtracking	Balance method

The equivalent equations made when backtracking are:
$$2q + 3 = 11$$
$$2q = 8$$
$$q = 4$$

Equal
$$2q + 3 = 11$$

Remove three counters from both sides (-3).

Equal
$$2q = 8$$

Halve the contents of each pan ($\div 2$).

Equal
$$q = 4$$

Solving equations

In both backtracking and the balance method:
- **the operations and the order in which they are performed to solve the equation are the same**
- **the equivalent equations created on the way to the solution are the same**
- **the aim is to get the variable by itself on one side of the equation.**

Substitution can be used to check that the answer to an equation you have solved is correct. Once substituted, if the LHS equals the RHS, then the answer is correct. If the LHS does not equal the RHS, then the answer is incorrect and you should try again.

tlvd-4687

WORKED EXAMPLE 12 Solving equations using inverse operations on both sides

Solve the following equations using inverse operations and by performing the same operations on both sides.

a. $4y + 3 = 31$

b. $\dfrac{t}{3} - 5 = 47$

c. $3(p - 11) + 9 = 57$

Check the answer by substitution.

THINK	WRITE
a. 1. The last operation performed when building $4y + 3 = 31$ was $+ 3$. Subtract 3 from both sides and then simplify.	**a.** $4y + 3 = 31$ $4y + 3 - 3 = 31 - 3$ $4y = 28$
2. To get y by itself, divide both sides by 4 and then simplify.	$\dfrac{4y}{4} = \dfrac{28}{4}$ $\dfrac{^1 4y}{^1 4} = \dfrac{^7 28}{^1 4}$ $y = 7$
3. Check the answer by substituting the value of $y = 7$ in the given equation.	$4y + 3 = 31$ $4 \times (7) + 3 = 31$ $28 + 3 = 31$ $31 = 31$ As LHS = RHS, the answer is correct.
b. 1. The last operation performed when building $\dfrac{t}{3} - 5 = 47$ was $- 5$. Add 5 to both sides and then simplify.	**b.** $\dfrac{t}{3} - 5 = 47$ $\dfrac{t}{3} - 5 + 5 = 47 + 5$ $\dfrac{t}{3} = 52$
2. To get t by itself, multiply both sides by 3 and then simplify.	$\dfrac{t}{3} \times \dfrac{3}{1} = 52 \times 3$ $\dfrac{t}{^1 3} \times \dfrac{^1 3}{1} = 156$ $t = 156$

3. Check the answer by substituting the value of $t = 156$ in the given equation.

$$\frac{t}{3} - 5 = 47$$
$$\frac{(156)}{3} - 5 = 47$$
$$52 - 5 = 47$$
$$47 = 47$$

As LHS = RHS, the answer is correct.

c. 1. The last operation performed when building $3(p - 11) + 9 = 57$ was $+9$. Subtract 9 from both sides and then simplify.

c.
$$3(p - 11) + 9 = 57$$
$$3(p - 11) + 9 - 9 = 57 - 9$$
$$3(p - 11) = 48$$

2. The last operation performed when building $3(p - 11) = 48$ was $\times 3$. Divide both sides by 3 and then simplify.

$$\frac{3(p - 11)}{3} = \frac{48}{3}$$
$$\frac{{}^{1}\cancel{3}(p - 11)}{{}^{1}\cancel{3}} = \frac{{}^{16}\cancel{48}}{{}^{1}\cancel{3}}$$
$$p - 11 = 16$$

3. The last operation performed when building $p - 11 = 16$ was -11. Add 11 to both sides and then simplify.

$$p - 11 + 11 = 16 + 11$$
$$p = 27$$

4. Check the answer by substituting the value of $p = 27$ in the given equation.

$$3(p - 11) + 9 = 57$$
$$3(27 - 11) + 9 = 57$$
$$3 \times 16 + 9 = 57$$
$$57 = 57$$

As LHS = RHS, the answer is correct.

9.6 Exercise

Students, these questions are even better in jacPLUS

 Receive immediate feedback and access sample responses

 Access additional questions

 Track your results and progress

Find all this and MORE in jacPLUS

1. In your own words, describe how to solve equations using inverse operations and by performing the same operations on both sides.

2. **WE12a** Solve the following equations by using inverse operations and performing the same operations on both sides.

   a. $2a = 18$ **b.** $b - 4 = 48$ **c.** $\dfrac{c}{8} = 2$ **d.** $-2d = 16$

3. Solve the following equations by using inverse operations and performing the same operations on both sides.

   a. $-e = 5$ **b.** $4f = 16$ **c.** $g + 7 = 11$ **d.** $\dfrac{u}{3} = 4$

4. Determine the first operation that would be undone when solving the following equations.

   a. $3a + 9 = 87$ **b.** $(b - 7) + 3 = 5$ **c.** $\dfrac{(c - 2)}{5} + 11 = 8$ **d.** $\dfrac{3(2d + 7)}{11} = -1$

5. **WE12b** Solve the following equations by using inverse operations and performing the same operations on both sides.

   a. $2x + 4 = 10$　　　　　b. $3p - 2 = 7$　　　　　c. $7a + 15 = 57$　　　　　d. $11a - 13 = 112$

6. Solve the following equations by using inverse operations and performing the same operations on both sides.

   a. $\dfrac{c}{2} + 6 = 28$　　　b. $\dfrac{d+8}{7} = 1$　　　c. $4(e + 7) = 32$　　　d. $\dfrac{k-10}{3} = 12$

7. a. Without using a calculator, solve $4(a + 2) = 30$ by dividing both sides by 4 first.
   b. Without using a calculator, solve $4(a + 2) = 30$ by expanding the brackets first and then solving.
   c. Explain what you notice about the solutions to parts **a** and **b**.
   d. Explain when it is simpler to expand the brackets first instead of dividing both sides by the coefficient of the brackets.

8. Solve the following equations.

   a. $3(a + 6) = 24$　　　　b. $2(b - 7) = 22$　　　　c. $4(c + 7) = 51$　　　　d. $7(d - 15) = 61$

9. **WE12c** Solve the following equations by using inverse operations and performing the same operations on both sides.

   a. $3(a - 7) - 2 = 23$　　　　b. $\dfrac{b+8}{2} + 9 = 30$　　　　c. $\dfrac{2(a+2)}{4} = 2$

   d. $5\left(\dfrac{d+12}{10}\right) = 100$　　　e. $\dfrac{4(e-2)}{6} + 7 = 19$　　　f. $\dfrac{8f}{10} + 6 = 14$

   Check the answer by substitution.

10. Simplify the following equations by collecting like terms, and then solve.

    a. $4x + 3x + 7 = 21$　　　　b. $25 + 3b + 10 - 4b = 15$　　　　c. $5c - 6c + 17 = 38$

11. In each of the following:

    a. calculate the value of the variable
    b. redraw each shape showing the length of each side in centimetres.

    i.
    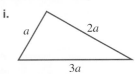
    Perimeter = 15 cm

    ii.
    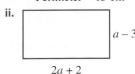
    Perimeter = 25 cm

    iii.
    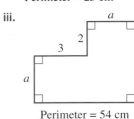
    Perimeter = 54 cm

12. The cost of catching a taxi can be calculated using the formula $C = 3.10 + 1.45k$, where C is the total cost of the taxi in dollars and k is the number of kilometres travelled.

a. Calculate the distance travelled in a taxi if the total cost of the trip is:

  i. $40.80
  ii. $5.20
  iii. $25.30

b. Calculate the amount it would cost to catch a taxi from your home to school.

9.7 Rearranging formulas

LEARNING INTENTION

At the end of this subtopic you should be able to:
• rearrange formulas using backtracking and inverse operations.

9.7.1 Rearranging formulas using backtracking

A formula is a mathematical expression that shows the relationship between different quantities.

> **Area of a triangle:**
>
> $$A = \frac{1}{2}bh$$
>
> **where b = the length of the base**
> **h = the height**
>
> **Speed:**
>
> $$s = \frac{d}{t}$$
>
> **where d = distance**
> **t = time**

It is possible to calculate the value of any variable in the formula as long as you are given the value of all the other variables in the formula. The variable by itself on one side of the formula is known as the **subject of the formula**. The subject of the formula $A = \frac{1}{2}bh$ is A; the subject of the formula $s = \frac{d}{t}$ is s.

Using backtracking techniques, it is possible to rearrange formulas to make any variable the subject of the formula.

For example, the formula $s = \frac{d}{t}$ can be rearranged to:

$$d = st \quad \text{(multiplying both sides with } t)$$
$$t = \frac{d}{s} \quad \text{(multiplying both sides with } \frac{t}{s})$$

WORKED EXAMPLE 13 Rearranging the formula using backtracking

Rearrange the following formula to make the variable a the subject.

$$v = at + u$$

THINK

1. Think about how the formula was built up from a and draw a flow chart representing this.

WRITE

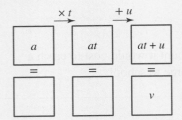

2. Backtrack through the flow chart one step at a time: the last step was $+u$, so you need to subtract u.

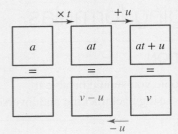

3. The other step was $\times t$, so you need to divide by t.

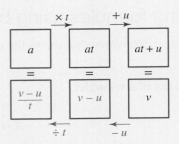

4. Write the new equation with a as the subject.

$$a = \frac{v - u}{t}$$

9.7.2 Rearranging formulas using inverse operations

Inverse operations may also be used to rearrange formulas.

WORKED EXAMPLE 14 Rearranging a formula using inverse operations

The area of a triangle is given by the formula $A = \dfrac{1}{2}bh$. Rearrange the formula to make the variable b the subject of the formula.

THINK

1. Write the formula and decide which operation to undo first.

2. Undo $\times \dfrac{1}{2}$ (the same as $\div 2$) by multiplying both sides by its reciprocal, 2, and then simplifying.

WRITE

$$A = \frac{1}{2}bh$$

$$A \times 2 = \frac{1}{2}bh \times 2$$

$$2A = bh$$

3. Undo $\times h$ by dividing both sides by h and then simplifying.	$\dfrac{2A}{h} = \dfrac{bh}{h}$
	$\dfrac{2A}{h} = b$
4. Rewrite the equation, placing the subject b on the LHS.	$b = \dfrac{2A}{h}$

Note: In Worked example 14, b was multiplied by both $\dfrac{1}{2}$ and h (same order of operation), so you can also divide both sides by h before multiplying both sides by 2.

Remember

The inverse operations with variables work the same way as with numbers; the inverse of $\times b$ is $\div b$, etc.

9.7 Exercise

1. Express in your own words what is meant by *formula*.

2. State the subject of each formula.

   a. $y = mx + c$ b. $E = mc^2$ c. $P = 2L + 2W$ d. $c^2 = a^2 + b^2$

3. State the subject of each formula.

   a. $d = 4.9t^2$ b. $F = \dfrac{5C}{9} + 32$ c. $v = \dfrac{m}{d}$ d. $E = \dfrac{mv^2}{2}$

4. State the subject of each formula.

   a. $S = 180(n - 2)$ b. $r^2 = x^2 + y^2$ c. $I = \dfrac{V}{R}$ d. $T = a + d(n - 1)$

5. **WE13** Rearrange the formula in the corresponding part of question 2 to make the given variable the subject of the formula.

   a. x b. m c. W d. b^2

6. Rearrange the formula in the corresponding part of question 3 to make the given variable the subject of the formula.

   a. t^2 b. C c. m d. m

7. Rearrange the formula in the corresponding part of question 4 to make the given variable the subject of the formula.

   a. n b. y^2 c. V d. n

8. **WE14** The area of the trapezium shown is given by the formula $A = h\left(\dfrac{a+b}{2}\right)$.

   Rearrange the formula to make:

   a. a the subject of the formula
   b. b the subject of the formula
   c. h the subject of the formula.

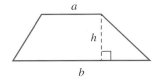

9. Answer the following questions.

   a. Write the formula for the perimeter of the rectangle shown.
   b. Rearrange the formula to make l the subject of the formula.
   c. Rearrange the formula to make w the subject of the formula.

10. **MC** A rearranged version of the formula $a = \dfrac{b - dc}{e}$ is:

    A. $e = ab - dc$

    B. $d = ae - \dfrac{b}{c}$

    C. $b = ae + dc$

    D. $c = \dfrac{ae - b}{d}$

    E. $ae + b = dc$

11. The total surface area of a cylinder can be found using the formula $A = 2\pi r(r + h)$, where r is the radius and h is the height of the cylinder.

    a. Rewrite the formula to make h the subject of the formula.
    b. Use your formula to calculate the height of a cylinder with a radius of 3 cm and a total surface area of 160 cm².

12. Use the formula $\text{speed} = \dfrac{\text{distance}}{\text{time}}$ $\left(s = \dfrac{d}{t}\right)$ to calculate:

    a. the speed of a car that travels 123 km in 2.5 h
    b. the distance covered by a car moving at 91 km/h for 2.25 h
    c. the time it takes a car to travel 334 km when it is travelling at 103 km/h.

13. You have agreed to take on a babysitting job for your mother's friend. You will be paid according to the following formula:

$$P = 10 + 8h$$

   where P is the total pay for one evening's babysitting ($) and h is the number of hours you spend babysitting on that evening.

   a. On the first evening, you babysat for 4 hours. Calculate the amount you earned.
   b. On another occasion, you earned a total of $66. Determine how long you spent babysitting.

14. The area of a circle is found using the formula $A = \pi r^2$.

    a. Rearrange the formula so that it can be used to calculate the radius of a circle if you are given the area of the circle.
    b. Use your rearranged formula to calculate the radii of the two clocks shown.

Area = 1018 cm²   Area = 452 cm²

15. The formula for working out how much water a rectangular swimming pool of uniform depth can hold is given by the formula $V = lwd$, where V is the volume of water (m³), l is the length of the swimming pool (m), w is the width of the swimming pool (m) and d is the depth of the swimming pool (m).

a. Calculate the amount of water the swimming pool can hold if its length is 14 m, its width is 8 m and its depth is 1.8 m when full.

b. Rearrange the formula to make d the subject.

c. Determine the depth of the water if the volume was 179.2 m².

d. When the water is at the depth you found in part **c**, determine how many centimetres it is below the top of the pool.

9.8 Applications of equations and formulas

LEARNING INTENTION

At the end of this subtopic you should be able to:
- construct equations for worded problems
- solve equations for worded problems.

9.8.1 Solving worded problems

To solve problems presented in words, turn the problem into an equation, solve the equation and then use the solution to answer the problem.

Steps to solve worded problems

1. Read the question carefully and identify the unknown variables.

2. Construct an appropriate equation using your variable.

3. Solve the equation.

4. Check your solution to see if it makes sense in the context of the question.

5. Write the answer to the question.

WORKED EXAMPLE 15 Constructing an appropriate equation and then solving it

The ages of three sisters add to 35 years. If the eldest is 4 years older than the middle sister and the youngest is 2 years younger than the middle sister, calculate the ages of the three sisters.

THINK	WRITE
1. Read the question carefully and identify the unknown variables.	Let m be the middle sister's age. Then the eldest sister is $m + 4$ years old and the youngest sister is $m - 2$ years old.
2. The sum of the sisters' ages is 35 years. Construct an appropriate equation using your variable.	Middle sister's age + eldest sister's age $\quad$ + youngest sister's age $= 35$ years $m + m + 4 + m - 2 = 35$
3. Solve the equation by: • subtracting 2 from both sides. • dividing both sides by 3.	$3m + 2 = 35$ $3m + 2 - 2 = 35 - 2$ $3m = 33$ $\dfrac{3m}{3} = \dfrac{33}{3}$ $m = 11$
4. Check your solution to see if it makes sense in the context of the question. It is reasonable for the middle sister to be 11 years old, which is the solution to the equation. This makes the eldest sister $11 + 4 = 15$ years old and the youngest sister $11 - 2 = 9$ years old. The ages of the three sisters add to 35 years.	If $m = 11$, LHS $= 3m + 2 \qquad$ RHS $= 35$ LHS $= 3 \times 11 + 2$ LHS $= 33 + 2$ LHS $= 35$ LHS $=$ RHS Therefore, $m = 11$ is a solution.
5. Write the answer as a sentence.	The ages of the three sisters are 15, 11 and 9.

WORKED EXAMPLE 16 Constructing and evaluating an equation

A gardener constructs a fence with three strands of wire around a rectangular garden bed. The gardener bought a roll of wire that was 100 m long and had 4 m left over. If the length of the garden bed was 2 m longer than its width, calculate the dimensions of the garden bed.

THINK	WRITE
1. Read the question carefully and identify the unknown variables.	Let $w =$ the width of the garden bed. The length will be $w + 2$.

2. There are three strands of wire around the garden bed and 4 m left over. Construct an equation to represent this using the variables.

$$\text{One strand (perimeter)} = 2(w) + 2(w + 2)$$
$$= 2w + 2w + 4$$
$$= 4w + 4$$
$$\text{For three strands} = 3(4w + 4)$$
$$= 12w + 12$$
$$\text{Add leftover wire} = 12w + 12 + 4$$
$$= 12w + 16$$
$$\text{Total length of wire} = 100$$
$$\text{Equation: } 12w + 16 = 100$$

3. Solve the equation by:
- subtracting 16 from both sides
- dividing both sides by 12.

$$12w + 16 - 16 = 100 - 16$$
$$12w = 84$$
$$\frac{12w}{12} = \frac{84}{12}$$
$$w = 7$$

4. Check your answer to see if it makes sense in the context of the question.
If $w = 7$, the length is $w + 2 = 9$. One strand of wire would be $14 + 18 = 32$ m long. Three strands would be 96 m long. That leaves 4 m left of the 100 m of wire. So the answer is reasonable.

If $w = 7$,
$$\text{LHS} = 12w + 16 \qquad \text{RHS} = 100$$
$$\text{LHS} = 12(7) + 16$$
$$\text{LHS} = 100$$
$$\text{LHS} = \text{RHS}$$
Therefore, $w = 7$ is a solution.

5. Write the answer as a sentence.

The garden bed is 7 m wide and 9 m long.

9.8 Exercise

Students, these questions are even better in jacPLUS

 Receive immediate feedback and access sample responses

 Access additional questions

Track your results and progress

Find all this and MORE in jacPLUS

1. In your own words, write a list of steps required to solve a problem written in words.

2. **WE15** A pair of shoes and matching bag cost $175. If the shoes cost $50 more than the bag, calculate the prices of the bag and the shoes.

3. A mother is five times as old as her son. Their ages add to 36 years. Calculate the ages of the mother and the son.

4. Your friend is 7 cm taller than you. The total of your heights is 290 cm. Calculate the height of your friend.

5. If a number is halved and then increased by 12, it will be six more than the original number. State the original number.

6. **WE16** Three brothers get an average of $12 pocket money per week. The eldest brother gets $5 more than the middle brother, while the youngest gets $2 less than the middle brother. Calculate the amount of money each brother gets per week.

7. Three sisters have an average age of 15. The eldest sister is 5 years older than the middle sister. The middle sister is 5 years older than the youngest sister. Calculate the ages of the three sisters.

8. During the walkathon around the school oval, you walk five laps more than your friend while your sister walks 10 laps more than you. The three of you raise $96 at a rate of $3 per lap. Determine the number of laps completed by each person.

9. You and your friend have a total of $45 dollars. If your friend has $2 more than four times the amount of money you have, determine how much money you have.

10. a. The sum of three consecutive numbers is 57. Determine the three numbers. (*Hint:* Let $n =$ the smallest number.)
    b. The sum of three consecutive odd numbers is 81. Determine the three numbers.
    c. The sum of four consecutive even numbers is 284. Determine the four numbers.
    d. Four consecutive multiples of 3 add to 130. If the smallest number is n, the next multiple of 3 will be $n + 3$. Determine the four numbers.
    e. Three consecutive multiples of 5 add to 195. Determine the three numbers.

11. a. The result of adding 1 to three times a number is the same as adding 2 to the same number. Determine the number.
    b. The result of adding 1 to two times a number and multiplying by 3 is the same as adding 2 to the number and multiplying by 5. Determine the number.

12. A rectangle has an area of 243 cm^2. One side is three times longer than the other. Calculate the dimensions of the rectangle.

13. There is a smaller rectangle that is exactly the same shape as the one in question 12. It has a perimeter of 16 cm.
    a. Calculate its dimensions.
    b. Calculate its area.

14. The two rectangles shown below have the same area but different dimensions.

a. Calculate the missing side lengths.
b. Calculate the area of each rectangle.

9.9 Review

9.9.1 Summary

doc-38050

Hey students! Now that it's time to revise this topic, go online to:

Access the topic summary   **Review** your results   **Watch teacher-led** videos   **Practise questions with** immediate feedback

Find all this and MORE in jacPLUS

9.9 Exercise

Multiple choice

1. **MC** The next three numbers in the pattern 14, 25, 36 are:
   - **A.** 46, 56, 66
   - **B.** 41, 52, 63
   - **C.** 45, 54, 63
   - **D.** 47, 58, 69
   - **E.** 43, 52, 61

2. **MC** The flow chart for $3(x-4)$ is:
   - **A.** $x \xrightarrow{\times 3} \square \xrightarrow{-4} \square$
   - **B.** $x \xrightarrow{-4} \square \xrightarrow{\times 3} \square$
   - **C.** $x \xrightarrow{\times 3} \square \xrightarrow{+4} \square$
   - **D.** $x \xrightarrow{-4} \square \xrightarrow{+3} \square$
   - **E.** $x \xrightarrow{+4} \square \xrightarrow{+3} \square$

3. **MC** The inverse operations from the flow chart 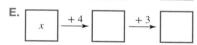 are:
   - **A.** subtract 7, then multiply by 5.
   - **B.** multiply by 5, then subtract 7.
   - **C.** add 7, then divide by 5.
   - **D.** subtract 7, then divide by 5.
   - **E.** subtract 5, then divide by 7.

4. **MC** Using inverse operations, the solution for x in $5x - 12 = 8$ is:
   - **A.** 6
   - **B.** 5
   - **C.** 4
   - **D.** 3
   - **E.** 7

5. **MC** Using inverse operations, the solution for p in $\dfrac{p-5}{3} = 2$ is:
   - **A.** 11
   - **B.** 6
   - **C.** 5
   - **D.** 3
   - **E.** 1

6. **MC** Using inverse operations, the solution for x in $4x - 10 = 6x - 18$ is:
   - **A.** 2
   - **B.** 3
   - **C.** 4
   - **D.** 5
   - **E.** 6

7. **MC** Rearranging the formula $E = \dfrac{1}{2}mv^2$ with m as the subject results in:
   - **A.** $m = \dfrac{1}{2}Ev^2$
   - **B.** $m = 2Ev^2$
   - **C.** $m = \dfrac{2E}{v^2}$
   - **D.** $m = \dfrac{E}{2v^2}$
   - **E.** $m = \dfrac{v^2}{2E}$

8. **MC** Select which of the following is a balanced equation of $3x + 2 = 5x - 3$.
   A. $9x + 6 = 15x - 9$
   B. $9x + 5 = 15x - 6$
   C. $6x + 4 = 10x - 1$
   D. $12x + 6 = 20x - 12$
   E. $6x + 2 = 5x - 3$

9. **MC** Let n represent the number of bars of chocolate in a packet. If you buy four packets and your friend buys six packets, the total number of bars of chocolate that you have between you could be written as:

   A. $n + 4$
   B. $4n + 6$
   C. $n + 6$
   D. $4n + 6n$
   E. $n + 6n$

10. **MC** If your friend told you that 'a number was multiplied by 7 and then 8 was subtracted', it could be written as:

    A. $8 - 7n$ B. $7n - 8$ C. $7(n - 8)$ D. $8n - 7$ E. $8(n - 7)$

Short answer

11. Build an expression by following the instructions on the flow chart below.

$$\boxed{m} \xrightarrow{+5} \boxed{} \xrightarrow{\times 2} \boxed{} \xrightarrow{-3} \boxed{}$$

12. Construct a flow chart for each of the following expressions.

    a. $4(x - 3) + 4$ b. $\dfrac{2(x - 8)}{5} + 6$

13. Use a flow chart and backtracking to solve each of the following equations.

    a. $2g + 7 = 11$ b. $\dfrac{3x - 9}{2} = 12$ c. $7(x + 2) - 8 = 27$

14. For the following pan balance scale:

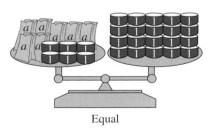

Equal

    i. write the equation represented by the scale
    ii. calculate the value of the variable.

15. Use inverse operations to solve the following equations.

    a. $4a = 24$ b. $\dfrac{c}{3} = 21$ c. $2d - 6 = 14$

16. Use inverse operations to solve the following equations.

    a. $\dfrac{e + 2}{2} = 15$ b. $3(f - 2) = 24$ c. $\dfrac{3(g + 4)}{5} = 6$

Extended response

17. Use inverse operations to solve the following equations.

a. $5w - 2 = 3w + 8$

b. $4e + 3 = 5e - 7$

c. $3 - r = 2r + 9$

18. Rearrange the following formulas to make the variable in brackets the subject.

a. $e = \dfrac{1}{2}ab^2$ (b)

b. $t = \dfrac{r - u}{m}$ (r)

c. $q^2 = \dfrac{2e + p}{i}$ (p)

19. Solve the following problems by writing an equation first and then solving it.

a. The sum of three consecutive numbers is 81. Determine the three numbers.

b. Three children are spaced 3 years apart. If the oldest is three times as old as the youngest, determine how old they are.

20. If you have twice as much money as your brother but only a third as much as your sister, determine how much you each have if all the money adds up to $36.

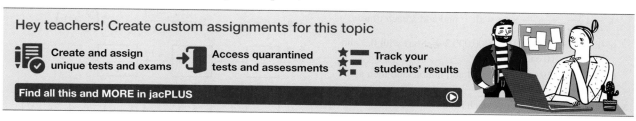

Hey teachers! Create custom assignments for this topic

Create and assign unique tests and exams

Access quarantined tests and assessments

Track your students' results

Find all this and MORE in jacPLUS

Answers

Topic 9 Equations

9.2 Number patterns and pronumerals

9.2 Exercise

1. Try subtracting or adding adjacent numbers or every second or third number.
2. a. 14, 16, 18 — Start with 6 and add 2 each time.
   b. 15, 14, 13 — Start with 19 and subtract 1 each time.
   c. 125, 150, 175 — Start with 25 and add 25 each time.
   d. 5, 2.5, 1.25 — Start with 80 and divide by 2 each time.
   e. $4\frac{1}{2}, 5\frac{1}{2}, 6\frac{1}{2}$ — Start with $\frac{1}{2}$ and add 1 each time.
   f. $\frac{1}{48}, \frac{1}{96}, \frac{1}{192}$ — Start with $\frac{1}{3}$ and add 1 each time.
   g. 0.000 03, 0.000 003, 0.000 000 3 — Start with 0.3 and divide by 10 each time.
   h. 79, 159, 319 — Start with 4 and double and then add 1 each time.
3. a. Start with 1 and double each time.
   b. Start with 12 and subtract 3 each time.
   c. Start with 1 and add 1 each time.
4. a. A letter that stands for a number
   b. Otherwise, we will not know what it represents.
5. Sample responses are available in the worked solutions in the online resources.
6. a. True
   b. True. They can contain more than one variable.
   c. False
   d. True
   e. False
   f. True
7. a. $35t$ b. $t - 1$ c. $t + 5$ d. $\frac{t}{2}$
8. a. A   b. C   c. B   d. B
   e. C
9. a. $n + 3$ b. $n - 5$ c. $2n + 6$
   d. $7n + 52$ e. $3n - 15$
10. a. $m - 8, m - 2, m + 1, m + 3$
    b. $n - 15, n - 10.3, n + 25$
11. Answers will vary. Examples are given.
    a. Let d be the number of dogs at my house. My friend has 2 more dogs at their house than I do ($d + 2$), while another friend has 1 fewer dog at their house than I do ($d - 1$).
    b. Let t be the number of textbooks in my schoolbag. My friend has double that number in their schoolbag ($2t$) while another friend has 1 more textbook than I do in their schoolbag ($t + 1$).
    c. Let y be the number of dollars in my wallet. My friend has 3 times that number of dollars in their wallet ($3y$), while another friend has 10 times that number of dollars in their wallet ($10y$).

12. a. $T - X$ b. $\$4B$ c. $45G$ d. $\$\frac{T}{5}$
13. a. $2n$ b. $4n$ c. $8n$
14. a. $2 \times c + 4$ b. $3 \times c + 4$ c. $2 \times c + 2$
    d. $2 \times c + 3$
15. a. $a + 6$ b. $a - 2$

9.3 Building expressions

9.3 Exercise

1. Flow charts keep track of the steps used in an expression and are particularly useful where a lot of steps are involved.
2. a. $2n + 1$ b. $3\left(\frac{k}{2} + 1\right)$ c. $\frac{2m - 7}{5}$
3. a. $-c + \frac{5}{8}$ b. $\frac{3(a - 12)}{6}$ c. $7\left(\frac{s}{2} + 1\right)$
4. a.
   b.
   c.
   d.
5. a. Multiply by 2
   b. Multiply by 4
   c. Multiply by -1
6. a. Multiply by 2
   b. Raise y to a power of 2 or multiply by y
   c. Raise k to a power of 3
7. a. Divide by 7   b. Divide by 27
   c. Add 7   d. Multiply by 19
   e. Add 7   f. Divide by 4
8. a.
   b.
   c.
   d.
9. a.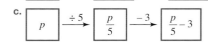
   b.

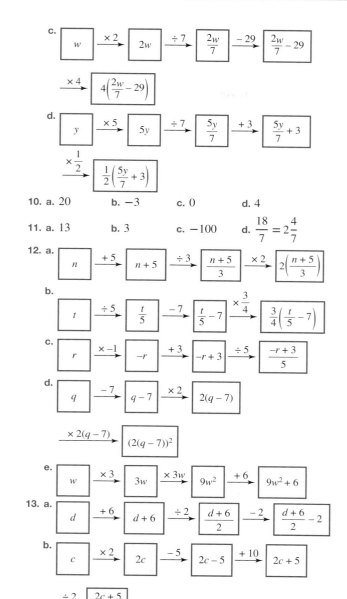

c.

w $\xrightarrow{\times 2}$ $2w$ $\xrightarrow{\div 7}$ $\dfrac{2w}{7}$ $\xrightarrow{-29}$ $\dfrac{2w}{7} - 29$

$\xrightarrow{\times 4}$ $4\left(\dfrac{2w}{7} - 29\right)$

d.

y $\xrightarrow{\times 5}$ $5y$ $\xrightarrow{\div 7}$ $\dfrac{5y}{7}$ $\xrightarrow{+3}$ $\dfrac{5y}{7} + 3$

$\xrightarrow{\times \frac{1}{2}}$ $\dfrac{1}{2}\left(\dfrac{5y}{7} + 3\right)$

10. a. 20 **b.** −3 **c.** 0 **d.** 4

11. a. 13 **b.** 3 **c.** −100 **d.** $\dfrac{18}{7} = 2\dfrac{4}{7}$

12. a.

n $\xrightarrow{+5}$ $n+5$ $\xrightarrow{\div 3}$ $\dfrac{n+5}{3}$ $\xrightarrow{\times 2}$ $2\left(\dfrac{n+5}{3}\right)$

b.

t $\xrightarrow{\div 5}$ $\dfrac{t}{5}$ $\xrightarrow{-7}$ $\dfrac{t}{5} - 7$ $\xrightarrow{\times \frac{3}{4}}$ $\dfrac{3}{4}\left(\dfrac{t}{5} - 7\right)$

c.

r $\xrightarrow{\times -1}$ $-r$ $\xrightarrow{+3}$ $-r+3$ $\xrightarrow{\div 5}$ $\dfrac{-r+3}{5}$

d.

q $\xrightarrow{-7}$ $q-7$ $\xrightarrow{\times 2}$ $2(q-7)$

$\xrightarrow{\times 2(q-7)}$ $(2(q-7))^2$

e.

w $\xrightarrow{\times 3}$ $3w$ $\xrightarrow{\times 3w}$ $9w^2$ $\xrightarrow{+6}$ $9w^2 + 6$

13. a.

d $\xrightarrow{+6}$ $d+6$ $\xrightarrow{\div 2}$ $\dfrac{d+6}{2}$ $\xrightarrow{-2}$ $\dfrac{d+6}{2} - 2$

b.

c $\xrightarrow{\times 2}$ $2c$ $\xrightarrow{-5}$ $2c-5$ $\xrightarrow{+10}$ $2c+5$

$\xrightarrow{\div 2}$ $\dfrac{2c+5}{2}$

c.

n $\xrightarrow{-5}$ $n-5$ $\xrightarrow{\times 2}$ $2(n-5)$ $\xrightarrow{-3}$ $2(n-5)-3$

14. Answers will vary. A sample response is provided in the worked solutions in the online resources.

15. $64\,\text{m}^2$

9.4 Backtracking to solve equations

9.4 Exercise

1. Work backwards through the flow chart, starting with the last box; work against the arrows using inverse operations.

2. a. To undo a step where you have added, you subtract.

b. To undo a step where you have subtracted, you add.

c. To undo a step where you have multiplied, you divide.

d. To undo a step where you have divided, you multiply.

3. a. i. $2x + 7 = -29$

    ii. $x = -18$

b. i. $2(x + 7) = 32$

    ii. $x = 9$

c. i. $3\left(\dfrac{k}{3} + 7\right) = -93$

    ii. $k = -114$

4. a. i. $\dfrac{2e}{5} + 11 = 17$

    ii. $e = 15$

b. i. $\dfrac{a + 7}{8} + 9 = -4$

    ii. $a = -111$

c. i. $\dfrac{-2w + 8}{3} = -10$

    ii. $w = 19$

5. a. $x = -7$ **b.** $x = 6$ **c.** $m = 9$

    d. $x = -3$ **e.** $x = -17$ **f.** $h = 9$

6. a. $k = 27$ **b.** $s = 16$ **c.** $r = \dfrac{14}{5}$ **d.** $x = -59$

7. a. i. $x = 42 = 21$

    ii. $x = -21$. The unknown number is -21.

b. i. $6x = 120$

    ii. $x = 20\,\text{m}$. The paddock is 20 m wide.

8. a. i. $6x = 180°$

    ii. $x = 30°$. The angles are $30°, 60°$ and $90°$.

b. i. $\dfrac{5x + 7}{3} = 10$

    ii. $x = \dfrac{23}{5} = 4.6$. The numbers are 4.6, 6.6 and 18.8.

9.

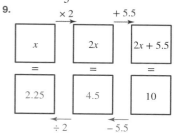

$x = 2.25$. One loaf of bread costs $2.25.

10.

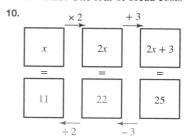

$x = 11$. The friends are 11 and 14 years old.

11.

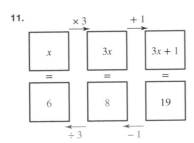

$x = 6$. I ran 6 km yesterday and 13 km today.

12. a. Your friend should have added 8 to 20 (under the third box), rather than divide 20 by 2. Therefore, your friend should have written 28 under the second box instead of 10. The correct answer is $x = 14$.

b. Your friend should have divided 16 under the third box by 4, rather than subtract 4 from 16. Therefore your friend should have written 4 under the second box instead of 12. The correct answer is $x = 0$.

9.5 Balancing equations

9.5 Exercise

1. You have to make the same change to the other side.

2. a. i. $4r = 12$

　　ii. $r = 3$

　b. i. $2n + 3 = 7$

　　ii. $n = 2$

　c. i. $3t + 5 = 8$

　　ii. $t = 1$

　d. i. $4p + 1 = 13$

　　ii. $p = 3$

3. a. i.

　　ii. $q = 3$

　b. i.

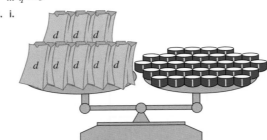

　　ii. $d = 3$

　c. i.

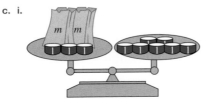

　　ii. $m = 2$

d. i.

　ii. $p = 1$

e. i.

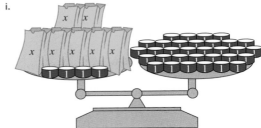

　ii. $x = 4$

f. i.

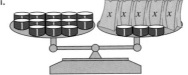

　ii. $x = 2$

4. a. $10y = 30$　　　　**b.** $y - 5 = -2$

　c. $\dfrac{y}{2} = \dfrac{3}{2}$　　　　**d.** $y + 28 = 31$

5. a. $-2a = -10$　　　**b.** $\dfrac{4}{5}a = 4$

　c. $a - \dfrac{2}{3} = \dfrac{13}{3}$　　**d.** $0.3a = 1.5$

6. Answers will vary. An example is $3x = 3$.

7. a. $m + 16 = 17$　　　**b.** $n - 5 = 3$

　c. $2p + 6 = 12$　　　**d.** $12m = 48$

8. a. $2n = 6$　　　　　**b.** $2p = 30$

　c. $12p + 28 = 8$　　　**d.** $5m - \dfrac{3}{3} = \dfrac{-1}{3}$

9. Answers will vary. Examples include $2x + 1 = 9$ and $12x + 6 = 54$.

10. Two equations are equivalent if they have the same answer.

11. a. i. Add 4 to both sides.

　　ii. Add 6 to both sides.

　　iii. Multiply both sides by 3.

　　iv. Multiply both sides by 2.

　　v. Add 2 to both sides, then multiply both sides by 4.

　　vi. Multiply both sides by 4, then add 2 to both sides.

　b. They have subtraction and division in the expressions, and these cannot be represented on a pan balance scale.

　c. In part **v**, the variable is divided by 4 and then 2 is subtracted from the result. In part **vi**, 4 is subtracted from the variable and then the result is divided by 2.

　d. i. $x = 14$　　**ii.** $x = 9$　　**iii.** $w = 21$

　　iv. $e = 18$　　**v.** $q = 48$　　**vi.** $q = 42$

12. a. $5x = 3$

$x = \dfrac{3}{5}$. One bag of sugar weighs $\dfrac{3}{5}$ kg.

b. $4x + 2.70 = 9.90$

$x = 1.8$. One chocolate bar costs $1.80.

c. $3x + 500 = 920$

$x = 140$. One apple weighs 140 g.

9.6 Solving equations

9.6 Exercise

1. Undo, on both sides of the equation, the last operation performed on the variable when the equation was built. Then undo, on both sides of the equation, the second last operation performed on the variable when the equation was built. Continue until the variable is left on one side of the equation by itself.

2. a. $a = 9$ **b.** $b = 52$
c. $c = 16$ **d.** $d = -8$

3. a. $e = -5$ **b.** $f = 4$
c. $g = 4$ **d.** $u = 12$

4. a. Subtract 9.   **b.** Subtract 3.
c. Subtract 11.   **d.** Multiply by 11.

5. a. $x = 3$ **b.** $p = 3$
c. $a = 6$ **d.** $a = \dfrac{125}{11}$

6. a. $c = 44$ **b.** $d = -1$
c. $e = 1$ **d.** $k = 46$

7. a. $a = \dfrac{11}{2}$ or 5.5

b. $a = \dfrac{11}{2}$ or 5.5

c. The answers are the same.

d. Expanding the brackets first when there is more than one set of brackets allows you to collect like terms and simplify the expression.

8. a. $a = 2$ **b.** $b = 18$
c. $c = 5.75$ **d.** $\dfrac{166}{7}$

9. a. $\dfrac{46}{3}$ **b.** $b = 34$ **c.** $a = 2$
d. $d = 188$ **e.** $e = 20$ **f.** $f = 10$

10. a. $x = 2$ **b.** $b = 20$ **c.** $c = -21$

11. a. i. $a = 2.5$ **ii.** $a = 4.5$ **iii.** $a = 11$

b. i.

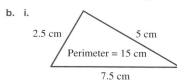

ii.

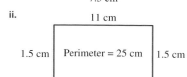

iii.

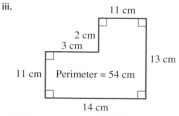

12. a. i. 26 km   **ii.** 1.45 km   **iii.** 15.31 km

b. Answers will vary. Sample responses can be found in the worked solutions in the online resources.

9.7 Rearranging formulas

9.7 Exercise

1. A formula is a mathematical expression that shows the relationship between different variable quantities.

2. a. y **b.** E **c.** P **d.** c^2

3. a. d **b.** F **c.** v **d.** E

4. a. S **b.** r^2 **c.** I **d.** T

5. a. $x = \dfrac{y - c}{m}$ **b.** $c = \sqrt{\dfrac{E}{m}}$

c. $W = \dfrac{P - 2L}{2}$ **d.** $b = \sqrt{c^2 - a^2}$

6. a. $t = \sqrt{\dfrac{d}{4.9}}$ **b.** $C = \dfrac{9(F - 32)}{5}$

c. $m = dv$ **d.** $m = \dfrac{2E}{v^2}$

7. a. $n = \dfrac{S}{180} + 2$ **b.** $y = \sqrt{r^2 - x^2}$

c. $V = IR$ **d.** $n = \dfrac{T - a}{d} + 1$

8. a. $a = 2\left(\dfrac{A}{h}\right) - b$ **b.** $b = 2\left(\dfrac{A}{h}\right) - a$

c. $h = \dfrac{2A}{a + b}$

9. a. $P = 2l + 2w$ **b.** $l = \dfrac{P - 2w}{2}$

c. $w = \dfrac{P - 2l}{2}$

10. C

11. a. $h = \dfrac{A}{2\pi r} - r$ **b.** Height $= 5.5$ cm

12. a. $s = 49.2$ km/h   **b.** $d = 204.75$ km
c. $t = 3.24$ h

13. a. $42   **b.** 7 hours

14. a. $r = \sqrt{\dfrac{A}{\pi}}$ **b.** 12 cm and 18 cm

15. a. $V = 201.6$ m^3 **b.** $d = \dfrac{V}{lw}$

c. $d = 1.6$ m   **d.** 20 cm

9.8 Applications of equations and formulas

9.8 Exercise

1. Turn the problem into an equation, solve the equation and use the solution to answer the problem.

2. Bag: $62.50; shoes: $112.50

3. Mother: 30

   Son: 6

4. 148.5 cm

5. 12

6. Eldest: $16; middle: $11; youngest: $9

7. 10, 15 and 20

8. You walked 9 laps, your friend walked 4 laps and your sister walked 19 laps.

9. $8.60

10. a. 18, 19, 20   b. 25, 27, 29
    c. 68, 70, 72, 74   d. 28, 31, 34, 37
    e. 60, 65, 70

11. a. $\dfrac{1}{2}$ b. 7

12. 9 cm and 27 cm

13. a. 2 cm and 6 cm

    b. 12 cm^2

14. a. $x = 36$; side lengths are 12 and 72.

    b. 504

9.9 Review

9.9 Exercise

Multiple choice

1. D   2. B   3. C   4. C   5. A
6. C   7. C   8. A   9. D   10. B

Short answer

11. $2(m + 5) - 3$

12. a.

| x | $\xrightarrow{-3}$ | $x - 3$ | $\xrightarrow{\times 4}$ | $4(x - 3)$ | $\xrightarrow{+4}$ | $4(x - 3) + 4$ |

b.

| x | $\xrightarrow{-8}$ | $x - 8$ | $\xrightarrow{\times 2}$ | $2(x - 8)$ | $\xrightarrow{\div 5}$ | $\dfrac{2(x - 8)}{5}$ | $\xrightarrow{+6}$ | $\dfrac{2(x - 8)}{5} + 6$ |

13. a. 2   b. 11   c. 3

14. a. $7a + 6 = 20$ b. 2

15. a. 6   b. 63   c. 10

16. a. 28   b. 10   c. 6

Extended response

17. a. 5   b. 10   c. −2

18. a. $b = \sqrt{\dfrac{2e}{a}}$ b. $r = tm + u$ c. $p = q^2 i - 2e$

19. a. 26, 27 and 28   b. 3, 6 and 9 years old

20. You = $8
    Sister = $24
    Brother = $4

10 Collecting and classifying data

LEARNING SEQUENCE

Fully worked solutions for this topic are available online.

10.1 Overview

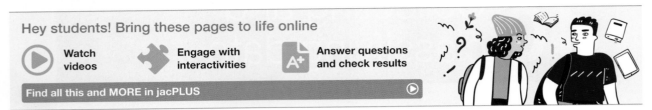

10.1.1 Introduction

The world we live in produces an ever-increasing amount of data. Being able to understand and interpret these large sets of data is a valuable skill in the 21st century. To be able to interpret data, you need to first know how to collect it. The Australian Bureau of Meteorology collects data about the weather, analyses it, and then uses that knowledge to predict what the weather will be like tomorrow, next week and even next year. The Australian Bureau of Statistics runs a census night every five years to collect data about the population. This data is then analysed so future decisions can be made by the government.

Social media companies collect data about how you use their platforms. This data is analysed and used to adapt these platforms to the way you use them (and to help these companies maximise profits). This demonstrates the importance of understanding how to collect and then interpret data to make important decisions.

KEY CONCEPTS

This topic covers the following key concepts from the VCE Mathematics Study Design:
- collection and representation of data in diagrammatic, tabular and graphical forms and the features, conventions and terminology used in these processes
- construction of charts, tables and graphs to represent data
- interpretation of data to summarise and communicate findings and possible conclusions.

Source: VCE Mathematics Study Design (2023–2027) extracts © VCAA; reproduced by permission.

10.2 Collecting data

10.2.1 Collecting data by census

Data collection is a process in which data is collected to obtain information and draw conclusions about issues of concern regarding a given population. A **population** consists of a complete group of people, objects, events, etc. with at least one common characteristic. Any subset of a population is called a **sample**. Data is collected using either a **census** of the entire population or a **survey** of a sample of the population.

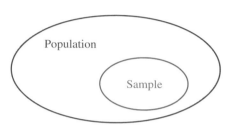

Census

A census is conducted by official bodies at regular time intervals on a set date. In Australia, a census of the entire population and housing is conducted by the **Australian Bureau of Statistics (ABS)** every five years on **census night**. The latest census in Australia occurred in August 2021.

Procedure for conducting a census

A census consists of three main stages:

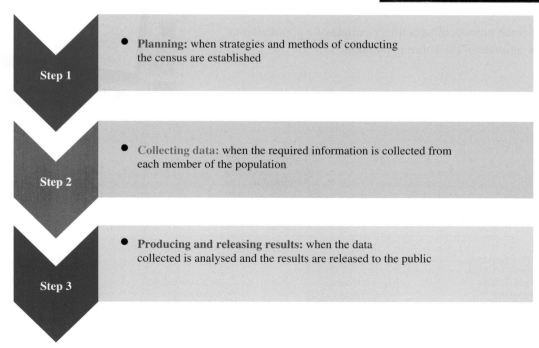

Step 1 • **Planning:** when strategies and methods of conducting the census are established

Step 2 • **Collecting data:** when the required information is collected from each member of the population

Step 3 • **Producing and releasing results:** when the data collected is analysed and the results are released to the public

WORKED EXAMPLE 1 Deciding whether to use a census or a survey

For each of the following situations, state whether the data should be collected using a census or a survey. Give reasons for your choice.

a. All students in a school are asked to state their method of travel to and from school.
b. Data is collected from every third person leaving a shopping centre on the time spent shopping.
c. Data is collected from 500 households regarding their average monthly water usage.

THINK	WRITE
a. 1. Determine whether the group considered represents an entire population or a sample.	**a.** All students in a school represent the whole student population of the school.
2. State whether a census or a survey is required.	As this represents the entire student population for the given school, a census is required.
b. 1. Determine whether the group considered represents an entire population or a survey.	**b.** As only every third person is involved in the data collection, this represents a sample of the whole population of people leaving the shopping centre.
2. State whether a census or a survey is required.	As this represents a sample, a survey is required.
c. 1. Determine whether the group considered represents an entire population or a survey.	**c.** Only 500 households are involved in the data collection, so this represents a sample of all the households.
2. State whether a census or a survey is required.	As this represents a sample, a survey is required.

10.2.2 Collecting data by surveys and sampling

As already stated in the introduction of this chapter, a survey of a sample of a given population is considered when a census cannot be used.

Sampling is the process of selecting a sample of a population to provide an estimate of the entire population.

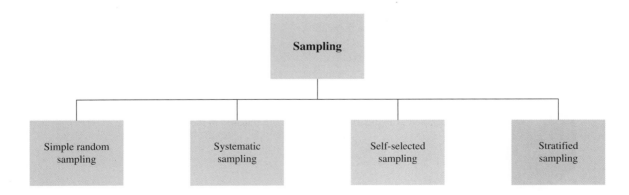

Simple random sampling

Simple random sampling is the basic method of selecting a sample. This type of sampling ensures that each individual in a population has an equal chance of being selected for the sample.

> ### Formula to calculate the sample size
>
> **The sample size can be calculated using the formula:**
>
> $$S = \sqrt{N}, \text{ where } N \text{ is the size of the population}$$

Consider a population of 25 Year 11 Mathematics students and a sample of 5 students. A basic method of choosing the sample is by assigning each Year 12 student a unique number from 1 to 25, writing each number on a piece of paper, placing all the papers in a box or a bowl, shaking them well and then choosing 5 pieces of paper. The students who correspond to the numbers drawn will be a sample of the Year 11 student population.

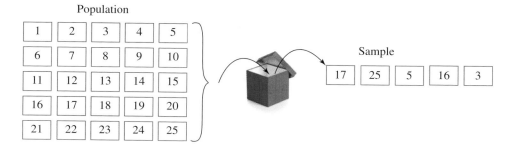

Random number generator using calculators or computer software

Calculators and computer software have a random number generator that makes the process of selecting a sample a lot easier. An Excel worksheet generates random numbers using the RAND( ) or RANDBETWEEN(a, b) command.

The RAND( ) command generates a random number between 0 and 1. Depending on the size of the sample, these generated numbers have to be multiplied by n, where n is the size of the population.

	A	B	C	D	E
1	0.433821				
2	0.78391				
3	0.547901				
4	0.892612				
5	0.390466				
6	0.264742				
7	0.690003				
8	0.070899				
9	0.876409				
10	0.976012				

RANDBETWEEN(a, b) generates a random number from a to b. For the population of Year 12 Mathematics students, the sample of five students can be generated by using **RANDBETWEEN(1, 25)**.

	A	B	C	D	E	F
1	9					
2	3					
3	25					
4	3					
5	17					

If the same number appears twice, a new number will need to be generated in its place to achieve five different numbers.

tlvd-4690

WORKED EXAMPLE 2 Selecting a sample

Select a sample of 12 days between 1 December and 31 January:
a. by hand
b. using a random number generator.

THINK

a. 1. Write every member of the population on a piece of paper.

2. Fold all papers and put them in a box.

3. Select the required sample. A sample of 12 days is required, so randomly choose 12 papers from the box.

4. Convert the numbers into the data represented.

5. Write the sample selected.

b. 1. Assign each member of the population a unique number.

WRITE

a. There are 31 days in December and 31 days in January. 62 pieces of paper will be required to write down each day.
$$1/12, 2/12, \ldots, 30/01, 31/01$$
Note: Alternatively, each day could be assigned a number, in order, from 1 to 62.

Ensure that the papers are folded properly so no number can be seen.

Sample: 15, 9, 41, 12, 1, 7, 36, 13, 26, 5, 50, 48

15 represents 15/12, 9 represents 9/12, 41 represents 10/01, 12 represents 12/12, 1 represents 1/12, 7 represents 7/12, 36 represents 5/01, 13 represents 13/12, 26 represents 26/12, 5 represents 5/12, 50 represents 19/01, 48 represents 17/01.

Sample: 15/12, 9/12, 10/01, 12/12, 1/12, 7/12, 5/01, 13/12, 26/12, 5/12, 19/01, 17/01

b. Assign each day a number, in order, from 1 to 62:
$$1/12 = 1, 2/12 = 2, \ldots, 31/01 = 62$$

2. Open a new Excel worksheet and use the RANDBETWEEN(1, 62) command to generate the random numbers required.

	A	B	C	D	E	F
1	29					
2	27					
3	17					
4	26					
5	4					
6	45					
7	41					
8	25					
9	10					
10	28					
11	16					
12	8					

Note: Some of the numbers may be repeated. For this reason, more than 12 numbers should be generated. Select the first 12 unique numbers.

3. Convert the numbers into the data represented.

29 represents 29/12, 27 represents 27/12, 17 represents 17/12, 26 represents 26/12, 4 represents 4/12, 45 represents 14/01, 41 represents 10/01, 25 represents 25/12, 10 represents 10/12, 28 represents 28/12, 16 represents 16/12, 8 represents 8/12.

4. Write the sample selected.

Sample: 29/12, 27/12, 17/12, 26/12, 4/12, 14/01, 10/01, 25/12, 10/12, 28/12, 16/12, 8/12

Systematic sampling

Systematic sampling or **systematic random sampling** requires a starting point chosen at random with members of the population chosen at regular intervals.

For example, if a sample of 5 students out of 30 is needed for a survey, to ensure that each member of the population has an equal chance of being chosen, divide the whole population by the size of the sample (in this case, $30 \div 5 = 6$). So, the starting point will be a random number from 1 to 6 and then every 6th student will be chosen.

Starting point interval formula

The starting point interval is chosen using the formula:

$$\text{maximum starting point} = N - I(s - 1)$$

where N is the size of the population

s is the size of the sample

I is the whole number of $\dfrac{N}{s}$.

Using a systematic random sampling method, a researcher wants to choose 8 people from a population of 59 people. State one such sample.

THINK	WRITE
1. Write a list of all the people in the population.	For the purpose of this example, we are going to use the initials of the people. GF, AS, TG, YH, ID, BK, CD, YT, UE, OM, LP, HI, BT, SJ, FR, IV, BX, MN, IT, UM, WK, FA, FP, ST, VP, AA, AR, BT, OD, PM, LC, RP, PO, GV, BF, TP, AD, IP, CR, AN, OO, FC, LM, LC, AF, AK, PY, SF, GH, VS, MR, CB, FJ, RM, CS, KF, GS, WM, FX
2. Choose an interval.	$\dfrac{59}{8} = 7.37$ The researcher could choose either every 7th person or every 8th person.
3. Choose the starting point.	Let $I = 7$. Maximum starting point $= N - I(s-1)$ $\qquad\qquad\qquad\qquad = 59 - 7 \times (8-1)$ $\qquad\qquad\qquad\qquad = 59 - 49$ $\qquad\qquad\qquad\qquad = 10$ The starting point will be a number from 1 to 10. Let this number be 4.
4. Form the sample.	The sample will be formed by the 4th, 11th, 18th, 25th, 32nd, 39th, 46th and 53rd person. GF, AS, TG, **YH**, ID, BK, CD, YT, UE, OM, **LP**, HI, BT, SJ, FR, IV, BX, **MN**, IT, UM, WK, FA, FP, ST, **VP**, AA, AR, BT, OD, PM, LC, **MR**, PO, GV, BF, TP, AD, IP, **CR**, AN, OO, FC, LM, LC, AF, **AK**, PY, SF, GH, VS, RP, CB, **FJ**, RM, CS, KF, GS, WM, FX
5. Write the sample selected.	YH, LP, MN, VP, MR, CR, AK, FJ

Self-selected sampling

Self-selected sampling is used when the members of a population are given the choice to participate in research. In this type of sampling, the researcher chooses the **sampling strategy**, such as:
- asking for volunteers to participate in the trial of a drug
- a website asking customers to answer a short questionnaire.

Self-selected sampling requires two steps:

For example, if you were asked to form a sample of 20 students out of 200 Year 11 students, the two steps would be:
- advertising your study around the school
- accepting the Year 11 students who showed an interest in participating in your study and rejecting any students who are not in Year 11.

Step 1 • Publicising the needs of the study to potential participants

Step 2 • Accepting or rejecting the applicants offering to participate in the study

A pharmaceutical company wants to trial a new flu vaccine. The study requires adults aged between 20 and 80 years. State the steps required to form a self-selected sample of 100 participants.

THINK	WRITE
1. State the ways to publicise the study.	Possible ways to publicise the study are: • online (social media or email) • TV • radio • hospitals
2. State the characteristics required for accepting or rejecting the applicants.	Accept 100 adults aged between 20 and 80 years old. Reject any others.

Stratified sampling

Stratified sampling is used when there are variations in the characteristics of a population. This method requires the population to be divided into subgroups called **strata**.

For example, if the Year 12 students of a secondary college were asked about their favourite movie, the preferences could be different for boys and girls. For this reason, the sample would have to reflect the same proportion of boys and girls as the actual population.

Calculating sample size for subgroups

$$\text{sample size for each subgroup} = \frac{\text{sample size}}{\text{population size}} \times \text{subgroup size}$$

Once the sample size of each subgroup has been determined, random sampling or systematic sampling can be used to form the sample required.

Calculate the number of female students and male students required to be part of a sample of 25 students if the student population is 652, with 317 male students and 335 female students.

THINK	WRITE
1. State the formula for determining the sample size.	$\text{Sample size for each subgroup} = \dfrac{\text{sample size}}{\text{population size}} \times \text{subgroup size}$
2. Calculate the sample size for each subgroup.	$\text{Sample size for female students} = \dfrac{25}{652} \times 335$ $= 13$ $\text{Sample size for male students} = \dfrac{25}{652} \times 317$ $= 12$
3. Write the answer.	The sample should contain 13 female students and 12 male students.

10.2 Exercise

1. **WE1** For each of the following situations, state whether the data should be collected using a census or a survey. Give reasons for your choice.
   a. An online business recorded the time that its customers spent searching for a product.
   b. 50% of the Year 12 students at a secondary college were asked to state their preference for either online tutorials or face-to-face tutorials.
   c. The heights of all the patients at a local hospital were recorded.

2. For each of the following, state which situation represents a sample and which represents a population. Give reasons for your choice.
   a. The number of absences of all Year 12 students in a school and the number of absences of all students in the school
   b. The number of passenger airplanes landing at an airport and the total number of airplanes landing at the same airport

3. State whether a census or a survey is required to collect the data in the following statistical investigations. Give reasons for your answer.
   a. Water savings in 150 suburbs across the country.
   b. Highest educational level of all people in Australia.
   c. Roll marking at home room in a secondary college.
   d. Asking every fifth person leaving the theatre whether they liked the play.
   e. Asking every customer in a car dealership to fill in a questionnaire.

4. **WE2** Select a sample of 5 students to participate in a relay from a group of 26 students:
   a. by hand
   b. using a random number generator.

5. A small business has 29 employees. The owner of the business has decided to survey 8 employees on their opinion about the working conditions. Select the required sample:
   a. by hand
   b. using a random number generator.

6. Calculate the approximate sample size of a population of 23 500 people.

7. Generate a sample of 10 members from a population of 100 people using the Excel worksheet random number generator and the following command:
   a. RAND( )
   b. RANDBETWEEN(0, 100)

8. **WE3** A car manufacturer wants to test 5 cars from a lot of 32 cars. Form one possible sample by using a systematic random sampling method.

9. A quality control officer is conducting a survey of 46 products. She decides to sample 9 products using a systematic random sampling method. Form two possible samples by using systematic random sampling methods.

10. Stratified sampling is used in the following surveys. Calculate how many members are needed from each subgroup.

    a. A school has a population of 127 VCE students, where 61 students are in Year 11. A sample of 11 students is chosen to represent all the VCE students in an interstate competition.
    b. The local cinema is running a promotional movie screening. A sample of 10 people from those attending the screening will be chosen to receive a free pass to the next movie screening. If the total number of people to attend the promotional screening is 138, with 112 adults and 26 children, calculate the number of children and adults who will receive a free movie pass in order to keep a proportional sampling.

11. **WE4** The local council is seeking volunteers to participate in a study on how to improve the local public library. State the steps required to form a self-selected sample of 50 participants.

12. A gym instructor is conducting a study on the effects of regular physical exercise on the wellbeing of teenagers. State what steps are required to form a self-selected sample of 30 participants.

13. Census At School Australia is a project that collects information from Australian students using a questionnaire developed by the Australian Bureau of Statistics. It is not compulsory for students to participate in this census; however, students can volunteer to participate. State what type of sampling is used in this project.

14. A farmer has 46 black sheep and 29 white sheep. He wants to try a new diet on a sample of 10 sheep.

    a. State what type of sampling the farmer is using.
    b. Calculate the number of black sheep and white sheep required for this trial in order to give a proportional representation of both types of sheep.

15. A manufacturing company of mobile phones is testing two batches of 100 phones each for defects. In the first batch, every 5th phone is checked, while in the second batch, every 7th phone is checked.

    a. State the method of sampling that is used for this testing.
    b. Determine the number of phones tested for defects in the two batches.
    c. Calculate the maximum starting point for each batch.

16. **WE5** Calculate the sample size of the subgroups in a sample of 33 people from a population of 568 males and 432 females.

17. Calculate the sample size of the subgroups in a sample of 15 cars from a population of 179 white cars and 215 black cars.

18. At Guntawang Secondary College, a survey is conducted to collect information from the 230 students enrolled.
    a. Calculate the appropriate sample size for this population.
    b. Explain how the sample could be chosen.

19. Generate a sample of 12 members from a population of 140 people using the Excel worksheet random number generator and the following command:
    a. RAND( )   b. RANDBETWEEN(1, 140).

20. For a population of 800 people:
    a. calculate the size of the sample required
    b. use the command RANDBETWEEN(1, 800) to generate the sample.

10.3 Simple survey procedures

LEARNING INTENTION

At the end of this subtopic you should be able to:
- identify the population type for a survey
- determine the appropriate questions for a questionnaire
- identify different types of questions: open-ended, closed and partially closed.

10.3.1 Surveys

The procedures and methods of conducting surveys are of great importance; if done wrong, they could provide misleading results.

In a survey, the data is collected from a sample of a given population. The characteristics of the population are estimated from the characteristics of the sample using statistical procedures.

The members of a population could be any entities, such as people, animals, organisations and businesses.

For example, if you were interested in knowing the students' opinion about a school policy, the target population would be all the students in the school. On the other hand, if you were interested in finding out the students' opinion about learning to drive, the target population would be all the students aged 16 or over.

WORKED EXAMPLE 6 Determining the target population of a survey

A company is conducting three surveys in order to obtain information required to improve its employees' work conditions and lifestyle. Define the target population in each of the surveys.
a. A survey about the time spent by employees to travel to work.
b. A survey about the amount of money spent weekly on childcare.
c. A survey about the amount of time spent using a computer.

THINK	WRITE
a. Define the target population.	a. All the company's employees.
b. Define the target population.	b. All employees with children.
c. Define the target population.	c. All employees required to use a computer for work.

Questionnaires

Surveys can be administered using paper or electronic questionnaires, face-to-face interviews or by telephone. The most common type of survey is the traditional paper survey in the form of a questionnaire. A **questionnaire** is a list of questions used to collect data from the survey's participants. Designing questionnaires is probably one of the most important tasks when administering a survey.

Some dos and don'ts for wording questions in a survey are shown in the table below.

👍	👎
Use simple and easy to understand language.	Avoid abbreviations and jargon. E.g. VCE (Victorian Certificate of Education), ABS (Australian Bureau of Statistics), HSC (Higher School Certificate), etc.
Use short sentences. E.g. 'There are many types of movies available in cinemas that people watch. Which type do you think most represents your preference?' could be replaced by a simple, straightforward question such as 'What type of movie do you prefer?'	Avoid vague or ambiguous questions by asking precise questions.
Ask for one piece of information at a time and avoid double-barrelled questions. E.g. 'How satisfied are you with your job and your boss?' is a double-barrelled question, putting two pieces of information together.	Avoid double negatives. E.g. 'Would you say that your Mathematics teacher is not unqualified?' is an affirmation; however, some people might not be aware of this and the question might look confusing.
Ask questions that are impartial and do not suggest the 'correct' answer.	Avoid leading questions. E.g. 'Do you agree with the new school council structure? Yes or No.' could be replaced with a question such as 'Do you agree or disagree with the new school council structure? 1. Strongly agree, 2. Agree, 3. Disagree, 4. Strongly disagree.'
Give consideration to sensitive issues like privacy and ethics. E.g. Questions about issues such as income, religious beliefs, political views and gender orientation are sensitive topics and they have to be stated in a sensitive manner.	Avoid or eliminate bias. E.g. 'How would you rate your Mathematics textbook: Excellent, Good or Average?' The bias is in the absence of a negative option. Questions with 'always' and 'never' can also bias participants.

WORKED EXAMPLE 7 Explaining whether questions are appropriate for a questionnaire

Consider the following types of questions and explain why they are not appropriate for a questionnaire. For every question state a more appropriate alternative.

a. **How many TV sets do you own?**

b. **How many times did you purchase lunch from the school's cafeteria last year?**

c. **What is your yearly income?**

THINK	WRITE
a. 1. State any concerns about the question.	**a.** The question assumes that all respondents own a TV set. Some respondents might have a TV set in their household but not own it.
2. State a possible replacement question or questions.	'Do you own a TV set? If yes, how many TV sets do you own?'
b. 1. State any concerns about the question.	**b.** It is hard to understand what period 'last year' covers. The question assumes that all students buy lunch from the school's cafeteria. It might be hard to recall the exact number of times.
2. State a possible replacement question or questions.	'Did you buy lunch at the school's cafeteria last school year (or over the last 12 months)? If yes, state the approximate number of times.' Alternatively, provide categories like 0–5, 6–10, 10–15 etc.
c. 1. State any concerns about the question.	**c.** People usually don't like to state their exact income. Broad categories are more likely to be answered honestly.
2. State a possible replacement question or questions.	'What category best describes your annual income? Less than $25 000, $25 000– $50 000, $50 000– $75 000, etc.'

Types of questions

Questions have to be designed keeping in mind the relevance to the survey. The data collected has to be clear and easy to analyse.

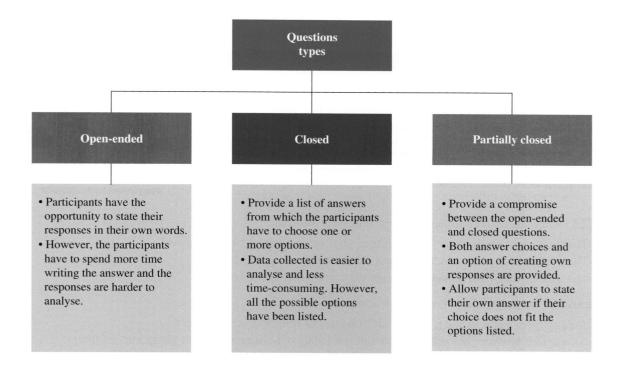

Questions types

Open-ended
- Participants have the opportunity to state their responses in their own words.
- However, the participants have to spend more time writing the answer and the responses are harder to analyse.

Closed
- Provide a list of answers from which the participants have to choose one or more options.
- Data collected is easier to analyse and less time-consuming. However, all the possible options have been listed.

Partially closed
- Provide a compromise between the open-ended and closed questions.
- Both answer choices and an option of creating own responses are provided.
- Allow participants to state their own answer if their choice does not fit the options listed.

WORKED EXAMPLE 8 Identifying types of questions

State whether the following questions are open-ended, partially closed or closed.
a. What mode of transport do you use when travelling to school?
b. What mode of transport do you use to travel to school? Multiple options may be selected.
   1. Walk   2. Bus   3. Tram   4. Bike   5. Car
c. What mode of transport do you use to travel to school?
   1. Walk   2. Bus   3. Tram   4. Bike   5. Car   6. Other [   ]

THINK	WRITE
a. 1. Determine if a written answer is required or possible options are available.	a. The question requires the participant to write their own answer.
2. State what type of question it is.	Open-ended
b. 1. Determine if a written answer is required or possible options are available.	b. The question has possible options available. However, it does not exhaust all possible options.
2. State what type of question it is.	Closed
c. 1. Determine if a written answer is required or possible options are available.	c. The question has possible options available. However, it does allow the participant to write their own option.
2. State what type of question it is.	Partially closed

 Resources

 Interactivities Planning a questionnaire (int-3810)
Questionnaires (int-3809)

10.3 Exercise

Students, these questions are even better in jacPLUS

 Receive immediate feedback and access sample responses

 Access additional questions

 Track your results and progress

Find all this and MORE in jacPLUS ▶

1. **WE6** Define the target population in the following surveys.

   a. The time spent by the Year 12 students in your school completing homework.
   b. The time spent by the students in your school completing homework.
   c. The time spent by the students in Australia completing homework.

2. Define the target population of the following samples.

   a. A dairy farm has selected a sample of 20 bottles of milk to test for quality.
   b. An internet provider has selected a sample of 150 people to survey about the quality of their internet service.

3. Maya is conducting a survey on the preferred way of doing research for statistical investigations. State the target population if she decides to survey:

   a. a sample of 20 Year 11 students
   b. a sample of 30 students from Years 11 and 12
   c. a sample of 50 students from the whole school.

4. **WE7** Consider the following types of questions and explain why they are not appropriate for a questionnaire. For every question, state a possible acceptable replacement.

   a. What brand of mobile phone do you have?
   b. Do you agree or disagree that drinking alcohol on the beach is not permitted but smoking is permitted?
   c. What church do you attend on Sundays?

5. Consider the following types of questions and explain why they are not appropriate for a questionnaire. For every question, state a possible acceptable replacement.

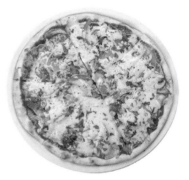

   a. Do you like pizza, pasta or both?
   b. Are your parents not unclear in their expectations of you?

6. **WE8** State whether the following questions are open-ended, partially closed or closed.

   a. How would you rate the quality of the meat at our supermarket?
      1. Excellent   2. Good   3. Average   4. Poor
   b. What is your favourite brand of cereal?
   c. Can you improve your grades in Mathematics? Yes or No.
   d. What steps have you taken to improve your Mathematics grades?
      1. Completing more homework
      2. Asking the teacher for assistance more often
      3. Spending more time understanding the concepts
      4. Other — insert your answer here: _____

7. Write one question of each of the following types.

   a. Open-ended
   b. Closed
   c. Partially closed

8. Design an open-ended question, a partially closed question and a closed question for a questionnaire surveying the students in your year level about the internet search engine they use.

9. The following questions are not properly worded for a questionnaire. Explain why, and rewrite them using the guidelines in this subtopic.
   a. Do you catch a bus to work, or a tram?
   b. Which of these is the most important issue facing teenagers today? Circle one answer.
      The environment, binge drinking, teenage pregnancy
   c. Do you feel better now that you have stopped smoking?
   d. How much money do you earn per week?

10. State if the following questions are open-ended, partially closed or closed.
    a. What is your country of birth?
       ☐ Australia ☐ Other: _____
    b. Do you have an account on a social networking site?
       ☐ Yes ☐ No
    c. What is your preferred way of spending weekends?

11. The ABS conducts a census every five years. To monitor changes that might occur between these times, surveys are conducted on samples of the population. The ABS selects a representative sample of the population and interviewers are allocated particular households. It is important that no substitutes occur in the sampling. The interviewer must persevere until the selected household supplies the information requested. It is a legal requirement that selected households cooperate.

    The following questionnaire is reproduced from the ABS website www.abs.gov.au. It illustrates the format and types of questions asked by an interviewer collecting data regarding employment from a sample.

Q.1.	I WOULD LIKE TO ASK ABOUT LAST WEEK, THAT IS, THE WEEK STARTING MONDAY THE ... AND ENDING (LAST SUNDAY THE .../YESTERDAY).	
Q.2.	LAST WEEK DID ... DO ANY WORK AT ALL IN A JOB, BUSINESS OR FARM?	
	Yes	☐ Go to Q.5
	No	☐
	Permanently unable to work	☐ No More Questions
	Permanently not intending to work (if aged 65+ only)	☐ No More Questions
Q.3.	LAST WEEK DID ... DO ANY WORK WITHOUT PAY IN A FAMILY BUSINESS?	
	Yes	☐ Go to Q.5
	No	☐
	Permanently not intending to work (if aged 65+ only)	☐ No More Questions
Q.4.	DID ... HAVE A JOB, BUSINESS OR FARM THAT ... WAS AWAY FROM BECAUSE OF HOLIDAYS, SICKNESS OR ANY OTHER REASON?	
	Yes	☐
	No	☐ Go to Q.13
	Permanently not intending to work (if aged 65+ only)	☐ No More Questions
Q.5.	DID ... HAVE MORE THAN ONE JOB OR BUSINESS LAST WEEK?	
	Yes	☐
	No	☐ Go to Q.7
Q.6.	THE NEXT FEW QUESTIONS ARE ABOUT THE JOB OR BUSINESS IN WHICH ... USUALLY WORKS THE MOST HOURS.	

Q.7. DOES ... WORK FOR AN EMPLOYER, OR IN ... OWN BUSINESS?

Employer ☐

Own business ☐ **Go to Q.10**

Other/Uncertain ☐ **Go to Q.9**

Q.8. IS ... PAID A WAGE OR SALARY, <u>OR</u> SOME OTHER FORM OF PAYMENT?

Wage/Salary ☐ **Go to Q.12**

Other/Uncertain ☐

Q.9. WHAT ARE ... (WORKING/PAYMENT) ARRANGEMENTS?

Unpaid voluntary work ☐ **Go to Q.13**

Contractor/Subcontractor ☐

Own business/Partnership ☐

Commission only ☐

Commission with retainer ☐ **Go to Q.12**

In a family business without pay ☐ **Go to Q.12**

Payment in kind ☐ **Go to Q.12**

Paid by the piece/item produced ☐ **Go to Q.12**

Wage/salary earner ☐ **Go to Q.12**

Other ☐ **Go to Q.12**

Q.10. DOES ... HAVE EMPLOYEES (IN THAT BUSINESS)?

Yes ☐

No ☐

Q.11. IS THAT BUSINESS INCORPORATED?

Yes ☐

No ☐

Q.12. HOW MANY HOURS DOES ... USUALLY WORK EACH WEEK IN (THAT JOB/THAT BUSINESS/ALL ... JOBS)?

1 hour or more ☐ **No More Questions**

Less than 1 hour/no hours ☐

Insert occupation questions if required

Insert industry questions if required

Q.13. AT ANY TIME DURING THE LAST 4 WEEKS HAS ... BEEN LOOKING FOR FULL-TIME OR PART-TIME WORK?

Yes, full-time work ☐

Yes, part-time work ☐

No ☐ **No More Questions**

Q.14. AT ANY TIME IN THE LAST 4 WEEKS HAS ...

Written, phoned or applied in person to an employer for work? ☐

Answered an advertisement for a job? ☐

Looked in newspapers? ☐

Checked factory notice boards, or used the touchscreens at Centrelink offices?

AT ANY TIME IN THE LAST 4 WEEKS HAS ...

Been registered with Centrelink as a jobseeker? ☐

Checked or registered with an employment agency? ☐

Done anything else to find a job?	☐	
Advertised or tendered for work	☐	
Contacted friends/relatives	☐	
Other	☐ **No More Questions**	
Only looked in newspapers	☐ **No More Questions**	
None of these	☐ **No More Questions**	

Q.15.	**IF … HAD FOUND A JOB COULD … HAVE STARTED WORK LAST WEEK?**	
	Yes	☐
	No	☐
	Don't know	☐

Remaining questions are only required if Duration of Unemployment is needed for output or to derive the long-term unemployed.

Q.16.	**WHEN DID … BEGIN LOOKING FOR WORK?**	
	Enter Date	
	Less than 2 years ago	...../...../..... **DD MM YY**
	2 years or more ago	...../...../..... **DD MM YY**
	5 years or more ago	...../...../..... **DD MM YY**
	Did not look for work	☐

Q.17.	**WHEN DID … LAST WORK FOR TWO WEEKS OR MORE?**	
	Enter Date	
	Less than 2 years ago	...../...../..... **DD MM YY**
	2 years or more ago	...../...../..... **DD MM YY**
	5 years or more ago	...../...../..... **DD MM YY**
	Has never worked (for two weeks or more)	☐ **No More Questions**

Reading the questionnaire carefully, you will note that, although the questions are labelled 1 to 17, there are only 15 questions requiring answers (two are introductory statements to be read by the interviewer). Because of directions to forward questions, no individual would be asked all 15 questions.

a. State how many questions could be asked of those who have a job.
b. State how many questions could be answered by unemployed individuals.
c. State how many questions apply to those not in the labour force.

12. Choose a topic of interest to you and conduct a survey.

a. Design an interview questionnaire of a similar format to the ABS survey, using directions to forward questions.
b. Decide on a technique to select a representative sample of the students in your class.
c. Administer your questionnaire to this sample.
d. Collate your results.
e. Draw conclusions from your results.

f. Prepare a report which details the:

  i. aim of your survey
  ii. design of the survey
  iii. sample selection technique
  iv. results of the survey collated in table format
  v. conclusions.

10.4 Classifying data

LEARNING INTENTION

At the end of this subtopic you should be able to:
- classify different types of data as categorical or numerical
- identify the difference between nominal and ordinal data
- identify the difference between discrete and continuous data.

10.4.1 Types of data

Statistics is the science of collecting, organising, presenting, analysing and interpreting data. The information collected is called **data**.

The information in the flowchart shown can be used to determine the type of data being considered.

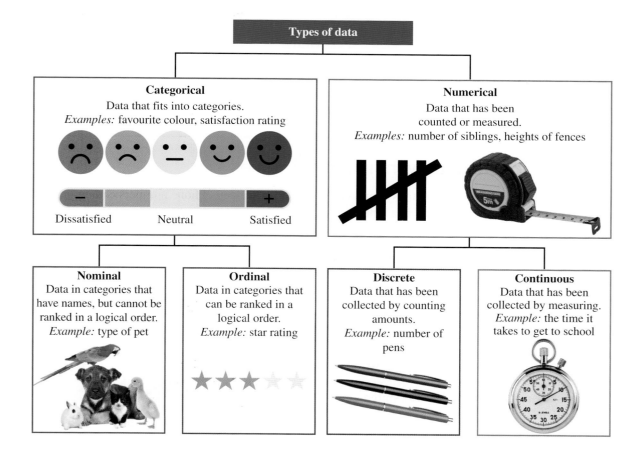

WORKED EXAMPLE 9 Classifying data as numerical discrete or numerical continuous

Classify the following data as either numerical discrete or numerical continuous.
a. A group of people is asked the question: 'How many cars does your family have?'
b. Students are asked to complete the statement: 'My height is ...'

THINK	WRITE
a. 1. Determine whether the data can be counted or measured.	a. The data can be counted.
2. State what type of data it is.	Numerical discrete
b. 1. Determine whether the data can be counted or measured.	b. The data can be measured.
2. State what type of data it is.	Numerical continuous

WORKED EXAMPLE 10 Classifying data as categorical nominal or categorical ordinal

Classify the following data as either categorical nominal or categorical ordinal.
a. A group of people is asked the question: 'What is your opinion about the service in your local supermarket?'
b. Students are asked to complete the statement: 'My favourite colour is ...'

THINK	WRITE
a. 1. Determine whether the data can be ordered.	a. Yes
2. State what type of data it is.	Categorical ordinal
b. 1. Determine whether the data can be ordered.	b. No
2. State what type of data it is.	Categorical nominal

10.4.2 Categorical data with numerical values

There are instances when categorical data is expressed by numerical values. 'Phone numbers ending in 5' and 'postcodes' are examples of categorical data represented by numerical values.

tlvd-4691

WORKED EXAMPLE 11 Classifying data as categorical or numerical

Classify the following data as either categorical or numerical.
a. A set of data is collected about bank account numbers.
b. A group of people are asked: 'How many minutes do you exercise per day?'
c. A group of people are asked: 'What is the area of your backyard?'
d. Students are asked to complete the statement: 'The subject I like the most is ...'

THINK	WRITE
a. 1. Determine whether the data can be counted or measured.	**a.** Neither
2. State what type of data it is.	Categorical
b. 1. Determine whether the data can be counted or measured.	**b.** The data can be measured.
2. State what type of data it is.	Numerical
c. 1. Determine whether the data can be counted or measured.	**c.** The data can be measured.
2. State what type of data it is.	Numerical
d. 1. Determine whether the data can be counted or measured.	**d.** Neither
2. State what type of data it is.	Categorical

10.4 Exercise

Students, these questions are even better in jacPLUS

 Receive immediate feedback and access sample responses   **Access additional questions**   **Track your results and progress**

Find all this and MORE in jacPLUS

1. **WE9** Classify the following data as either numerical discrete or numerical continuous.

   a. The amount of time spent on homework per night.
   b. The marks in a test.

2. Explain in your own words why 'the number of pieces of fruit sold' is numerical discrete data while 'the quantity (in kg) of fruit sold' is numerical continuous data.

3. Classify the following data as either discrete or continuous.

   a. Data is collected about the water levels in a dam over a period of one month.
   b. An insurance company is collecting data about the number of work injuries in ten different businesses.
   c. 350 students are selected by a phone company to answer the question: 'How many pictures do you take per week using your mobile phone?'
   d. Customers in a shopping centre are surveyed on the number of times they visited the shopping centre in the last 12 months.
   e. 126 people participated in a 5-km charity marathon. Their individual times are recorded.
   f. A group of students are asked to measure the perimeter of their bedrooms. This data is collected and used for a new project.

4. **WE10** Classify the following data as either categorical nominal or categorical ordinal.

   a. The brands of cars sold by a car yard over a period of time
   b. The opinion about the quality of a website: poor, average or excellent

5. Explain in your own words why 'favourite movie' is categorical nominal data while 'the rating for a movie' is categorical ordinal data.

6. Classify the following data as either ordinal or nominal.

   a. The members of a football club are asked to rate the quality of the training grounds as 'poor', 'average' or 'very good'.

   b. A group of students are asked to state the brand of their mobile phone.

   c. A fruit and vegetable shop is conducting a survey, asking the customers, 'Do you find our products fresh?' Possible answers are 'Never', 'Sometimes', 'Often', 'Always'.

   d. A chocolate company conducts a survey with the question: 'Do you like sweets?'. The possible answers are 'Yes' and 'No'.

7. **WE11** Classify the following data as either numerical or categorical.

   a. Data collected on 'whether the subject outcomes 1, 2 and 3 have been completed'.

   b. The number of learning outcomes per subject.

8. Explain in your own words why 'country phone codes' is categorical data while 'the number of international calls' is numerical data.

9. Classify the following data as either numerical or categorical.

   a. The amount of time spent at the gym per week.

   b. The opinion about the quality of a website: poor, average or excellent.

10. Explain in your own words why 'the lengths of pencils in a pencil case' is numerical data while 'the colours of pencils' is categorical data.

11. **MC** A train company recorded the time by which trains were late over a period of one month. The data collected is:

    A. numerical discrete.

    B. categorical nominal.

    C. categorical ordinal.

    D. numerical continuous.

    E. none of the above.

12. Classify the following data as either numerical or categorical.

    a. The average time spent by customers in a homeware shop is recorded.

    b. Mothers are asked: 'What brand of baby food do you prefer?'

    c. An online games website asked the players to state their gender.

    d. A teacher recorded the lengths of his students' long jumps.

10.5 Categorical data

10.5.1 One-way or frequency tables

Categorical data can be displayed in tables or graphs.

One-way tables or **frequency distribution tables** are used to organise and display data in the form of frequency counts. The **frequency** of a score is the number of times the score occurs in the set of data.

Frequency distribution tables display one variable only. For large amounts of data, a **tally** column helps make accurate counts of the data. The frequency distribution table shown displays the tally and the frequency for a given data set.

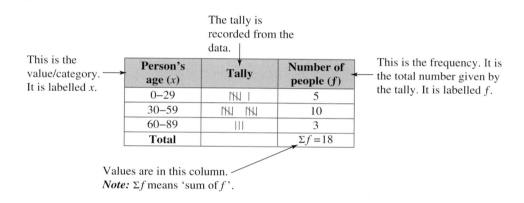

The tally is recorded from the data.

This is the value/category. It is labelled x.

Person's age (x)	Tally	Number of people (f)
0–29	⦀⦀ I	5
30–59	⦀⦀ ⦀⦀	10
60–89	III	3
Total		$\Sigma f = 18$

This is the frequency. It is the total number given by the tally. It is labelled f.

Values are in this column.
Note: Σf means 'sum of f'.

tlvd-4692

WORKED EXAMPLE 12 Constructing a frequency distribution table

Alana is organising the Year 11 formal and has to choose three types of drinks out of the five available: cola, lemonade, apple juice, orange juice and soda water. She decided to survey her classmates to determine their favourite drinks. The data collected is shown below.

orange juice	cola	cola	cola	apple juice	lemonade
lemonade	cola	apple juice	lemonade	orange juice	orange juice
lemonade	cola	apple juice	orange juice	cola	soda water
orange juice	cola	soda water	orange juice	cola	cola

By constructing a frequency table, determine the three drinks Alana should choose for the formal.

THINK

1. Draw a frequency table with three columns and the headings 'Drink', 'Tally' and 'Frequency'.

2. Write all the possible outcomes in the first column and 'Total' in the last row.

3. Fill in the 'Tally' column.

4. Fill in the 'Frequency' column. Ensure the table has a title also.

5. Write your answer as a sentence.

WRITE

Drink	Tally	Frequency

Drink	Tally	Frequency
Apple juice		
Cola		
Lemonade		
Orange juice		
Soda water		
Total		

Drink	Tally	Frequency
Apple juice	|||	
Cola	|||| ||||	
Lemonade	||||	
Orange juice	|||||	
Soda water	||	
Total		

Favourite drinks for a Year 11 class		
Drink	Tally	Frequency
Apple juice	|||	3
Cola	|||| ||||	9
Lemonade	||||	4
Orange juice	|||||	6
Soda water	||	2
Total		24

Alana should choose cola, orange juice and lemonade for the Year 11 formal.

10.5.2 Two-way frequency tables

Two-way tables are used to examine the relationship between two related categorical variables. Because the entries in these tables are frequency counts, they are also called **two-way frequency tables**.

The table shown is a two-way table that relates two variables. It displays the relationship between age groups and party preference.

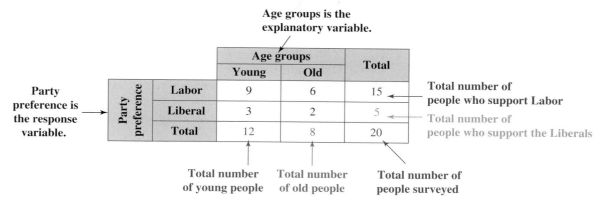

Age groups is the explanatory variable.

Party preference is the response variable.

		Age groups		Total
		Young	Old	
Party preference	Labor	9	6	15
	Liberal	3	2	5
	Total	12	8	20

Total number of people who support Labor

Total number of people who support the Liberals

Total number of young people

Total number of old people

Total number of people surveyed

One variable is **explanatory** (age groups) and the other variable is **response** (party preference).

The explanatory variable is the variable that can be manipulated to produce changes in the response variable. The response variable changes as the explanatory variable changes. Its outcome is a response to the input of the explanatory variable.

WORKED EXAMPLE 13 Displaying the data in a two-way frequency table

A sample of 40 people was surveyed about their income level (low or high). The sample consisted of 15 young people (25–45 years old) and 25 old people (above 65 years old). 7 young people were on a low-level income while 6 old people were on a high-level income. Display this data in a two-way table and determine the number of young people on a high-level income.

THINK

1. State the two categorical variables.

2. Decide where the two variables are placed.

3. Construct the two-way table and label both the columns and the rows.

WRITE

Age groups: young and old people
Income levels: low and high

The explanatory variable is 'Age groups' — columns.
The response variable is 'Income level' — rows.

Income levels for young and old people			
	Age groups		Total
	Young	Old	
Income level — Low			
High			
Total			

4. Fill in the known data:
   - 7 young people are on a low-level income.
   - 6 old people are on a high-level income.
   - There are 15 young people in total and 25 old people in total.
   - The total number of people in the sample is 40.

Income levels for young and old people

		Age groups		Total
		Young	Old	
Income level	Low	7		
	High		6	
	Total	15	25	40

5. Determine the unknown data.

Young people on high-level income: $15 - 7 = 8$
Old people on low-level income: $25 - 6 = 19$
Total number of people on a low-level income:
$7 + 19 = 26$
Total number of people on a high-level income:
$8 + 6 = 14$

6. Fill in the table.

Income levels for young and old people

		Age groups		Total
		Young	Old	
Income level	Low	7	19	26
	High	8	6	14
	Total	15	25	40

7. Write the answer as a sentence.

The total number of young people on a high-level income is 8.

10.5.3 Two-way relative frequency tables

Consider the party preference table shown. As the number of young and old people is different, it is difficult to compare the data.

Party preference for young and old people

		Age groups		Total
		Young	Old	
Party preference	Labor	9	6	15
	Liberal	3	2	5
	Total	12	8	20

Nine young people and six old people prefer Labor; however, there are more young people surveyed than old. In order to accurately compare the data, we need to convert these values in a more meaningful way as proportions or percentages.

Formula for relative frequency

The frequency given as a proportion is called relative frequency.

$$\text{relative frequency} = \frac{\text{score}}{\text{total number of scores}}$$

The relative frequency table for the party preference data is:

<table>
<tr><td colspan="5">Party preference for young and old people</td></tr>
<tr><td colspan="2" rowspan="2"></td><td colspan="2">Age groups</td><td rowspan="2">Total</td></tr>
<tr><td>Young</td><td>Old</td></tr>
<tr><td rowspan="6">Party preference</td><td>Labor</td><td>$\frac{9}{20}$</td><td>$\frac{6}{20}$</td><td>$\frac{15}{20}$</td></tr>
<tr><td>Liberal</td><td>$\frac{3}{20}$</td><td>$\frac{2}{20}$</td><td>$\frac{5}{20}$</td></tr>
<tr><td>Total</td><td>$\frac{12}{20}$</td><td>$\frac{8}{20}$</td><td>$\frac{20}{20}$</td></tr>
</table>

In simplified form using decimal numbers, the relative frequency table for party preference becomes:

<table>
<tr><td colspan="5">Party preference for young and old people</td></tr>
<tr><td colspan="2" rowspan="2"></td><td colspan="2">Age groups</td><td rowspan="2">Total</td></tr>
<tr><td>Young</td><td>Old</td></tr>
<tr><td rowspan="3">Party preference</td><td>Labor</td><td>0.45</td><td>0.30</td><td>0.75</td></tr>
<tr><td>Liberal</td><td>0.15</td><td>0.10</td><td>0.25</td></tr>
<tr><td>Total</td><td>0.60</td><td>0.40</td><td>1.00</td></tr>
</table>

Note: In a relative frequency table, the total value is always 1.

Formula for percentage relative frequency

The frequency written as a percentage is called percentage relative frequency. We can use the following formula to calculate % relative frequency:

$$\text{percentage relative frequency} = \frac{\text{score}}{\text{total number of scores}} \times 100\%$$

or

$$\text{percentage relative frequency} = \text{relative frequency} \times 100\%$$

tlvd-4693

WORKED EXAMPLE 14 Calculating the percentage relative frequency

Consider the party preference data discussed previously, with the relative frequencies shown.

		Party preference for young and old people		
		Age groups		
		Young	Old	Total
Party preference	Labor	0.45	0.30	0.75
	Liberal	0.15	0.10	0.25
	Total	0.60	0.40	1.00

Calculate the percentage relative frequency for all entries in the table out of the total number of people. Discuss your findings.

THINK

1. Write the formula for % relative frequency.

2. Substitute all scores into the percentage relative frequency formula.

WRITE

Percentage relative frequency = relative frequency × 100%

		Party preference for young and old people		
		Age groups		
		Young	Old	Total
Party preference	Labor	0.45 × 100	0.30 × 100	0.75 × 100
	Liberal	0.15 × 100	0.10 × 100	0.25 × 100
	Total	0.60 × 100	0.40 × 100	1.00 × 100

3. Simplify.

		Party preference for young and old people		
		Age groups		
		Young	Old	Total
Party preference	Labor	45	30	75
	Liberal	15	10	25
	Total	60	40	100

4. Discuss your findings.

Of the total number of people, 75% prefer Labor and 25% prefer the Liberals. 60% of people are young and 40% of people are old. 45% of people who prefer Labor are young and 30% of people who prefer the Liberals are old. 15% of people who prefer the Liberals are young and 10% of people who prefer the Liberals are old.

10.5.4 Grouped column graphs

Grouped column graphs are used to display the data for two or more categories. They are visual tools that allow for easy comparison between various categories of data sets. The frequency is measured by the height of the column. Both axes have to be clearly labelled, including scales and units, if used. The title should explicitly state what the grouped column graph represents.

Consider the data recorded in the two-way distribution table shown.

		Income levels for young and old people		
		Age groups		Total
		Young	Old	
Income level	Low	7	19	26
	High	8	6	14
	Total	15	25	40

The grouped column graph for this data is shown. Notice that all rectangles have the same width.

The colour or pattern of the rectangles differentiates between the two categories, 'Young' and 'Old'.

Making comparisons

- It is helpful to change the table to a percentage frequency table to compare the income levels of old and young people.
- We first calculate the percentage relative frequency for all the entries in the two-way table out of the total number of people.

		Income levels for young and old people		
		Age groups		Total
		Young %	Old %	
Income level	Low	$\frac{7}{40} \times 100 = 17.5$	$\frac{19}{40} \times 100 = 47.5$	$\frac{26}{40} \times 100 = 65.0$
	High	$\frac{8}{40} \times 100 = 20.0$	$\frac{6}{40} \times 100 = 15.0$	$\frac{14}{40} \times 100 = 35.0$
	Total %	$\frac{15}{40} \times 100 = 37.5$	$\frac{25}{40} \times 100 = 62.5$	$\frac{40}{40} \times 100 = 100$

From these percentages we can conclude that:
- 17.5% of the people surveyed are young on a low income and 47.5% of the people are old on a low income.
- 20% of the total number of people surveyed are young on a high income while 15% are old on a high income.
- 37.5% of the people surveyed are young while 62.5% are old.
- 65% of the people surveyed are on a low income while 35% are on a high income.

Another way of comparing the results of this survey is to calculate percentages out of the total numbers in columns and in rows.

Note: This method of comparison is more applicable when data has even or close to even numbers of each age group surveyed.

Row 1: Percentages of young people and old people on low incomes out of the total number of people on a low income

$$\text{Young on low income} = \frac{\text{young on low income}}{\text{total people on low income}} \times 100\%$$

$$= \frac{7}{26} \times 100\%$$

$$= 26.9\%$$

$$\text{Old on low income} = \frac{\text{old on low income}}{\text{total people on low income}} \times 100\%$$

$$= \frac{19}{26} \times 100\%$$

$$= 73.1\%$$

These percentages indicate that 26.9% of the people on a low income are young, while 73.1% are old.

Row 2: Percentages of young and old people on high incomes out of the total number of people on a high income

$$\text{Young on high income} = \frac{\text{young on high income}}{\text{total people on high income}} \times 100\%$$

$$= \frac{8}{14} \times 100\%$$

$$= 57.1\%$$

$$\text{Old on high income} = \frac{\text{old on high income}}{\text{total people on high income}} \times 100\%$$

$$= \frac{6}{14} \times 100\%$$

$$= 42.9\%$$

These percentages indicate that 57.1% of the people on a high income are young while 42.9% are old.

Column 1: Percentages of young people on low and high income out of the total number of young people surveyed

$$\text{Young on low income} = \frac{\text{young on low income}}{\text{total number of young}} \times 100\%$$

$$= \frac{7}{15} \times 100\%$$

$$= 46.7\%$$

$$\text{Young on high income} = \frac{\text{young on high income}}{\text{total number of young}} \times 100\%$$

$$= \frac{8}{15} \times 100\%$$

$$= 53.3\%$$

These percentages indicate that 46.7% of the young people surveyed are on a low income while 53.3% are on a high income.

Column 2: Percentages of old people on low and high income out of the total number of old people surveyed

$$\text{Old on low income} = \frac{\text{old on low income}}{\text{total number of old}} \times 100\%$$
$$= \frac{19}{25} \times 100\%$$
$$= 76\%$$

$$\text{Old on high income} = \frac{\text{old on high income}}{\text{total number of old}} \times 100\%$$
$$= \frac{6}{25} \times 100\%$$
$$= 24\%$$

These percentages indicate that 76% of the old people surveyed are on a low income while 24% of old people are on a high income.

WORKED EXAMPLE 15 Drawing a grouped column graph displaying the percentage relative frequencies of the given data

The data shown is part of a random sample of 25 students who participated in a questionnaire run by the Australian Bureau of Statistics through Census@School. The aim of the questionnaire was to determine how often the students use the internet for schoolwork.

Draw a grouped column graph displaying the percentage relative frequencies of the data set given.

		Use of internet for schoolwork		
		Place		Total
		School	Home	
Time spent	Rarely	2	8	10
	Sometimes	4	4	8
	Often	6	1	7
	Total	12	13	25

THINK

1. Draw a labelled set of axes with accurate scales. Ensure the graph has a title.

WRITE/DRAW

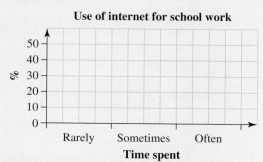

---

2. Calculate the % relative frequency for the data needed.

		Use of the internet for schoolwork		
		Place		Total
		Young	Old	
Time spent	Rarely	$\frac{2}{25} \times 100 = 8$	$\frac{8}{25} \times 100 = 32$	$\frac{10}{25} \times 100 = 40$
	Sometimes	$\frac{4}{25} \times 100 = 16$	$\frac{4}{25} \times 100 = 16$	$\frac{8}{25} \times 100 = 32$
	Often	$\frac{6}{25} \times 100 = 24$	$\frac{1}{25} \times 100 = 4$	$\frac{7}{25} \times 100 = 28$
	Total	$\frac{12}{25} \times 100 = 48$	$\frac{13}{25} \times 100 = 52$	100

3. Draw the sets of corresponding rectangles. The graph can be drawn vertically.

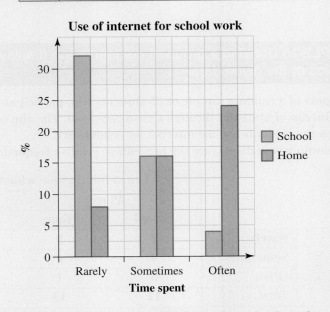

4. The graph can also be drawn horizontally.

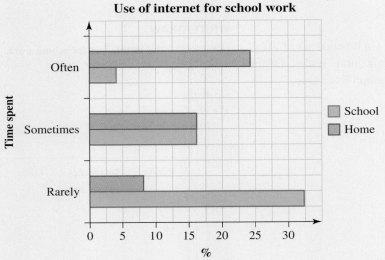

10.5.5 Misuses of grouped column graphs

The column graph is a powerful tool for displaying categorical data. However, a badly constructed column graph could be misleading or deceiving. The graph shown displays issues that could give a false impression about the data represented:

- The horizontal axis jumps from June 2014 to June 2017 while the rest of the scale represents equal intervals of one year. This gives the impression that the increase between the first two columns is a lot bigger than the increases between the rest of the columns.
- The overlapping drawing of the woman on the columns gives the impression that the June 2018 column has a higher value than it actually has.

A more accurate representation of this data is shown in the following histogram.

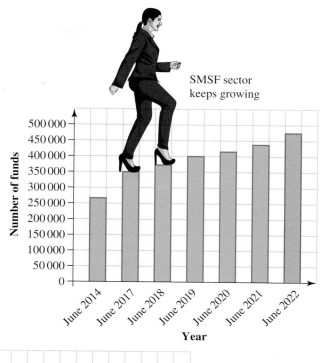

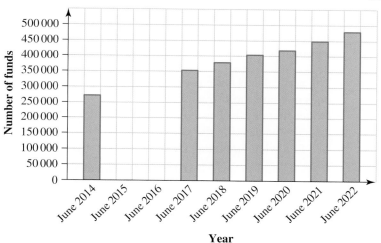

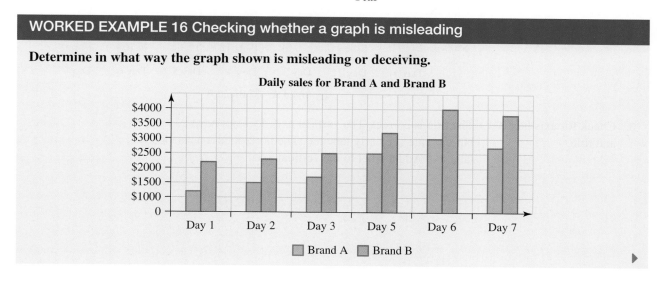

WORKED EXAMPLE 16 Checking whether a graph is misleading

Determine in what way the graph shown is misleading or deceiving.

THINK	**WRITE**
1. Check the starting point of the vertical axis.	The vertical axis does not have an even scale. The first incremental increase is $1000 (from $0 to $1000) whereas the following incremental increases are all of $500 (e.g. from $1000 to $1500). This gives the impression that the sales of Brand B are a lot higher than those of Brand A. If we compare the first two columns, it looks like the sales of Brand B are more than double the sales of Brand A. This assumption is incorrect. The following graph has a scale showing even incremental increases of $500.

As can be seen from the graph, in reality, the sales of Brand B are roughly double those of Brand A only on day 1.

| 2. Check the accuracy of the scales on both axes. | The vertical axis has equally spaced scales. The horizontal axis does not have the correct scale.
Notice that day 4 sales are missing. This makes the jump from the sales on day 3 to the sales on day 5 look a lot higher.
The following graph includes day 4 (taken from original data). |

| 3. Check for axis labels and title. | The title is correct. The vertical axis does not have a label, but its scale is clearly labelled. The horizontal axis is clearly labelled. Add a label to the vertical axis. |

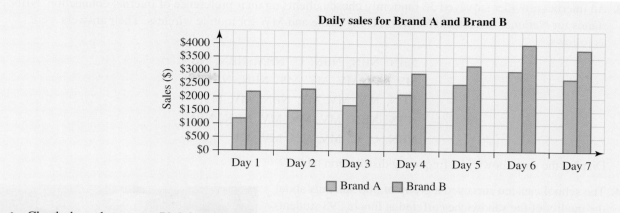

Daily sales for Brand A and Brand B

4. Check the colours or patterns of the columns.

Pink is used for one brand and blue for the other brand. Because of this choice of colour, the pink columns tend to stand out more than the blue columns. Care has to be taken when choosing the colours or patterns of the columns so that there is no extreme discrepancy.

on Resources

Interactivity Create a bar graph (int-6493)

10.5 Exercise

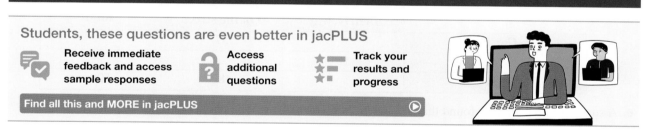

Students, these questions are even better in jacPLUS

Receive immediate feedback and access sample responses

Access additional questions

Track your results and progress

Find all this and MORE in jacPLUS

1. **WE12** The form of transport for a group of 30 students is given below.

car, car, bicycle, tram, tram, tram, car, walk, walk, walk,

tram, walk, car, walk, walk, car, walk, bicycle, bicycle, walk,

bicycle, tram, walk, tram, tram, car, car, walk, tram, tram

Construct a frequency distribution table to display this data.

2. A group of 24 students was asked 'Which social network do you use more, MyPage or FlyBird?' Their answers were recorded as follows.

MyPage, MyPage, FlyBird, MyPage, FlyBird, MyPage,

MyPage, MyPage, FlyBird, MyPage, MyPage, FlyBird,

FlyBird, MyPage, MyPage, FlyBird, FlyBird, MyPage,

MyPage, MyPage, MyPage, FlyBird, MyPage, FlyBird

Draw a frequency table to display this data.

3. An internet provider surveyed 35 randomly chosen clients on their preference of internet connection. NBN stands for National Broadband Network, C for Cable and MW for mobile wireless. Their answers were as follows.

NBN, NBN, NBN, C, NBN, NBN, NBN,

MV, MV, MV, MV, MV, NBN, NBN,

NBN, MV, MV, MV, NBN, MV, MV,

NBN, NBN, MV, MV, MV, NBN, NBN,

NBN, C, C, MV, MV, MV, MV

Display the data shown in a frequency distribution table.

4. The school canteen surveyed a group of 500 students about the quality of the sandwiches offered at lunch. 298 students were of the opinion that the sandwiches offered at the school canteen were very good, 15 students thought they were of poor quality, 163 students thought they were excellent, while the rest said they were average.
Construct a frequency distribution table for this set of data.

5. **WE13** Determine the missing values in the two-way table shown.

		Parking place and car theft		
		Parking type		Total
		In driveway	On the street	
Theft	Car theft		37	
	No car theft	16		
	Total		482	500

6. A survey of 263 students found that 58 students who owned a mobile phone also owned a tablet, while 11 students owned neither. The total number of students who owned a tablet was 174.
Construct a complete two-way table to display this data and calculate all the unknown entries.

7. The numbers of the Year 11 Maths students at Senior Secondary College who passed and failed their exams are displayed in the table shown.

		Year 11 students' Mathematics results		
		Result		Total
		Pass	Fail	
Subject	Foundation Mathematics		23	
	General Mathematics		49	74
	Mathematical Methods	36	62	
	Specialist Mathematics	7		25
	Total	80		

a. Complete the two-way table.
b. Determine the number of Year 11 students who study Mathematics at Senior Secondary College.
c. Calculate the total number of students enrolled in Foundation Mathematics.

8. George is a hairdresser who keeps a record of all his customers and their hairdressing requirements. He made a two-way table for last month and recorded the numbers of customers who wanted a haircut only, and those who wanted a haircut and colour.

   a. Design a two-way table that George would need to record the numbers of young and old customers who wanted a haircut only, and those who wanted a haircut and colour.
   b. George had 56 old customers in total and 2 young customers who wanted a haircut and colour. Place these values in the table from part a.
   c. George had 63 customers during last month, with 30 old customers requiring a haircut and colour. Fill in the rest of the two-way table.

9. The manager at the local cinema recorded the movie genres and the ages of the audience at the Monday matinee movies. The total audience was 156 people. A quarter of the audience were children under the age of 10, 21 were teenagers and the rest were adults. There were 82 people who watched the drama movie, 16 teenagers who watched comedy, and 11 people under the age of 10 who watched the drama movie.

   a. Construct a complete two-way table.
   b. Determine the total number of adults who watched the comedy movie.
   c. Calculate the number of children.

10. **WE14** Consider the data given in the two-way table shown.

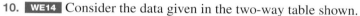
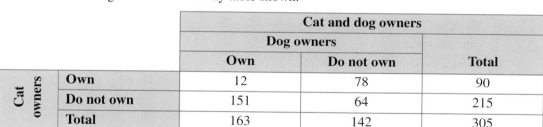

		Cat and dog owners		
		Dog owners		
		Own	Do not own	Total
Cat owners	Own	12	78	90
	Do not own	151	64	215
	Total	163	142	305

   a. Calculate the relative frequencies for all the entries in the two-way table given. Give your answers correct to 3 decimal places or as fractions.
   b. Calculate the percentage relative frequencies for all the entries in the two-way table out of the total number of pet owners.

11. Using the two-way table from question 7, construct:

   a. the relative frequency table, correct to 4 decimal places
   b. the percentage relative frequency table.

12. Consider the data given in the two-way table shown.

		Swimming attendance		
		People		
		Adults	**Children**	**Total**
Day of the week	**Monday**	14	31	45
	Tuesday	9	17	26
	Wednesday	10	22	32
	Thursday	15	38	53
	Friday	18	45	63
	Saturday	27	59	86
	Sunday	32	63	95
	Total	125	275	400

a. Calculate the relative frequencies and the percentage relative frequencies for all the entries for adults out of the total number of adults.

b. Calculate the relative frequencies and the percentage relative frequencies for both adults and children out of the total number of people swimming on Wednesday.

13. A ballroom studio runs classes for adults, teenagers and children in Latin, modern and contemporary dances. The enrolments in these classes are displayed in the two-way table shown.

		Dance class enrolments			
		People			
		Adults	**Teenagers**	**Children**	
Dance	**Latin**			23	75
	Modern		34	49	96
	Contemporary	36		62	
	Total	80			318

a. Complete the two-way table.

b. Construct a relative frequency table for this data set.

c. Construct a percentage relative frequency table for this data set.

14. **WE15** Bianca manages a bookstore. She recorded the number of fiction and non-fiction books she sold over two consecutive weeks in the two-way table below.

		Types of books sold in a fortnight		
		Week		**Total**
		Week 1	**Week 2**	
Type of book	**Fiction**	51	73	124
	Non-fiction	37	28	65
	Total	88	101	189

Construct a grouped column graph for this data.

15. Construct a grouped column graph for the data displayed in the following two-way table.

		Type of exercise enjoyed by different age groups		
		People		Total
		Teenagers	Adults	
Type of exercise	Running	12	10	22
	Walking	8	15	23
	Total	20	25	45

16. **MC** The horizontal column graph below displays data about school sector enrolments in each of the Australian states collected by the ABC in 2017.

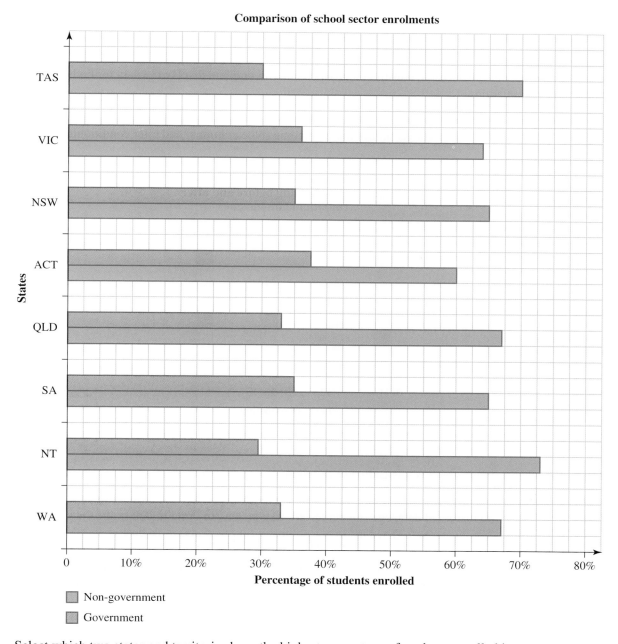

Comparison of school sector enrolments

Percentage of students enrolled

States

Non-government
Government

Select which two states and territories have the highest percentage of students enrolled in government schools.

A. VIC and NSW   **B.** VIC and ACT   **C.** NSW and QLD   **D.** VIC and QLD   **E.** VIC and SA

17. Human blood is grouped in 8 types: O+, O−, A+, A−, B+, B−, AB+ and AB−. The following % relative frequency table shows the percentage of Australians and Chinese who have a particular blood type.

Blood type	Percentage of Australia's population	Percentage of China's population
O+	40	47.7
O−	9	0.3
A+	31	27.8
A−	7	0.2
B+	8	18.9
B−	2	0.1
AB+	2	5.0
AB−	1	0.01

Source: Wikipedia; http://en.wikipedia.org/wiki/ABO_blood_group_system

Use a spreadsheet or otherwise to construct a grouped column graph to display the data for both Australia's and China's populations.

18. **WE16** The graph shown represents the relationship between job satisfaction of a group of people surveyed and their years of schooling before conditioning on income (blue columns) and after conditioning on income (pink rectangles).

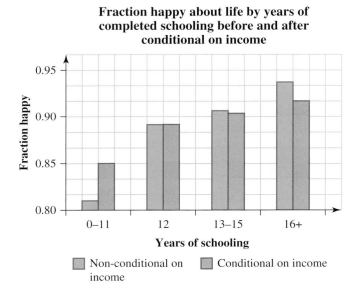

Fraction happy about life by years of completed schooling before and after conditional on income

Explain in what way this graph is misleading or deceiving.

19. a. Car insurance companies calculate premiums according to the perceived risk of insuring a person and their vehicle. From historical data, age is a perceived 'risk factor'. In order to determine the premium of their car insurance policies, Safe Drive insurance company surveyed 300 of its customers.

They determined three age categories: category A for young drivers, category B for mature drivers, and category C for older drivers. They then recorded whether the customer had a traffic accident within the last year. Their records showed that 113 drivers from category B and 17 drivers out of 47 in category C did not have a traffic accident within the last year.

If 75 drivers had a traffic accident last year and 35 of the drivers were in category A, construct a complete two-way table for this data set.

b. State how many mature drivers had a traffic accident last year.

c. Determine the total number of drivers who did not have a traffic accident last year.

d. Use a spreadsheet or otherwise to construct a grouped column graph for this data.

20. Explain in what way this graph is misleading or deceiving.

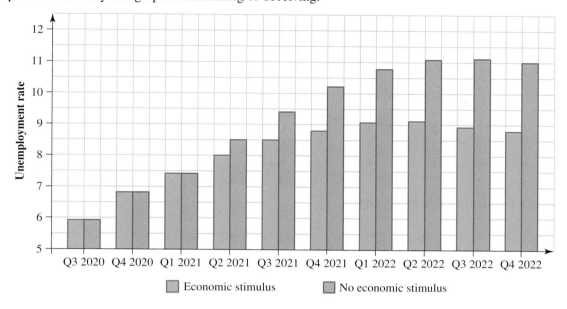

10.6 Numerical data

LEARNING INTENTION

At the end of this subtopic you should be able to:
- display and interpret numerical data in frequency tables
- construct frequency tables for ungrouped and grouped data
- construct a dot plot for a data set
- construct a histogram for grouped and ungrouped data
- construct a stem-and-leaf plot for ten- and five-interval data sets.

10.6.1 Displaying numerical data

The visual display of data is an important tool in communicating the information collected. Numerical data can be displayed in tables or graphical representations.

Frequency distributions

Frequency distributions (also known as **frequency distribution tables** or **frequency tables**) for numerical data are constructed in the same way as for categorical data, given that they display frequency counts in both cases.

Ungrouped numerical data

For **ungrouped data**, each score is recorded separately. For large sets of data, it is helpful to use tally marks.

> ▶ **WORKED EXAMPLE 17 Constructing a frequency table for data**

For his statistics project, Radu surveyed 50 students on the number of books they had read over the past six months. The data recorded is listed below.

3	2	0	6	4	3	3	4	2	2
2	2	1	2	1	1	5	4	4	3
3	2	3	3	3	5	0	1	1	1
4	2	2	3	2	3	2	2	2	1
2	1	1	5	4	4	3	3	4	2

Construct a frequency table for this data.

THINK

1. Draw a frequency table with three columns to display the variable (number of books), tally and frequency.
   - The data has scores between 0 as the lowest score and 6 as the highest score.
   - The last cell in the first column is labelled 'Total'.

WRITE

Number of books	Tally	Frequency
0		
1		
2		
3		
4		
5		
6		
Total		

2. Complete the 'Tally' and 'Frequency' columns.

Number of books	Tally	Frequency
0				2
1	卌 ||||	9
2	卌 卌 卌	15
3	卌 卌 ||	12
4	卌 |||	8
5					3
6			1
Total		50

3. Check the frequency total with the total number of data.   50

Grouped data

Grouped data is data that has been organised into groups or class intervals. It is used for very large amounts of data and for continuous data.

Care has to be taken when choosing the class sizes for the data collected. The most common class sizes are 5 or 10 units. A class interval is usually written as 5– < 10, meaning that all scores between 5 and 10 are part of this group, including 5 but excluding 10.

WORKED EXAMPLE 18 Constructing a frequency table for grouped data

The data below displays the maximum daily temperature in Melbourne for every day in a month of summer.

Date	Maximum temp (°C)	Date	Maximum temp (°C)	Date	Maximum temp (°C)
1	27.4	11	29.1	21	22.4
2	20.7	12	34.3	22	28.3
3	21.4	13	35.1	23	38.3
4	18.9	14	22.9	24	27.6
5	18.2	15	24.8	25	21.9
6	21.2	16	23.1	26	22.4
7	30.6	17	22	27	31.2
8	37	18	27.8	28	21.7
9	21.1	19	32.6	29	23.3
10	21.4	20	23.5	30	21.9
				31	24.7

Source: Copyright Commonwealth Bureau of Meterology

Construct a frequency table for this data.

THINK

1. Draw a frequency table with three columns to display the variable (temperature (°C) in this case), tally and frequency.
   - The data has scores between 18.2 °C as the lowest score and 38.3 °C as the highest score. This leads us to five groups of class size 5 °C between 18 °C and 43 °C.
   - The last cell in the first column is labelled 'Total'.

WRITE

Temperature (°C)	Tally	Frequency
18– <23		
23– <28		
28– <33		
33– <38		
38– <43		
Total		

2. Complete the 'Tally' and 'Frequency' columns.

Temperature (°C)	Tally	Frequency
18– <23	JHT JHT IIII	14
23– <28	JHT III	8
28– <33	JHT	5
33– <38	III	3
38– <43	I	1
Total		31

3. Check the frequency total with the total number of data.   31

10.6.2 Dot plots

A dot plot is a graphical representation of numerical data made up of dots. A dot plot can also be used to represent categorical data, with dots being stacked in a column above each category. All data are displayed using identical equally spaced dots, where each dot represents one score of the variable. The height of the column of dots represents the frequency for that score.

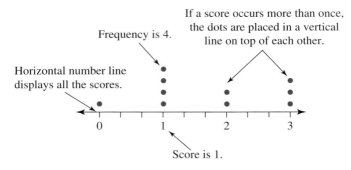

tlvd-4695

WORKED EXAMPLE 19 Construct a dot plot for a data set

At an electronics store, customers were surveyed about the number of TV sets they have in their household. The data collected from 20 customers is displayed in the following frequency table.

Number of TV sets	Frequency
0	1
1	3
2	5
3	8
4	2
5	1
Total	20

Construct a dot plot for this set of data.

THINK

1. Draw a labelled horizontal line for the variable 'Number of TV sets'.
   - Start from the minimum score, 0, and continue until you've placed the maximum score on the line (5 in this example).
   - The spaces between scores have to be equal.

2. Place the dots for each score.
   - The frequency for 0 is 1. Place one dot above score 0.
   - The frequency for 1 is 3. Place three dots above score 1.
   - Continue until you've placed all data above the line.
   - Include a title.

WRITE/DRAW

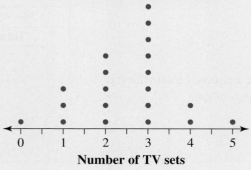

10.6.3 Histograms

A histogram is a graphical representation of numerical data. It is very similar to the column graph used to represent categorical data, because the data is displayed using rectangles of equal width. However, although there is a space before the first column, there are no spaces between columns in a histogram. Each column represents a different score of the same variable.

The vertical axis always displays the values of the frequency.

Features

Ungrouped data	Grouped data
The value of the variable is written in the middle of the column under the horizontal axis.	The ends of the class interval are written under the ends of the corresponding columns.

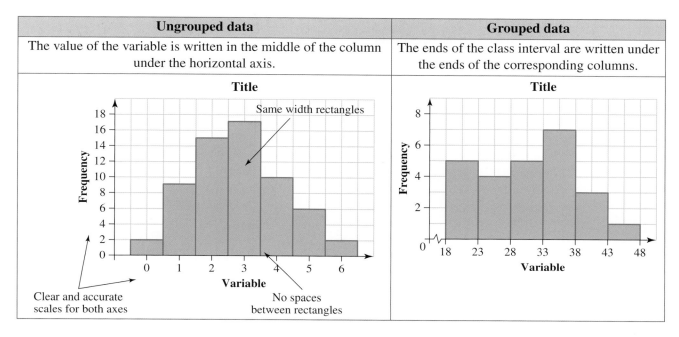

WORKED EXAMPLE 20 Constructing a histogram for numerical discrete data

Draw a histogram for the data given in the frequency table below.

Number of books read over six months	Frequency
0	2
1	9
2	15
3	12
4	8
5	3
6	1
Total	50

THINK	WRITE/DRAW

THINK

1. Draw a labelled set of axes:
   - Place 'Frequency' on the vertical axis.
   - Place 'Number of books' on the horizontal axis.
   - Ensure the histogram has a title.
   - The maximum frequency is 18, so the vertical scale should have ticks from 0 to 18. The ticks could be written by ones (a bit cluttered) or by twos.
   - The data is ungrouped, so the ticks are in the middle of each column.
   - Allow for a space before drawing the columns of the histogram.

2. Draw the columns.

WRITE/DRAW

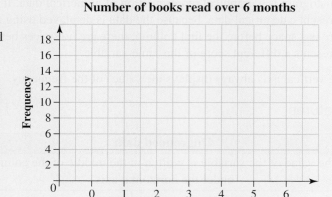

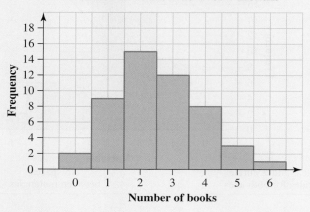

The data displayed in Worked example 20 is numerical discrete data. For numerical continuous data, we have to group scores in intervals. The intervals must be equal in size and must not overlap.

tlvd-4696

WORKED EXAMPLE 21 Constructing a histogram for numerical continuous data

Using the frequency table of maximum daily temperatures in Melbourne in December 2021, construct a histogram to represent this set of data.

Temperature (°C)	Frequency
18− < 23	14
23− < 28	8
28− < 33	5
33− < 38	3
38− < 43	1
Total	31

THINK

1. Draw a labelled set of axes:
   - Place 'Frequency' on the vertical axis.
   - Place 'Temperature (°C)' on the horizontal axis.
   - Ensure the histogram has a title.
   - The maximum frequency is 14, so the vertical scale should have ticks from 0 to 16. The ticks could be written by ones (a bit cluttered) or by twos.
   - The data is grouped so the ticks should be at the edges of each column.
   - Allow for a space before drawing the columns of the histogram.

2. Draw the columns.

WRITE/DRAW

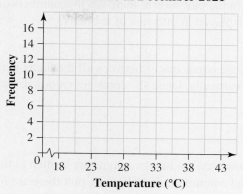

Maximum daily temperatures in Melbourne in December 2021

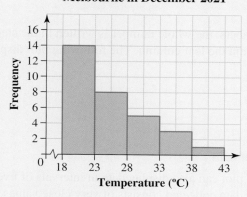

Maximum daily temperatures in Melbourne in December 2021

10.6.4 Stem-and-leaf plots

A stem-and-leaf plot is a graphical representation of grouped numerical data. The stem-and-leaf plot and the histogram shown are graphical representations of the same set of data. The two graphical representations have similar shapes. However, a stem-and-leaf plot displays all the data collected, while a histogram loses the individual scores.

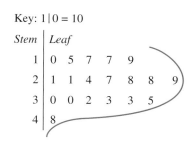

Key: 1|0 = 10

Stem	Leaf
1	0 5 7 7 9
2	1 1 4 7 8 8 9
3	0 0 2 3 3 5
4	8

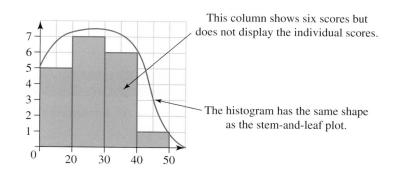

This column shows six scores but does not display the individual scores.

The histogram has the same shape as the stem-and-leaf plot.

Ten-unit intervals

A stem-and-leaf plot has two columns marked 'Stem' and 'Leaf'. For two-digit numbers, the first digit, the ten, is written in the stem column; the second digit, the unit, is written in the leaf column. In the plot shown, there are five numbers displayed: 31, 37, 35, 48 and 42. The data is displayed in intervals of ten units.

Key: $3|1 = 31$

Stem	Leaf
3	1 7 5
4	8 2

Ensure that there are equal spaces between all the digits in the leaf column and that all numbers line up vertically. Once all scores are recorded, they have to be placed in an increasing order.

Key: $3|1 = 31$

Stem	Leaf
3	1 5 7
4	8 8

For three-digit numbers, the first two digits are written in the first column and the third digit in the second column. In the following plot there are four numbers: 120, 125, 126 and 134.

Key: $3|1 = 31$

Stem	Leaf
12	0 5 6
13	4

For decimal numbers, the whole number is written in the stem column and the decimal digits are written in the leaf column. The following plot displays the data: 2.7, 2.8, 3.5, 3.6, 3.9, 4.1.

Key: $2|7 = 2.7$

Stem	Leaf
2	7 8
3	5 6 9
4	1

Five-unit intervals

A stem-and-leaf plot can also display data in intervals of five units. This situation occurs when the data collected consists of very close values.

For example, the set of data 11, 12, 12, 13, 14, 15, 17, 18, 21, 22, 26, 26, 27, 29 is better represented in five-unit intervals. In this case, the stem will consist of the values 1, 1*, 2 and 2*.

Key: $2|3 = 23$
$2^*|6 = 26$

Stem	Leaf
1	1 2 2 3 4
1*	5 7 8
2	1 2
2*	6 6 7 9

The first 1 represents the data values from 10 to 14 inclusive, while 1* represents the data values from 15 to 19 inclusive.

The first 2 represents the data values from 20 to 24 inclusive, while 2* represents the data values from 25 to 29 inclusive.

tlvd-4697

WORKED EXAMPLE 22 Construct a stem-and-leaf plot using ten- and five-unit intervals

The heights of 20 students are listed below. All measurements are given in centimetres. Construct a stem-and-leaf plot for this set of data:

a. using ten-unit intervals   **b. using five-unit intervals.**

156 172 162 164 174 151 150 169 171 169
167 161 153 155 165 172 148 166 169 158

THINK

a. **1.** Draw a horizontal line and a vertical line, as shown, and label the two columns stem-and-leaf.

a.

Stem	Leaf

2. Write the values in the stem part of the graph using intervals of 10 units.
As the values of the data are three-digit numbers, we write the first two digits in the stem column in increasing order. The lowest value is 148 and the highest value is 174. The stem will have the numbers 14, 15, 16 and 17.

Stem	Leaf
14	
15	
16	
17	

3. Write the values in the leaf part of the graph. Start with the first score, 156, and place 6 to the right of 15.
The next value is 172. Place 2 to the right of 17.

Stem	Leaf
14	
15	6
16	
17	2

4. Continue until all data is displayed.

Stem	Leaf								
14	8								
15	6	1	0	3	5	8			
16	2	4	9	9	7	1	5	6	9
17	2	4	1	2					

5. Rewrite all data in increasing order and add a key.

Key: 14|8 = 148

Stem	Leaf								
14	8								
15	0	1	3	5	6	8			
16	1	2	4	5	6	7	9	9	9
17	1	1	2	2	4				

b. **1.** Rewrite the stem part of the graph using intervals of five units.

b.

Stem	Leaf
14	
14*	
15	
15*	
16	
16*	
17	
17*	

2. Write the values in the leaf part of the graph and add a key.

Key: $15|1 = 151$
$15*|8 = 158$

Stem	Leaf
14*	8
15	0 1 3
15*	5 6 8
16	1 2 4
16*	5 6 7 9 9 9
17	1 2 2 4

 Resources

 Interactivities Create a histogram (int-6494)
Stem plots (int-6242)
Create stem plots (int-6495)
Dot plots, frequency tables and histogram and bar charts (int-6243)

10.6 Exercise

Students, these questions are even better in jacPLUS

 Receive immediate feedback and access sample responses

 Access additional questions

Track your results and progress

Find all this and MORE in jacPLUS ▶

1. **WE17** A group of 25 Year 11 students was surveyed by a psychologist about how many hours of sleep they each have per night. The set of scores below shows this data.

6   6   7   8   5   10   8   8   9   6   4   7   7   6   5   10   7   8   6   8   9   7   5   9   7

Display this data in a frequency distribution table.

2. Using the random sampler of 30 students from the Census@School (Australian Bureau of Statistics website), Eva downloaded the responses of 30 Year 11 students across Australia to the question, 'What is the length, in cm, of your foot without a shoe?' Measurements are given to the nearest centimetre.

28	27	28	23	27	29	24	26	29	26
23	23	27	25	28	24	26	28	24	28
22	26	23	26	29	21	23	25	24	26

Display this data in a frequency distribution table.

3. **WE18** The times of entrants in a charity run are listed below. The times are given in minutes.

34	29	57	45	26	40	19	28	33	37	39	46
18	52	19	36	28	19	20	19	54	38	38	51
19	21	53	36	37	25	22	30	25	34	18	17

Display this data in a frequency distribution table.

4. The data set shown represents the number of times last month a randomly chosen group of 30 people spent their weekends away from home.

4 3 2 0 0 0 1 2 0 1 0 0 2 1 1

4 1 1 0 0 0 2 1 1 0 2 1 3 3 0

Construct a frequency distribution table for this set of data, including a column for the tally.

5. For research purposes, 36 Tasmanian giant crabs were measured and the widths of their carapaces are given below.

35.3	34.9	36.0	35.4	37.0	34.2
35.9	35.5	35.0	36.2	35.7	33.1
36.5	36.4	37.2	35.4	34.9	35.1
34.8	35.8	35.2	35.6	37.1	36.7
35.0	34.2	36.3	33.9	35.2	34.1
36.5	36.4	36.6	34.3	35.8	35.7

Width, cm

Construct a frequency table for this set of data.

6. **WE19** The following frequency distribution table displays the data collected by MyMusic website about the number of songs downloaded by 40 registered customers per week.

Number of songs	Frequency
2	12
3	8
4	7
5	6
6	3
7	1
Total	40

Construct a dot plot for this data.

7. Daniel asked his classmates 'How many pairs of shoes do you have?'
   He collected the data and listed it in the following frequency table.

Pairs of shoes	Frequency
3	9
4	5
5	3
6	2
7	3
8	2
9	1
Total	**25**

Construct a dot plot for this set of data.

8. The frequency table shown displays the ages of 30 people who participated in a spelling contest.

Age	Frequency
11	3
12	6
13	8
14	7
15	3
16	2
17	1

Construct a dot plot for this set of data.

9. The data given represents the results in a typing test, in characters per minute, of a group of 24 students in a beginners class.

36 39 40 38 41 39 37 40
41 38 34 35 39 38 37 36
39 41 39 38 42 40 39 37

Construct a dot plot for this set of data.

10. **WE20** The following frequency distribution table displays the data collected by a Year 11 student about the number of emails his friends send per day.

Number of emails	Frequency
0	1
1	12
2	17
3	5
4	2
5	3
Total	**50**

Construct a histogram for this data.

11. 'In how many languages can you hold an everyday conversation?' is one of the questions in the Census@School questionnaire. Using the random sampler from the Census@School (Australian Bureau of Statistics) website, Brennan downloaded the responses of 100 randomly chosen students. He displayed the data in the frequency distribution table shown.

Number of languages	Frequency
1	127
2	50
3	13
4	4
5	3
6	2
7	1
Total	200

Construct a histogram for this data.

12. A group of 24 students was given a task that involved searching the internet to find how to construct an ungrouped histogram. The times, in minutes, they spent searching for the answer was recorded.

10	15	18	12	19	13	14	10
16	18	12	15	18	12	13	12
15	12	12	10	15	11	10	16

Draw an ungrouped histogram for this set of data.

13. **WE21** A phone company surveyed 50 randomly chosen customers about the number of times they check their email account on weekends.

5	10	2	17	8	8	13	6	3	0
11	1	6	10	1	4	0	11	5	12
0	3	4	5	6	5	14	9	4	10
14	9	2	16	3	9	8	10	5	11
7	2	5	6	1	2	2	12	9	17

a. Display the data collected in a frequency distribution table using class intervals of 3.
b. Construct a histogram for this data.

14. The following frequency distribution table shows the number of road fatalities in Victoria over a period of 12 months per age group, involving people between 30 and 70 years old.

Age group	Frequency
30 − < 40	47
40 − < 50	27
50 − < 60	41
60 − < 70	35
Total	210

Source: TAC road safety statistical summary, page B

Construct a histogram for this data.

15. The data below displays the responses of 50 Year 11 students who participated in a questionnaire related to the amount of money they earned from their part-time jobs over the previous week.

30	103	50	30	0	90	0	80	68	157
0	350	123	70	80	50	0	330	210	26
25	0	0	50	230	50	20	60	0	305
177	126	90	0	0	12	200	120	15	45
90	30	60	0	0	80	0	260	25	150

a. Display this data in a frequency table.
b. Draw a grouped histogram to represent this data.

16. **WE22** The BMI or body mass index is the number that shows the proportion between a person's mass and height. The formula used to calculate the BMI is:

$$\text{BMI} = \frac{\text{weight (kg)}}{\text{height (m)}}$$

The BMIs of 24 students are listed below.

25	24	18	16	29	23	20	21
21	17	19	23	25	24	32	22
22	23	30	18	23	20	19	24

Display this data in a stem-and-leaf plot.

17. The marks out of 50 obtained by a Year 11 Foundation Mathematics group of students on their exam are given below.

36	41	29	50	45	23	48	56
20	12	43	33	35	44	32	49
39	48	50	18	43	20	38	29

Display this data in a stem-and-leaf plot.

18. The manager of a timber yard started an inventory for the upcoming stocktake sale. The lengths of the first 30 timber planks recorded are given.

250	220	245	229	260	210	250	261	244	218
250	251	216	243	232	210	212	227	219	207
231	243	204	230	265	220	206	253	229	225

Display this data set in an ordered stem-and-leaf plot.

19. Consider the following set of data, which represents the heights, in cm, of 40 children under the age of 3.

58.9	54.6	62.9	45.7	49.6	62.8	43.2	56.7	69.0	65.3
57.8	59.4	66.7	49.6	72.7	35.2	72.8	75.6	32.9	56.7
54.8	45.2	69.5	47.3	68.6	67.4	63.4	61.5	52.6	48.0
56.3	78.4	39.9	75.3	45.2	56.9	66.3	55.4	63.8	72.3

Using a spreadsheet or otherwise, construct a grouped histogram for this set of data.

20. The reaction times using their non-dominant hand of a group of people are listed below. The times are given in deciseconds (a tenth of a second).

48	39	40	36	35	37	46	41	43	39	35
50	59	37	32	39	42	43	46	33	38	30
41	29	36	38	51	42	34	39	36	37	41

Using a spreadsheet or otherwise, construct a grouped histogram for this data set.

10.7 Comparing and interpreting data

LEARNING INTENTION

At the end of this subtopic you should be able to:
- display data as a back-to-back stem-and-leaf plot.

10.7.1 Back-to-back stem-and-leaf plots

Back-to-back stem-and-leaf plots allow us to compare two sets of data on the same graph.

A back-to-back stem-and-leaf plot has a central stem with leaves on either side.

The stem-and-leaf plot shown gives the ages of members of two teams competing in a bowling tournament.

Key: $5\,|\,5 = 55$

Team A	Stem	Team B
1	4	
	5	5 7 8 9
	6	0 2 3 5 7
6 5 4 3	7	0 1 2
8 6 5 4 3 2 1	8	8
	9	0

WORKED EXAMPLE 23 Displaying data as a back-to-back stem-and-leaf plot

The following data was collected from two Year 11 Maths classes that completed the same test. The total mark available was 100.

Class 1	84	90	86	95	92	81	83	97	88	99	79	100	85	82	97
Class 2	90	55	48	62	70	58	63	67	72	59	60	88	57	65	71

Draw a back-to-back stem-and-leaf plot.

THINK	WRITE/DRAW
1. To create a back-to-back stem-and-leaf plot, determine the highest and lowest values of each set of data to help you decide on a suitable scale for the stems.	The highest value is 100 and the lowest value is 48, so the stems should be 4, 5, 6, 7, 8, 9, 10.
2. Put the data for each class in order from lowest to highest value.	**Class 1:** 79, 81, 82, 83, 84, 85, 86, 88, 90, 92, 95, 97, 97, 99, 100 **Class 2:** 48, 55, 57, 58, 59, 60, 62, 63, 65, 67, 70, 71, 72, 88, 90

3. Create a back-to-back stem-and-leaf plot.

Key: $4\,|\,8 = 48$

Leaf (class 1)	Stem	Leaf (class 2)
	4	8
	5	5 7 8 9
	6	0 2 3 5 7
9	7	0 1 2
8 6 5 4 3 2 1	8	8
9 7 7 5 2 0	9	0
0	10	

10.7.2 Choosing appropriate data representations

To display similar sets of data, appropriate tabular and/or graphical representations are selected to enable comparisons of similarities and differences.

Side-by-side bar charts or back-to-back bar charts are used when comparing multiple sets of categorical data.

Back-to-back stem-and-leaf plots and dot plots are used to compare distributions of numerical data.

WORKED EXAMPLE 24 Choosing an appropriate data representation for a data set

A sample of New South Wales residents were asked to state the environmental issue that was most important to them. The results were sorted by the age of the people surveyed. The results were as follows.

	Environmental issue		
Age	Reducing pollution	Conserving water	Recycling rubbish
Under 30	52	63	41
Over 30	32	45	23

a. State the type of data display that you would use to compare the data sets.
b. Construct a side-by-side bar chart.
c. State what number of people surveyed were over 30.
d. Determine the percentage of under-30s surveyed who listed conserving water as the most important environmental issue.

THINK

a. Determine what type of data is being compared.

b. Construct a side-by-side column graph for the data.

c. Total the response of the over-30s.

WRITE/DRAW

a. Categorical data is being compared. A side-by-side bar chart or back-to-back bar chart would be the most suitable data display.

b.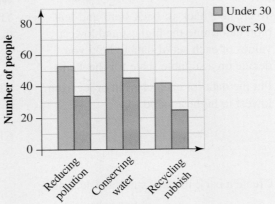

Important environmental issues for NSW residents

c. For the over-30s, there were 32 responses for reducing pollution, 45 for conserving water and 23 for recycling rubbish.
The total number of people over 30 who were surveyed is $32 + 45 + 23 = 100$.

---

d. To calculate the percentage of under-30s who listed conserving water as an important issue, use the formula:

$$\text{percentage} = \frac{\text{frequency}}{\text{total frequency}} \times 100\%.$$

d. $\text{Percentage} = \dfrac{63}{52 + 63 + 41} \times 100\%$

$$= \frac{63}{156} \times 100\%$$

$$= 40.4\%$$

Approximately 40% of residents under 30 listed conserving water as the most important environmental issue.

 Resources

 Interactivity Back-to-back stem plots (int-6252)

10.7 Exercise

Students, these questions are even better in jacPLUS

 Receive immediate feedback and access sample responses

 Access additional questions

 Track your results and progress

Find all this and MORE in jacPLUS

---

1. The two dot plots below display the latest Maths test results for two Year 11 classes. The results show the marks out of 20.

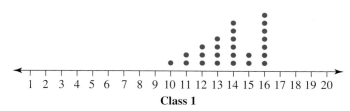

Class 1

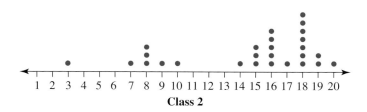

Class 2

a. Determine the number of students in each class.
b. For each class, determine the number of students who scored 15 out of 20 on the test.
c. For each class, state the number of students who scored more than 10 on the test.
d. Use the dot plots to describe the performance of each class on the test.

2. **WE23** The following data shows the ages of adult and children players at a ten-pin bowling centre. Draw a back-to-back stem-and-leaf plot of the data.

Adult: 20, 36, 16, 38, 32, 18, 19, 21, 25, 45, 29, 60, 31, 21, 16, 38, 52, 43, 17, 28, 23, 23, 43, 17, 22, 23, 32, 34

Children: 21, 23, 30, 16, 31, 46, 15, 17, 22, 17, 50, 34, 65, 25, 27, 19, 15, 43, 22, 17, 22, 16, 48, 57, 54, 23, 16, 30, 18, 21, 28, 35

3. The comparisons between the battery lives of two mobile phone brands are shown in the back-to-back stem-and-leaf plot below. State which mobile phone brand has the better battery life. Explain.

Key: 6 | 1 = 61 hours

Brand A	Stem	Brand B
8 8 7 5	0	7
9 7 4 1 0	1	0 5 5 5 7 9
2 2 2 1	2	0 2 2 6 7
8 6 4 2 0	3	0 2 4 6 8
	4	
	5	6
1	6	
	7	5

4. **WE24** A city newspaper surveyed a sample of Victorian residents about an upcoming state election on the issues most important to each age group. The results were as follows:

	Election issue					
Age	Marriage equality	Education	Refugees and immigration	Tax and superannuation	Health	Housing affordability
18–40	18	35	38	43	22	47
Over 40	12	38	19	61	41	32

a. State the type of data display you would use to compare the data sets.
b. State what number of people surveyed were over 40.
c. State the issue most important to the people aged over 40.
d. Determine the percentage of under-40s surveyed who listed marriage equality as the most important election issue. Give your answer to the nearest whole number.

5. Ten workers were required to complete two tasks. Their supervisor observed them and gave them a score for the quality of their work on each task, where higher scores indicated better-quality work.
The results are indicated in the side-by-side bar chart on the next page.

a. State which worker had the largest difference between scores for the two tasks.
b. State what number of workers received a lower score for task 2 than for task 1.

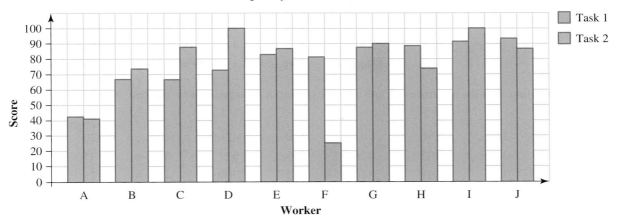

Score for quality of work on each task

6. The following graphical display summarises the ages of patients seen by two doctors in a medical surgery during one particular day.

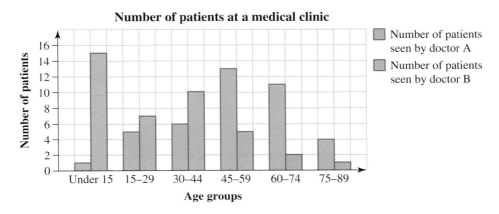

Number of patients at a medical clinic

a. Determine the number of patients aged under 15 who consulted doctor B.

b. Doctor A tends to consult patients aged over 45, whereas doctor B tends to consult patients aged under 45. State whether the statement is True or False.

7. Two households estimate the electricity consumed by different household appliances and devices as follows.

Appliance or device	Household 1 electricity consumption (%)	Household 2 electricity consumption (%)
Water heater	29	32
Refrigerator	18	21
Stove/cooktop	21	24
Washing machine	6	10
Lighting	8	7
Computer	7	2
Audiovisual equipment	2	2
Air-conditioner	6	2
Heating	3	3

Assuming that both households use the same amount of electricity overall, answer True or False to the following statements.

a. Household 1's computer uses more electricity than household 2's computer.

b. Household 2 uses less electricity in heating and air-conditioning than household 1.

c. Household 2's washing machine uses less electricity than household 1.

8. The daily number of hits a fashion blogger gets on her new website over 3 weeks is:

126	356	408	404	420	425	176
167	398	433	446	419	431	189
120	431	390	495	454	215	117

At the same time, the daily number of hits a healthy lifestyle blogger gets on his new website over 3 weeks is:

240	156	462	510	420	474	520
225	402	426	563	621	339	195
320	621	340	495	700	415	371

a. Compare the two data sets using an appropriate graphical display.
b. Comment on the two data sets.

9. The winning times in seconds for the women's and men's 100-metre sprint in the Olympics are shown below.

Year	Women's 100-m sprint	Men's 100-m sprint
1928	12.2	10.8
1932	11.9	10.3
1936	11.5	10.3
1948	11.9	10.3
1952	11.5	10.4
1956	11.5	10.5
1960	11.0	10.2
1964	11.4	10.0
1968	11.0	9.9
1972	11.07	10.14
1976	11.08	10.06
1980	11.60	10.25
1984	10.97	9.99
1988	10.54	9.92
1992	10.82	9.96
1996	10.94	9.84
2000	10.75	9.87
2004	10.93	9.85
2008	10.78	9.69

a. Display the winning times for women and men using a stem-and-leaf plot.
b. Determine if there is a large difference in winning times. Explain your answer.

10. The following data sets show the rental price (in $) of two-bedroom apartments in two different suburbs of Castlemaine.
Suburb A

215	225	211	235	244	210	215	210	256	207
200	200	242	225	231	205	240	205	235	200

Suburb B

| 235 | 245 | 231 | 232 | 240 | 280 | 280 | 270 | 255 | 275 |
| 275 | 285 | 245 | 265 | 270 | 255 | 260 | 258 | 251 | 285 |

a. Draw a back-to-back stem-and-leaf plot to compare the data sets.

b. Compare and contrast the rental price in the two suburbs.

11. A side-by-side bar chart shows the distribution of road fatalities over a year.

Distribution of fatalities for 12 months, by age and road user class

Legend: Driver, Passenger, Pedestrian, Motorcyclist, Pedal cyclist

a. Determine which age group had the most passenger and pedestrian fatalities. Give a reason why this might be.

b. The government wants to introduce a road campaign. Determine which age groups the government should focus on. Explain your answer.

12. The horizontal side-by-side bar chart shows a monthly comparison of road fatalities.

a. Using the data from 2020 and 2021, determine which year had the most road fatalities.

b. State which year has had the most fatalities from January to March.

c. Explain why there were 44 fatalities in the month of April 2021.

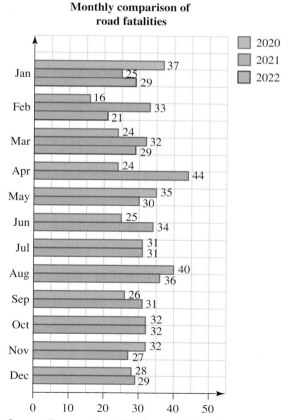

Monthly comparison of road fatalities

Legend: 2020, 2021, 2022

Jan	37, 25, 29
Feb	16, 33, 21
Mar	24, 32, 29
Apr	24, 44
May	35, 30
Jun	25, 34
Jul	31, 31
Aug	40, 36
Sep	26, 31
Oct	32, 32
Nov	32, 27
Dec	28, 29

Source: Transport for NSW, Centre for Road Safety

10.8 Review

10.8.1 Summary

doc-38051

Hey students! Now that it's time to revise this topic, go online to:

Access the topic summary

Review your results

Watch teacher-led videos

Practise questions with immediate feedback

Find all this and MORE in jacPLUS

10.8 Exercise

Multiple choice

1. **MC** Select the data collection situation for which a census is required.

   A. The employment status of a group of 1200 people over 18.
   B. The employment status of all people living in Australia at a given time.
   C. The employment status of a group of women aged between 30 and 40 years.
   D. Temperatures in Australia from June 2017 to June 2018.
   E. The favourite subject of Year 11 students at your school.

2. **MC** Categorical data can be graphically represented using a:

   A. histogram.
   B. dot plot.
   C. stem-and-leaf plot.
   D. column graph.
   E. line graph.

3. **MC** An example of numerical discrete data is the:

   A. depth of water in a dam over a period of one month.
   B. number of children enrolled in 20 high schools.
   C. daily average temperatures in April 2022 at a meteorological station.
   D. weekly average time a group of 46 students spends doing household chores.
   E. monthly average rainfall during December.

4. **MC** The Australian Bureau of Statistics collects data on the price of unleaded petrol. The type of data being collected is:

   A. nominal.
   B. ordinal.
   C. discrete.
   D. continuous.
   E. binomial.

5. **MC** Select which of the following can be classified as categorical ordinal data:

   A. Someone's favourite colour.
   B. Someone's favourite football team.
   C. Someone's opinion on how well their car was serviced.
   D. Someone's favourite car.
   E. Someone's favourite food.

6. **MC** At a hospital nursing station, the following information is available about a patient. From the following, select the information that is ordinal.

   A. Temperature: 36.7 °C
   B. Blood type: A
   C. Blood pressure: 120/80 mm Hg
   D. Response to treatment: excellent
   E. Weight: 65 kg

7. **MC** Select the graph that represents the data as a grouped vertical column graph.

A.

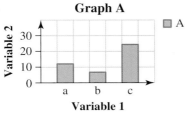

Graph A

B.

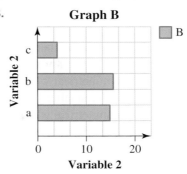

Graph B

C.

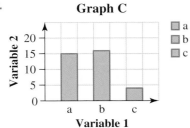

Graph C

D.

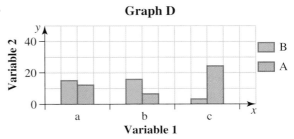

Graph D

E.

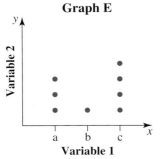

Graph E

8. **MC** Consider the following set of data representing the heights, in cm, of 40 children under the age of 3.

58.9	54.6	62.9	45.7	49.6	62.8	43.2	56.7	69.0	65.3
57.8	59.4	66.7	49.6	72.7	35.2	72.8	75.6	32.9	56.7
54.8	45.2	69.5	47.3	68.6	67.4	63.4	61.5	52.6	48.0
56.3	78.4	39.9	75.3	45.2	56.9	66.3	55.4	63.8	72.3

A suitable graphical representation for the data set given is:

A. a histogram with ungrouped data.
B. a histogram with grouped data.
C. a stem-and-leaf plot.
D. both **B** and **C**.
E. a line graph.

9. **MC** The data sets displayed in the back-to-back stem-and-leaf plot shown can be described by:

   A. The data set on the right is symmetrical while the data set on the left tails off as the numbers get larger.
   B. The data set on the left is symmetrical while the data set on the right tails off for smaller data values.
   C. The data set on the left is symmetrical while the data set on the right tails off as the numbers get larger.
   D. The data set on the right is symmetrical while the data set on the left tails off for smaller data values.
   E. The data set on the right is asymmetrical while the data set on the left tails off for larger values.

Left	Stem	Right
	10	0 3 4 4 5
5 2	11	0 1 2 2 2 3 5 8 9
8 8 7 5 1 1	12	0 1 4 6 6 9
7 5 5 4 3 2 1 0	13	3 3 7 8 9
9 8 8 6 4 2 0	14	3 5 6 8
9 6 6 2	15	1 6 8
7 3 3	16	0 2
	17	0 4
	18	6

Stem units: 10

10. **MC** In a group of 165 people aged 16–30 years, 28 people aged 23–30 years have a learner driver licence while 15 people aged 16–22 years have a probationary driver licence. Out of the 117 people who have a licence, 59 are learners.

Type of driver licence for people aged 16–30 years			
	Age groups		
	16–22 years	23–30 years	Total
Driver licence — L	31	28	59
P	15	43	58
None	27	21	48
Total	73	92	165

If the total number of people aged 16–22 years surveyed was 73, the number of people aged 23–30 who have a probationary driver licence is:

A. 92
B. 58
C. 64
D. 43
E. 15

Short answer

11. For each of the following, state whether a census or a survey has been used.

   a. Two hundred people in a shopping centre are asked to nominate the supermarket where they do most of their grocery shopping.
   b. To determine the most popular new car on the road, 500 new car buyers are asked the make and model of the car they purchased.
   c. To determine the most popular new car on the road, the make and model of every new car registered is recorded.
   d. To determine the average mark in the Mathematics half-yearly examination, every student's mark is recorded.
   e. To test the quality of tyres on a production line, every 100th tyre is road-tested.

12. One hundred teenagers were surveyed about their favourite type of music genre. The data was organised into a frequency table.

Music genre	Frequency
Hip hop/rap	28
Pop/R&B/soul	27
Rock	26
Country	3
Blues/jazz	4
Classical	2
Alternative	10

a. State whether the data is numerical or categorical.
b. State the data type of 'music genre'.
c. Calculate what percentage of teenagers preferred rock music.

13. A road worker recorded the number of vehicles that passed through an intersection in the morning and in the afternoon and summarised the data in a two-way table. A total of 47 heavy vehicles passed through the intersection during the day and 126 light vehicles in the afternoon.

Of the 238 vehicles that passed through the intersection in the morning, 199 were light vehicles.

a. Construct a two-way table for this set of data.
b. Determine the number of vehicles that passed through the intersection during the whole day.
c. Determine the number of heavy vehicles that passed through the intersection in the afternoon.

14. The numbers of the immediate family members of the teachers in a small school are given.

5 6 3 8 2 2 3 4 3 4 6 7
2 5 4 5 4 6 1 3 4 5 5 4

a. Present this set of data in a frequency table.
b. Present this data as a dot plot.

15. Students from a Year 11 class were asked about their favourite subjects and the following data were recorded.

Maths	English	PE	Science	PE
Art	Maths	Science	English	Science
Cooking	PE	English	Cooking	Maths
PE	Art	Science	Maths	Art
Science	Cooking	Art	PE	PE

a. Put the data in a frequency table and record the number of students who were surveyed.
b. Calculate the percentage of students who preferred Maths.
c. State the most popular subject. Calculate the percentage of students who preferred this subject.
d. State what type of data this is.

16. The following graph shows the number of people in selected occupations in Victoria.

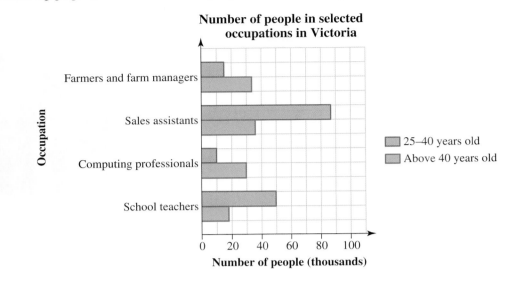

Number of people in selected occupations in Victoria

25–40 years old
Above 40 years old

a. Use the graph to estimate the number of people aged 25–40 who were sales assistants in Victoria at the time the data was collected.
b. Estimate the number of school teachers in Victoria.
c. State in which occupation there are about 30 000 people above the age of 40.
d. State in which occupation the total number of workers is about 40 000.

Extended response

17. Calculate the percentage relative frequency, correct to 1 decimal place, for all the entries in the two-way table given.

		Eye colour and hair colour			
		Eye colour			Total
		Adults	Teenagers	Children	
Hair colour	Light	49	38	26	113
	Dark	106	12	29	147
	Total	155	50	55	260

18. Consider the grouped bar graph shown displaying the numbers of people lodging visa applications in Australia, by sex and age.

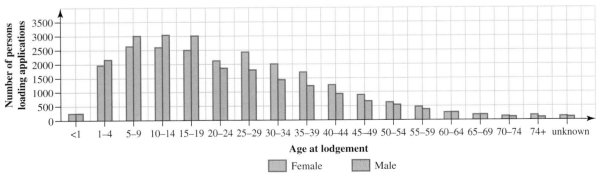

Female   Male

Notes:
1. Visas counted include subclass 200 (Refugee), 201 (In-Country Special Humanitarian Program), 202 (Global Special Humanitarian Program), 203 (Emergency Rescue) and 204 (Woman at Risk).
2. Data was extracted from Departmental systems.

a. Determine the number of males in the age bracket of 30–34 years who lodged a visa application.
b. Determine the number of children aged 1–4 years for whom a visa application was lodged.
c. State in what age brackets the number of female applicants was equal to the number of male applicants.
d. State the age bracket that has the highest proportion of female applicants.

19. A coffee bar serves either skim, reduced-fat or whole milk in coffees. The coffees sold on a particular day are shown in the table below, sorted by the type of milk and the weight range of the customers.

		Weight range	
		Underweight	Overweight
Type of milk	**Skim**	87	124
	Reduced fat	55	73
	Whole	112	49

a. State what number of coffees were sold on this day.
b. Represent this data in a graphical display.
c. Determine the percentage of underweight customers who asked for skim milk in their coffees. Give your answer to the nearest whole number.
d. Determine the percentage of coffees sold that contained reduced-fat milk. Give your answer to the nearest whole number.
e. If this was the daily trend of sales for the coffee bar, determine the percentage of the coffee bar's customers you would expect to be overweight. Give your answer to the nearest whole number.

20. The local childcare centre, HappyTodd, is attended by 37 children aged 2–4 years. In order to provide an appropriate diet for these toddlers, their weights, in kilograms, were recorded at enrolment.

14.5	12.4	16.3	13.0	13.6	12.9	15.4	14.7
14.3	14.9	13.8	15.2	15.4	16.1	13.2	13.8
13.5	13.4	15.1	12.7	14.2	13.6	14.4	14.1
14.7	14.8	14.2	14.3	13.5	13.1	14.6	15.3
14.2	15.5	15.7	15.8	16.5			

a. Display this data in a grouped frequency table using intervals of 0.5 kg including a column for the midpoint of the class intervals.
b. Construct a histogram to represent this data set.
c. Estimate the average weight at enrolment of the toddlers at HappyTodd childcare centre by first adding a column for fx in the frequency distribution table.
d. State the minimum and the maximum weights, and calculate the range of this data set.

Answers

Topic 10 Collecting and classifying data

10.2 Collecting data

10.2 Exercise

1. a. Census; the entire population of the online business
   b. Survey. 50% of the Year 12 student population is a sample of the population
   c. Census; the entire population of the hospital's patients

2. a. The number of absences of all Year 12 students is a sample since it is only one part of the school. The number of absences of all students in the school is a population, since all students are included.
   b. The number of passenger airplanes landing at an airport is a sample since it is only one group of airplanes that land at an airport. The total number of airplanes landing at an airport is a population since it includes all airplanes that land at an airport.

3. a. Survey; a sample of 150 suburbs
   b. Census; the entire population of Australia
   c. Census; the entire home room
   d. Survey; sample of the population who viewed the play
   e. Census; all the customer population

4. a, b Any sample with 5 different numbers between 1 and 26

5. a, b Any sample with 8 different numbers between 1 and 29

6. 153 people

7. a. Any sample with 10 different numbers between 1 and 100
   b. Any sample with 10 different numbers between 1 and 100

8. 1 to 8 if selecting every sixth car, or 1 to 4 if selecting every seventh car

9. From 1 to 6 selecting every fifth product

10. a. 5 Year 11 and 6 Year 12 students
    b. 8 adults and 2 children

11. Publicising includes advertising at the local library, local radio station, letter drop in the surrounding households and advertisements at the local shopping centres.
    All applicants could be accepted as there are no restrictions set on the study.

12. Publicising using advertisements in the local paper, or at the local gym

13. Self-selected sampling

14. a. Stratified sampling
    b. 6 black sheep and 4 white sheep

15. a. Systematic sampling
    b. 20 phones in the first batch; 14 phones in the second batch
    c. Maximum starting point is 5 in the first batch when selecting every fifth phone, and 9 in the second batch when selecting every seventh phone.

16. 19 males and 14 females

17. 7 white cars and 8 black cars

18. a. 15 students
    b. Either using a random number generator or using systematic sampling

19. a. Any sample with 12 different numbers between 1 and 140
    b. Any sample with 12 different numbers between 1 and 140

20. a. 28
    b. Any sample with 28 different numbers between 1 and 800

10.3 Simple survey procedures

10.3 Exercise

1. a. All Year 12 students
   b. All students in the school
   c. All students in Australia

2. a. All the milk bottles at the dairy farm
   b. All the customers of the internet provider

3. a. The Year 11 students
   b. The Years 11 and 12 students
   c. All students in the school

4. a. Assumes all people have mobile phones. A possible replacement question is 'Do you own a mobile phone? If yes, what brand do you have?'
   b. Double-barrelled question. It should be separated into two questions: 'On a scale of 1 (strongly agree) to 5 (strongly disagree), should drinking alcohol be permitted at the beach? Should smoking be banned at the beach?'
   c. Assumes that all people go to church on Sundays and all people are of a certain religion. A possible replacement question is 'Do you attend a place of prayer?
      • Never   • Rarely
      • Sometimes   • Often
      • Regularly'

5. a. A leading question.
      'Do you like pizza, pasta, both or neither?'
   b. Double negative question.
      'Are your parents' expectations of you clear?'

6. a. Closed   b. Open-ended
   c. Closed   d. Partially closed

7. a. Sample responses can be found in the worked solutions in the online resources.
   b. Sample responses can be found in the worked solutions in the online resources.
   c. Sample responses can be found in the worked solutions in the online resources.

8. Possible answers:
   Open-ended question: 'What is your preferred search engine when surfing the internet?'
   Partially closed question: 'The internet search engine I prefer to use is:
   ☐ Firefox ☐ Google ☐ Safari ☐ Other ....'
   Closed question: 'The internet search engine I prefer to use is:
   ☐ Firefox ☐ Google ☐ Safari ☐ None of these
   ☐ I don't use the internet'

9. **a.** Double-barrelled question. Possible replacement: 'What mode of transport do you use to go to work?

☐ Bus ☐ Tram ☐ Other ....'

b. Some people might have other opinions, such as smoking or lack of social skills or too much technology.

Possible replacement: 'In your opinion, what is the most important issue facing teenagers today?'

c. Leading question suggesting the desired answer. Possible replacement: 'How do you feel now after stopping smoking?'

d. Too personal. Possible replacement:

Your annual income is:

☐ less than $30 000
☐ $30 000–$50 000
☐ $50 000–$70 000
☐ $70 000–$90 000
☐ more than $90 000

10. **a.** Partially closed

b. Closed

c. Open-ended

11. See sample responses in the worked solutions.

12. See sample responses in the worked solutions.

10.4 Classifying data

10.4 Exercise

1. **a.** Numerical continuous

b. Numerical discrete

2. The number of fruits is discrete data as it is represented by counting numbers. The quantity of fruit sold is continuous data because it represents a measurement; there is always another value between any two values.

3. **a.** Continuous   **b.** Discrete
   c. Discrete   **d.** Discrete
   e. Continuous   **f.** Continuous

4. **a.** Categorical nominal

b. Categorical ordinal

5. 'Favourite movies' identifies a category; it cannot be ordered. 'The rating for a movie' can be ordered in categories like poor, average, very good and excellent.

6. **a.** Ordinal   **b.** Nominal
   c. Ordinal   **d.** Nominal

7. **a.** Categorical nominal

b. Numerical discrete

8. Although country phone codes are numerical, they are neither countable nor measurable. They represent categories. The number of international calls is numerical data as it is countable.

9. **a.** Numerical continuous

b. Categorical ordinal

10. Data collected on the lengths of pencils is a measurement, while data collected on the colours of pencils can be represented in colour categories.

11. D

12. **a.** Numerical   **b.** Categorical
    c. Categorical   **d.** Numerical

10.5 Categorical data

10.5 Exercise

1.

Mode of transport to school		
Mode of transport	Tally	Frequency
Car	$\overline{||||}$ ||	7
Bicycle	||||	4
Walk	$\overline{||||}$ $\overline{||||}$	10
Tram	$\overline{||||}$ ||||	9
Total		30

2.

Favourite social network		
Social network	Tally	Frequency
MyPage	$\overline{||||}$ $\overline{||||}$ $\overline{||||}$	15
FlyBird	$\overline{||||}$ ||||	9
Total		24

3.

Internet connection		
Type of internet connection	Tally	Frequency
C	|||	3
NBN	$\overline{||||}$ $\overline{||||}$ $\overline{||||}$	15
MW	$\overline{||||}$ $\overline{||||}$ $\overline{||||}$ ||	17
Total		35

4.

Opinion	Frequency
Excellent	163
Very good	298
Average	24
Poor	15
Total	500

5.

Parking place and car theft			
	Parking type		
	In driveway	On the street	Total
Car theft	2	37	39
No car theft	16	445	461
Total	18	482	500

6.

Mobile phones and tablets			
	Tablets		
	Own	**Do not own**	**Total**
Mobile phones Own	58	78	136
Do not own	116	11	127
Total	174	89	263

7. a.

Year 11 students' Mathematics results			
	Result		
	Pass	**Fail**	**Total**
Subject Foundation Mathematics	12	23	35
General Mathematics	25	49	74
Mathematical Methods	36	62	98
Specialist Mathematics	7	18	25
Total	80	152	232

b. 232 students

c. 35 students

8. a.

Hairdressing requirements per age group			
	Age group		
	Young	**Old**	**Total**
Hairdressing requirement Haircut			
Haircut and colour			
Total			

b.

Hairdressing requirements per age group			
	Age group		
	Young	**Old**	**Total**
Hairdressing requirement Haircut			
Haircut and colour	2		
Total		56	

c.

Hairdressing requirements per age group			
	Age group		
	Young	**Old**	**Total**
Hairdressing requirement Haircut only	5	26	31
Haircut and colour	2	30	32
Total	7	56	63

9. a.

Monday matinee audience per movie genre			
	Movie genres		
	Drama	**Comedy**	**Total**
Audience type Adults	66	30	96
Teenagers	5	16	21
Children	11	28	39
Total	82	74	156

b. 30

c. 39

10. a. Relative frequencies

Cat and dog owners			
	Dog owners		
	Own	**Do not own**	**Total**
Cat owners Own	0.039	0.256	0.295
Do not own	0.495	0.210	0.705
Total	0.534	0.466	1.000

b. Percentage relative frequencies

Cat and dog owners (%)			
	Dog owners		
	Own (%)	**Do not own (%)**	**Total**
Cat owners Own (%)	3.9	25.6	29.5
Do not own (%)	49.5	21.0	70.5
Total	53.4	46.6	100.0

11. a. Relative frequency table

		Year 11 students' Mathematics results		
		Result		**Total**
		Pass	**Fail**	
Subject	**Foundation Mathematics**	0.0517	0.0991	0.1509
	General Mathematics	0.1078	0.2112	0.3190
	Mathematical Methods	0.1552	0.2672	0.4224
	Specialist Mathematics	0.0302	0.0776	0.1078
	Total	0.3448	0.6552	1.0000

b. Percentage relative frequency table

		Year 11 students' Mathematics results		
		Result		**Total (%)**
		Pass (%)	**Fail (%)**	
Subject	**Foundation Mathematics (%)**	5.17	9.91	15.09
	General Mathematics (%)	10.78	21.12	31.90
	Mathematical Methods (%)	15.52	26.72	42.24
	Specialist Mathematics (%)	3.02	7.76	10.78
	Total (%)	34.48	65.52	100.00

Note: Due to rounding, some of the percentages written do not add up to the actual value in some of the total columns. To have the correct values in the total columns, ensure that the actual values (no rounding) are summed up.

12. a. Relative frequencies

Swimming attendance	
Day of the week	**Adults**
Monday	0.112
Tuesday	0.072
Wednesday	0.08
Thursday	0.12
Friday	0.144
Saturday	0.216
Sunday	0.256

b. Percentage relative frequencies

Swimming attendance	
Day of the week	**Adults (%)**
Monday	11.2
Tuesday	7.2
Wednesday	8
Thursday	12.2
Friday	14.4
Saturday	21.6
Sunday	25.6

Adults: relative frequency = 0.3125; percentage relative frequency = 31.25%
Children: relative frequency = 0.6875; percentage relative frequency = 68.75%

13. a.

		Dance class enrolments			
		People			**Total**
		Adults	**Teenagers**	**Children**	
Dance	**Latin**	31	21	23	75
	Modern	13	34	49	96
	Contemporary	36	49	62	147
	Total	80	104	134	318

b. Relative frequency table

		Dance class enrolments			
		People			**Total**
		Adults	**Teenagers**	**Children**	
Dance	**Latin**	0.10	0.07	0.07	0.24
	Modern	0.04	0.11	0.15	0.30
	Contemporary	0.11	0.15	0.19	0.46
	Total	0.25	0.33	0.42	1.00

Note: Due to rounding, some of the percentages written do not add up to the actual value in some of the total columns. To have the correct values in the total columns, ensure that the actual values (no rounding) are summed up.

c. Percentage relative frequency table

		Dance class enrolments (%)			
		People			**Total**
		Adults (%)	**Teenagers (%)**	**Children (%)**	
Dance	**Latin (%)**	10	7	7	24
	Modern (%)	4	11	15	30
	Contemporary (%)	11	15	19	46
	Total (%)	25	33	42	100

14.

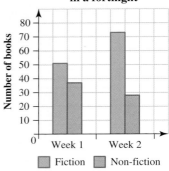

Types of books sold in a fortnight

15.

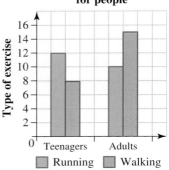

Types of fitness exercise for people

16. B

17. See the figure at the bottom of the page.*

18. The vertical axis starts at 0.80 with no break shown. The differences between columns look exaggerated with a non-zero baseline. The percentage difference is quite minimal — approximately 7% between the people who never finished high school and the people who have a postgraduate degree. Here is the graph with a baseline of zero:

Job satisfaction and years of schooling

19. a.

		Drivers and traffic accidents			
		Drivers			
		Category A	**Category B**	**Category C**	
		Young drivers	**Mature drivers**	**Older drivers**	**Total**
Event	**Traffic accident**	35	10	30	75
	No traffic accident	95	113	17	225
	Total	130	123	47	300

b. 10

c. 225

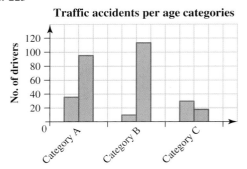

Traffic accidents per age categories

20. The vertical axis starts at 5 with no break shown. The differences between columns look exaggerated with a non-zero baseline. This makes the differences between the columns look greater than they actually are.

*17.

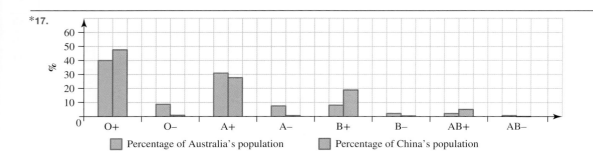

10.6 Numerical data

10.6 Exercise

1.

Hours of sleep	Tally	Frequency
4			1
5					3
6						5
7							6
8						5
9					3
10				2
Total		25

2.

Foot length	Tally	Frequency
21			1
22			1
23						5
24						4
25				2
26								6
27					3
28						5
29					3
Total		30

3.

Time	Tally	Frequency
17 − < 22	|||| ||||	10
22 − < 27	||||	4
27 − < 32	||||	4
32 − < 37	||||	5
37 − < 42	||||		6
42 − < 47	||	2
47 − < 52			1
52 − < 57	|||	3
57 − < 62			1
Total		36

4.

Weekends	Tally	Frequency
0											11
1										9
2						5
3					3
4				2
Total		30

5.

Width	Tally	Frequency
33.0 − < 34.0	||	2
34.0 − < 35.0	|||| ||	7
35.0 − < 36.0	|||| |||| ||||	15
36.0 − < 37.0	|||| ||||	9
37.0 − < 38.0	|||	3
Total		36

6.

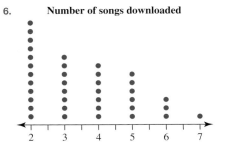

Number of songs downloaded

Number of songs

7.

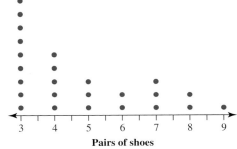

Number of pairs of shoes

Pairs of shoes

8.

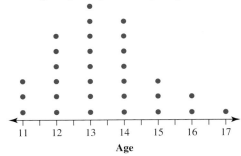

Ages of participants in a spelling contest

Age

9.

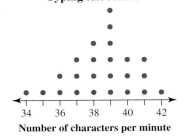

Typing test results

Number of characters per minute

10.

Number of emails sent per day by Year 11 students

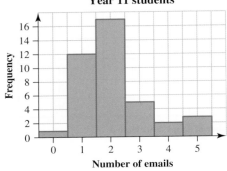

11.

Number of languages spoken

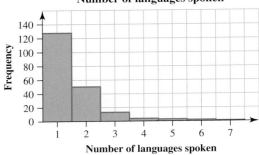

12. The figure at the bottom of the page.*

13. a.

Email check	Frequency
0 – < 3	11
3 – < 6	12
6 – < 9	8
9 – < 12	11
12 – < 15	5
15 – < 18	3

b.

Email check per weekend

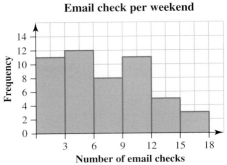

14.

Road fatalities in Victoria

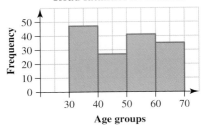

15. a.

Money earned	Frequency
0 – < 50	22
50 – < 100	14
100 – < 150	4
150 – < 200	3
200 – < 250	3
250 – < 300	1
300 – < 350	2
350 – < 400	1
Total	50

*12.

Time spent searching the internet

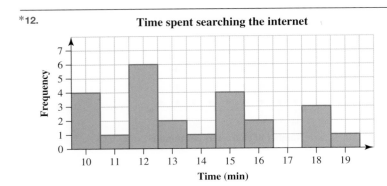

b.

Weekly wage from part-time jobs

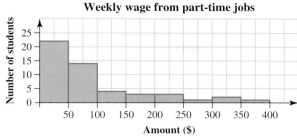

16. Key: 1|2 = 12
 1*|6 = 16

Stem	Leaf
1*	6 7 8 8 9 9
2	0 0 1 1 2 2 3 3 3 3 4 4 4
2*	5 5 9
3	0 2

17. Key: 1|2 = 12

Stem	Leaf
1	2 8
2	0 0 3 9 9
3	2 3 5 6 8 9
4	1 3 3 4 5 8 8 9
5	0 0 6

18. Key: 20|4 = 204

Stem	Leaf
20	4 6 7
21	0 0 2 6 8 9
22	0 0 5 7 9 9
23	0 1 2
24	3 3 4 5
25	0 0 0 1 3
26	0 1 5

19.

Heights of children under the age of 3

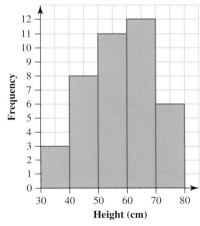

20.

Reaction time with non-dominant hand

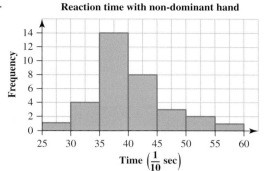

10.7 Comparing and interpreting data

10.7 Exercise

1. a. Class 1 : 25; class 2 : 27

 b. Class 1 : 2; class 2 : 3

 c. Class 1 : 24; class 2 : 20

 d. Eleven students in class 2 scored better than students in class 1, but 6 students scored worse than those in class 1. Class 2 results are more widely spread than class 1 results. The median for class 1 was $\frac{14}{20}$, which is lower than the median for class 2, $\frac{16}{20}$.

2. Key: 1 | 6 = 16 years

Leaf (adults)	Stem	Leaf (children)
9 8 7 7 6 6	1	5 5 6 6 6 7 7 7 8 9
9 8 5 3 3 3 2 1 1 0	2	1 1 2 2 2 3 3 5 7 8
8 8 6 4 2 2 1	3	0 0 1 4 5
5 5 3	4	3 6 8
2	5	0 4 7
0	6	5

3. Brand B seems to have a better battery life, as brand A has more batteries that have a battery life of less than 10 hours. The median of brand A is 21 hours and is lower than that of brand B, which is 22 hours.

4. a. Side-by-side bar chart

 b. 203

 c. Tax and superannuation

 d. 9%

5. a. Worker F **b.** 4 (A, F, H and J)

6. a. 14 **b.** True

7. a. True **b.** True **c.** False

8. a.

Number of hits	Fashion blogger	Healthy lifestyle blogger
100−<150	3	0
150−<200	3	2
200−<250	1	2
250−<300	0	0
300−<350	0	3
350−<400	3	1
400−<450	9	4
450−<500	2	3
500−<550	0	2
550−<600	0	1
600−<650	0	2
650−<700	0	0
700−<750	0	1

See the figure at the bottom of the page.*

b. Answers will vary. Example answer:
The fashion blogger had 400−<450 hits on her website on 9 days in the 3-week period. The healthy lifestyle blogger got a wider spread of hits. He received over 400 hits for 11 days in the 3-week period. His website seems to be gaining in popularity as its spread falls over a higher maximum and minimum than the spread of the fashion blog.

9. a. See the table at the bottom of the page.*

b. Answers will vary. There is not a large difference within each gender, but there is a large difference in time between the two genders.

10. a. Key: 23 | 1 = $231

Suburb A	Stem	Suburb B
7 5 5 0 0 0	20	
5 5 1 0 0	21	
5 5	22	
5 5 1	23	1 2 5
4 2 0	24	0 5 5
6	25	1 5 5 8
	26	0 5
	27	0 0 5 5
	28	0 0 5 5

b. Answers will vary. Suburb A has lower rent than suburb B. Suburb B is a more expensive suburb to rent an apartment. The median rental price in suburb A is $215; the median rental price in suburb B is $259.

11. a. 70+ age group. Answers about the reason will vary. Elderly people don't drive as much and tend to walk or be passengers.

---

*8. a.

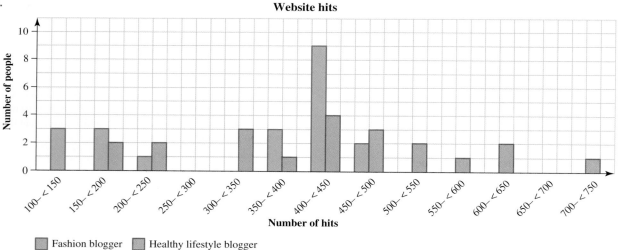

Website hits

*9. a.

Key: 9 | 69 = 9.69

Women's sprint	Stem	Men's sprint
	9	69
	9	84 85 87 90 92 96 99
	10	00 06 14 20 25 30 30 30 40
	10	50 80
97 94 93 82 78 75 54		
40 08 07 00 00	11	
90 90 60 50 50 50	11	
20	12	

b. Answers will vary. The road campaign could focus on:
- elderly people as pedestrians — being alert when crossing a road
- motorcyclists in the 30–49 age group, not on probationary licence — risk taking
- driving for all age groups — awareness and safety for all drivers.

12. a. 2021

b. 2021

c. Answers will vary but could include that more people are on the roads during the Easter break and school holidays.

10.8 Review

10.8 Exercise

Multiple choice

1. B
2. D
3. B
4. D
5. C
6. D
7. D
8. A
9. C
10. D

Short answer

11. a. Survey b. Survey c. Census
    d. Census e. Survey

12. a. Categorical b. Categorical, nominal
    c. 26%

13. a.

	Number of vehicles		
	Type of vehicle		
	Heavy vehicles	**Light vehicles**	**Total**
Morning	39	199	238
Afternoon	8	126	134
Total	47	325	372

(Time of the day)

b. 372 vehicles

c. 8 heavy vehicles

14. a.

Family members	Frequency
1	1
2	3
3	4
4	6
5	5
6	3
7	1
8	1

b. Draw a horizontal line for the number of family members. Place the numbers 1 to 8 evenly spaced along the line. Place the dots for each score.

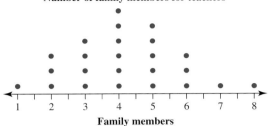

Number of family members for teachers

Family members

15. a.

Subject	Tally	Frequency (f)
Maths	IIII	4
Art	IIII	4
Cooking	III	3
PE	IIIII I	6
Science	IIIII	5
English	III	3
Total		25

b. 16%

c. PE; 24%

d. Nominal (categorical)

16. a. 87 000

b. 68 000

c. Computing professionals

d. Computing professionals

Extended response

17.

		Eye colour and hair colour			
		Eye colour			
		Adults	**Teenagers**	**Children**	**Total**
Hair colour	**Light**	18.8%	14.6%	10%	43.5%
	Dark	40.8%	4.6%	11.1%	56.5%
	Total	59.6%	19.2%	21.1%	100%

18. a. 1400

b. 4200

c. Babies under 1 year of age, 60–64 year olds and 65–69 year olds

d. 5–9 years of age

19. a. 500

b.

Coffee sales

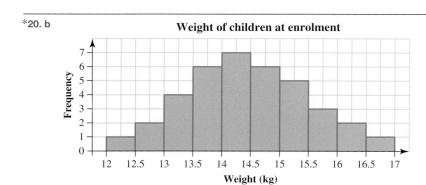

(Legend: Underweight, Overweight)

Number of people (y-axis): 0, 20, 40, 60, 80, 100, 120, 140

Type of milk (x-axis): Skim, Reduced fat, Whole

c. 34%

d. 26%

e. 49%

b. See the figure at the foot of the page.*

c.

Weight class (kg)	Midpoint (kg)	Frequency	fx (kg)
12.0– < 12.5	12.25	1	12.25
12.5– < 13.0	12.75	2	25.50
13.0– < 13.5	13.25	4	53.00
13.5– < 14.0	13.75	6	82.50
14.0– < 14.5	14.25	7	99.75
14.5– < 15.0	14.75	6	88.50
15.0– < 15.5	15.25	5	76.25
15.5– < 16.0	15.75	3	47.25
16.0– < 16.5	16.25	2	32.50
16.5– < 17.0	16.75	1	16.75
Total		37	534.25

Mean = 14.4 kg

d. Minimum weight = 12.4 kg, maximum weight = 16.5 kg, range = 4.1 kg

20. a.

Weight (kg) class	Midpoint (kg)	Frequency
12.0– < 12.5	12.25	1
12.5– < 13.0	12.75	2
13.0– < 13.5	13.25	4
13.5– < 14.0	13.75	6
14.0– < 14.5	14.25	7
14.5– < 15.0	14.75	6
15.0– < 15.5	15.25	5
15.5– < 16.0	15.75	3
16.0– < 16.5	16.25	2
16.5– < 17.0	16.75	1
Total		37

*20. b

Weight of children at enrolment

Frequency (y-axis): 0, 1, 2, 3, 4, 5, 6, 7

Weight (kg) (x-axis): 12, 12.5, 13, 13.5, 14, 14.5, 15, 15.5, 16, 16.5, 17

11 Financial systems and income payments

LEARNING SEQUENCE

Fully worked solutions for this topic are available online.

11.1 Overview

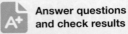

11.1.1 Introduction

There are many famous sayings about money. You have probably heard 'money doesn't grow on trees', 'money can't buy happiness' and 'a fool and his money are soon parted'. Most of us have to be conscious of where our money comes from and where it goes. Understanding the basic principles of finance will be very helpful in your life. Some of you will already be earning money from a part-time casual job. Do you keep a record of your spending? Do you have your own savings account? Many of you may already be saving for a new games console, your first car or your dream holiday.

In this topic, you will investigate different kinds of employment as well as the different investment options for saving your money. Every branch of industry and business, whether large or small, international or domestic, will have to pay their employees either an annual salary or a wage based on an hourly rate. Financial investments are another way for businesses and individuals to make money — it is important to understand how investments work to be able to decide whether an investment is a good idea or not.

A sound knowledge of financial mathematics is essential in a range of careers, including financial consultancy, accountancy, business management and pay administration.

KEY CONCEPTS

This topic covers the following key concepts from the VCE Mathematics Study Design:
- personal financial services and information such as borrowing, bills and banking
- income calculations including rates of pay and pay slips
- personal taxation and superannuation
- taxation as a duty to community and contribution to government
- fees and interest
- cost structures and related information associated with financial transactions.

Source: VCE Mathematics Study Design (2023–2027) extracts © VCAA; reproduced by permission.

11.2 Purchasing methods

11.2.1 Cash purchases

Buying goods with cash is the most straightforward type of purchase you can make. The buyer owns the goods outright and no further payments are necessary. Some retailers or services offer a discount if you pay with cash.

WORKED EXAMPLE 1 Calculating amount of payment if paying with cash

A plumber offers a 5% discount for customers who pay with cash. Calculate the amount a customer would be charged if they paid in cash and the fee before the discount was $139.

THINK	WRITE
1. Determine the percentage of the fee that the customer will pay after the discount is taken into account.	$100\% - 5\% = 95\%$
2. Multiply the fee before the discount by the percentage the customer will pay. Turn the percentage into a fraction.	$139 \times 95\% = 139 \times \dfrac{95}{100}$
3. Evaluate the amount to be paid.	$= 132.05$
4. Write the answer.	The customer will be charged $132.05.

11.2.2 Credit and debit cards

Credit cards

A **credit card** is an agreement between a financial institution (usually a bank) and an individual that the financial institution will loan an amount of money up to a pre-approved limit to the individual. It can be used to pay for transactions until the amount of debt on the credit card reaches the agreed limit of the credit card.

If a customer pays off the debt on their credit card within a set period of time after purchases are made, known as the interest-free period, they will pay no interest on the debt. Otherwise, they will pay a high-interest rate on the debt (usually 20–30% p.a.), with the interest calculated monthly. Customers are obliged to pay at least a minimum monthly amount off the debt — for example, 3% of the balance.

Customers are often charged an annual fee for using credit cards, but they can also earn rewards such as frequent flyer points for major airlines or discounts at certain retailers.

11.2.3 Debit cards

Debit cards are usually linked to bank accounts, although they can also be preloaded with set amounts of money. When a customer uses a debit card, the money is debited directly from their bank account or from the preloaded amount.

If a customer tries to make a transaction with a debit card that exceeds the balance in their bank account, then either their account will become overdrawn (which typically incurs a fee from the banking facility), or the transaction will be declined.

Online payment systems

Online payment systems such as PayPal, Apple Pay and Google Pay allow people to send money to others. They can be linked to your bank account, debit card or credit card in order to send and receive money to and from others. These online payment systems are designed to be convenient and to add an extra level of security and fraud prevention when purchasing items online.

tlvd-4808

WORKED EXAMPLE 2 Calculating the total amount of interest over a three-month period

Heather has a credit card that charges an interest rate of 19.79% p.a. She tries to ensure that she always pays off the full amount at the end of the interest-free period, but an expensive few months over the Christmas holidays leaves the outstanding balance on her card at $635, $427 and $155 for three consecutive months.

Calculate the total amount of interest Heather has to pay over the three-month period. Give your answer correct to the nearest cent.

THINK

1. Use the simple interest formula to determine the amount of interest charged each month.

WRITE

1st month:
$$I = \frac{Prn}{100}$$
$$= \frac{635 \times 19.79 \times \frac{1}{12}}{100}$$
$$\approx 10.47$$

2nd month:
$$I = \frac{Prn}{100}$$
$$= \frac{427 \times 19.79 \times \frac{1}{12}}{100}$$
$$\approx 7.04$$

3rd month:
$$I = \frac{Prn}{100}$$
$$= \frac{155 \times 19.79 \times \frac{1}{12}}{100}$$
$$\approx 2.56$$

2. Calculate the sum of the interest for the three months.

$10.47 + 7.04 + 2.56 = 20.07$

3. Write the answer.

Heather has to pay $20.07 in interest over the three-month period.

11.2.4 Personal loans

A **personal loan** is a loan made by a lending institution to an individual. A personal loan will usually have a fixed interest rate attached to it, with the interest paid by the customer calculated on a reduced balance. This means that the interest for each period will be calculated on the amount still owing, rather than the original amount of the loan.

WORKED EXAMPLE 3 Calculating the outstanding balance on a personal loan

Francis takes out a loan of $3000 to help pay for a business management course. The loan has a fixed interest rate of 7.75% p.a. and Francis agrees to pay back $275 a month. Assuming that the interest is calculated before Francis's payments, calculate the outstanding balance on the loan after Francis's third payment. Give your answer correct to the nearest cent.

THINK

1. Calculate the interest payable for the first month of the loan.

WRITE

$$I = \frac{Prn}{100}$$
$$= \frac{3000 \times 7.75 \times \frac{1}{12}}{100}$$
$$\approx 19.38$$

2. Calculate the total value of the loan before Francis's first payment.

$3000 + $19.38 = $3019.38

3. Calculate the total value of the loan after Francis's first payment.

$3019.38 − $275 = $2744.38

4. Calculate the interest payable for the second month of the loan.

$$I = \frac{Prn}{100}$$
$$= \frac{2744.38 \times 7.75 \times \frac{1}{12}}{100}$$
$$\approx 17.72$$

5. Calculate the total value of the loan before Francis's second payment.

$2744.38 + $17.72 = $2762.10

6. Calculate the total value of the loan after Francis's second payment.

$2762.10 − $275 = $2487.10

7. Calculate the interest payable for the third month of the loan.

$$I = \frac{Prn}{100}$$
$$= \frac{2487.1 \times 7.75 \times \frac{1}{12}}{100}$$
$$\approx 16.06$$

8. Calculate the total value of the loan before Francis's third payment.

$2487.10 + $16.06 = $2503.16

9. Calculate the total value of the loan after Francis's third payment.

$2503.16 − $275 = $2228.16

10. Write the answer.

The outstanding balance of the loan after Francis's third payment is $2228.16.

11.2.5 The effective rate of interest

A **buy now–pay laterplan**, or hire purchase, can be used when a customer wants to make a purchase but doesn't have the means to pay up-front. It usually works by paying a small amount up-front, and then paying weekly or monthly instalments.

The interest rate of a buy now–pay later plan can be determined by using the simple interest formula. However, the actual interest rate will be higher than that calculated, as these calculations don't take into account the reducing balance owing after each payment has been made.

The **effective rate of interest** can be used to give a more accurate picture of how much interest is actually charged on a buy now–pay later plan. To determine this, we can use the following formula.

Formula for the effective rate of interest

$$R_{ef} = \frac{2400I}{P(m+1)}$$

where R_{ef} is the effective rate of interest, I is the total interest paid, P is the principal (the cash price minus the deposit) and m is the number of monthly payments.

---

tlvd-4809

A furniture store offers its customers the option of purchasing a $2999 bed and mattress by paying $500 up-front, followed by 12 monthly payments of $230.
a. Calculate the amount a customer will pay in total if they choose the offered buy now–pay later plan.
b. Calculate the effective rate of interest for the buy now–pay later plan correct to 2 decimal places.

THINK

a. 1. Determine the total amount to be paid under the buy now–pay later plan.

2. Write the answer.

b. 1. Calculate the total amount of interest paid.

2. Calculate the principal (the cash price minus the deposit).

3. Identify the components of the formula for the effective rate of interest.

4. Substitute the information into the formula and determine the effective rate of interest.

5. Write the answer.

WRITE

a. Total payment $= 500 + 12 \times 230$
$$= 500 + 2760$$
$$= 3260$$

The total amount paid under the buy now–pay later plan is $3260.

b. $I = 3260 - 2999$
$$= 261$$

$P = 2999 - 500$
$$= 2499$$

$I = 261$
$P = 2499$
$m = 12$

$R_{ef} = \dfrac{2400I}{P(m+1)}$

$$= \dfrac{2400 \times 261}{2499(12+1)}$$

$$= 19.28\% \text{ (to 2 decimal places)}$$

The effective rate of interest for the time purchase plan is 19.28%.

11.2 Exercise

Unless otherwise directed, where appropriate give all answers correct to 2 decimal places or the nearest cent.

1. **WE1** An electrician offers a discount of 7.5% for customers who pay by cash. Determine the amount customers will pay in cash if the charge before the discount being applied is:

   a. $200
   b. $312
   c. $126.

2. **MC** George runs a pet-care service looking after cats and dogs on weekend afternoons. He charges a fee of $20 per pet plus $9 per hour. He also gives his customers a 6% discount if they pay in cash.
   Charlene asks George to look after her two cats between 1 pm and 5 pm on a Saturday afternoon. From the following options, select the amount she has to pay if she pays in cash.

   A. $33.85
   B. $52.65
   C. $71.45
   D. $72.95
   E. $73.85

3. **WE2** Barney is struggling to keep control of his finances and starts to use his credit card to pay for purchases. At the end of three consecutive months, his outstanding credit card balance is $311.55, $494.44 and $639.70 respectively. If the interest rate on Barney's credit card is 22.75% p.a., calculate how much interest he is charged for the three-month period.

4. Dani uses her credit card while on an overseas trip and returns with an outstanding balance of $2365.24 on it. Dani can only afford to pay the minimum monthly balance of $70.96 off her credit card before the interest-free period expires.

   a. Dani's credit card charges an interest rate of 24.28% p.a. Calculate the amount she will be charged in interest for the next month.
   b. If Dani spent $500 less on her overseas trip, determine how much the interest would be reduced by. (*Note:* Assume that Dani still pays $70.96 off her credit card.)

5. **WE3** Petra takes out a loan of $5500 to help pay for a business management course. The loan has a fixed interest rate of 6.85% p.a. and Petra agrees to pay back $425 a month. Assuming that the interest is calculated before Petra's payments, calculate the outstanding balance on the loan after her third payment.

6. Calculate the total amount of interest paid on a $2500 personal loan if the rate is 5.5% p.a. and $450 is paid off the loan each month. (Assume that the interest is calculated before the monthly payments.)

7. Shawna takes out a personal loan of $4000 to help support her brother in a new business venture. The loan has a fixed interest rate of 9.15% calculated on the reduced monthly balance, and Shawna agrees to pay $400 back per month.

   a. Calculate the amount of interest Shawna will pay over the lifetime of the loan.
   b. Shawna's brother's business goes well, and he is able to pay her back the $4000 after 1 year with 30% interest. Calculate the total amount Shawna earned from taking out the loan.

8. **WE4** A car dealership offers its customers the buy now–pay later option to purchase a $13 500 car by paying $2500 up-front, followed by 36 monthly payments of $360.

   a. Calculate the amount a customer will pay in total if they choose the buy now–pay later plan.
   b. Calculate the effective rate of interest for the buy now–pay later plan.

9. Georgie is comparing purchasing plans for the latest 4K television. The recommended retail price of the television is $3500. She goes to three stores and they offer her the following buy now–pay later plans.

   - Store 1: $250 up-front + 12 monthly payments of $300
   - Store 2: 24 monthly payments of $165
   - Store 3: $500 up-front + 6 monthly payments of $540

   a. Calculate the total amount payable for each purchase plan.
   b. State which purchase plan has the lowest effective rate of interest.

10. **MC** A new outdoor furniture set normally priced at $1599 is sold for an up-front fee of $300 plus 6 monthly instalments of $240. The effective rate of interest is:

   A. 27.78%
   B. 30.23%
   C. 31.51%
   D. 37.22%
   E. 43.42%

11. Drew has a leak in his water system and gets quotes from 5 different plumbers to try to find the best price for the job. From previous experience he believes it will take a plumber 90 minutes to fix his system. Calculate approximately how much each plumber will charge to help Drew decide who to hire.

   - Plumber A: A call-out fee of $5121.84 plus an hourly charge of $80, with a 5% discount for payment in cash
   - Plumber B: A flat fee of $200 with no discount
   - Plumber C: An hourly fee of $130, with a 10% discount for payment in cash
   - Plumber D: A call-out fee of $70 plus an hourly fee of $90, with an 8% discount for payment in cash
   - Plumber E: An hourly fee of $120 with no discount

12. An electrical goods store allows purchasers to buy any item priced at $74 907.6 or more for a 10% deposit, with the balance payable 6 months later at a simple interest rate of 7.64% p.a. Calculate the final cost of each of the following items under this arrangement.

   a. An entertainment system priced at $1265
   b. An oven priced at $1450
   c. A refrigerator priced at $2018
   d. A washing machine priced at $3124

13. Elise gets a new credit card that has an annual fee of $100 and earns 1 frequent flyer point per $1 spent. In her first year using the card she spends $27 500 and has to pay $163 in interest. Elise exchanged the frequent flyer points for a gift card to her favourite store, which values each point as being worth 0.8 cents.
   Determine if using the credit card over the year was a profitable investment. Explain.

14. Michelle uses all of the $12 000 in her savings account to buy a new car worth $25 000 on a buy now–pay later scheme. The purchase also requires 24 monthly payments of $750.

   a. Calculate the amount Michelle paid in total for the car.

   Michelle gets a credit card to help with her cash flow during this 24-month period, and over this time her credit card balance averages $215 per month. The credit card has an interest rate of 23.75% p.a.

   b. Calculate the amount of interest Michelle pays on her credit card over this period.
   c. In another 18 months Michelle could have saved the additional $13 000 she needed to buy the car outright. Calculate the amount she saved by choosing to save this money first.

15. Javier purchases a new kitchen on a buy now–pay later plan. The kitchen usually retails for $24 500, but instead Javier pays an up-front fee of $5000 plus 30 monthly instalments of $820.

   a. Calculate the amount Javier pays in total.
   b. Calculate the effective rate of interest of the buy now–pay later plan.

   If Javier paid an up-front fee of $10 000, he would only have to make 24 monthly instalments of $710.

   c. Calculate the amount Javier would save by going for the second plan.
   d. Calculate the effective rate of interest of the second plan.

16. Some loans calculate the interest each month after a monthly repayment has been paid. This is called a reducing balance loan.

   a. Using a spreadsheet, calculate the time it will take to pay back a $10 000 loan with an interest rate of 6.55% p.a. on a reducing monthly balance when paying back $560 per month.
   b. Calculate the amount of interest that is payable over the lifetime of the loan.

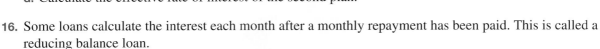

11.3 Financial systems — permanent employment

11.3.1 Salaries, wages and penalty rates

Employees may be paid for their work in a variety of ways. Most receive either a **wage** or a **salary**.

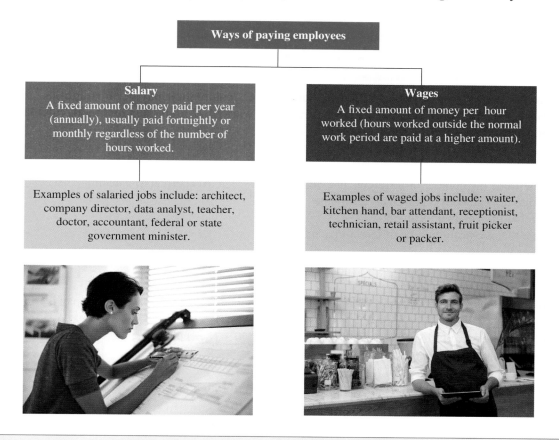

Ways of paying employees

Salary
A fixed amount of money paid per year (annually), usually paid fortnightly or monthly regardless of the number of hours worked.

Wages
A fixed amount of money per hour worked (hours worked outside the normal work period are paid at a higher amount).

Examples of salaried jobs include: architect, company director, data analyst, teacher, doctor, accountant, federal or state government minister.

Examples of waged jobs include: waiter, kitchen hand, bar attendant, receptionist, technician, retail assistant, fruit picker or packer.

Key points

- **Normal working hours in Australia are *38 hours* per week.**
- **There are *52 weeks* in a year.**
- **There are *26 fortnights* in a year (this value will be slightly different for a leap year).**
- **There are *12 months* in a year.**

Overtime is paid when a wage earner works more than the regular hours each week. These additional payments are often referred to as penalty rates. Penalty rates are usually paid for working on weekends, public holidays or at night.

The extra hours are paid at a higher hourly rate, normally calculated at either time and a half or double time.

Calculating overtime

The overtime hourly rate is usually a multiple of the regular hourly rate. Some examples of overtime rates include:
- **1.5 × regular hourly wage** (*time and a half*)
- **2 × regular hourly wage** (*double time*)
- **2.5 × regular hourly wage** (*double time and a half*)

Regular hourly rate
$25.00

→

Overtime hourly rate
1.5 × regular hourly rate
1.5 × 25.00 = $37.50

WORKED EXAMPLE 5 Calculating salary paid per fortnight

a. A bank employee earns a salary of $61 000 per year. Calculate their pay per fortnight.
b. A fast-food employee is paid $15.50 per hour. If they worked 72 hours last fortnight, calculate the amount they were paid.

THINK	WRITE
a. 1. Determine the number of fortnights in a year: divide the number of days in a year by the number of days in a fortnight. Note that the result will be different in a leap year (366 days).	**a.** $\dfrac{365}{14} \approx 26.07$ (correct to 2 decimal places)
2. Divide the annual salary by 26.07 and round to the nearest cent.	$61\,000 \div 26.07 \approx 2339.85$
3. Write the answer as a sentence.	The bank employee's fortnightly pay is $2339.85.
b. 1. The worker is paid $15.50 for each hour of work, so multiply this amount by the number of hours worked.	**b.** $15.50 \times 72 = 1116.00$
2. Write the answer as a sentence.	Last fortnight, the fast-food employee was paid $1116.00.

Note: We will use the approximations of 52.14 weeks and 26.07 fortnights in a year in this text.

---

tlvd-4810

WORKED EXAMPLE 6 Calculating wage earned over a weekend

A hospitality worker is paid time and a half for working on Saturday nights and double time for working on Sundays or public holidays. The normal hourly pay rate is $15.50. If the worker worked 5 hours on Saturday and 6 hours on Sunday, calculate the amount of money they earned over the weekend.

THINK	WRITE
1. Calculate the hourly rate of pay for Saturday. Time-and-a-half means × 1.5.	$15.50 \times 1.5 = \$23.25$ per hour
2. Calculate the amount they were paid for Saturday.	$23.25 \times 5 = \$116.25$

3. Calculate the hourly rate of pay for Sunday. double time means × 2.

$15.50 \times 2 = \$31$ per hour

4. Calculate the amount paid for Sunday.

$31 \times 6 = \$186$

5. Write the answer as a sentence.
   Note: This is the amount they earned before tax was taken out.

The amount earned for the weekend is $\$116.25 + 186 = \$302.25$.

11.3.2 Superannuation

Employees may receive an **allowance** to complete certain tasks, such as to use their own tools at work, or to work in certain conditions. Common allowances include uniforms, tools and equipment, travel costs, car and phone costs, and first-aid training.

Allowances are included in wages and are documented separately on pay sheets.

All workers in Australia get an additional sum of money for their retirement. This sum is called **superannuation**. From 2017, the law requires employers to pay an additional 9.5% of annual salary into a recognised superannuation fund. The amount is based on usual earnings before tax and is calculated at each pay cycle.

Some workers choose to contribute additional funds from their wages to increase their superannuation. There are tax incentives, such as paying lower tax rates on superannuation lump sums, to encourage workers to save for their retirement.

Additional government financial support, known as the **pension**, is also available for some people who have retired. The amount a retired person receives in the pension depends on their personal financial security and wealth.

WORKED EXAMPLE 7 Calculating the amount of superannuation paid by the employer each fortnight

A worker's hourly rate is $28.75 and he works 38 hours each week. The worker is paid fortnightly. Calculate the amount of superannuation the employer pays on his behalf each fortnight.

THINK	WRITE
1. Calculate the fortnightly wage. Note that if the employee works 38 hours per week, then he works 76 (38×2) hours per fortnight.	$76 \times 28.75 = \$2185.00$
2. Calculate 9.5% of his fortnightly wage. Remember to convert 9.5% to a decimal by dividing by 100.	$9.5\% \times 2185 = 0.095 \times 2185$ $\approx \$207.58$
3. Write the answer as a sentence.	The employer pays $207.58 into the superannuation fund each fortnight.

11.3.3 Annual leave loading

Some workers will receive an extra payment on top of the 4-week annual leave pay. It is usually 17.5% of their normal pay for 4 weeks. It will depend on the working agreement workers have with their employers if they receive an **annual leave loading**.

The purpose of annual leave loading is to compensate workers who are unable to earn additional money through overtime while on leave, and to help cover the costs associated with taking holidays.

tlvd-4811

WORKED EXAMPLE 8 Calculating the amount paid for annual leave loading before tax

A worker's annual salary is \$58 056. Calculate the amount, in dollars, paid for annual leave loading before tax.

THINK	WRITE
1. Calculate the weekly salary by dividing the annual salary by 52.14.	$\dfrac{58\,056}{52.14} \approx \$1113.46$
2. Calculate the wage for 4 weeks.	$\$1113.46 \times 4 = \$4453.84$
3. Calculate 17.5% of the 4-week wage. Remember to convert 17.5% to a decimal by dividing by 100.	$17.5\% \times 4453.84 = 0.175 \times 4453.84$ ≈ 779.42
4. Write the answer as a sentence.	The amount paid for annual leave loading, before tax, is \$779.42.

 Resources

 Interactivity Special rates (int-6068)

11.3 Exercise

Students, these questions are even better in jacPLUS

 Receive immediate feedback and access sample responses

 Access additional questions

 Track your results and progress

Find all this and MORE in jacPLUS

1. A lawyer is offered a job with a salary of either \$74 000 per year or \$40 per hour. Assuming that they work 80 hours every fortnight, determine which pay is greater.

2. **MC** An hourly wage of \$32.32 for 77.5 hours per fortnight results in a fortnightly pay of:
   A. \$1252.40 B. \$2504.80 C. \$842.58 D. \$1292.80

3. **WE5** Calculate the wage for the following hourly rates and hours worked.
   a. \$19.75 for 74.75 hours b. \$24.85 for 45.75 hours c. \$45.30 for 35.25 hours

4. A worker earns \$20.40 an hour. They need to earn a minimum of \$700 each week before tax to buy food, pay rent and bills, and have some money for entertainment. Calculate the minimum number of hours they have to work each week.
   Give your answer correct to 2 decimal places.

5. **WE6** A retail worker is paid time and a half for working on Saturday and double time for working on Sundays or public holidays. The normal pay rate is $18.75. The worker worked over a long weekend, with the Monday being a public holiday. The hours worked for Saturday, Sunday and Monday (public holiday) were 8, 5 and 6 hours respectively.
Calculate the amount of money they earned over the weekend.

6. Over the last four weeks, a person has worked 35, 36, 34 and 41 hours. If she earns $24.45 per hour, calculate the amount they earn for each of the two fortnights.

7. **WE7** A worker's hourly rate is $29.45 and they work 25.75 hours each week. The worker is paid fortnightly. Calculate the amount of superannuation the employer pays on her behalf each fortnight.

8. A school principal earns an annual salary of $155 750.
   a. Calculate the amount they earn each month.
   b. Calculate their superannuation fund payment if the fund is paid 9.5% of her annual salary.

9. A salary earner makes $62 000 per year.
   a. Calculate the amount they earned each month.
   b. Calculate their superannuation fund payment each month if he receives 9.5% superannuation.
   c. Calculate the total amount deposited into the fund in a year.

10. **WE8** A worker's annual salary is $85 980. Calculate the amount, in dollars, paid for annual leave loading before tax.

11. Some employers offer a superannuation bonus scheme. They pay you 9.5%, then for every additional 1% of your salary that you contribute, they match it with a further 1%. For example, if you contribute 2% of your salary to superannuation, you will receive a total of 9.5% + 2% + 2% = 13.5% put into your fund.
A software salesperson makes $70 000 per year and decides to contribute 3% of their salary into superannuation.

   a. Calculate the salesperson's contribution from their own salary into their superannuation each year.
   b. The salesperson's employer matches their superannuation contribution. In total, calculate the amount they receive into their superannuation fund each year.

12. There is a proposal to have the same penalty rates for both Saturday and Sunday of time and a half. Currently workers receive time and a half to work on Saturdays and double time to work on Sundays.
A waiter receives penalty rates and their hourly rate is $17.75.

   a. Calculate the percentage, to the nearest whole number, which they will be worse off by if the penalty rates change.
   b. Calculate the percentage, to 2 decimal places, by which their hourly rate will need to increase to ensure they receive the same amount they currently earn for working both Saturdays and Sundays, assuming they work the same number of hours each day.

13. A factory worker receives an hourly rate of $28.40 to work a standard 38-hour week. They receive overtime of time and a half for any hour worked above 38 hours.

    a. On average, they work 42.5 hours each week. Calculate their weekly wage before tax.

    b. The worker is offered a salary of $68 000. Determine if they will be better off remaining on a wage or taking the salary offer.
    Justify your answer using calculations.

14. An employer is proposing a new working agreement: the removal of the 17.5% annual leave loading, an increase of 5% in annual wages, and 10.5% superannuation contributions. Employees are considering whether to accept the agreement.
    Determine if they should accept the agreement. Justify your answer using calculations.

15. A secretary's current annual salary is $56 890. They are offered two packages:
Package 1: a pay increase of 1.5% each year for 3 years
Package 2: a superannuation contribution of 10.75% for 5 years
By calculating their salary each year for 5 years and the additional amount of superannuation contributions that will be deposited into their superannuation fund, explain which package they should choose.

11.4 Financial systems — taxation and deductions

LEARNING INTENTION

At the end of this subtopic you should be able to:
- calculate personal taxation, GST and superannuation
- calculate and interpret different income-related taxes, payments and deductions and their impact on income, such as pay scales, personal tax, withholding tax, PAYE and PAYG.

11.4.1 The purpose of taxation

Taxation is a means by which state and federal governments raise revenue for public services, welfare and community needs by imposing charges on citizens, organisations and businesses.

Services include education, health, pensions for the elderly, unemployment benefits, public transport and much more.

Tax file numbers

A tax file number (TFN) is a personal reference number for every tax-paying individual, company, funds and trusts. Tax file numbers are valid for life and are issued by the Australian Taxation Office (ATO).

Income tax

Income tax is a tax levied on people's financial income. It is deducted from each fortnightly or monthly pay.

The amount of income tax is based upon **total income** and **tax deductions**, which determines a worker's **taxable income**.

Formula to calculate taxable income

taxable income = total income − tax deductions

The calculation of income tax is based upon an income tax table. The income tax table at the time of writing is:

Taxable income	Tax on this income
0–$18 200	Nil
$18 201–$45 000	19c for each $1 over $18 200
$45 001–$120 000	$5092 plus 32.5c for each $1 over $45 000
$120 001–$180 000	$29 467 plus 37c for each $1 over $120 000
$180 001 and over	$51 667 plus 45c for each over $180 000

Note: The income tax table is subject to change.

Tax deductions

Workers who spend their own money for work-related expenses are entitled to claim the amount spent as tax deductions. Tax deductions are recorded in the end-of-financial-year tax return. The deductions are subtracted from the taxable income, which lowers the amount of money earned and hence reduces the amount of tax to be paid.

What can be claimed as tax deductions is determined by the Australian Taxation Office. Some examples of deductions that can be claimed include using your car to travel to work-related events, purchasing materials, union fees, donations made to charities and using a home office. Deductions must be work-related expenses and evidence, such as receipts, must be provided.

tlvd-4812

WORKED EXAMPLE 9 Calculating the income tax payable

Calculate the income tax payable by the teacher who earns a salary of $67 400. Tax deductions for the teacher are $4240 and the teacher earned $1240 in bank interest.

THINK	WRITE
1. Determine the total income by adding the salary and bank interest together.	$67 400 + 1240 = \$68 640$
2. Subtract the tax deductions from the total income.	Taxable income $= \$68 640 − 4240$ $= \$64 400$
3. Determine the tax bracket based on the taxable income of $64 400.	Taxable income is between $45 000 and $120 000.
4. Determine the amount of tax to be paid.	$5092 plus 32.5c for every $1 over $45 000
5. Determine the amount of taxable income over $45 000 by subtracting $45 000 from the teacher's taxable income.	$\$64 400 − \$45 000 = \$19 400$
6. Determine the tax rate amount as 32.5% of $19 400. Remember to convert 32.5% to a decimal.	$0.325 \times \$19 400 = \$6305$
7. Add the 'plus amount' to calculate the payable tax.	$\$5092 + \$6305 = \$11 397$
8. Write the answer as a sentence.	The total tax payable by the teacher is $11 397.

11.4.2 The Medicare levy

Australian residents have access to health care through Medicare, which is partly funded by taxpayers through the payment of a **Medicare levy**. The current Medicare levy is 2% of taxable income, and may be reduced if the taxable income is below a certain amount.

In addition to the Medicare levy, residents who do not have adequate private health care may also be required to pay a Medicare levy surcharge.

WORKED EXAMPLE 10 Calculating the amount of Medicare levy

A plumber's taxable income for the financial year is **$65 850**. They claim **$5680** in work-related expenses (deductions) and they also have private health insurance.
Determine the amount they have to pay for the Medicare levy.

THINK	WRITE
1. Calculate the plumber's taxable income.	Taxable income = $65 850 − $5680 = $60 170
2. Calculate 2% of the taxable income. Remember to convert 2% to a decimal.	2% × 60 170 = 0.02 × 60 170 = $1203.40
3. Write the answer as a sentence.	They will pay $1203.40 for the Medicare levy.

11.4.3 Pay As You Go (PAYG) tax

The Australian Taxation Office (ATO) administers a **Pay As You Go (PAYG) tax**, which is a withholding tax system. It requires employers to calculate the amount of income tax to withhold from employees. The amount withheld is determined by taxation tables provided by the ATO.

This amount is withheld from the employee's regular earnings (gross pay) and contributes towards the employee's tax to be paid at the end of the financial year.

> **Formula to calculate net pay**
>
> **net pay = gross pay − tax withheld**

At the end of the financial year, workers submit a tax return which lists their deductions and all money earned.
- If the total amount of tax paid is less than is required, the worker will have to pay the difference.
- If the amount of tax paid over the year is more than is required, the worker receives a refund.

Australian workers who provide their tax file number and do not work another job claim the tax-free threshold. This means that they do not pay tax on the first $18 200 earned in the financial year.

A factory worker is paid $19.94 per hour. Each week they work 38 hours, with an additional 4 hours overtime paid at time and a half. Using the following taxation table, calculate their net pay (the amount they receive each week).

Weekly earnings ($)	Tax withheld ($)	Weekly earnings ($)	Tax withheld ($)
856.00	133.00	881.00	141.00
857.00	133.00	882.00	142.00
858.00	133.00	883.00	142.00
859.00	134.00	884.00	142.00
860.00	134.00	885.00	143.00
861.00	134.00	886.00	143.00
862.00	135.00	887.00	143.00
863.00	135.00	888.00	144.00
864.00	135.00	889.00	144.00
865.00	136.00	890.00	144.00
866.00	136.00	891.00	145.00
867.00	136.00	892.00	145.00
868.00	137.00	893.00	145.00
869.00	137.00	894.00	146.00
870.00	137.00	895.00	146.00
871.00	138.00	896.00	146.00
872.00	138.00	897.00	147.00
873.00	138.00	898.00	147.00
874.00	139.00	899.00	147.00
875.00	139.00	900.00	148.00
876.00	139.00		
877.00	140.00		
878.00	140.00		
879.00	141.00		
880.00	141.00		

THINK

1. Calculate the weekly wage for normal hours.
2. Calculate the overtime.
3. Calculate the total weekly wage.
4. Using the table, locate the weekly wage and read the amount of tax withheld.
5. Calculate the net pay by subtracting the tax withheld from the total wages.

WRITE

$38 \times 19.94 = \$757.72$

$19.94 \times 1.5 \times 4 = \$119.64$

$\$757.72 + 119.64 = \$877.36$

Weekly wage is $877; tax withheld is $140.

Weekly income $= 877.36 - 140$
$\qquad\qquad\qquad = \$737.36$

Students, these questions are even better in jacPLUS

Receive immediate feedback and access sample responses

Access additional questions

Track your results and progress

Find all this and MORE in jacPLUS

1. **WE9** Calculate the income tax payable by a baker whose salary is $51 260. The baker's tax deductions are $2120. The baker earned $1850 in bank interest. Refer to the income tax table provided in section 11.4.1.

2. **WE10** Calculate the Medicare levy for the following taxable incomes.
   a. $60 400
   b. $77 300
   c. $89 400
   d. $108 423

3. **MC** A shop worker earns $18.95 per hour. They work 52 hours in a fortnight. Select the amount that is closest to the amount of Medicare levy they will be expected to pay at the end of the financial year.
   Note: Assume there are 26.07 fortnights in a year.
   A. $20
   B. $51
   C. $200
   D. $514
   E. $510

4. **WE11** A garage mechanic is paid by the hour.
   a. If they work 84 hours in a fortnight and are paid $22.50 an hour, calculate their gross pay (amount before tax).
   b. Using the taxation table shown, calculate their net pay.

Fortnightly earnings ($)	Tax withheld ($)	Fortnightly earnings ($)	Tax withheld ($)
1832.00	306.00	1878.00	322.00
1834.00	308.00	1880.00	324.00
1836.00	308.00	1882.00	324.00
1838.00	308.00	1884.00	324.00
1840.00	310.00	1886.00	326.00
1842.00	310.00	1888.00	326.00
1844.00	310.00	1890.00	326.00
1846.00	312.00	1892.00	328.00
1848.00	312.00	1894.00	328.00
1850.00	314.00	1896.00	330.00
1852.00	314.00	1898.00	330.00
1854.00	314.00	1900.00	330.00
1856.00	316.00	1902.00	332.00
1858.00	316.00	1904.00	332.00
1860.00	316.00	1906.00	332.00
1862.00	318.00	1908.00	334.00
1864.00	318.00	1910.00	334.00
1866.00	318.00	1912.00	334.00
1868.00	320.00	1914.00	336.00
1870.00	320.00	1916.00	336.00
1872.00	320.00	1918.00	336.00
1874.00	322.00	1920.00	338.00
1876.00	322.00		

5. For the mechanic in question **4**, any hours worked over 82 hours per fortnight are paid overtime at a rate of time and a half. Adjust their gross and net pay accordingly.

6. Using the taxation table provided:

Taxable income	Tax on this income
0–$18 200	Nil
$18 201–$45 000	19c for each $1 over $18 200
$45 001–$120 000	$5092 plus 32.5c for each $1 over $45 000
$120 001–$180 000	$29 467 plus 37c for each $1 over $120 000
$180 001 and over	$51 667 plus 45c for each over $180 000

   i. write down the percentage tax payable for the following annual salaries
   ii. hence, calculate the amount of tax payable.

     a. $37 500   **b.** $15 879   **c.** $85 670   **d.** $131 000

7. An apprentice electrician is paid $17.14 per hour. They are paid the normal rate for the first 38 hours worked in any week and then the overtime rate of $25 for hours worked over 38 hours. In one week, they worked 45 hours. Using the taxation table, calculate their net wage for the week (the amount they receive).

Weekly earnings ($)	Amount to be withheld ($)
816.00	119.00
817.00	119.00
818.00	119.00
819.00	120.00
820.00	120.00
821.00	120.00
822.00	121.00
823.00	121.00
824.00	121.00
825.00	122.00
826.00	122.00
827.00	122.00
828.00	123.00
829.00	123.00
830.00	123.00

Weekly earnings ($)	Amount to be withheld ($)
831.00	124.00
832.00	124.00
833.00	125.00
834.00	125.00
835.00	125.00
836.00	126.00
837.00	126.00
838.00	126.00
839.00	127.00
840.00	127.00
841.00	127.00
842.00	128.00
843.00	128.00
844.00	128.00
845.00	129.00

     a. $1500   **b.** $2000   **c.** $2500   **d.** $4000

8. The table shows the tax payable for fortnightly wages. Using the table, calculate the tax payable on the following fortnightly wages by workers who claim the tax-free threshold.

Fortnightly income ($)	Tax to be withheld		Fortnightly income ($)	Tax to be withheld	
	With tax-free threshold ($)	No tax-free threshold ($)		With tax-free threshold ($)	No tax-free threshold ($)
1000	7	262	2000	366	608
1100	96	296	2200	434	676
1200	118	330	2500	540	780
1500	192	434	3000	712	968
1800	296	538	4000	1086	1358

9. Calculate the percentage of tax payable for the incomes from question **8** by workers who don't claim the tax-free threshold. Give your answers to 2 decimal places where necessary.

10. A nurse's gross annual salary is $58 284 and they are paid monthly.
    a. Calculate their gross monthly salary.
    b. Using the tax table shown, determine the amount of tax withheld from their monthly salary.

Monthly earnings ($)	Tax withheld ($)	Monthly earnings ($)	Tax withheld ($)
4792.67	949.00	4870.67	979.00
4797.00	953.00	4875.00	979.00
4801.33	953.00	4879.33	979.00
4805.67	958.00	4883.67	984.00
4810.00	958.00	4888.00	984.00
4814.33	958.00	4892.33	984.00
4818.67	962.00	4896.67	988.00
4823.00	962.00	4901.00	988.00
4827.33	962.00	4905.33	988.00
4831.67	966.00	4909.67	992.00
4836.00	966.00	4914.00	992.00
4840.33	966.00	4918.33	997.00
4844.67	971.00	4922.67	997.00
4849.00	971.00	4927.00	997.00
4853.33	971.00	4931.33	1001.00
4857.67	975.00	4935.67	1001.00
4862.00	975.00	4940.00	1001.00
4866.33	975.00		

c. Determine the percentage of tax, to 2 decimal places, they pay each month.
d. Calculate the Medicare levy they pay.

11. An accountant pays $1719.40 in Medicare levy.
    a. If they have private health insurance, calculate their taxable income for the year.
    b. Calculate their tax payable for the financial year. Use the tax table provided in subtopic 11.4.1.

12. A hairdresser receives a weekly wage of $930.60, which includes 38 hours of normal pay plus 4 hours of overtime, paid at time and a half.
    a. Explain how their hourly rate can be determined, and hence calculate the hourly rate.
    b. Determine their annual salary if they work an average of 40 hours a week (38 hours at normal pay and 2 hours overtime). Assume 52 weeks in 1 year.
    c. Using your answer from part **b**, determine the amount of tax payable at the end of the financial year.
    d. The hairdresser forgot to claim $1980 in deductions and is looking forward to receiving $1980 in a tax refund. By recalculating the amount of tax payable at the end of the financial year based on their new taxable income, explain why they won't receive $1980 as tax refund.

13. A truck driver earns $24.07 per hour for working a 76-hour fortnight. An overtime rate of time and a half is paid for additional hours worked over the 76 hours during Monday to Friday, and double time is paid for hours worked on weekends (Saturday and Sunday). Assume 52 weeks in 1 year.

   Over a fortnight, the truck driver worked 80 hours Monday–Friday and 15 hours over the weekend.

   a. Calculate the gross fortnightly wage of the truck driver.

Fortnightly earnings ($)	Tax withheld ($)	Fortnightly earnings ($)	Tax withheld ($)
2682.00	602.00	2712.00	612.00
2684.00	602.00	2714.00	614.00
2686.00	604.00	2716.00	614.00
2688.00	604.00	2718.00	614.00
2690.00	604.00	2720.00	616.00
2692.00	606.00	2722.00	616.00
2694.00	606.00	2724.00	616.00
2696.00	606.00	2726.00	618.00
2698.00	608.00	2728.00	618.00
2700.00	608.00	2730.00	618.00
2702.00	608.00	2732.00	620.00
2704.00	610.00	2734.00	620.00
2706.00	610.00	2736.00	620.00
2708.00	610.00	2738.00	622.00
2710.00	612.00	2740.00	622.00

   b. Show that the truck driver's net fortnightly pay was $2089.84 by using the PAYG tax table shown above.
   c. If the truck driver claims $2486 in deductions, calculate their taxable income, and hence calculate the amount of payable tax for the year by using the tax table provided in section 11.4.1.
   d. Calculate the Medicare levy the driver is expected to pay, given that he has private health insurance.

14. A computer technician's annual salary is $67 374. They claim $1580 in deductions and receive $225 in dividends from shares.

   a. The technician is paid monthly. Calculate their gross monthly salary.
   b. Their employer withholds $1200 PAYG tax each month. Calculate the total amount of tax withheld for the year.
   c. Determine the tax payable by the computer technician for the financial year. Hence, determine whether they receive a tax refund or are required to pay more tax, and state the amount.

15. The Australian tax system is a tiered system, as shown in the table.

Taxable income	Tax on this income
0–$18 200	Nil
$18 201–$45 000	19c for each $1 over $18 200
$45 001–$120 000	$5092 plus 32.5c for each $1 over $45 000
$120 001–$180 000	$29 467 plus 37c for each $1 over $120 000
$180 001 and over	$51 667 plus 45c for each over $180 000

A dentist's annual salary is $97 605.

a. Explain why the tax payable is not found by calculating 32.5% of $97 605.

b. Explain how the additional 'plus amount' of $19 822 is calculated.

16. A mechanical engineer pays $14 812 in tax.

a. Explain how their taxable income can be determined.

b. Determine their taxable income.

c. After the financial year they realise that they forgot to claim $985 in tax deductions. Determine their taxable income and the tax refund they should expect.

11.5 Income payments — commission, piecework and royalties

LEARNING INTENTION

At the end of this subtopic you should be able to:
- calculate commissions and royalties
- calculate payment for piecework.

11.5.1 Commission

A **commission** is paid to a person when an item is sold. For example, when real estate agents sell houses, they are paid a commission.

Commissions are calculated as a percentage of the sale price. If no sales are made, no commission is received. A commission table is often used to determine the value of a commission.

tlvd-4813

WORKED EXAMPLE 12 Calculating the commission to be paid

When selling real estate, an agent is paid according to the table shown.

Sale price	Commission	Plus
Between 0 and $80 000	**2% of sale price**	**0**
Between $80 001 and $140 000	**1.5% of the amount over $80 000**	**$1600**
$140 000 and over	**1.1% of the amount over $140 000**	**$2500**

If a house is sold for $200 000, calculate the commission paid to the real estate agent.

THINK	WRITE
1. Since the amount of commission varies according to the sale price, determine which of the 3 ranges the sale falls into. $200\,000 > 140\,000$, so the price falls into range 3 ($140 000 and over). The commission for this range is based on the amount of the sale *over* $140 000. Calculate the amount over $140 000.	$\begin{array}{r} 200\,000 \\ -140\,000 \\ \hline 60\,000 \end{array}$
2. The commission is 1.1% of 60 000; write 1.1% as a decimal.	$1.1\% = \dfrac{1.1}{100} = 0.011$
3. Multiply 0.011 by the amount over $140 000 — that is, $60 000.	$\begin{aligned} 1.1\% \text{ of } 60\,000 &= 0.011 \times 60\,000 \\ &= 660 \end{aligned}$
4. The table also specifies an additional commission of $2500, due to the sale being over $140 000. Calculate the total of the commission.	$660 + 2500 = 3160$
5. Write the answer as a sentence.	The total commission is $3160.

11.5.2 Salary and commissions

In some industries, the rate of sales varies widely, so it is not practical to live on commissions only. For example, car salespeople are given a base salary in addition to a commission.

WORKED EXAMPLE 13 Calculating the total wage including salary and commission

A car salesperson earns a salary of $1200 per month plus a commission of 3% of the total sales that they make that month. In February, they sold cars worth $198 500. Calculate their total wage (salary + commission) for that month.

THINK	WRITE
1. Their commission is 3% of sales. Write 3% as a decimal.	$3\% = \dfrac{3}{100} = 0.03$
2. Calculate 3% of the total sales in order to determine the commission.	$\begin{aligned} 3\% \text{ of } 198\,500 &= 0.03 \times 198\,500 \\ &= 5955 \end{aligned}$
3. Their total wage is the commission plus their base salary. Add the commission of $5955 to the base salary of $1200.	$5955 + 1200 = 7155$
4. Write the answer as a sentence.	The total wage paid in February is $7155.

11.5.3 Piecework

A person paid by **piecework** is paid a fixed amount for each item produced. Often the work is done at home. There may be bonuses for faster workers.

WORKED EXAMPLE 14 Calculate the amount earned by piecework

A dressmaker gets paid $4.75 for every dress they sew. If they sewed 124 dresses last week, calculate the amount of money they earned.

THINK	WRITE
1. The dressmaker gets $4.75 for each dress, so multiply this amount by the number of dresses sewn.	$4.75 \times 124 = 589$
2. Write the answer as a sentence.	Last week the dressmaker was paid $589.00.

11.5.4 Royalties

A **royalty** is a payment made to an author, composer or creator for each copy of the work or invention sold. For example, if a pop star sold $2 million worth of music last year, she is entitled to some of the profits. Other people, such as managers, also get a share. Royalties are usually calculated as a percentage of the total sales (not total profit).

WORKED EXAMPLE 15 Calculating royalties

Last year, a new rock star sold music worth $2 156 320. Their recording contract specifies that they receive 2.4% royalty on the total sales. Determine the amount of money they earned last year.

THINK	WRITE
1. They earned 2.4% of the sales ($2 156 320). Write 2.4% as a decimal.	$2.4\% = \dfrac{24}{100}$ $= 0.024$
2. Calculate 2.4% of the total sales; multiply 0.024 by the total sales.	2.4% of $2\,156\,320 = 0.024 \times 2\,156\,320$ $= \$51\,751.68$
3. Write the answer as a sentence.	The rock star earned $51 751.68 in royalties.

on Resources

🧩 **Interactivities** Piecework (int-6069)

Commission and royalties (int-6070)

1. **WE12** Using the commission table shown, calculate the commission on the sale of a house for $945 000.

SOLD for $945 000

Sale price	Commission	Plus
Between $0 and $400 000	0.5% of sale price	0
Between $400 001 and $700 000	0.3% of the amount over $400 000	$2000 (0.5% of $400 000)
$700 001 and over	0.2% of the amount over $700 000	$2900 (0.5% of $400 000 + 0.3% of $300 000)

2. Using the commission table from question **1**, calculate the commission for the following sales.
   a. $330 000   b. $525 000   c. $710 000   d. $1 330 000

3. **MC** Using the commission table from question **1**, the commission on the sale of a $500 000 house would be:
   A. $1000   B. $2000   C. $2300   D. $2500   E. $2700

4. **WE13** A car salesperson earns a salary of $1400 per month plus a commission of 3.5% of the total sales they make that month. In April, they sold cars worth $155 000. Calculate their total wage (salary + commission) for that month.

5. **WE14** A shoemaker is paid $5.95 for each pair of running shoes they can make. If the shoemaker made 235 pairs of shoes last week, calculate the amount they were paid.

6. **MC** If a software engineer gets paid $3.40 for every line of computer code they write, determine how much they will make if they write 865 lines in a fortnight.
   A. $29 410.00   B. $2941.00   C. $294.10   D. $2329.00   E. $29 000

7. A dressmaker gets paid per item they sew. If they received $545 for sewing 45 items, determine the amount they were paid per item.

8. **WE15** Last year, a pop star sold $5 342 562 worth of music on their website and via music streaming services. Their contract calls for a 3.5% royalty on all sales. Determine the amount they earned last year.

9. **MC** If an author earned $25 560 on the basis of sales of $568 000, then the royalty payment is:
   A. 4.5%   B. 45%   C. 0.045%
   D. 0.45%   E. 0.0045%

10. An actor is paid a 2.5% royalty plus $300 000 cash to act in the latest blockbuster movie.

   a. Complete the following table of royalties.

Time	Jan–Mar	Apr–Jun	Jul–Aug	Sep–Dec
Sales	$123 400	$2 403 556	$432 442	$84 562
Royalty payment				

   b. Calculate the total amount the actor received, including the cash.

11. To keep matters simple, the winner of a television singing competition receives a royalty of 1% on sales of their first album. If this year's winner sells $4 563 453 worth of music, determine the royalty payment.

12. A new children's book author is offered a choice of $100 000 cash or a 4% royalty on sales. Calculate the sales needed for the royalty offer to match the cash offer.

13. A hot-shot used-car salesperson earns $1500 per month plus 6.2% commission on all sales. If they sold $243 540 worth of cars last month, calculate their total wage.

14. A songwriter receives royalties on the sales of their songs. They received $11 473.75 in royalties for the sale of $458 950 in songs. Explain how the royalty percentage they receive can be determined, and hence calculate the royalty percentage they receive.

15. A shoe manufacturer decides to offer their shoemakers two pay options:
   Option 1: $8.25 per pair of shoes made
   Option 2: receive a commission on the shoes sold

   A shoemaker can make 15 pairs of shoes each day (over 5 days) and each pair of shoes sells for $75.95. In one week the manufacturer will sell 250 pairs of shoes.
   Determine which option will earn the shoemaker the most money. Justify your answer using calculations.

16. A telecommunications salesperson receives a base wage of $450 per week plus 2.5% commission on all new telephone plans sold during the week. In one week, their earnings (before tax) were $502.25.

   a. Calculate the total amount (in dollars) of telephone plans they sold during the week.
   b. If each plan was $95, determine how many plans they sold.
   c. The salesperson is offered the choice to remain on a weekly base wage of $530 with an increase of 3.5% in commission on all sales, or to receive an annual salary of $37 500.
   Determine which option you would recommend they choose. Justify your answers using calculations.

11.6 Income payments and calculations

LEARNING INTENTION

At the end of this subtopic you should be able to:
- calculate percentage change
- calculate tax payable on gross income
- prepare a wage sheet.

11.6.1 Calculating percentage change

Percentage increase and decrease are examples of percentage change. It is the extent to which an amount grows or decreases.

Formula to calculate percentage change

$$\text{percentage change} = \frac{\text{difference between two numbers}}{\text{original number}} \times 100$$

A spreadsheet can be used to find the percentage change by inserting a formula that calculates the difference between the cell numbers, and then divides by the original number and multiplies by 100.

WORKED EXAMPLE 16 Calculating the percentage change of wages

A shop assistant receives a pay increase. Their original weekly pay was $545 and their new pay is $567. Using a spreadsheet, calculate the percentage change of their wage. Write your answer to the nearest whole per cent.

THINK

1. Set up a spreadsheet with headings 'original pay', 'new pay' and 'percentage change'.

2. Enter her original pay ($545) in cell A2 and her new pay ($567) in cell B2.

WRITE

	A	B	C
1	Original pay	New pay	Percentage change
2			
3			
4			
5			
6			

	A	B	C
1	Original pay	New pay	Percentage change
2	545	567	
3			
4			
5			
6			

3. In cell C2, enter the formula that calculates the difference between cells A2 and B2 (B2 − A2) and then divides by the original pay: (B2 − A2)/A2, and finally converts to a percentage by multiplying by: (B2 − A2)/A2 ∗ 100. Be sure to place = before your equation.

	A	B	C
1	Original pay	New pay	Percentage change
2	545	567	= (B2−A2)/A2*100
3			
4			
5			
6			

4. Write the answer as a sentence.

The percentage change of her wage is 4% (a 4% pay rise).

11.6.2 Calculating tax payable

Calculating the tax payable on gross incomes can be found using a spreadsheet and inserting formulas that determine the amount of tax payable for each tax tier. (See the table in subtopic 11.4.1 for the tax tiers.)

WORKED EXAMPLE 17 Calculating the amount of tax payable on gross income

A law clerk's annual salary is \$45 675. She claims \$450 in tax deductions and receives \$250 in interest from investments. Using a spreadsheet, calculate the amount of tax payable for the financial year by using the tax table provided in subtopic 11.4.1.

THINK

WRITE

1. Create a spreadsheet with headings 'gross income', 'interest', 'deductions' and 'taxable income'. Enter the values.

	A	B	C	D
1	Gross income	Interest	Deductions	Taxable income
2	45 675	250	450	
3				
4				
5				
6				

2. In cell D2 calculate the taxable income: add the interest (\$250) to the annual salary (\$45 675) and subtract the deductions (\$450) by writing the formula = A2 + B2 − C2.

D2			fx = A2 + B2 − C2
	A	B	C	D
---	---	---	---	---
1	Gross income	Interest	Deductions	Taxable income
2	45 675	250	450	45 475
3				
4				
5				
6				

3. Insert a column to represent the tax bracket that the law clerk falls into.

	A	B	C	D	E
1	Gross income	Interest	Deductions	Taxable income	Tax tier (45 001–120 000)
2	45 675	250	450	45 475	
3					
4					
5					
6					

4. Insert a formula to calculate the tax payable for this tax tier.

   Type the following formula into
   E2: $= 5092 + (D2 - 45\,000)^* \, 0.325$

5. Calculate the tax payable.

	A	B	C	D	E
1	Gross income	Interest	Deductions	Taxable income	Tax tier (45 001–120 000)
2	45 675	250	450	45 475	5246.375
3					
4					
5					
6					

6. Write the answer to the nearest cent.

   The tax payable by the law clerk is $5246.38.

11.6.3 Preparing a wage sheet

A wage sheet shows a list of workers and details of their earnings. These include net wages, pay rates, gross deductions and allowances. A wage sheet may look similar to the one shown.

Employee	Pay rate ($)	Normal hours worked	Overtime 1.5	Penalty rate 1.5	Penalty rate 2	Allowance	Gross pay ($)	Tax withheld ($)	Net pay ($)
Dave	19.50	38	2	5		25.5	945.75	163.00	782.75
Jose	21.80	32	4	3.5	5		1253.50	271.00	982.50
Neeve	25.70	25		4		15	1002.30	183.00	819.30
Niha	22.45	38				10	853.10	131.00	722.10
Yen	29.85	36			6		1074.60	208.00	866.60

A pay sheet (or pay slip) shows individual workers' details, including the hours worked, pay rate, net wage, tax deduction, leave days and superannuation paid.

A wage sheet can be set up in a spreadsheet containing formulas that perform the necessary calculations.

WORKED EXAMPLE 18 Completing a wage sheet

Using a spreadsheet, complete the following wage sheet.

Employee	Pay rate ($)	Normal hours worked	Overtime 1.5	Penalty rate 1.5	Penalty rate 2	Allowance ($)	Gross pay ($)	Tax withheld ($)	Net pay ($)
Honura	38.95	38	4		6	15		620	
Skye	28.40	38	3	5				289	
Beau	19.15	35		5				234	

THINK

1. Set up a spreadsheet.

2. Calculate the pay for hours worked: Pay for normal hours formula (e.g. 38*38.95)

Overtime pay calculation (e.g. 4*1.5*38.95)

Pay for working on penalty rates (e.g. 0*1.5*38.95 + 6*2*38.95).

3. Calculate gross pay (= hours worked plus allowances).

4. Complete the table by filling the formula in column H down for all workers.

5. Calculate the net pay by deducting tax withheld.

6. Write the answer.

WRITE

	A	B	C	D	E	F	G	H	I	J
1 2 3	Employee	Pay rate ($)	Normal hours worked	Overtime 1.5	Penalty rate 1.5	2	Allowance ($)	Gross pay ($)	Tax withheld ($)	Net pay ($)
4	Honura	38.95	38	4		6	15		620	
5	Skye	28.40	38	3	5				289	
6	Beau	19.15	35		5				234	

B4*C4

D4*D3*B4

E4*E3*B4 + F4*F3*B4

In cell H4 enter the following formula:
H4 = B4*C4 + D4*D3*B4 + E4*E3*B4 + F4*F3*B4 + G4

	A	B	C	D	E	F	G	H	I	J
1 2 3	Employee	Pay rate ($)	Normal hours worked	Overtime 1.5	Penalty rate 1.5	2	Allowance ($)	Gross pay ($)	Tax withheld ($)	Net pay ($)
4	Honura	38.95	38	4		6	15	2196.20	620	
5	Skye	28.40	38	3	5			1420	289	
6	Beau	19.15	35		5			813.88	234	

In cell J4 type = H4 – I4

	A	B	C	D	E	F	G	H	I	J
1 2 3	Employee	Pay rate ($)	Normal hours worked	Overtime 1.5	Penalty rate 1.5	2	Allowance ($)	Gross pay ($)	Tax withheld ($)	Net pay ($)
4	Honura	38.95	38	4		6	15	2196.20	620	1576.20
5	Skye	28.40	38	3	5			1420	289	1131
6	Beau	19.15	35		5			813.88	234	579.88

1. **WE16** Calculate the percentage change for the following using a spreadsheet. Give your answers correct to 2 decimal places.

   a. Original tax paid is $450 and the new tax paid is $435.
   b. A person receives a $45 pay rise. The fortnightly pay was originally $1680.
   c. The original price of an item was $1580 and after two years it is worth $1375.
   d. After a $115 pay increase, a person's new monthly pay is $5890.

2. A bus driver receives a pay increase. Their original weekly pay was $875 and her new pay is $897. Using a spreadsheet, calculate the percentage change of their wage. Give your answer correct to 2 decimal places.

3. A steel worker receives a pay increase of 5.5%. If their original weekly pay (before tax) is $1023.45, calculate their new weekly pay.

4. **WE17** A fast-food manager's annual salary is $48 131. They claim $980 in tax deductions and receives $500 in interest from investments. Using a spreadsheet, calculate the amount of tax payable for the financial year by using the tax table provided in subtopic 11.4.1.

5. Using a spreadsheet, calculate the tax payable by the following workers.

Employee	Gross salary ($)	Interest ($)	Deductions ($)	Taxable income	Tax free (0–18 200) 0c	Tax tier (18 201– 45 000) 19c	Tax tier (45 001– 120 000) 32.5c	Tax tier (120 001– 180 000) 37c	Tax tier (180 000+) 45c
Stan	52 895	350							
Bett	65 845		650						
Xiao	36 080	125							
Mohamed	75 040	870	1500						

6. Workers in a textile factory are paid per item. The number of items 4 employees made for the week are shown.

	Rate/item	Monday	Tuesday	Wednesday	Thursday	Friday
Harry	$4.75	25	26	24	30	20
Theo	$6.50	19	18	16	20	15
Maria	$3.50	35	40	32	29	32
Marcia	$7.95	22	18	25	21	15

Using a spreadsheet, calculate each employee's weekly earnings, before tax.

7. **WE18** Using a spreadsheet, complete the following wage sheet.

Employee	Pay rate ($)	Normal hours worked	Overtime 1.5	Penalty rate 1.5	Penalty rate 2	Allowance	Gross pay ($)	Tax withheld ($)	Net pay ($)
Rex	15.85	28		7	7			124	
Tank	22.15	35		5		15		167	
Gert	30.10	32	1.5		8	19		367	

8. A pay sheet for an individual employee is shown. Complete the pay sheet using a spreadsheet, and hence state the employee's net weekly pay and amount of superannuation paid into their superannuation fund.

Entitlements	Unit	Rate	Total
Wages for ordinary hours worked	30 hours	$35.05	$
Total ordinary hours = 30 hours			
Penalty (double time)	5 hours	$70.10	$
		Gross payment	$

Deductions	
Taxation	$325
Total deductions	
Net payments	

Employer superannuation contribution	
Contribution	$

9. A worker receives a 4.5% pay increase. Their new weekly pay after the pay rise is $1095.

   a. Calculate their annual salary, assuming 52.14 weeks in a year.
   b. Calculate their annual salary before the pay rise.

10. A small business owner employs a manager who receives a monthly salary, 2 office workers who are paid an hourly rate and 3 other workers who are paid per item constructed.
The following table shows the work information of the 6 employees for one week.

Employee	Hours worked	Hourly rate	Items	Rate/item	Annual salary
Sara (manager)	40 hours	–	–	–	$55 850
Troy (office worker)	25	$16.85	–	–	
Helga	30	$22.50	–	–	
Nina	20	–	19	$4.25	
Bill	15	–	15	$5.75	
Max	35	–	28	$4.95	

Using a spreadsheet, construct a weekly wage sheet that shows the earnings, before tax, of the employees. Assume 52 weeks in 1 year.

11. The taxation table is shown.

Taxable income	Tax on this income
0–$18 200	Nil
$18 201–$45 000	19c for each $1 over $18 200
$45 001–$120 000	$5092 plus 32.5c for each $1 over $45 000
$120 001–$180 000	$29 467 plus 37c for each $1 over $120 000
$180 001 and over	$51 667 plus 45c for each over $180 000

Using a spreadsheet, explain how the 'plus' calculations of $5092, $29 467 and $51 667 are found for the different tax levels in the table.

12. A waiter is paid an hourly rate of $21.50. They receive penalty rates of time and a half for working Saturdays and double time for working Sundays and public holidays.

Over the Christmas period, they worked 5 hours on Friday night, 8 hours on Saturday, 9 hours on Sunday and 7.5 hours on Boxing Day (a public holiday). Their employer withholds $185 in tax. Prepare a pay sheet using a spreadsheet to represent his weekly pay.

13. A teacher's annual salary is $94 961. They decide to reduce their time to 0.8 (this means they receive 0.8 of their salary and work 4 days out of 5), claiming that they will be on the lower tax level and pay less in tax overall, so their net fortnightly pay will not be much lower.

By calculating the tax payable for both the full-time salary and the 0.8 salary, determine how accurate their claim is.

14. Each year, an IT technician moves up one level on a pay scale. The current pay scale is shown.

The technician is currently on level 1–3. At the start of the following year, they move up the pay scale to level 1–4.

Level	Annual salary
1–1	$51 758
1–2	$53 258
1–3	$55 333
1–4	$57 508
1–5	$60 088

a. Determine the percentage change in their wage correct to 2 decimal places.

A new working agreement is reached, which increases the pay level of technicians on levels 1 and 2 by 3.75%.

b. Determine the percentage change in the technician's wage from their original 1–3 wage, correct to 2 decimal places.

c. After the pay increase of 3.75%, another technician moves to level 2–3 with an annual salary of $65 880. The overall percentage change in their wage is 8.5%. Determine the salary at level 2–2 before and after the pay increase.

15. A new taxation system is being proposed.

Taxable income	Tax on this income
0–$19 500	Nil
$19 501–$65 000	21 c for each $1 over $19 500
$65 001–$125 000	A plus 35.5 c for each $1 over $65 000
$125 001–$180 000	B plus 37 c for each over $125 000
$180 001 and over	C plus 45 c for each $1 over $180 000

a. Calculate the values of A, B and C using a spreadsheet.

b. Write down the formula that would calculate the tax payable for a taxable income within the tax level $65 001–$125 000.

c. By comparing this proposed taxation table with the current table, state whether this would be a fairer taxation system. Justify your answer using calculations.

11.7 Review

doc-38052

11.7.1 Summary

Hey students! Now that it's time to revise this topic, go online to:

Access the topic summary

Review your results

Watch teacher-led videos

Practise questions with immediate feedback

Find all this and MORE in jacPLUS

11.7 Exercise

Multiple choice

1. **MC** A new racing bike priced at $3600 is sold for an up-front payment fee of $300 plus 15 monthly instalments of $280. The effective rate of interest is:

   A. 14.36%   B. 40.91%   C. 69.94%   D. 204.61%   E. 25%

2. **MC** Meredith walks dogs on the weekends. She charges $14.00 per dog plus $6.00 an hour. She offers her clients a 5% discount for paying in cash. The amount she would charge for someone paying cash to walk 3 dogs for 2 hours is:

   A. $51.30   B. $131.10   C. $54   D. $2.70   E. $48

3. **MC** The energy usage and rates for a household are as shown. There is a 25% discount for paying the bill before the due date. If the bill was paid before the due date, determine the amount paid.
   (Remember to include GST.)

Electricity

Tariff	Bill days	Current reading	Previous reading	Total usage (kWh)	Charge/rate (c/kWh)
All day	90	78 250	77 682		25.08
Service to property	90			$1.105/day	

Gas

Tariff	Bill days	Current reading	Previous reading	Total usage (kWh)	Charge/rate (c/MJ)
All day	90	8920	8138		2.452
Service to property	90			$0.985/day	

   A. $107.82   B. $241.90   C. $262.29   D. $288.52   E. $242.29

4. **MC** A car can travel 485 km on 68 litres of fuel. The fuel consumption in L/100 km for the car is:

   A. 0.14   B. 7.01   C. 7.13   D. 14.02   E. 14.05

5. **MC** Select the term that represents receipt of payment per item constructed.

   A. Wages   B. Piecework   C. Commission   D. Superannuation   E. Royalty

6. **MC** In a leap year, an annual salary of $56 200 results in a fortnightly pay of:

    A. $2161.54   **B.** $2155.73   **C.** $2149.73   **D.** $2150.80   **E.** $2152.80

7. **MC** A bookkeeper receives a pay rise. Their original weekly pay was $758 and their new weekly pay is $775. The percentage change of their pay is approximately:

    A. 0.0219%   **B.** 0.024%   **C.** 2.19%   **D.** 2.24%   **E.** 20.4%

8. **MC** A librarian pays $1137.60 in Medicare levy. Their annual salary is:

    A. $1137.60   **B.** $2200.75   **C.** $56 880   **D.** $37 920   **E.** $46 880

9. **MC** According to the manufacturer's guidelines, a vehicle has its four tyres replaced 3 times in 9 years. If the vehicle travels an average of 17 500 km each year, select the number of kilometres (on average) the driver gets a new set of tyres.

    A. 157 500 km   **B.** 52 500 km   **C.** 17 500 km
    D. 50 000 km   **E.** 55 200 km

10. **MC** A household's daily water usage is 1700 litres. The total water bill for 60 billing days is $392.12, which includes $187.10 for water services. The charge per kilolitre is:

    A. $2.01/kL   **B.** $2.10/kL   **C.** $20.1/kL   **D.** $2.05/kL   **E.** $3.01/kL

Short answer

11. A nurse is paid $28.50 per hour for an 80-hour fortnight. Assuming 26.07 fortnights per year, calculate:

    a. the nurse's annual salary
    b. the nurse's annual superannuation (assuming 9.5% superannuation)
    c. the nurse's weekly pay.

12. A novelist gets a royalty payment of 2.9% of gross sales of their book. Complete the following table.

	January	**February**	**March**	**April**
Gross sales	$45 000	$125 000	$320 000	
Royalty				$1508

13. A vehicle is purchased for $18 570 on finance of 5.5% per annum for 3 years. A 10% deposit is paid and there are no other administration costs. Stamp duty payable on the purchase of a vehicle is as follows, if the vehicle is worth:

    $100 000 or less — 3% ($3 per $100 or part thereof)
    more than $100 000 — 5% ($5 per $100 or part thereof).
    Calculate, correct to the nearest cent:

    a. the stamp duty
    b. the monthly instalments.

14. There is a 12% discount for customers who pay their energy bill before the due date. The energy usage of a customer is:
    - electricity: 875 k Wh at 19.02c/k Wh
    - gas: 720 MJ at 2.034c/MJ.

    The customer is also charged a $112 fixed fee for service to the property.
    If the customer pays before the due date, calculate the amount they paid, including GST.

15. Calculate the amount of superannuation that is paid each fortnight (assume 9.5% superannuation) for the following annual salaries:

a. $67 899   **b.** $98 765   **c.** $101 010   **d.** $123 456

16. Employees at a cafe are paid penalty rates for working on weekends. The normal hourly rate is $18.50. Penalty rates for Saturday are time and a half, and double time for Sundays and public holidays.

Four employees worked on the long weekend. The hours worked are shown in the table.

Employee	Saturday	Sunday	Monday (public holiday)
Tran	5	8	
Gus		6	
Meg	3.5		7
Warren	4		6

Calculate the earnings before tax for each employee.

17. A factory worker's hourly rate is $21.50. They work 38 hours a week with 4 hours overtime paid at time and a half. The tax withheld each week is $195. They claim $25 a week for the uniform and earn $120 in interest for the year. (Assume 52.14 weeks in the year.)

a. Calculate the taxable income and Medicare levy.

b. Explain why the worker should expect to receive a tax refund. Support your explanation with calculations.

18. The fortnightly pay slip for an employee is shown.

Entitlements	Unit	Rate	Total
Wages for ordinary hours worked	76 hours	$20.95	$
Total ordinary hours = 76 hours			
Overtime	15 hours	$31.40	$
Travel allowance	350 km	$0.66/km	$
Gross payments			$

Deductions	
Taxation	$430
Total deductions	
Net payment	

a. Using a spreadsheet, calculate the fortnightly pay.

b. Assuming they receive the same fortnightly pay for the year, calculate their annual taxable income. Assume 26.07 fortnights in the year.

c. Determine if the employee will receive a tax refund or have to pay more in tax.

Extended response

19. A window cleaner has the opportunity to move from casual to part-time employment. Workers employed on a casual basis do not receive annual or sick leave. All employees in the cleaning business receive 9.5% superannuation.
The window cleaner's current casual hourly rate is $25.76 and the part-time hourly rate is $23.70. Determine if they should move to part-time employment.
Justify your answer by calculating their annual salary based on working 38 hours each week at the normal hourly rate and their annual leave loading.

20. Business groups are seeking a change to the taxation system. Their proposal is as follows:

Taxable income	Tax on this income
0–$29 000	Nil
$29 001–$85 000	35c for each $1 over $29 000
$85 001–$175 000	A plus 40c for each $1 over $85 000
$175 001	B plus 58c for each $1 over $175 000

a. Calculate the values of A and B.

b. Compare the tax payable for the following annual salaries under the new proposed system and the current taxation table.

   i. $25 890   **ii.** $45 870   **iii.** $67 950   **iv.** $81 940   **v.** $195 870

c. Based on your calculations from part **b**, explain how workers will be better or worse off under the proposed new taxation system.

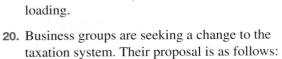

Hey teachers! Create custom assignments for this topic

Create and assign unique tests and exams

Access quarantined tests and assessments

Track your students' results

Find all this and MORE in jacPLUS ▶

Answers

Topic 11 Financial systems and income payments

11.2 Purchasing methods

11.2 Exercise

1. a. $185　　b. $288.60　　c. $116.55
2. C
3. $27.41
4. a. $46.42　　b. $10.32
5. $4312.44
6. $38.42
7. a. $176.94　　b. $1023.06
8. a. $15 460　　b. 11.56%
9. a. Store 1: $3850
   Store 2: $3960
   Store 3: $3740
   b. Store 2
10. D
11. Plumber C
12. a. $1308.49　　b. $1499.85
    c. $2087.38　　d. $3231.40
13. No, Elise loses $43.
14. a. $30 000　　b. $102.13
    c. $5102.13
15. a. $29 600　　b. 20.25%
    c. $2560　　d. 16.82%
16. a. 19 months　　b. $550.42

11.3 Financial systems — permanent employment

11.3 Exercise

1. The wage (hourly rate) is the greater pay ($3200 per fortnight compared to $2838.51).
2. B
3. a. $1476.31　　b. $1136.89　　c. $1596.83
4. 34.31 hours
5. $637.50
6. First fortnight: $1735.95
   Second fortnight: $1833.75
7. $144.08
8. a. $12 979.17
   b. $14 796.25 annually, $1233.02 monthly
9. a. $5166.67　　b. $490.83　　c. $5890
10. $1154.31
11. a. $2100　　b. $10 850
12. a. They will be worse off by 25% for every hour they work on Sunday but no worse off for working on Saturdays.
    b. 16.65%

13. a. $1270.90
    b. Their weekly salary would be $1304.18, which is higher than their hourly rate calculation ($1270.90). They should take the salary offer.
14. The employees should accept the new agreement. The new agreement is 15.5% more than the base salary compared to 10.8% more under the current agreement.
15. They will receive the largest amount over 5 years by choosing package 1 ($322 826.55 compared to $315 028.40).

11.4 Financial systems — taxation and deductions

11.4 Exercise

1. $7038.75
2. a. $1208　　b. $1546
   c. $1788　　d. $2168.46
3. D
4. a. $1890　　b. $1564
5. Gross: $1912.50
   Net: $1578.50
6. a. i. 19% on amounts over $18 201
      ii. $3667
   b. i. Nil　　　　　ii. $0
   c. i. 32.5% for amounts over $45 000 + 5 092
      ii. $18 309.75
   d. i. 37% on amounts over $120 000 + 29 467
      ii. $33 537
7. $704.32
8. a. $192　　b. $366　　c. $540　　d. $1086
9. a. 28.93%　　b. 30.4%　　c. 31.2%　　d. 33.95%
10. a. $4857　　b. $975　　c. 20.07%　　d. $1165.68
11. a. $85 970　　b. $18 407.25
12. a. $21.15
    Working 4 hours overtime equates to the same pay as working 6 hours normal time ($4 \times 1.5 = 6$). Calculate the number of 'normal' hours worked ($38 + 6 = 44$) and divide into wage.
    b. $45 091.80
    c. $5121.84
    d. $4733.24
    They won't receive $1980 as a tax refund because the tax payable is a percentage of their taxable income. Their taxable income was reduced by $1980 but the tax payable was reduced by 32.5% of 1980 (0.325×1980).
13. a. $2695.84
    b. $2695.84 − $606 = $2089.84
    c. $12 438.90
    d. $1352.12
14. a. $5614.50
    b. $14 400
    c. They will receive a refund of $2476.82.

15. a. When calculating tax using this tiered system, the taxable salary of $97 605 is split into the different tax-bracket amounts, which are found by the difference between the lower and upper limits, e.g. 45 000 − 18 200 = $26 800. See the table at the bottom of the page.*
This is much lower than 32.5% × $97 605 = $31 721.63.

b. The dentist's taxable salary falls into the third tax bracket, 32.5% for amounts over $45 000 plus $5092. The $5092 is the tax amount from the previous tax bracket (19% × $26 800), as shown.

16. a. Determine the tax bracket their taxable income falls within. Since they paid more than $5092 and less than $29 467, their taxable income is within the third tax bracket: $5092 + 32.5% × (taxable income − $45 000).

b. $74 907.69

c. Refund: $320.13

11.5 Income payments — commission, piecework and royalties

11.5 Exercise

1. $3390

2. a. $1650   **b.** $2375   **c.** $2920   **d.** $4160

3. C

4. $6825

5. $1398.25

6. B

7. $12.11

8. $186 989.67

9. A

10. a. See the table at the bottom of the page.*

b. $376 099

11. $45 634.53

12. $2 500 000

13. $16 599.48

14. The royalty percentage can be found by dividing the royalty amount by the total sale amount and then multiplying by 100. The songwriter's royalty percentage is 2.5%.

15. Option 1:
$15 \times 5 \times 8.25 = \$618.75$
Option 2:
$250 \times 75.95 \times 0.035 = \$664.56$
Option 2 is the better option.

16. a. $2090

b. 22

c. If they sell 22 new telephone plans each week, taking the salary option of $37 500 per annum would give them a higher weekly wage.

11.6 Income payments and calculations

11.6 Exercise

1. a. 3.33%   **b.** 2.68%   **c.** 12.97%   **d.** 1.99%

2. 2.51%

3. $1079.74

4. $5953.58

5. Stan = $7771.63, Bett = $11 655.38, Xiao = $3420.95, Mohamed = $14 650.25

6. Harry = $593.75, Theo = $572.00, Maria = $588.00, Marcia = $802.95

7. See the table at the bottom of the page.*

---

*15. a.

Tax bracket	Amount of salary ($)	Tax ($)	Tax paid ($)
0–$18 200	18 200	0	
$18 201–$45 000	26 800	19	0.19 × 26 800 = 5092
$45 001–$120 000	52 605	32.5	0.325 × 52 605 = 17 096.63
Total	**$97 605**		**Total = $22 188.63**

*10. a.

	Jan–Mar	Apr–Jun	Jul–Aug	Sept–Dec
Sales	$123 400	$2 403 556	$432 442	$84 562
Royalty payment	$3085.00	$60 088.90	$10 811.05	$2114.05

*7.

Employee	Pay rate ($)	Normal hours worked	Overtime 1.5	Penalty rate 1.5	Penalty rate 2	Allowance	Gross pay ($)	Tax with held ($)	Net pay ($)
Rex	15.85	28		7	7		832.13	124	708.13
Tank	22.15	35		5		15	956.38	167	789.38
Gert	30.10	32	1.5		8	19	1531.53	367	1164.53

8.

Entitlements	Unit	Rate	Total
Wages for ordinary hours worked	30 hours	$35.05	$1051.50
Total ordinary hours = 30 hours			
Penalty (double time)	5 hours	$70.10	$350.50
		Gross payment	**$1402.00**

Deductions		
Taxation		$325
	Total deductions	**$325**
	Net payment	**$1077.00**

Employer superannuation contribution	
Contribution	$133.19

9. a. $57 093.30   **b.** $54 634.74

10. Sara = $1074.04, Troy = $421.25, Helga = $675.00, Nina = $80.75, Bill = $86.25, Max = $138.60

11. The 'plus' calculations are found by adding the tax payable in the previous brackets.

12. See the table at the bottom of the page.*

13. The claim is incorrect. Although the teacher will pay $6171.47 less in tax, reducing their salary by 20% has an overall reduction of $18 922.20 in pay, so they will have $12 819.73 less pay.

14. a. 3.93%

   b. 7.83%

   c. Before pay increase = $60 718.89
      After pay increase = $62 995.85

15. a. A = $9555, B = $30 855, C = $51 205

   b. 9555 + 0.355 × (taxable salary − 65 000)

c. Workers who previously earned between $45 000 and $65 000 will pay less tax in the new system; however, workers previously earning between $65 000 and $120 000 will pay 3% more in tax. Workers who previously earned between $18 200 and $45 000 will pay 1% more in tax, while workers earning more than $180 000 will pay less. So on balance, the middle earners will pay less but low-income earners will pay more and high-income earners will pay less; therefore, it does not seem to be a fairer system.

11.7 Review

11.7 Exercise

Multiple choice

1. B
2. A
3. D
4. D
5. B
6. C
7. D
8. C
9. B
10. A

Short answer

11. **a.** $59 439.60   **b.** $5646.76   **c.** $1140

12. See the table at the bottom of the page.†

13. **a.** $557.10   **b.** $558.88

14. $283.69

15. **a.** $247.43   **b.** $359.90
    c. $368.08   **d.** $449.88

---

*12.

Hourly rate	Normal hours	Saturday 1.5	Sunday 2	Public holiday 2	Gross pay	PAYG tax	Net pay
21.5	5	8	9	7.5	$1075.00	$185.00	$890.00

†12.

	January	February	March	April
Gross sales	$45 000	$125 000	$320 000	$\frac{1508}{0.029} = \$52\,000$
Royalty	2.9% × $45 000 = 0.029 × $45 000 = $1305	2.9% × $125 000 = 0.029 × $125 000 = $3625	2.9% × $320 000 = 0.029 × $320 000 = $9280	$1508

16. See the table at the bottom of the page.*

17. a. $962.82

   b. They have paid $3091.60 more in tax than they are required to. Therefore, they will receive a tax refund.

18. a.

Entitlements	Unit	Rate	Total
Wages for ordinary hours worked	76	$20.95	$76 \times 20.95 = $1592.20
Total ordinary hours = 78 hours			
Overtime	15	$31.40	$31.40 \times 15 = $471
Travel allowance	350	$0.66	$350 \times 0.66 = $231.00
	Gross payment		1592.20 + 471 + 231 = $2294.20
Deductions			
Taxation		$430	
	Total deductions		**$430**
	Net payment		2294.20 − 430 = $1864.20

   b. $59 809.79

   c. Their PAYG tax is more than the required tax. Therefore, they will get a return of 11 210.10 − 9905.11 = $1304.99.

Extended response

19. Casual employment will give them more money overall ($52 048.64 compared to $51 599.99 including annual leave loading) based on working 38 hours each week for 52.14 weeks. However, they will not have sick or annual leave. Considering taking annual leave, **wo**rking part-time would be the better deal. They will also have sick leave and more job security.

20. a. A: $19 600
       B: $55 600

b. See the table at the bottom of the page.*

c. Under the proposed taxation system, lower income earners will pay less in tax due to an increase in the upper limit of the tax-free threshold from $18 200 to $25 890. Higher income earners will pay more due to a higher percentage tax rate (58% compared to 45% currently). Middle income workers will pay slightly more under the proposed system due to an increase in the tax rate (35% compared to 19% and 32.5% under the current system).

---

*16.

Employee	Saturday	Sunday	Monday (public holiday)	Gross wage
Rate	$27.75	$37	$37	
Tran	$5	$8		$5 \times 27.75 + 8 \times 37 = $434.75
Gus		$6		$6 \times 37 = $222
Meg	$3.5		$7	$3.5 \times 27.75 + 7 \times 37 = $356.13
Warren	$4		$6	$4 \times 27.75 + 6 \times 37 = $333

*20. b.

	Taxable income	Tax payable (current)	Tax payable (proposed)
i	$25 890	$0.19 \times (25\,890 − 18\,200) = $1461.10	$0
ii	$45 870	$5092 + 0.325 \times (45\,870 − 45\,000) = $5374.75	$0.35 \times (45\,870 − 29\,000) = $5904.50
iii	$67 950	$5092 + 0.325 \times (67\,950 − 45\,000) = $12\,550.75	$0.35 \times (67\,950 − 29\,000) = $13\,632.50
iv	$81 940	$5092 + 0.325 \times (81\,940 − 45\,000) = $17\,097.50	$0.35 \times (81\,940 − 29\,000) = $18\,529
v	$195 870	$51\,667 + 0.45 \times (195\,870 − 180\,000) = $58\,808.50	$55\,600 + 0.58 \times (195\,870 − 175\,000) = $67\,704.60

12 Length and area

LEARNING SEQUENCE

Fully worked solutions for this topic are available online.

12.1 Overview

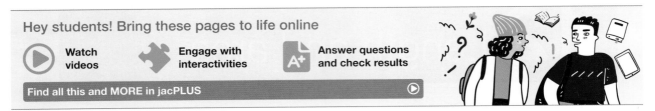
12.1.1 Introduction

We all use measurement every day, often without even realising that we're doing so. For example, if you estimate the time that it'll take you to finish a Mathematics exercise, that's a measure of time, and if you count the number of steps you take to walk to the milk bar, that's a measure of distance. Understanding measurement allows us to understand the world around us, from the size of bacteria to the time it would take us to travel to the stars.

Calculating the perimeter and area of two-dimensional shapes is one of the oldest problems in mathematics, with the ancient Egyptians and Babylonians discovering various methods to solve problems involving perimeter and area. The knowledge of calculating the perimeters and areas of shapes is applied in several fields such as mapping, architecture, and surveying. Likewise, by knowing how to calculate the surface area of shapes, we're able to know the amount of fabric we need to make a tent or how much paint we need to paint a new living room.

KEY CONCEPTS

This topic covers the following key concepts from the VCE Mathematics Study Design:
- standard metric units and measures, including common derived metric measures
- reading and interpretation of scales on digital and analogue instruments
- estimation and approximation strategies.

Source: VCE Mathematics Study Design (2023–2027) extracts © VCAA; reproduced by permission.

12.2 SI units, length and estimation

LEARNING INTENTION

At the end of this subtopic you should be able to:
- identify the International System of Units (SI)
- convert between different units of length
- estimate length using different types of measuring tools.

12.2.1 International System of Units (SI)

The International System of Units (SI) was introduced so that a consistent unit of measurement is used in all countries. There are seven basic SI units, but the most familiar are the first four in the following table.

International System of Units (SI)		
Physical quantity	**Abbreviation**	**Unit**
Length	**m**	**metre**
Time	**s**	**second**
Mass	**kg**	**kilogram**
Temperature	**K**	**kelvin**
Current	A	ampere
Luminous intensity	cd	candela
Amount of substance	mol	mole

Using these basic SI units, many other units can be formed. A few of these are:
- area – m^2 (metres squared)
- volume – m^3 (metres cubed)
- capacity – l (litres).

Units of length

Units of length are used to describe the dimensions of an object, such as its length, width and height, or the distance between two points. The metric system is based on the number 10. The base unit of length in the metric system is the metre.

The following figures show the most commonly used units of length, together with their abbreviations and approximate examples.

1. **Kilometre (km)**

The distance travelled in one minute by a car travelling at the speed limit of 60 kilometres per hour

2. **Metre (m)**

The length of an adult's step while walking

3. Centimetre (cm)

The width of each of your fingers

4. Millimetre (mm)

The width of a wire in this computer chip

When measuring lengths, you should use a unit that gives a reasonable value (not too large and not too small).

Converting between units of length

The relationship between the metric units of length can be used to convert a measurement from one unit to another.

The main metric units of length are related as follows:

$$1\,\text{km} = 1000\,\text{m} \quad 1\,\text{m} = 100\,\text{cm} \quad 1\,\text{cm} = 10\,\text{mm}$$

Unit conversion for length

Units of length can be converted as shown in the following diagram. The numbers next to each arrow are called conversion factors.

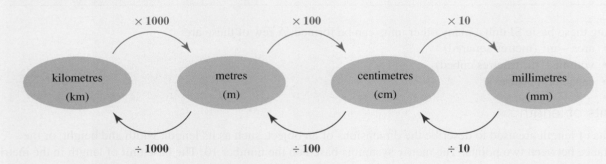

Note: **When converting from a larger unit to a smaller unit, multiply by the conversion factor. When converting from a smaller unit to a larger unit, divide by the conversion factor.**

---

tlvd-4941

WORKED EXAMPLE 1 Converting between units of length

Convert the following lengths to the units shown.
a. 0.739 km = _____ m **b. 53 250 mm = _____ m**

THINK	WRITE
a. 1. To convert from km to m, multiply by 1000.	a. 0.739×1000
2. When multiplying by 1000, move the decimal point 3 places to the right. Make sure the new unit is stated.	$= 739$ Thus: $0.739\,\text{km} = 739\,\text{m}$

b.
1. To convert from mm to m, divide by 1000.
2. When dividing by 1000, move the decimal point 3 places to the left. Make sure the new unit is stated.

b. $53\,250 \div 1000$
$= 53.25$
Thus:
$53\,250 \text{ mm} = 53.25 \text{ m}$

WORKED EXAMPLE 2 Converting and adding units of length

Calculate the value of 7 m 36 cm + 5 m 45 cm.

THINK

1. Convert both values to metres by dividing the centimetre part of the measurement by 100 and adding it to the metre part of the measurement.

2. Add the values in metres.

3. Write the answer using the correct unit.

WRITE

$7 \text{ m } 36 \text{ cm} = 7 + 36 \div 100$
$= 7 + 0.36$
$= 7.36 \text{ m}$
$5 \text{ m } 45 \text{ cm} = 5 + 45 \div 100$
$= 5 + 0.45$
$= 5.45 \text{ m}$

$7.36 + 5.45 = 12.81$

12.81 m

12.2.2 Measuring tools

A variety of tools can be used to help measure length.
- A ruler can be used to measure small objects.
- A tape measure can be used to measure longer objects or distances.
- A car's odometer can record long measurements, such as the distance between two towns.
- A picture with a scale, such as a map or a microscope drawing, can be used to measure very large or very small lengths.

When using a ruler or a tape measure to measure length, it is important to ensure that the zero line of the scale markings is always placed at the start of the length being measured.

Other measuring tools are:

Thermometer — to measure temperature

Scales — to measure mass

Stopwatch — to measure time

tlvd-4942

WORKED EXAMPLE 3 Determining the measurement marked on a ruler by the arrow

State the measurement marked by the arrow in each of the following. Record your answer in the unit indicated in the brackets.

a.
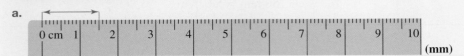
(mm)

b.

(cm)

THINK	WRITE
a. 1. From left to right, count how many centimetres the arrow has passed.	**a.** 1 cm
2. Multiply that number by 10 to account for the number of mm in each cm (1 cm = 10 mm).	$1 \times 10 = 10$ mm
3. Add the extra millimetres (mm).	$10 + 5 = 15$ mm
b. 1. Count the number of centimetres the arrow has passed. The whole numbers represent the number of cm.	**b.** 2 cm
2. The extra 4 mm represent $\dfrac{4}{10}$ cm or 0.4 cm.	0.4 cm
3. Add the whole and part cm. Write the answer.	$2 + 0.4 = 2.4$ cm

 Resources

 Interactivity Converting units of length (int-4011)

12.2 Exercise

Students, these questions are even better in jacPLUS

 Receive immediate feedback and access sample responses

 Access additional questions

 Track your results and progress

Find all this and MORE in jacPLUS

1. **MC** The SI unit for mass is:
   A. litre.   B. gram.   C. metre.   D. kilogram.   E. pound.

2. **MC** Select which of the following is *not* one of the seven basic SI units.
   A. Kelvin   B. Kilometre   C. Second   D. Gram   E. Kilogram

3. **MC** Select which of the following is a measurement tool used to measure length.
   A. Stopwatch   B. Scales   C. Tape measure   D. Thermometer   E. Ammeter

4. **MC** Select which of the following would *not* measure one of the seven basic SI units.
   A. Thermometer   B. Scales   C. Measuring jug   D. Tape measure   E. Stopwatch

5. Select the most suitable metric unit of length to measure the following distances.

   a. Height of the tallest building

   b. Size of the television

   c. The length of the Yarra River

   d. The width of the cable connector

6. **WE1** Convert the following lengths to the units shown.

   a. 0.283 km = _____ m
   b. 520 cm = _____ m
   c. 0.418 m = _____ cm
   d. 450 m = _____ km
   e. $5\dfrac{1}{2}$ km = _____ m
   f. 78 459 mm = _____ m

7. **MC** The length 375 cm expressed in millimetres is equal to:

   A. 0.375 mm   B. 3.75 mm   C. 37.5 mm   D. 3750 mm   E. 0.0375 mm

8. **MC** The distance 95.6 kilometres expressed in metres is equal to:

   A. 0.956 m   B. 956 m   C. 9560 m   D. 95 600 m   E. 95.60 m

9. **MC** The length 875 mm is equivalent to:

   A. 0.875 m   B. 0.0875 m   C. 0.875 km   D. 0.008 75 km   E. 8.75 m

10. Convert the following lengths into the units shown.

    a. 0.27 km = _____ m   b. 73 500 mm = _____ m   c. 257 mm = _____ cm   d. 1.3 km = _____ cm

11. **WE2** Calculate the value of each of the following.

    a. 5 m 67 cm + 9 m 51 cm
    b. 2 m 88 cm + 3 m 20 cm

12. **WE3** State the measurement marked by the arrow in each of the following. Record your answer in the unit indicated in brackets.

a.

(cm)

b.

(mm)

c.

(mm)

d.

(cm)

13. For each of the following line segments:

    i. estimate its length
    ii. use a ruler to measure its length
    iii. comment on the accuracy of your estimate.

    a. _____
    b. _____
    c. _____
    d. _____
    e. _____
    f. _____

14. a. State whether the measurements you obtained by using a ruler in question **13** are exact or approximate.
    b. Explain how you can be sure that the ruler you used is accurate.

15. List three situations in which you would use each of the following measuring tools in everyday life.

    a. A 30-cm ruler that includes millimetre markings
    b. A 100-cm ruler that has only centimetre markings
    c. A 2-m dressmaker's tape that includes centimetre markings
    d. A 100-m measuring tape that has metre and centimetre markings

16. Explain why it would not be wise to use millimetres to measure the distance between Melbourne and Perth.

17. The combined height of two students is 3.62 m. If one student is 1.85 m tall, calculate the height of the other student in centimetres.

18. Arrange each of the following in ascending order.

   a. 0.15 km, 135 m, 2400 cm

   b. 25 cm, 120 mm, 0.5 m

   c. 9 m, 10 000 mm, 0.45 km

   d. 32 000 cm, 1200 m, 1 km

19. Calculate the value of each of the following in centimetres.

   a. 15 mm + 5 mm   **b.** 1.5 m + 40 mm

   c. 995 mm + 1.2 m   **d.** 5.67 cm + 1156 mm − 0.25 m

20. Calculate the value of each of the following.

   a. 15 cm 15 mm + 27 cm 86 mm   **b.** 3 km 65 m + 79 km 38 m

   c. 66 cm 15 mm − 26 cm 5 mm   **d.** 125 m 49 cm − 37 m 79 cm

21. The heights of Melbourne's second tallest building, Eureka Tower, and Queensland's tallest building, Q1 Tower, measure 0.297 km and 322.5 m respectively.

   a. Name the taller building.

   b. Calculate the height difference, in metres, between the buildings.

   c. Calculate the sum of the heights of both buildings, in metres.

22. Consider the graph of the average height of the players in each of the AFL clubs in centimetres.

Average height per club

Club	Average height (cm)
Gold Coast 1	188.61
Melbourne 2	188.54
Collingwood 3	188.48
Fremantle 4	188.33
Richmond 5	188.27
GWS 6	188.25
Carlton 7	188.25
Brisbane 8	188.06
Port Adelaide 9	187.93
Geelong 10	187.65
North Melbourne 11	187.59
West Coast 12	187.50
Western Bulldogs 13	187.33
Sydney 14	187.20
Hawthorn 15	187.13
St Kilda 16	187.00
Essendon 17	186.94
Adelaide 18	186.40

Determine:

   a. the average height of Collingwood players, in metres

   b. the difference between the average height of Carlton and Sydney players, in centimetres

   c. the difference between the tallest and smallest teams' average heights, in centimetres

   d. the league's average height

   e. West Coast's average height in millimetres

   f. two clubs whose difference in average height is 5 mm.

12.3 Perimeters of familiar and composite shapes

LEARNING INTENTION

At the end of this subtopic you should be able to:
- calculate the perimeter of a given shape
- apply Pythagoras' theorem to calculate the perimeter of a shape
- calculate the circumference of a circle.

12.3.1 Perimeter

The **perimeter** of a plane (flat) figure is the distance around the outside of the figure. If the figure has straight edges, the perimeter can be found by simply adding all the side lengths. Ensure that all lengths are in the same unit.

Formula for the perimeter of a shape

To calculate the perimeter of a shape, change all lengths to the same unit and add them.

$$\text{perimeter} = a + b + c + d + e$$

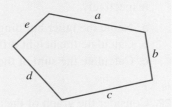

WORKED EXAMPLE 4 Calculating the perimeter of a shape

Calculate the perimeters of the following shapes.

a.
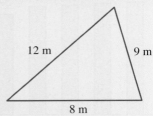

b.
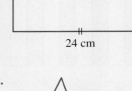

c.
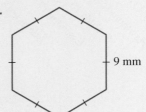

d.
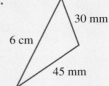

THINK

a. 1. Add the side lengths of the triangle.

2. Write the answer as a sentence.

WRITE

a. $P = 12 + 9 + 8$
$\quad = 29$

The perimeter is 29 m.

b. 1. Add the side lengths of the rectangle.

b. $P = 24 + 15 + 24 + 15$
$= 78$

2. Write the answer as a sentence.

The perimeter is 78 cm.

c. 1. Add the side lengths of the hexagon.

c. $P = 9 + 9 + 9 + 9 + 9 + 9$
$= 54$

2. Write the answer as a sentence.

The perimeter is 54 mm.

d. 1. Convert all lengths into the same unit.

d. 6 cm = 60 mm

2. Add the side lengths of the triangle, in millimetres.

$P = 60 + 30 + 45$
$= 135$

3. Write the answer as a sentence.

The perimeter is 135 mm.

12.3.2 Pythagoras' theorem

Pythagoras' theorem can be used to calculate an unknown side length of a right-angled triangle if two of the sides are known.

Pythagoras' theorem

Pythagoras' theorem states that:

$$c^2 = a^2 + b^2$$

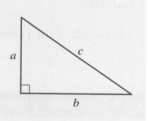

tlvd-4943

WORKED EXAMPLE 5 Calculating the perimeter of a shape using Pythagoras' theorem

Calculate the perimeter of the triangle.

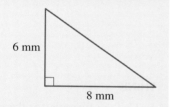

THINK

1. To calculate the perimeter, we need to know all the side lengths. We can use Pythagoras' theorem to determine the unknown side of the right-angled triangle.

2. Substitute the known values into the formula, remembering that c represents the hypotenuse, then solve for c. Take the square root of both sides to undo the square.

WRITE

$c^2 = a^2 + b^2$

$c^2 = a^2 + b^2$
$c^2 = 6^2 + 8^2$
$c^2 = 36 + 64$
$c^2 = 100$
$c = \sqrt{100}$
$c = 10$ mm

3. To calculate the perimeter, add the side lengths.

$$P = 10 + 6 + 8$$
$$= 24$$

4. Write the answer.

The perimeter is 24 mm.

12.3.3 Circumference

The **circumference** (C) is the term used for the perimeter of a circle.

Formula for the circumference of a circle

The circumference of a circle is given by the formula:

$$C = 2\pi r$$
or
$$C = \pi D$$

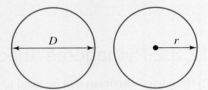

The diameter (D) of a circle is the distance from one side of the circle to the other, passing through the centre. The radius (r) of a circle is the distance from the centre of the circle to the outside, and it is half the length of the diameter.

The diameter is twice the length of the radius. That is, $D = 2r$ or $r = \dfrac{D}{2}$.

WORKED EXAMPLE 6 Calculating the circumference of a circle to 2 decimal places

Calculate the circumference of the following circle to 2 decimal places.

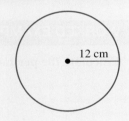

12 cm

THINK	WRITE
1. Use the formula for the circumference of a circle in terms of the radius.	$C = 2\pi r$
2. Substitute the radius into the equation and solve for C using the π key on your calculator.	$C = 2 \times \pi \times 12$ $= 75.3982$ ≈ 75.40 cm
3. Write the answer to 2 decimal places.	The circumference is 75.40 cm.

12.3.4 Perimeter of a sector

An **arc** is part of the circumference of a circle.

An **arc length** of a circle is calculated by determining the circumference of a circle and multiplying by the fraction of the angle that it forms at the centre of the circle.

Formula for arc length

The length of an arc in a circle can be calculated using the formula:

$$\text{arc length, } l = \frac{\theta}{360} \times 2\pi r$$

$$= \frac{\theta}{180}\pi r$$

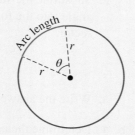

where r = radius of the circle and θ = the angle that the ends of the arc make with the centre of the circle (in degrees).

A sector is the portion of a circle enclosed by an arc and two radii.

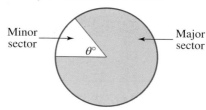

The perimeter of a sector is found by adding the length of the arc and the lengths of two radii.

Formula for the perimeter of a sector

The perimeter of a sector can be calculated using the formula:

$$\text{perimeter of a sector} = \text{arc length} + 2r$$

$$= \frac{\theta}{180}\pi r + 2r$$

where r = the radius of the circle and θ = the angle the ends of the arc make with the centre of the circle (in degrees).

WORKED EXAMPLE 7 Calculating the perimeter of a sector to 2 decimal places

Calculate the perimeter of each of the following shapes to 2 decimal places.

a.

2.4 cm

b.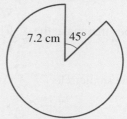

7.2 cm   45°

THINK

a. 1. Determine the angle that the ends of the arc make with the centre.

WRITE

a. $\theta = 360° - 90°$
  $= 270°$

2. Substitute the values for the radius and the angle into the formula for the perimeter of a sector.	Perimeter $= \dfrac{\theta}{180} \times \pi r + 2r$ $= \dfrac{270}{180} \times \pi \times 2.4 + 2 \times 2.4$ $= 11.3097... + 4.8$ ≈ 16.11 cm
3. Write the answer as a sentence.	Perimeter $= 16.11$ cm
b. 1. Determine the angle the ends of the arc make with the centre.	**b.** $\theta = 360° - 45°$ $= 315°$
2. Substitute the values for the radius and the angle into the formula for the perimeter of a sector.	Perimeter $= \dfrac{\theta}{180} \times \pi r + 2r$ $= \dfrac{315}{180} \times \pi \times 7.2 + 2 \times 7.2$ $= 39.5841... + 14.4$ ≈ 53.98 cm
3. Write the answer as a sentence.	Perimeter $= 53.98$ cm

12.3.5 Perimeter of composite shapes

The perimeter of a **composite shape** is the same as a regular perimeter; it is the total distance around the outside of a shape.

WORKED EXAMPLE 8 Calculating the perimeter of a composite shape to 2 decimal places

Calculate the perimeter of the following shape to 2 decimal places.

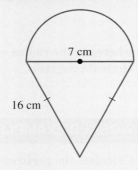

THINK	WRITE
1. Determine the different shapes that make the composite shape.	The shape is made up of a semicircle and two lengths of 16 cm.
2. Calculate the arc length of the semicircle. Recall that $r = \dfrac{D}{2}$.	Arc length $= \dfrac{180}{180} \times \pi \times \dfrac{7}{2}$ $= 10.995\,574$ cm
3. Calculate the perimeter of the shape by adding all three sides.	$P = 10.995\,574 + 16 + 16$ $= 42.995\,574$ cm ≈ 43.00 cm
4. Write the answer as a sentence.	The perimeter is 43.00 cm.

12.3 Exercise

Students, these questions are even better in jacPLUS

Receive immediate feedback and access sample responses

Access additional questions

Track your results and progress

Find all this and MORE in jacPLUS

1. Calculate the perimeter of each of the following rectangles.

a.

8 cm

b.

10 cm

6 cm

c.
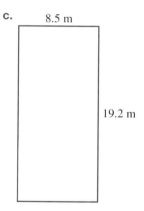
8.5 m

19.2 m

d.
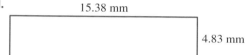
15.38 mm

4.83 mm

2. Calculate the perimeter of each of the following triangles.

a.
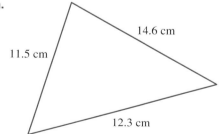
14.6 cm

11.5 cm

12.3 cm

b.
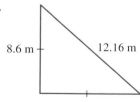
8.6 m   12.16 m

c.

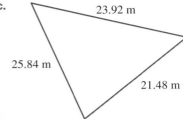

23.92 m

25.84 m

21.48 m

d.

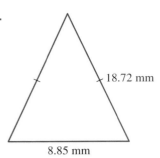

18.72 mm

8.85 mm

3. Calculate the perimeter of each of the following shapes.

a.

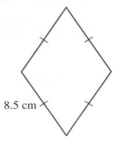

8.5 cm

b.

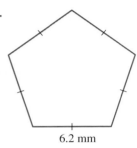

6.2 mm

c.

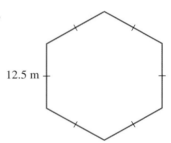

12.5 m

d.

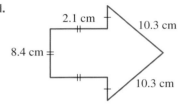

2.1 cm

10.3 cm

8.4 cm

10.3 cm

4. **WE4** Calculate the perimeter of the each of following shapes.

a.

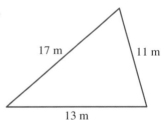

17 m

11 m

13 m

b.

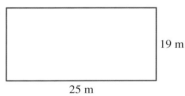

19 m

25 m

c.

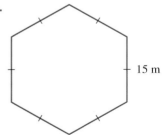

15 m

d.

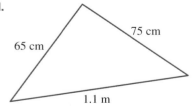

75 cm

65 cm

1.1 m

5. Determine the perimeter of the following.

   a. A square table of length 1.2 metres

   b. A rectangular yard 23 m by 37 m

   c. An octagon of side length 25 cm

6. Calculate the perimeter of each of the following shapes.

a.

85 cm

2.6 m

b.

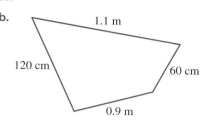

1.1 m

120 cm

60 cm

0.9 m

7. **WE5** Calculate the perimeter of each of the following shapes, correct to 2 decimal places where necessary.

a.

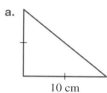

10 cm

b.

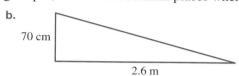

70 cm

2.6 m

8. **WE6** Calculate the circumference of each of the following circles, correct to 2 decimal places.

a.

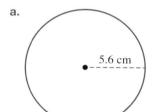

5.6 cm

b.

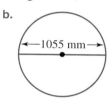

1055 mm

9. **WE7** Calculate the perimeter of each of the following shapes, correct to 2 decimal places.

a.

12 cm

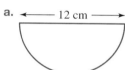

b.

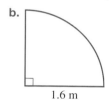

1.6 m

c.

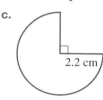

2.2 cm

10. Calculate the perimeter of each of the following shapes to 2 decimal places.

a.

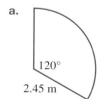

120°

2.45 m

b.

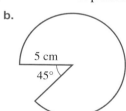

5 cm

45°

11. Calculate the perimeter of each of the following shapes to 2 decimal places.

a.

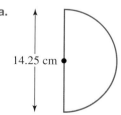

14.25 cm

b.

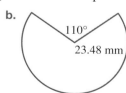

110°

23.48 mm

12. Calculate the perimeter of each of the following shapes to 2 decimal places.

a.

115°
2.3 cm

b.

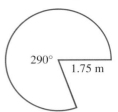

290°   1.75 m

c.

8.7 m
80°

d.

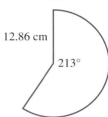

12.86 cm
213°

13. Calculate the perimeter of each of the following shapes, correct to 2 decimal places where necessary.

a.

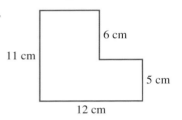

6 cm
11 cm
5 cm
12 cm

b.

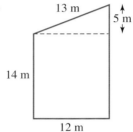

13 m
5 m
14 m
12 m

c.

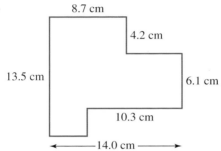

8.7 cm
4.2 cm
13.5 cm
6.1 cm
10.3 cm
14.0 cm

d.

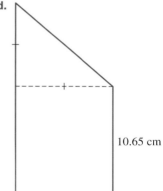

10.65 cm
9.56 cm

e.

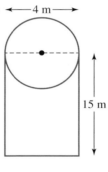

4 m
15 m

f.

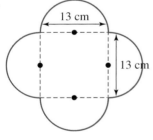

13 cm
13 cm

14. **WE8** Calculate the perimeter of the following shape, to 2 decimal places.

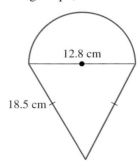

12.8 cm
18.5 cm

15. Calculate the perimeter of the following shape, to 1 decimal place.

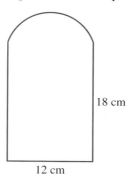

18 cm

12 cm

16. Calculate the radius (in centimetres) of a trundle wheel that travels one metre in one revolution.

17. The minute hand on a clock is 22 centimetres in length. Calculate the distance travelled by the tip of the hand in 3 hours.

18. The straight lengths of a running track measure 80 metres. If the half-circles at either end of the track are equal in length, answer the following questions, giving your answers correct to 2 decimal places.

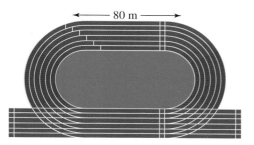

— 80 m —

 a. Calculate the radius of the end semicircles if the inside length of the track is exactly 400 metres in length.

 b. Determine how much further in front of the runner in lane 1 must the runner in lane 2 start to ensure both runners complete 400 metres, if each lane is 1 metre in width. Assume each runner runs in the centre of the lane.

12.4 Units of area and estimation

LEARNING INTENTION

At the end of this subtopic you should be able to:
- convert area units
- estimate the area of an irregular shape
- calculate the area of a rectangle.

12.4.1 Conversion of area units

The **area** of a two-dimensional shape is the amount of space enclosed by the shape. A square centimetre is the amount of space enclosed by a square with a length 1 cm and width 1 cm.

Some common metric units of area and their abbreviations are square kilometres (km^2), square metres (m^2), square centimetres (cm^2) and square millimetres (mm^2).

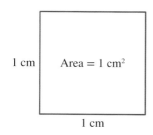

1 cm   Area = 1 cm^2

1 cm

When converting area units, you need to take into account that the shape is two-dimensional, so the units will be squared.

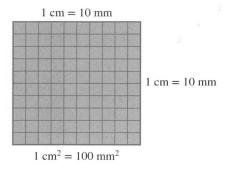

1 cm = 10 mm

1 cm = 10 mm

$1 \text{ cm}^2 = 100 \text{ mm}^2$

To convert area units:
- follow the conversion of length
- square the conversion.

Consider, for example, converting 1 cm^2 to mm^2.

To convert from cm to mm, multiply by 10. Since area is two-dimensional, we need to square the conversion, so multiply by $10^2 = 100$.

Therefore, $1 \text{ cm}^2 = 1 \times 100 \text{ mm}^2$
$= 100 \text{ mm}^2$

Unit conversion for area
The following diagram shows how to convert between metric units of area.

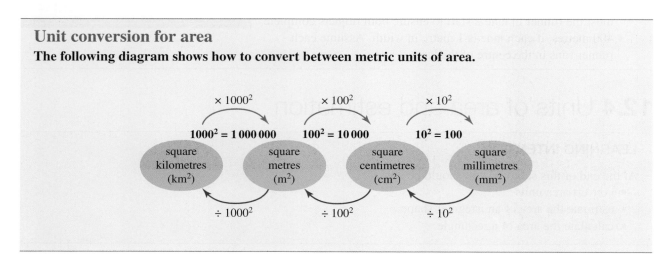

For example, to convert 54 km^2 to cm^2:

$$54 \text{ km}^2 = 54 \times 1000^2 \times 100^2$$
$$= 540\,000\,000\,000 \text{ cm}^2$$

Another common unit is the **hectare** (ha), a $100 \text{ m} \times 100 \text{ m}$ square equal to $10\,000 \text{ m}^2$, which is used to measure small areas of land.

Hectares

$1 \text{ ha} = 10\,000 \text{ m}^2$

$100 \text{ ha} = 1 \text{ km}^2$

Convert the following units of area.

a. $56\,cm^2 = $ _____ mm^2 b. $78\,700\,cm^2 = $ _____ m^2 c. $591\,ha = $ _____ km^2

THINK	WRITE
a. 1. To convert from cm^2 to mm^2, multiply by 10^2.	a. $56\,cm^2 = 56 \times 10^2\,mm^2$
2. Complete the calculation and write the answer.	$= 56 \times 100\,mm^2$ $= 5600\,mm^2$
b. 1. To convert from cm^2 to m^2, divide by 100^2.	b. $78\,700\,cm^2 = 78\,700 \div 100^2\,m^2$
2. Complete the calculation and write the answer.	$= 78\,700 \div 10\,000\,m^2$ $= 7.87\,m^2$
c. 1. To convert from ha to km^2, divide by 100.	c. $591\,ha = 591 \div 100\,km^2$
2. Complete the calculation and write the answer.	$= 5.91\,km^2$

12.4.2 Area of a rectangle

The area of a rectangle can be calculated by multiplying its length by its width.

Formula for the area of a rectangle

The area of a rectangle can be calculated using the formula:

$$A = l \times w$$

where $l = $ length and $w = $ width.

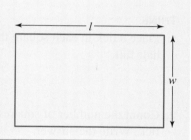

Calculate the area of the following shape in:
a. cm^2
b. m^2.

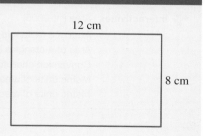

THINK	WRITE
a. 1. The shape is a rectangle, so use the formula for the area of a rectangle.	a. $A = l \times w$

2. Substitute $l = 12$ cm and $w = 8$ cm.

$$A = 12 \times 8$$
$$= 96$$

3. Write the answer.

$$\text{Area} = 96 \text{ cm}^2$$

b. 1. To convert cm^2 to m^2, divide by 100^2.

b. $96 \text{ cm}^2 = 96 \div 100^2 \text{ m}^2$
$$= 0.0096 \text{ m}^2$$

2. Write the answer.

$$\text{Area} = 0.0096 \text{ m}^2$$

12.4.3 Estimation of area

Follow the steps below to estimate the area of a shape where a square is not completely filled by the object.
- If more than half of the square is covered, count it in the area.
- If less than half of the square is covered, don't count it in the area.

WORKED EXAMPLE 11 Estimating the area of a shape

Estimate the area of the following shape, giving your answer in square units.

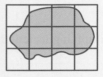

THINK

1. Place a tick in each square that is more than half full.

2. Count the number of squares that contain a tick. Write the answer.

WRITE

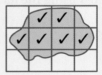

The area is approximately 6 square units.

 Resources

Interactivities Area (int-0005)
Area of a rectangle (int-4017)
Area of rectangles (int-3784)
Conversion chart for area (int-3783)
Metric units of area 1 (int-4015)
Metric units of area 2 (int-4016)

1. Name the most appropriate unit for measuring each of the following.
   a. Area of a floor rug
   b. Area of a fingerprint
   c. Area of a page from a novel
   d. Area of a painted wall in the lounge room
   e. Area of the surface of Uluru
   f. Area of a five-cent coin

2. Determine which metric unit of area you would use to measure each of the following areas.
   a. The palm of your hand
   b. A suburban house block
   c. A dairy farm
   d. Your fingernail
   e. Australia
   f. Your local football ground

3. Provide two examples of area (that have not already been mentioned) that could be measured in:
   a. mm^2
   b. cm^2
   c. m^2
   d. km^2.

4. **WE9** Convert the following units of area.
   a. $34\ cm^2 =$ _____ mm^2
   b. $14\,250\ cm^2 =$ _____ m^2
   c. $592\ ha =$ _____ km^2

5. Convert the following units of area.
   a. $0.18\ m^2 =$ _____ cm^2
   b. $0.0798\ km^2 =$ _____ m^2
   c. $374\,300\ m^2 =$ _____ ha

6. Convert each of the following measurements to the unit shown in brackets.
   a. $2.3\ cm^2\ \left(mm^2\right)$
   b. $2.57\ m^2\ \left(cm^2\right)$
   c. $470\ mm^2\ \left(cm^2\right)$
   d. $27\,000\ m^2\ \left(km^2\right)$
   e. $87\,500\ m^2\ (ha)$
   f. $17\,000\ cm^2\ \left(m^2\right)$

7. Convert each of the following measurements to the unit shown in the brackets.
   a. $51\,200\ m^2\ (ha)$
   b. $2.85\ ha\ \left(m^2\right)$
   c. $8380\ ha\ \left(km^2\right)$
   d. $12.6\ km^2\ (ha)$
   e. $0.23\ ha\ \left(m^2\right)$
   f. $623\,450\ m^2\ (ha)$

8. Convert each of the following measurements to the unit shown in the brackets.
   a. $0.048\ m^2\ \left(mm^2\right)$
   b. $0.0012\ km^2\ \left(cm^2\right)$
   c. $300\,800\ mm^2\ \left(m^2\right)$
   d. $4478\ cm^2\ \left(km^2\right)$

9. If each square is $1\ cm^2$, determine the area of each of the following shapes.
   a.
   b.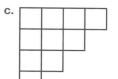
   c.
   d.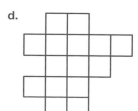

10. Calculate the area of each of the following shapes.

a.

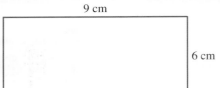

9 cm

6 cm

b.

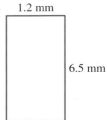

1.2 mm

6.5 mm

c.

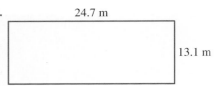

24.7 m

13.1 m

d.

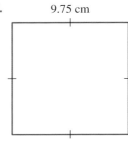

9.75 cm

11. `WE10` Calculate the area of each of the following shapes in the specified unit.

i. cm^2

ii. m^2

a.

2.36 m

1.45 m

b.

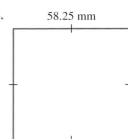

58.25 mm

c.

2.4 m

85 cm

d.

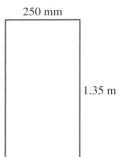

250 mm

1.35 m

12. `WE11` Estimate the area of each of the following shapes (giving your answer in units of squares).

a.

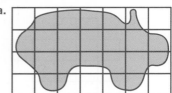

b.

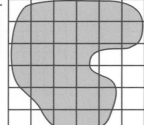

c.

13. Estimate the area of the banner held up by the club members. Use the squares shown in the diagram to help determine the area from one end to the other where each square represents 1 square metre.

14. Estimate the area of the following, by first estimating their dimensions.

    a. A singles tennis court
    b. A basketball court
    c. A netball court
    d. A standard door

15. If a cricket pitch has an area of 60.36 m² and it is 3 metres wide, calculate the length of the cricket pitch.

16. A rectangular photograph measuring 8 cm by 10 cm is enlarged so that its dimensions are doubled.

    a. Calculate the area of the original photograph.
    b. Calculate the dimensions of the new photograph.
    c. Calculate the area of the new photograph.
    d. Compare the area of the enlarged photo to the original photo and complete the following sentence.
    If the side lengths of a rectangle are doubled, its area is

    _____.

17. A rectangular flowerbed measures 25 m by 14 m. A gravel path 2 m wide surrounds it.

    a. Draw a diagram representing the flowerbed and path.
    b. Calculate the area of the flowerbed.
    c. Calculate the area of the gravel path.
    d. If gravel costs $7.50 per square metre, calculate the cost to cover the path.

18. A wall in a house is to be painted. The wall is rectangular and measures 9.5 metres by 3.2 m. The wall has two rectangular windows, 2.4 metres by 80 cm.

    a. Calculate the area of the wall to be painted in m².
    b. If the wall needs three coats, detemine how much paint is required to paint the wall in m².
    c. If 1 litre of paint covers 5 m², calculate the amount of paint required to paint the whole wall, to the nearest litre.
    d. If the paint costs $15.75 per litre, calculate the cost of the paint.

12.5 Areas of regular shapes

> **LEARNING INTENTION**
>
> At the end of this subtopic you should be able to:
> - calculate the area of a square, rectangle, triangle, trapezium and parallelogram
> - calculate the area of a circle and sector.

12.5.1 Areas of two-dimensional shapes

The areas of two-dimensional shapes can be calculated using the following formulas.

Shape	Diagram	Area formula
Square		$\begin{aligned} A_{\text{square}} &= l \times l \\ &= l^2 \end{aligned}$
Rectangle		$A_{\text{rectangle}} = l \times w$
Triangle		$A_{\text{triangle}} = \dfrac{1}{2} \times b \times h$
Trapezium		$A_{\text{trapezium}} = \dfrac{1}{2}(a+b) \times h$
Parallelogram		$A_{\text{parallelogram}} = b \times h$

tlvd-4945

WORKED EXAMPLE 12 Calculating the area of a shape

Calculate the area of each shape.

a.

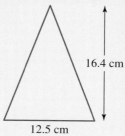

16.4 cm

12.5 cm

b.

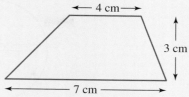

4 cm

3 cm

7 cm

THINK

a. 1. Identify the shape as a triangle and use the area of a triangle formula.

2. Substitute $b = 12.5$ cm and $h = 16.4$ cm.

3. Write the answer.

b. 1. Identify the shape as a trapezium and use the area of a trapezium formula.

2. Substitute $a = 4$ cm, $b = 7$ cm and $h = 3$ cm.

3. Write the answer.

WRITE

a. $A_{\text{triangle}} = \dfrac{1}{2} \times b \times h$

$A_{\text{triangle}} = \dfrac{1}{2} \times 12.5 \times 16.4$

$= 102.5 \text{ cm}^2$

The area is 102.5 cm^2.

b. $A_{\text{trapezium}} = \dfrac{1}{2}(a + b) \times h$

$A_{\text{trapezium}} = \dfrac{1}{2}(4 + 7) \times 3$

$= \dfrac{1}{2}(11) \times 3$

$= \dfrac{33}{2}$

$= 16.5 \text{ cm}^2$

The area is 16.5 cm^2.

12.5.2 Area of a circle

The area of a circle can be calculated using the following formula.

Formula for the area of a circle

The area of a circle can be calculated using the formula:

$$A_{\text{circle}} = \pi \times r^2,$$

where r = radius of the circle.

Note that $r = \dfrac{D}{2}$.

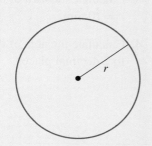

WORKED EXAMPLE 13 Calculating the area of a circle to 2 decimal places

Calculate the area of the following circle correct to 2 decimal places.

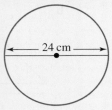

24 cm

THINK

1. Write the formula for the area of a circle.

2. Determine the radius of the circle.

3. Substitute $r = 12$ into the formula for the area of a circle and evaluate.

WRITE

$A_{circle} = \pi \times r^2$

The radius is half the length of the diameter, so:
$$r = \frac{24}{2}$$
$$= 12 \text{ cm}$$

$A_{circle} = \pi \times 12^2$
$= 144\pi$
$\approx 452.39 \text{ cm}^2$

---

tlvd-4946

WORKED EXAMPLE 14 Calculating the area of the shaded region to 2 decimal places

Calculate the area of the shaded region to 2 decimal places.

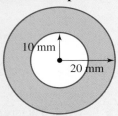

10 mm

20 mm

THINK

1. The shaded region is the area of the larger circle minus the area of the smaller circle.

2. Calculate the area of the larger circle with $r = 20$ mm.

3. Calculate the area of the smaller circle with $r = 10$ mm.

4. Calculate the area of the shaded region to 2 decimal places.

WRITE

$A_{shaded} = A_{large} - A_{small}$

$A_{large} = \pi \times (20)^2$
$= 400\pi$

$A_{small} = \pi \times (10)^2$
$= 100\pi$

$A_{shaded} = 400\pi - 100\pi$
$= 300\pi$
$\approx 942.48 \text{ mm}^2$

12.5.3 Area of a sector

The area of a sector can be calculated by finding the area of the circle and multiplying it by the fraction of the angle that the sector forms at the centre of the circle.

Formula for the area of a sector

The area of a sector can be calculated using the formula:

$$\text{area of a sector} = \frac{\theta}{360} \times \pi r^2$$

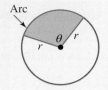

where:

r = radius of the circle
θ = angle that the sector makes with the centre of the circle, in degrees.

WORKED EXAMPLE 15 Calculating the area of a sector to 2 decimal places

Calculate the area of the shaded region to 2 decimal places.

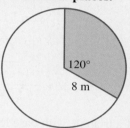

THINK	WRITE
1. Write the formula for the area of a sector.	Area of a sector $= \dfrac{\theta}{360} \times \pi r^2$
2. Substitute $r = 8$ cm and $\theta = 120°$.	Area of a sector $= \dfrac{120}{360} \times \pi \times 8^2$ $= 67.020\,643$ $\approx 67.02 \text{ cm}^2$
3. Write the answer.	The shaded area is 67.02 cm^2.

---

on Resources

Interactivities Area of circles (int-3788)
Area of parallelograms (int-3786)
Area of trapeziums 1 (int-3789)
Area of trapeziums 2 (int-3790)
Area of a sector (int-6076)
Area of a circle (int-4441)
Area of trapeziums (int-4442)

12.5 Exercise

1. Calculate the area of each of the following triangles.

   a.
   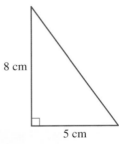
   8 cm
   5 cm

   b.
   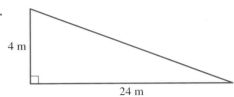
   4 m
   24 m

   c.
   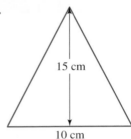
   15 cm
   10 cm

   d.
   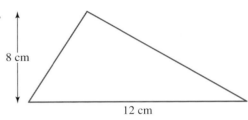
   8 cm
   12 cm

2. **MC** Select from the following the area of a triangle with base length of 12 cm and vertical height of 12 cm.

   A. 144 cm^2 **B.** 100 cm^2 **C.** 72 cm^2 **D.** 64 cm^2 **E.** 132 cm^2

3. Calculate the area of the following trapeziums.

   a.
   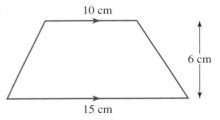
   10 cm
   6 cm
   15 cm

   b.
   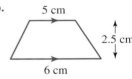
   5 cm
   2.5 cm
   6 cm

   c.
   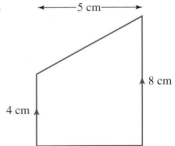
   5 cm
   8 cm
   4 cm

   d.
   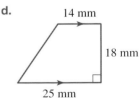
   14 mm
   18 mm
   25 mm

4. **MC** Select from the following the area of a parallelogram of length 25 cm and vertical height of 5 cm.

   A. 30 cm^2 **B.** 100 cm^2 **C.** 62.5 cm^2 **D.** 125 cm^2 **E.** 225 cm^2

5. Calculate the area of the following parallelograms.

a.

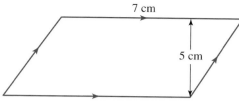

7 cm

5 cm

b.

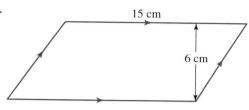

15 cm

6 cm

c.

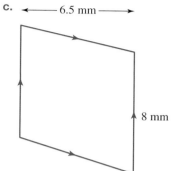

← 6.5 mm →

8 mm

d.

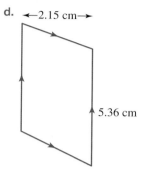

←2.15 cm→

5.36 cm

6. **WE12** Calculate the area of the following shapes.

a.

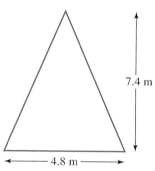

7.4 m

← 4.8 m →

b.

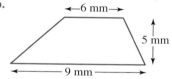

←6 mm→

5 mm

← 9 mm →

7. **WE13** Calculate the area of the following circles to 2 decimal places.

a.

5 cm

b.

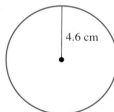

4.6 cm

c.

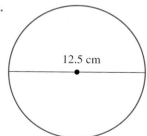

12.5 cm

d.

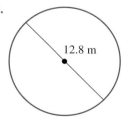

12.8 m

8. **WE14** Calculate the area of the shaded region to 2 decimal places.

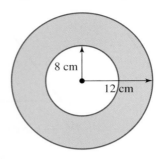

9. Calculate the area of the shaded region to 2 decimal places.

a.

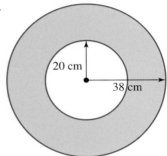

b.

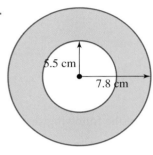

10. **WE15** Calculate the area of the shaded region to 2 decimal places.

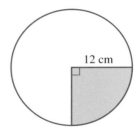

11. Calculate the area of the shaded region to 2 decimal places.

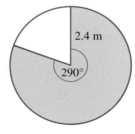

12. Calculate the area of the shaded region to 2 decimal places.

a.

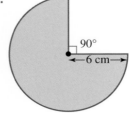

b.

7.2 m

140°

13. **MC** The area of a circle is equal to 113.14 mm². The radius of the circle is closest to:
    A. 36 mm
    B. 6 mm
    C. 2133 mm
    D. 7 mm
    E. 56.57 mm

14. Circular pizza trays come in three different sizes: small, medium and large.

    a. Calculate the area of each tray if the diameters are 20 cm, 30 cm and 40 cm respectively.
    b. Determine the areas of 50 trays each of the three sizes.
    c. A slice from the large pizza makes an angle from the centre of 45°. Calculate the area of the pizza slice.

15. A rugby pitch is rectangular and measures 100 m in length and 68 m in width.

    a. Calculate the area of the pitch.
    b. If the pitch was laid with instant turf, with each sheet measuring 4 m by 50 cm, calculate how many sheets of turf are required to cover the pitch.
    c. If each sheet costs $10.50, calculate the cost to cover the pitch.

16. A school is looking to build four netball courts side by side. A netball court measures 15.25 m wide by 30.5 m long, and a 3-m strip is required between each court and around the outside.

    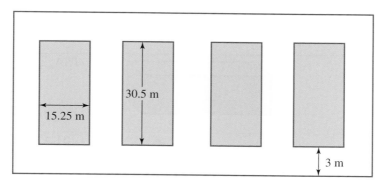

    a. Determine the area of each netball court.
    b. Calculate the total area, in metres, that the school would need for the four netball courts and their surrounding space.
    c. If the four netball courts are painted, determine the area that needs to be painted.
    d. If it costs $9.50 per square metre to paint the netball courts, calculate the cost to paint the four courts.

12.6 Surface areas of familiar prisms

LEARNING INTENTION

At the end of this subtopic you should be able to:
- calculate the total surface area of a prism
- calculate the total surface area of a cylinder and sphere.

12.6.1 Total surface area of a prism

The **total surface area** (TSA) of a three-dimensional (3D) object is the total area of each outer face of that object.

The total surface area of a 3D object can be determined by calculating the area of each individual face and adding these together.

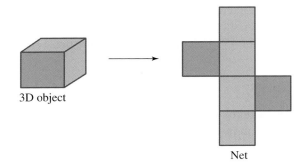

3D object

Net

Drawing the net of a solid can help you determine the individual shapes of the faces that make up the solid.

As with area, total surface area is measured using **squared units** (mm^2, cm^2, m^2, or km^2).

The total surface area of some common 3D objects can be calculated using appropriate formulas.

A **prism** is a 3D shape with identical opposite ends joined by straight edges. It can be sliced into identical slices called cross-sections.

Prism	Diagram	Total surface area formula
Cube		$TSA_{cube} = 6l^2$
Rectangular prism		$TSA_{rectangular\ prism} = 2(lh + lw + wh)$
Triangular prism		$TSA_{triangular\ prism} = bh + 2ls + bl$

Calculate the total surface area of the following rectangular prism.

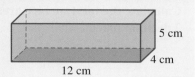

5 cm

4 cm

12 cm

THINK	WRITE
1. Identify the object as a rectangular prism, and write the appropriate total surface area formula.	$\text{TSA}_{\text{rectangular prism}} = 2(lh + lw + wh)$
2. Identify the dimensions of the object.	$l = 12$, $w = 4$ and $h = 5$
3. Substitute the length, width and height values into the TSA formula and calculate the TSA of the rectangular prism.	$\begin{aligned}\text{TSA}_{\text{rectangular prism}} &= 2(12 \times 5 + 12 \times 4 + 4 \times 5) \\ &= 2(60 + 48 + 20) \\ &= 2(128) \\ &= 256\end{aligned}$
4. Write the answer.	The total surface area of the rectangular prism is 256 cm^2.

12.6.2 Total surface area of a cylinder

Drawing the net of a cylinder can help determine the formula for the total surface area.

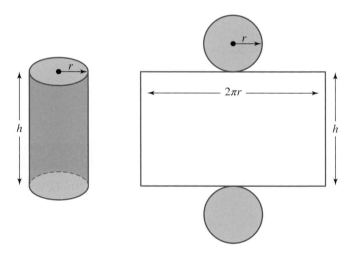

A cylinder is made up of 2 circles and a curved surface (which becomes a rectangle when rolled out and flattened).

Formula for the total surface area of a cylinder

The formula for the total surface area of a cylinder is:

$$\begin{aligned}\text{TSA}_{\text{cylinder}} &= 2\pi rh + 2\pi r^2 \\ &= 2\pi r(r + h)\end{aligned}$$

tlvd-4947

WORKED EXAMPLE 17 Calculating the total surface area of a cylinder

Calculate the total surface area of the cylinder shown correct to
2 decimal places.

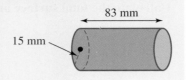

THINK	WRITE
1. Identify the object as a cylinder, and write the appropriate total surface area formula.	$\text{TSA}_{\text{cylinder}} = 2\pi rh + 2\pi r^2$
2. Identify the radius and the height of the cylinder.	$r = 15, \ h = 83$
3. Substitute the value of the radius and height into the TSA formula and calculate the TSA of the cylinder.	$\begin{aligned} \text{TSA}_{\text{cylinder}} &= 2 \times \pi \times 15 \times 83 + 2 \times \pi \times 15^2 \\ &= 2490\pi + 450\pi \\ &= 2940\pi \\ &\approx 9236.282\,401\,55 \end{aligned}$
4. Write the answer.	The total surface area of the cylinder is 9236.28 mm^2.

12.6.3 Total surface area of a sphere

A sphere is a round solid figure, such as a tennis ball, soccer ball or basketball.

Formula for the total surface area of a sphere

The formula for the total surface area of a sphere is:

$$\text{TSA}_{\text{sphere}} = 4\pi r^2$$

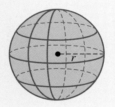

WORKED EXAMPLE 18 Calculating the total surface area of a sphere

Calculate the total surface area of a basketball with a radius of 15 cm. Give your answer correct to
2 decimal places.

THINK	WRITE
1. Identify the basketball as a sphere, and write the appropriate total surface area formula.	$\text{TSA}_{\text{sphere}} = 4\pi r^2$
2. Write the radius of the sphere.	$r = 15$
3. Substitute the value of the radius into the TSA formula and calculate the TSA of the basketball.	$\begin{aligned} \text{TSA}_{\text{sphere}} &= 4 \times \pi \times 15^2 \\ &= 4 \times \pi \times 225 \\ &= 900\pi \\ &= 2827.433\,388 \\ &\approx 2827.43 \text{ cm}^2 \end{aligned}$
4. Write the answer.	The total surface area of the basketball is 2827.43 cm^2.

 Resources

12.6 Exercise

1. State which of the shapes below are prisms.

a.

b.

c.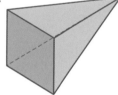

d.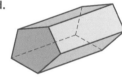

2. Calculate the total surface area of the cubes with the following side lengths.
   a. 4 cm
   b. 12.5 mm
   c. 0.85 m
   d. $7\frac{3}{4}$ cm

3. **WE16** Calculate the total surface area of this rectangular prism.

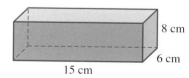

8 cm

6 cm

15 cm

4. Calculate the total surface area of each of the following rectangular prisms.

a.

33 cm
48 cm
72 cm

b.

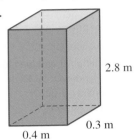

2.8 m
0.3 m
0.4 m

c.

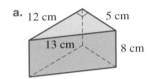

4 mm
15.5 mm
27 mm

d.

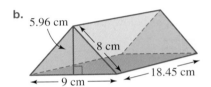

0.8 cm
2.8 cm
3.4 cm

5. For each of the following triangular prisms, calculate:

    i. the area of the triangular face
    ii. the area of each of the three rectangular faces
    iii. the total surface area.

a.

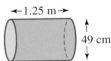

12 cm   5 cm
13 cm   8 cm

b.

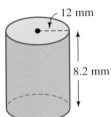

5.96 cm
8 cm
9 cm
18.45 cm

6. **WE17** Calculate the total surface area of each of the following cylinders, correct to 2 decimal places.

a.

←1.25 m→
49 cm

b.

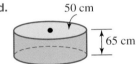

12 mm
8.2 mm

c.

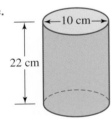

←10 cm→
22 cm

d.

50 cm
65 cm

7. Calculate the total surface area of this cylinder to 2 decimal places.

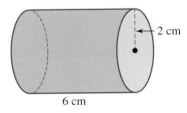

2 cm
6 cm

8. Calculate the side length of a cube with a total surface area of 384 m².

9. Calculate the total surface area of the following sphere, correct to 2 decimal places.

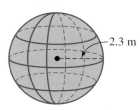

10. Calculate the total surface area of the following shapes, correct to 2 decimal places.

a.

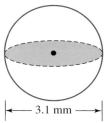

|← 3.1 mm →|

b.

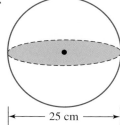

|← 25 cm →|

11. **WE18** Calculate the total surface area of a tennis ball with a diameter of 6 cm, correct to 2 decimal places.

12. **MC** A sphere has a total surface area of 125 cm^2. Calculate the radius of the sphere correct to 2 decimal places.

   A. 3.15 cm   **B.** 9.94 cm   **C.** 31.5 cm   **D.** 31.25 cm   **E.** 30.25 cm

13. Calculate the total surface area of the tin of tennis balls shown, correct to 2 decimal places, in:

   a. mm^2
   b. cm^2.

14. Calculate the surface area of the following shape, correct to 2 decimal places.

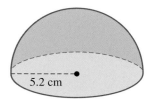

5.2 cm

12.7 Review

doc-38053

12.7.1 Summary

Hey students! Now that it's time to revise this topic, go online to:

 Access the topic summary

 Review your results

 Watch teacher-led videos

Practise questions with immediate feedback

Find all this and MORE in jacPLUS

12.7 Exercise

Multiple choice

1. **MC** The perimeter of the shape shown is closest to:
   - A. 174.35 cm
   - B. 208 cm
   - C. 3.10 m
   - D. 308 cm
   - E. 140.35 cm

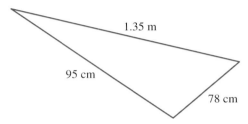

2. **MC** The perimeter of the shape shown, correct to 2 decimal places, is:
   - A. 10.80 cm
   - B. 22.62 cm
   - C. 24.16 cm
   - D. 16.96 cm
   - E. 19.66 cm

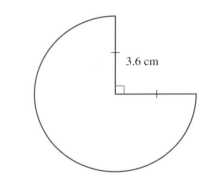

3. **MC** The area of the rectangle shown is:
   - A. 47.36 cm^2
   - B. 4.736 cm^2
   - C. 8.82 cm^2
   - D. 4.367 mm^2
   - E. 34.36 mm^2

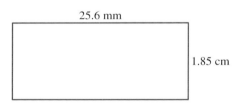

4. **MC** The area of the shape shown is:
   - A. 33.4 cm^2
   - B. 69.7 cm^2
   - C. 139.4 cm^2
   - D. 48.175 cm^2
   - E. 56.8 cm^2

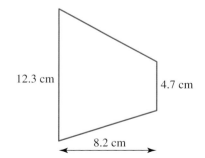

5. **MC** The area of the shape shown to 2 decimal places is:

A. 162.86 mm²

B. 155.52 mm²

C. 27.14 mm²

D. 135.72 mm²

E. 22.62 mm²

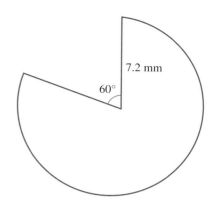

6. **MC** The area of the shape shown is closest to:

A. 97.474 cm²

B. 225.543 cm²

C. 218.237 cm²

D. 209.56 cm²

E. 194.948 cm²

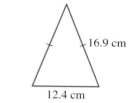

7. **MC** The total surface area of the rectangular prism shown is:

A. 221.45 cm²

B. 94.5 cm²

C. 255.55 cm²

D. 271.66 cm²

E. 244.24 cm²

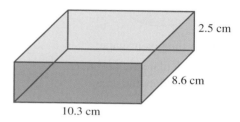

8. **MC** The total surface area of a sphere with a diameter of 5.35 m, to 2 decimal places, is closest to:

A. 359.68 m²

B. 22.48 m²

C. 89.92 m²

D. 33.62 m²

E. 19.38 m²

9. **MC** The area of the following shape, correct to 2 decimal places, is:

A. 37.38 cm²

B. 18.69 cm²

C. 33.18 cm²

D. 74.48 cm²

E. 33.78 cm²

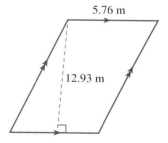

10. **MC** The area of the following shape, correct to 2 decimal places, is:

A. 15.17 cm²

B. 58.86 cm²

C. 130.32 cm²

D. 189.18 cm²

E. 198.18 cm²

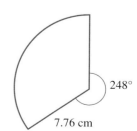

Short answer

11. Calculate the area of the following shape to 3 decimal places.

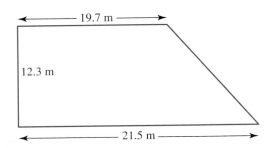

12. Calculate the total surface area of the following shapes to 2 decimal places.

a.

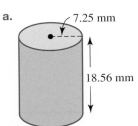

b.

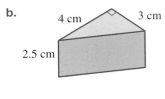

13. For the given shape:
   a. calculate the perimeter
   b. calculate the area.

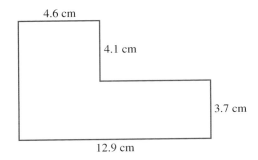

14. Calculate the total surface area of the following shape to 3 decimal places.

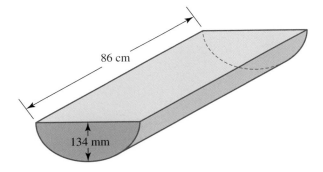

Extended response

15. A cube of length 18 cm has a hole with a 6-cm diameter drilled through its centre. Calculate the total surface area of the new shape to 2 decimal places.

16. A rectangular garden bed has a length of 15.35 m and width 9.52 m. An 85-cm wide path is built on the outside of the garden bed, reducing its size.
   a. Calculate the perimeter of the garden bed before the path was built.
   b. Calculate the perimeter of the garden bed after the path is built.
   c. Determine the area of the garden bed once the path is built.

17. Tennis is played on a rectangular court, with dimensions as shown.

    a. Determine the length and width of the tennis court.
    b. Calculate the perimeter of the tennis court.
    c. To mark all the lines shown with tape on the court, calculate the length of tape required.
    d. Determine the area of the tennis court.

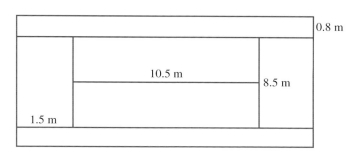

18. A rectangular prism can be made from the following net.

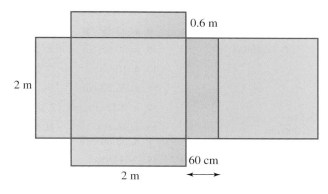

Calculate:

    a. the perimeter of the net
    b. the total surface area of the prism.

19. An Olympic swimming pool is 50 metres long, 25 metres wide and 1.8 metres deep.

    a. If Haylee swam 25 laps of the pool, determine how many metres she has swum.
    b. If you were to swim the 400-metre medley, calculate the number of laps you would swim.
    c. If you were to swim the longest Olympic pool event of 1500 metres, determine how many laps you would complete.
    d. Determine the surface area of the pool in m^2.
    e. Determine the surface area of the pool in cm^2.

20. A tin contains four tennis balls, each with a diameter of 6.3 cm. The tin has a height of 27.2 cm and a diameter of 73 mm.

    a. Determine the circumference of the lid.
    b. Calculate the total surface area of the outside of the tin.
    c. Calculate the total surface area of the four tennis balls.
    d. Determine the difference between the total surface area of the four balls and that of the tin.

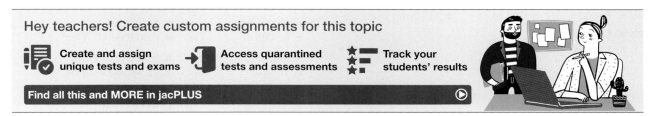

Answers

Topic 12 Length and area

12.2 SI units, length and estimation

12.2 Exercise

1. D
2. D
3. C
4. C
5. a. Metres   b. Centimetres
   c. Kilometres   d. Millimetres
6. a. 283 m   b. 5.2 m   c. 41.8 cm
   d. 0.45 km   e. 5500 m   f. 78.459 m
7. D
8. D
9. A
10. a. 270 m   b. 73.5 m
    c. 25.7 cm   d. 130 000 cm
11. a. 15.18 m   b. 6.08 m
12. a. 2 cm   b. 95 mm   c. 64 mm   d. 10 cm
13. a. ii. 50 mm   b. ii. 70 mm   c. ii. 31 mm
    d. ii. 80 mm   e. ii. 89 mm   f. ii. 24 mm
14. a. Approximate, since there is always a level of uncertainty
    in measuring
    b. Check it against a known length.
15. Answers will vary. Sample response is provided here.
    a. Measuring dimensions off a plan
    b. Drawing a line
    c. Measuring a person's waist size
    d. Measuring garden dimensions
16. Since the trip from Melbourne to Perth is a long way, you
    would usually use a large unit such as kilometres.
17. 177 cm
18. a. 2400 cm, 135m, 0.15 km
    b. 120 mm, 25 cm, 0.5 m
    c. 9 m, 10 000 mm, 0.45 km
    d. 32 000 cm, 1 km, 1 200 m
19. a. 2.0 cm   b. 154 cm
    c. 219.5 cm   d. 96.27 cm
20. a. 52 cm 1 mm or 52.1 cm
    b. 82 km 103 m or 82.103 km
    c. 41 cm
    d. 87 m 70 cm or 87.70 m
21. a. The Q1   b. 25.5 m   c. 619.5 m
22. a. 1.88 m
    b. 1.05 cm
    c. 2.21 cm
    d. 187.75 cm
    e. 1875 mm
    f. West Coast (1875 mm) and St Kilda (1870 mm)

12.3 Perimeters of familiar and composite shapes

12.3 Exercise

1. a. 32 cm   b. 32 cm   c. 55.4 m   d. 40.42 mm
2. a. 38.4 cm   b. 29.36 m   c. 71.24 m   d. 46.29 mm
3. a. 34 cm   b. 31 mm   c. 75 m   d. 50 cm
4. a. 41 m   b. 88 m   c. 90 cm   d. 250 cm
5. a. 4.8 m   b. 120 m   c. 200 cm
6. a. 690 cm   b. 3.8 m
7. a. 34.14 cm   b. 5.99 m
8. a. 35.19 cm   b. 3314.38 mm
9. a. 30.85 cm   b. 5.71 m   c. 14.77 cm
10. a. 10.03 m   b. 37.49 cm
11. a. 36.63 cm   b. 149.41 mm
12. a. 9.22 cm   b. 12.36 m   c. 29.55 cm   d. 73.53 cm
13. a. 46 cm   b. 58 m   c. 55 cm
    d. 53.94 cm   e. 40.28 m   f. 81.68 cm
14. 57.11 cm
15. 66.8 cm
16. 15.92 cm
17. 414.69 cm
18. a. 38.20 m   b. 6.30 m

12.4 Units of area and estimation

12.4 Exercise

1. a. m^2 b. mm^2 c. cm^2
   d. m^2 e. km^2 f. mm^2
2. a. cm^2 b. m^2 c. ha
   d. mm^2 e. km^2 f. m^2
3. There are many possible answers. Some examples are
   given.
   a. A computer chip or face of a small diamond
   b. A fridge magnet or dinner plate
   c. A block of land or surface of a concrete slab
   d. Tasmania or Kakadu National Park
4. a. $3400 \, mm^2$ b. $1.4250 \, m^2$ c. $5.92 \, km^2$
5. a. $1800 \, cm^2$ b. $79\,800 \, m^2$ c. 37.43 ha
6. a. $230 \, mm^2$ b. $25\,700 \, cm^2$ c. $4.7 \, cm^2$
   d. $0.027 \, km^2$ e. 8.75 ha   f. $1.7 \, m^2$
7. a. 5.12 ha   b. $28\,500 \, m^2$ c. $83.8 \, km^2$
   d. 1260 ha   e. $2300 \, m^2$ f. 62.345 ha
8. a. $48\,000 \, mm^2$ b. $12\,000\,000 \, cm^2$
   c. $0.3008 \, m^2$ d. $0.000\,000\,447\,8 \, km^2$
9. a. 5 squares = $5 \, cm^2$ b. 9 squares = $9 \, cm^2$
   c. 10 squares = $10 \, cm^2$ d. 15 squares = $15 \, cm^2$
10. a. $54 \, cm^2$ b. $7.8 \, mm^2$
    c. $323.57 \, m^2$ d. $95.0625 \, cm^2$

11. a. i. 34 220 cm^2 ii. 3.422 m^2

     b. i. 33.93 cm^2 ii. 0.003 393 m^2

     c. i. 20 400 cm^2 ii. 2.04 m^2

     d. i. 3375 cm^2 ii. 0.3375 m^2

12. a. 18 squares   b. 25 squares   c. 15 squares

13. 84 m^2

14. Answers will vary. Examples are:

     a. 200 m^2 b. 450 m^2 c. 450 m^2 d. 2 m^2

15. 20.12 m

16. a. 80 cm^2

     b. Width = 16 cm; length = 20 cm

     c. 320 cm^2

     d. Multiplied by 4 or quadrupled

17. a.

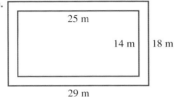

     b. 350 m^2

     c. 172 m^2

     d. $1290

18. a. 26.56 m^2 b. 79.68 m^2 c. 16 litres   d. $252

12.5 Areas of regular shapes

12.5 Exercise

1. a. 20 cm^2 b. 48 m^2 c. 75 cm^2 d. 48 cm^2

2. C

3. a. 75 cm^2 b. 13.75 cm^2 c. 30 cm^2 d. 351 mm^2

4. D

5. a. 35 cm^2 b. 90 cm^2 c. 52 mm^2 d. 11.524 cm^2

6. a. 17.76 m^2 b. 37.5 mm^2

7. a. 78.54 cm^2 b. 66.48 cm^2

     c. 122.72 m^2 d. 128.68 m^2

8. 251.33 cm^2

9. a. 3279.82 cm^2 b. 96.10 cm^2

10. 113.10 cm^2

11. 14.58 m^2

12. a. 84.82 cm^2 b. 99.53 m^2

13. B

14. a. Small = 314.16 cm^2; medium = 706.86 cm^2;

       large = 1256.64 cm^2

     b. 113 883 cm^2

     c. 157.08 cm^2

15. a. 6800 m^2

     b. 3400 sheets of turf

     c. $35 700

16. a. 465.125 m^2 b. 2774 m^2

     c. 1860.5 m^2 d. $17 674.75

12.6 Surface areas of familiar prisms

12.6 Exercise

1. a. Prism   b. Not a prism

     c. Not a prism   d. Prism

2. a. 96 cm^2 b. 937.5 mm^2

     c. 4.335 m^2 d. 360.375 cm^2

3. 516 cm^2

4. a. 14 832 cm^2 b. 4.16 m^2

     c. 1177 mm^2 d. 28.96 cm^2

5. a. i. 30 cm^2

       ii. 104 cm^2, 40 cm^2, 96 cm^2

       iii. 300 cm^2

     b. i. 26.82 cm^2

       ii. 147.6 cm^2, 147.6 cm^2, 166.05 cm^2

       iii. 514.89 cm^2

6. a. 2.30 m^2 b. 1523.04 mm^2

     c. 848.23 cm^2 d. 36 128.32 cm^2

7. 100.53 cm^2

8. 8 m

9. 66.48 m^2

10. a. 30.19 mm^2 b. 1963.50 cm^2

11. 113.10 cm^2

12. A

13. a. 72 452.98 mm^2 b. 724.5298 cm^2

14. 254.85 cm^2

12.7 Review

12.7 Exercise

Multiple choice

1. D   2. C   3. B   4. B   5. D

6. A   7. D   8. C   9. D   10. B

Short answer

11. 253.38 m^2

12. a. 1175.73 cm^2 b. 42 cm^2

13. a. 41.4 cm

     b. 66.59 cm^2

14. 0.65 m^2

Extended response

16. a. 49.74 m   b. 42.94 m   c. 106.743 m^2

17. a. 13.5 m and 10.1 m   b. 47.2 m

     c. 101.7 m   d. 136.35 m^2

18. a. 16.8 m   b. 12.8 m^2

19. a. 1250 m   b. 8 laps   c. 30 laps

     d. 1520 m^2 e. 15 200 000 cm^2

20. a. 22.93 cm   b. 707.50 cm^2

     c. 498.76 cm^2 d. 208.74 cm^2

21. 2226.74 cm^2

13 Volume, capacity and mass

LEARNING SEQUENCE

Fully worked solutions for this topic are available online.

13.1 Overview

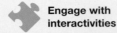

13.1.1 Introduction

People must measure! How much paint will you need to repaint your bedroom? How many litres of water will it take to fill the new pool? How many tiles do you need to order to retile the bathroom walls? How far is it from the North Pole to the South Pole? These are just a few examples where measurement skills are needed. Measurements in three dimensions include volumes of objects and capacity, such as the number of millilitres in a can of soft drink.

Measuring tools have advanced significantly in their capability to measure extremely small and extremely large amounts and objects, leading to many breakthroughs in medicine, engineering, science, architecture and astronomy.

In architecture, not all buildings are simple rectangular prisms. In our cities and towns, you will see buildings that are cylindrical in shape, buildings with domes, and even buildings that are hexagonal or octagonal in shape. Architects, engineers and builders all understand the relationships between these various shapes and how they are connected. Industrial and interior designers use the properties of plane figures, prisms, pyramids and spheres in various aspects of their work.

Have you ever wondered why tennis balls are sold in cylindrical containers? This is an example of manufacturers wanting to minimise the amount of waste in packaging. Understanding the concepts involved in calculating the surface area and volume of the common shapes we see around us is beneficial in many real-life situations.

KEY CONCEPTS

This topic covers the following key concepts from the VCE Mathematics Study Design:
- standard metric units and measures, including common derived metric measures
- estimation and approximation strategies

Source: VCE Mathematics Study Design (2023–2027) extracts © VCAA; reproduced by permission.

13.2 Volume of prisms and capacity

13.2.1 Volume of a right prism

The **volume** of a 3-dimensional object is the amount of space it takes up.

Volume is measured in units of mm^3, cm^3 and m^3.

Converting units of volume

The following chart is useful when converting between units of volume.

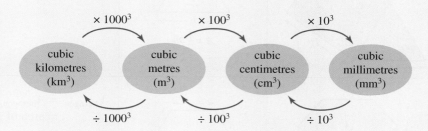

For example, $3\, m^3 = 3 \times 100^3 \times 10^3\, mm^3$
$$= 3\,000\,000\,000\, mm^3$$

Formula for the volume of a right prism

The volume of a prism can be calculated using the formula:

$$V_{\text{prism}} = A_{\text{base}} \times H$$

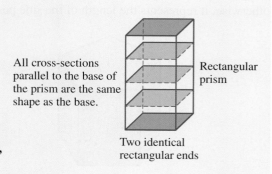

All cross-sections parallel to the base of the prism are the same shape as the base.

Rectangular prism

Two identical rectangular ends

where A is the area of the cross-section of the prism and H is the dimension that is at a right angle to the cross-section, often referred to as the height.

Shape	Diagram	Cross-section area	Volume formula
Cube		$A = l^2$	$\begin{aligned} V_{\text{square}} &= A_{\text{square}} \times H \\ &= \text{area of a square} \times \text{height} \\ &= l^2 \times l \\ &= l^3 \end{aligned}$
Rectangular prism		$A = lw$	$\begin{aligned} V_{\text{rectangle}} &= A_{\text{rectangle}} \times H \\ &= \text{area of a rectangle} \times \text{height} \\ &= lw \times H \end{aligned}$
Triangular prism		$A = \dfrac{1}{2}bh$	$\begin{aligned} V_{\text{triangle}} &= A_{\text{triangle}} \times H \\ &= \text{area of a triangle} \times \text{height} \\ &= \dfrac{1}{2}bh \times H \end{aligned}$
Hexagonal prism		$A = \dfrac{3\sqrt{3}}{2} \times a^2$	$\begin{aligned} V_{\text{hexagon}} &= A_{\text{hexagon}} \times H \\ &= \text{area of a hexagon} \times \text{height} \\ &= \dfrac{3\sqrt{3}}{2}a^2 \times H \end{aligned}$

If a prism is standing on a side that has the same shape as its cross-section, then H represents its height; otherwise, it represents the length of the side perpendicular to the base.

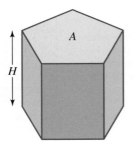

This pentagonal prism is standing on its cross-section, so H represents the height of the prism.

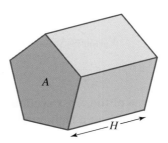

This is the same prism, but it is lying on its side. In this diagram, H represents the prism's length.

tlvd-4948

Calculate the volume of the following 3-dimensional shape.

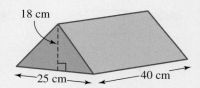

18 cm

25 cm — 40 cm

THINK

1. Write the formula for the volume of a prism.

2. This object has a triangular cross-section, so A represents the area of the triangle. Write the formula for the area of the triangle.

3. Identify the values of the pronumerals. (*Note:* Here, h is the height of the triangle, not of the prism.)

4. Substitute the values of the pronumerals into the formula and evaluate.

5. State the value of H. Since the prism is not standing on its triangular end, H represents the prism's length.

6. Substitute the values of A and H into the volume formula and evaluate.

7. Write the answer as a sentence, with correct units.

WRITE

$V = A \times H$

$A_{\text{triangle}} = \dfrac{1}{2} \times b \times h$

$b = 25 \, \text{cm}; h = 18 \, \text{cm}$

$A_{\text{triangle}} = \dfrac{1}{2} \times 25 \times 18$

$\qquad\quad = 225 \, \text{cm}^2$

$H = 40 \, \text{cm}$

$V = 225 \times 40$

$\quad = 9000 \, \text{cm}^3$

The volume of the prism is $9000 \, \text{cm}^3$.

13.2.2 Volume of a cylinder

You can calculate the volume of a cylinder using your knowledge of circles.

First calculate the area of the circle at the base of the cylinder using the formula $A = \pi r^2$, where r is the radius. Then multiply the area by the height (or length) of the cylinder.

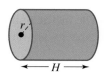

H

Formula for the volume of a cylinder

The formula used to calculate the volume of a cylinder is:

$$V_{\text{cylinder}} = AH$$
$$= \text{area of a circle} \times \text{height}$$
$$= \pi r^2 H$$

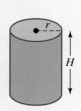

H

Calculate the volume of the cylinder correct to 2 decimal places.

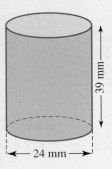

THINK	WRITE
1. Write the formula for the volume of a cylinder.	$V_{\text{cylinder}} = \pi r^2 H$
2. State the values of the pronumerals. (The value of r can be found by halving the given diameter.)	$r = \dfrac{D}{2} = \dfrac{24}{2} = 12 \, \text{mm}$ $H = 39 \, \text{mm}$
3. Substitute the values of the pronumerals into the formula and evaluate.	$V = \pi \times 12^2 \times 39$ $= \pi \times 5616$ $= 17\,643.18 \, \text{mm}^3$
4. Write the answer as a sentence, with correct units.	The volume of the cylinder is $17\,643.18 \, \text{mm}^3$.

13.2.3 Capacity

Capacity is another term for volume that is usually applied to the measurement of liquids and containers. The capacity of a container is the volume of liquid that it can hold. The standard measurement for capacity is the litre (L).

A container, such as a jar, has both volume and capacity, as it occupies space and can hold, contain or absorb something other than what it is made of. A solid block of wood has volume but not capacity, because it cannot hold or absorb anything.

The metric unit for volume, $1 \, \text{cm}^3$, was defined as having a capacity of $1 \, \text{mL}$. Therefore, units for volume and capacity are related, as shown below.

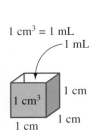

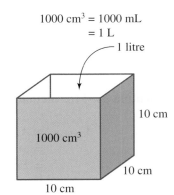

Converting units of capacity

The following chart can be used to convert between units of capacity:

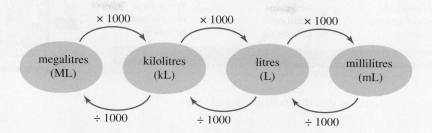

Converting between units of volume and capacity

The units of capacity and volume are related as follows:

$$1\,cm^3 = 1\,mL$$
$$1000\,cm^3 = 1000\,mL$$
$$= 1\,L$$
$$1\,m^3 = 1000\,L$$
$$= 1\,kL$$

---

tlvd-4949

WORKED EXAMPLE 3 Converting units of volume and capacity

Complete the following unit conversions.

a. $50\,mL = \underline{\hspace{1cm}}\,cm^3$ b. $150\,mL = \underline{\hspace{1cm}}\,L$ c. $0.35\,L = \underline{\hspace{1cm}}\,cm^3$

THINK

a. $1\,cm^3$ holds $1\,mL$.

b. There are $1000\,mL$ in $1\,L$. So, to convert from millilitres to litres, divide by 1000.

c. 1. First, convert $0.35\,L$ to mL. There are $1000\,mL$ in $1\,L$. So, to convert from litres to millilitres, multiply by 1000.

   2. Use the relationship $1\,cm^3 = 1\,mL$ to complete the conversion.

WRITE

a. $1\,mL = 1\,cm^3$, so $50\,mL = 50\,cm^3$.

b. $150\,mL = (150 \div 1000)\,L$
   $= 0.15\,L$

c. $0.35\,L = (0.35 \times 1000)\,mL$
   $= 350\,mL$

   $350\,mL = 350\,cm^3$
   $0.35\,L = 350\,cm^3$

---

on Resources

Interactivities Volume (int-6476)
                     Capacity (int-4024)

13.2 Exercise

1. **WE1** Calculate the volume of this three-dimensional shape.

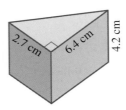

2. Calculate the volume of each of the following rectangular prisms.

   a.
   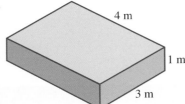

   b.
   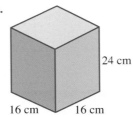

   c.
   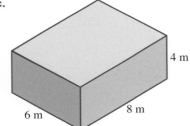

   d.
   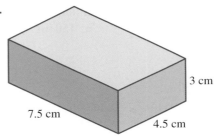

3. Calculate the volume of each of the following prisms.

   a.
   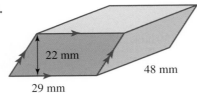

   b.
   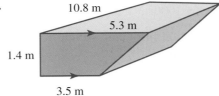

   c.
   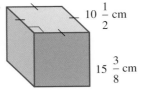

4. Calculate the volume of each of the following objects.

a.

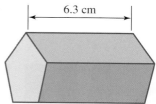

6.3 cm

Area of base = 15.4 cm²

b.

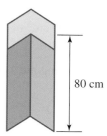

80 cm

Area of base = 1.2 m²

5. WE2 Calculate the volumes of the following shapes, correct to 2 decimal places.

a.

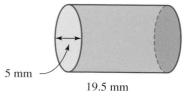

5 mm

19.5 mm

b.

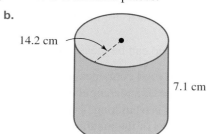

14.2 cm

7.1 cm

6. A diving pool in the shape of a rectangular prism has a length of 16 m, a width of 12 m and a depth of 4 m. Calculate the volume of the pool.

7. WE3 Complete the following unit conversions.

a. $3 \, L = \_\_\_ mL$

b. $0.524 \, L = \_\_\_ mL$

c. $425 \, mL = \_\_\_ L$

d. $0.04 \, L = \_\_\_ mL$

e. $2500 \, mL = \_\_\_ L$

f. $1.3 \, L = \_\_\_ mL$

8. Complete the following unit conversions.

a. $165 \, cm^3 = \_\_\_ mL = \_\_\_ L$

b. $6350 \, L = \_\_\_ mL = \_\_\_ cm^3$

c. $4.2 \, L = \_\_\_ mL = \_\_\_ cm^3$

9. Use the image shown to answer the following questions.

a. Calculate the volume of the fish tank in cm^3.

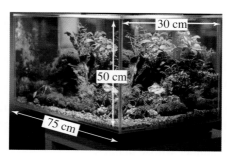

30 cm

50 cm

75 cm

b. If $1 \, cm^3$ is equivalent to 1 mL, determine the capacity of the tank in litres.

10. MC A triangular prism has a volume of $225 \, cm^3$. If the triangular base has an area of $25 \, cm^2$, the height is equal to:

A. 10 cm   **B.** 9 cm   **C.** 19 cm   **D.** 1 cm   **E.** 90 cm

11. MC The volume of a cylinder is $64 \, cm^3$. If the height is equal to 8 cm, the radius of the cylinder correct to 2 decimal places is:

A. 2.82 cm   **B.** 1.60 cm   **C.** 8 cm   **D.** 3.20 cm   **E.** 2.56 cm

12. A skateboard is packaged in a rectangular box that is 0.8 m long, 30 cm wide and 20 cm high.
   a. Calculate the volume of the box in cubic centimetres.
   b. If the skateboard takes up 7200 cm³ of the space in the box, determine how much of the space in the box is not occupied by the skateboard.

13. Standard measurements for cooking are as follows.
   1 cup = 250 mL
   1 tablespoon = 20 mL
   1 teaspoon = 5 mL
   In cooking quantities, determine:
   a. how many teaspoons there are in a tablespoon
   b. how many tablespoons there are in a cup
   c. how half a cup can be measured with only teaspoons and tablespoons
   d. how a quarter of a cup can be measured with only teaspoons and tablespoons.

14. Consider a set of three food containers of different sizes, each of which is in the shape of a rectangular prism.
   The smallest container is 6 cm long, 4 cm wide and 3 cm high. All the dimensions of the medium container are double those of the smallest one; all the dimensions of the largest container are triple those of the smallest one.
   a. Calculate the volume of the smallest container.
   b. State the dimensions and hence calculate the volume of:
      i. the medium container
      ii. the largest container.
   c. Calculate the ratio of the volumes of:
      i. the smallest container to the medium container
      ii. the smallest container to the largest container.
   d. Consider your answers to part c and use them to copy and complete the following:
      If all dimensions of a rectangular prism are increased by a factor of n, the volume of the prism is increased by a factor of _____.

15. Answer the following questions.
   a. Design, draw and label a prism that has a volume of 16 m³.
   b. Draw and label a second prism that also has a volume of 16 m³.

16. A block of cheese is in the shape of a triangular prism. The area of the triangle is 2.3 cm² and the length of the prism is 9.7 cm. If the price of cheese is $0.41 per cm³, calculate the price of the block.

17. A water tank is in the shape of a cylinder whose diameter is twice its height. If the tank holds 1000 litres, determine the dimensions of the tank, correct to 1 decimal place.

18. A cylinder fits exactly into a cube of side 8 cm.
   a. Determine the volume of the cylinder, correct to 2 decimal places.
   b. Determine the percentage of space inside the cube that is not occupied by the cylinder, correct to 2 decimal places.

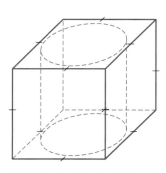

13.3 Volume of spheres and composite shapes

13.3.1 Volume of a sphere

The volume of a sphere of radius r is given by the following formula.

Volume of a sphere

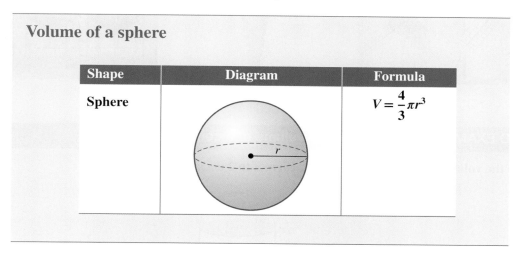

Shape	Diagram	Formula
Sphere		$V = \dfrac{4}{3}\pi r^3$

The volume of a sphere is measured using cubic units appropriate to the task (mm^3, cm^3, m^3 or km^3).

WORKED EXAMPLE 4 Calculating the volume of a sphere

Calculate the volume of the sphere shown correct to 2 decimal places.

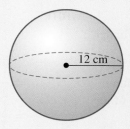

12 cm

THINK	WRITE
1. Write the formula for calculating the volume of a sphere.	$V_{\text{sphere}} = \dfrac{4}{3}\pi r^3$
2. Substitute the value of the radius into the equation and calculate.	$V_{\text{sphere}} = \dfrac{4}{3} \times \pi \times 12^3$ ≈ 7238.23
3. Write the answer with appropriate units.	$V_{\text{sphere}} = 7238.23 \text{ cm}^3$

13.3.2 Volume of composite shapes

The total volume of a **composite solid** can be calculated by determining the volume of each individual solid that it contains and adding them together to calculate the total volume.

Some examples of composite solids include:

tlvd-4950

WORKED EXAMPLE 5 Calculating the volume of a composite shape

Calculate the volume of the following object, correct to 2 decimal places.

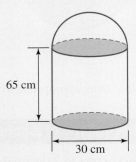

65 cm

30 cm

THINK	WRITE
1. The composite object is made up of a cylinder and a hemisphere.	
2. Write the formula for calculating the volume of a hemisphere and substitute the value of the radius into the formula. Calculate.	$V_{\text{hemisphere}} = \frac{1}{2} \times \frac{4}{3} \times \pi \times r^3$ $= \frac{1}{2} \times \frac{4}{3} \times \pi \times 15^3$ $= 2250\pi$
3. Write the formula for calculating the volume of a cylinder and substitute the values of the height and radius into the equation. Calculate.	$V_{\text{cylinder}} = \pi \times r^2 \times h$ $= \pi \times 15^2 \times 65$ $= 14\,625\pi$
4. Add the volumes of the hemisphere and the cylinder.	Total volume $= 2250\pi + 14\,625\pi$ $= 16\,875\pi$ $\approx 53\,014.38$
5 Write the answer using appropriate units.	$V_{\text{total}} \approx 53\,014.38 \text{ cm}^3$

1. **WE4** Calculate the volumes of the following shapes. Give your answer correct to the nearest whole number.

   a.

   17 mm

   b.

   8.2 mm

   c.

   38 mm

   d.

   32 cm

2. A sphere has a volume of 523.81 cm³. Calculate the radius of the sphere, correct to the nearest whole number.

3. The radius of the Earth is 6380 km. Assuming the Earth is a sphere, calculate its volume.

4. **MC** A hemispherical bowl has a volume of 2145.52 cm³. The diameter of the bowl is closest to:

   A. 20 cm   B. 8.0 cm   C. 16 cm   D. 512 cm   E. 10 cm

5. **WE5** Calculate the volume of the solid below, correct to the nearest cm³.

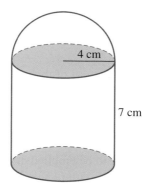

4 cm

7 cm

6. Calculate the volumes of the following objects, correct to 2 decimal places.

   a.

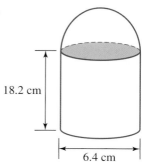

   18.2 cm

   6.4 cm

   b.

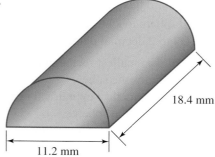

   18.4 mm

   11.2 mm

7. Calculate the capacity of the object shown in cm³. State the result to 4 significant digits.

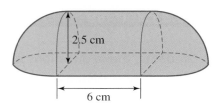
2.5 cm
6 cm

8. Calculate the volume of water that can completely fill the inside of this washing machine. Give your answer in litres, correct to 2 decimal places.

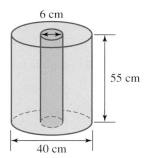
6 cm
55 cm
40 cm

9. This cylindrical steel water tank has a diameter of 2 m and a length of 3 m. Ignoring the thickness of the material, calculate:

a. the volume of the tank, correct to 3 decimal places
b. the capacity of the tank, to the nearest litre (1 m³ = 1000 L).

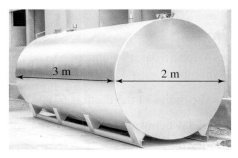
3 m
2 m

10. A house plan is drawn as shown. The house is going to be built on a concrete slab.

a. Calculate the area of the slab.
b. The slab is to be 15 cm thick. Calculate the volume of concrete needed for the slab. (*Hint:* Write 15 cm as 0.15 m.)
c. Concrete costs $180/m³ to pour. Calculate the cost of this slab.

11. Give some real-life examples of when it would be important to know how to calculate the volume of a sphere.

12. A spherical container with a radius of 15 cm³ is filled with water. A tap at the bottom of the container allows the water to be emptied at a rate of 80 cm³/s. Determine how long it would take to completely empty the container. Give your answer in seconds correct to 1 decimal place.

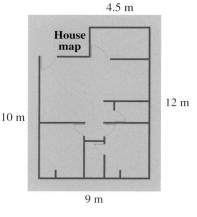
House map
4.5 m
12 m
10 m
9 m

13. The sculpture shown is to be packaged into a rectangular or cylindrical cardboard box.

a. Assuming each sphere touches the one next to it, calculate correct to 2 decimal places (where necessary):

  i. the volume of the smallest rectangular box that completely contains the spheres
  ii. the volume of the smallest cylindrical box that completely contains the spheres
  iii. the volume of space in your boxes not occupied by the spheres.

b. Use your calculations to justify which box you would choose.

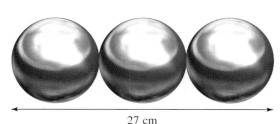
27 cm

14. A sphere is enclosed inside a cube of side length equal to the diameter of the sphere. Determine the percentage of the volume of the cube taken up by the sphere. Give your answer correct to 1 decimal place.

13.4 Mass

13.4.1 Units of mass

Mass describes how much matter makes up an object, and its standard unit of measurement is the kilogram (kg).

The mass of an object can be measured in milligrams (mg), grams (g), kilograms (kg) and tonnes (t). Just as with length and area, it is important to use the appropriate unit to measure the mass of an object.

Milligrams (mg)	Milligrams are used to measure the mass of a light object such as a strand of hair. The mass of a strand of hair is around 1 mg.	
Gram (g)	There are 1000 milligrams in 1 gram. Grams are commonly used when measuring non-liquid cooking ingredients. The mass of 1 cup of flour is about 150 g.	
Kilograms (kg)	There are 1000 grams in 1 kilogram. Kilograms are commonly used to measure larger masses, such as the mass of a person.	
Tonne (T)	There are 1000 kilograms in 1 tonne. Tonnes are used to measure the mass of very large objects such as a truck or boat.	

Conversion between units of mass

The relationship between the units of mass can be used to change a measurement from one unit to another.

$$1000 \text{ mg} = 1 \text{ g}$$
$$1000 \text{ g} = 1 \text{ kg}$$
$$1000 \text{ kg} = 1 \text{ t}$$

Converting between units of mass

The diagram shown can be used to assist in the conversion between different units of mass.

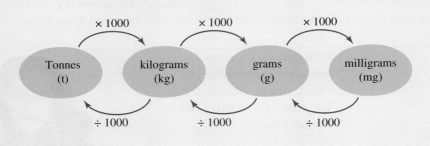

tlvd-4951

WORKED EXAMPLE 6 Converting mass measurements to different units

Convert the following mass measurements to the units specified.
a. 15.86 kg to grams   **b.** 13 650 mg to kilograms   **c.** 0.0071 t to kilograms

THINK

a. 1. To convert from kilograms to grams, multiply by 1000.

b. 1. To convert from milligrams to kilograms, divide by 1000 to convert to grams first.

   2. Once converted to grams, divide by 1000 again to convert to kilograms.

c. 1. To convert from tonnes to kilograms, multiply by 1000.

WRITE

a. $15.86 \text{ kg} = (15.86 \times 1000) \text{ g}$
              $= 15\,860 \text{ g}$

b. $13\,650 \text{ mg} = (13\,650 \div 1000) \text{ g}$
               $= 13.65 \text{ g}$

$13.65 \text{ g} = (13.65 \div 1000) \text{ kg}$
          $= 0.013\,65 \text{ kg}$

c. $0.0071 \text{ t} = (0.0071 \times 1000) \text{ kg}$
             $= 7.1 \text{ kg}$

1. State which unit of mass would be best used to describe the mass of each of the following.
   a. A full suitcase checked in for a flight
   b. A grass seed
   c. A family car
   d. A calculator
   e. A bag of apples
   f. A brick

2. Estimate the mass of each of the following objects and compare the actual values to your estimated values.
   a. A football
   b. A Maths textbook
   c. A can of soft drink

3. Estimate the mass of each of the following objects and compare the actual values to your estimated values.
   a. A video game cover
   b. A computer printer
   c. Your pen

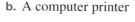

4. **WE6** Convert the following mass measurements to the units specified.
   a. 18.9 kg to grams
   b. 76 490 mg to kilograms
   c. 0.083 t to kilograms

5. Convert the following mass measurements to the units specified.
   a. 717 mg to grams
   b. 3867 g to tonnes
   c. 0.0084 t to grams

6. From your knowledge of household essentials, estimate the mass of the following items.

a.

b.

c.

d.

7. Change the following mass measurements to the units specified.

a. 2.4 g to milligrams

b. 46.7 t to kilograms

c. 2510 mg to grams

d. 82 g to kilograms

e. 0.03 kg to grams

f. 7893 g to milligrams

8. Change the following mass measurements to the units specified.

a. 25 781 mg to kilograms

b. $\frac{7}{8}$ kg to milligrams

c. 384.2 kg to tonnes

d. 0.075 kg to milligrams

e. 0.000 004 5 t to grams

f. 33 456 mg to kilograms

9. For Noah's party, 8 blocks of chocolate were purchased.

a. Calculate the total mass of the 8 blocks of chocolate if each chocolate block is 250 g.

b. The mass of one chocolate block is 250 g net. State what the term *net* means.

10. A family of 5 (2 parents and 3 children) are going on a holiday to Disneyland. The allowable limit for their luggage is 23 kg per person.
They pack light, with each parent taking 18 kg of luggage and each of the 3 children taking 12 kg. They hope to do some shopping during their holiday.
Calculate the maximum mass of their shopping so that they don't exceed the luggage allowance.

11. Christiana decided to go shopping to get some fruit and vegetables for the weekend. Complete the table shown to calculate the total cost to the nearest cent.

Item	Cost per kg	Cost
1.5 kg of bananas	$3.50	$1.5 \times \$3.50 = \$5.25$
3.4 kg of apples	$3.95	
1.2 kg of carrots	$2.80	
2.3 kg of grapes	$7.90	
Total cost		

12. If Rob's mass is 94.4 kg and his son Jack's is 67.8 kg, calculate the difference in their masses.

13. A group of contestants enter a weight-loss competition. Use the information given in the table to complete the table.

Name	Original mass (kg)	Loss	New mass (kg)
Matthew	112.3	16.7 kg	
Jack	133.8	24% of the original mass	
Sarah	93.6	$\frac{1}{6}$ of the original mass	
Jane	88.3	Jane now weighs $\frac{7}{9}$ of the original mass.	
Paul	105.6	13.9 kg	
Kelly	96.1	17.5% of the original mass	

Answer the following questions:

a. State who lost the most mass and how much they lost.
b. State who lost the least mass and how much they lost.
c. Place the contestants in descending order of their new mass.
d. Calculate the total mass loss of the six contestants.

14. The mass of an object is constant, but its weight changes depending on gravity. To calculate the weight (W) of an object, its mass (m) is multiplied by the value of the acceleration due to gravity (g). On the moon, the acceleration due to gravity is about $\frac{1}{6}$ of that on Earth (which is 9.8 m/s^2). The method for calculating the weight of a 100-kg object on the moon, measured in newtons (N), is shown below.

$$W = m \times g$$
$$= 100 \text{ kg} \times \left(\frac{1}{6} \times 9.8 \text{ m/s}^2 \right)$$
$$\approx 163.33 \text{ N}$$

Note: Weight is different to mass, as weight describes the gravitational force acting on an object, and is measured in newtons (N).

Use this information to calculate the weight of each of the following objects on the moon.

a. A 4500-kg boat   b. A 0.56-kg bottle   c. A 76.25-kg person   d. A 331-g iPad mini

13.5 Review

13.5.1 Summary

doc-38054

13.5 Exercise

Multiple choice

1. **MC** A mass of 259 grams is the same as:
   - **A.** 259 kg
   - **B.** 0.259 kg
   - **C.** 2.59 kg
   - **D.** 0.259 t
   - **E.** 2590 kg

2. **MC** If Jack buys seven 250-gram packets of chips for a group of friends who are coming around to watch a movie, the total mass of chips is:
   - **A.** 2.5 kg
   - **B.** 1.5 kg
   - **C.** 2.0 kg
   - **D.** 1.75 kg
   - **E.** 0.175 kg

3. **MC** A cube has a volume of 512 cm^3. The length of an edge of the cube is:
   - **A.** 64 cm^2
   - **B.** 16 cm
   - **C.** 8 cm
   - **D.** 2 cm
   - **E.** 16 cm^2

4. **MC** 45 mL is the same as:
   - **A.** 4500 cm^3
   - **B.** 45 cm^3
   - **C.** 450 cm^3
   - **D.** 4.5 cm^3
   - **E.** 0.45 cm^3

5. **MC** The volume of the following shape is closest to:
   - **A.** 85.68 cm^3
   - **B.** 134.59 cm^3
   - **C.** 1372.78 cm^3
   - **D.** 565.26 cm^3
   - **E.** 2010.62 cm^3

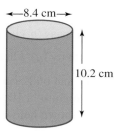

6. **MC** Select the correct statement from the following.
   - **A.** 4 km > 705 m
   - **B.** 1 T < 899 kg
   - **C.** 100 cm^3 = 1 L
   - **D.** 4500 cm < 4.5 km
   - **E.** 1 mm^2 = 100 cm^2

7. **MC** 12 500 mg converted to kg is:
   - **A.** 1.25 kg
   - **B.** 12.5 kg
   - **C.** 125 kg
   - **D.** 1250 kg
   - **E.** 0.125 kg

8. **MC** The volume of the following prism is closest to:

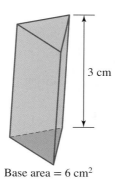

Base area = 6 cm^2

A. $18\,\text{cm}^2$ **B.** $9\,\text{cm}^3$ **C.** $54\,\text{cm}^3$ **D.** $18\,\text{cm}^3$ **E.** $1.8\,\text{cm}^3$

9. **MC** The volume of the shape shown is closest to:

A. $1767.15\,\text{cm}^3$
B. $883.57\,\text{cm}^3$
C. $14137.17\,\text{cm}^3$
D. $1325.36\,\text{cm}^3$
E. $7068.59\,\text{cm}^3$

15 cm

10. **MC** A spa is 1.5 m wide, 1.5 m long and 75 cm deep. The water required to fill the spa is closest to:

A. 1687.5 L
B. 1 687 500 L
C. 168.75 L
D. 16.875 L
E. 16 875 L

Short answer

11. Convert the following units as indicated.

a. $300\,\text{cm}^3 = $ _____ mL   **b.** $750\,\text{cm}^3 = $ _____ L
c. $3300\,\text{cm}^3 = $ _____ L   **d.** $25\,200\,\text{cm}^3 = $ _____ mL

12. You are hiking overnight and want to keep the weight of essential equipment below 3 kg before you add food and clothing. You have the following equipment: backpack 865 g, sleeping bag 1 kg, hiking stove 106 g and cookware 680 g.
Determine if the total weight of these items is under 3 kg.

13. Convert the following mass measurements to the units specified.

a. 717 mg to grams   **b.** 3867 g to tonnes   **c.** 0.0084 t to grams

14. Calculate the volume of each of the following objects.

a.

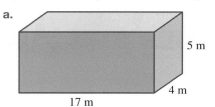

5 m
4 m
17 m

b.

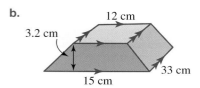

12 cm
3.2 cm
15 cm
33 cm

15. Calculate the volume of each of the following objects. Give your answers correct to 2 decimal places where necessary.

a.

22 mm

6 mm

6 mm

b.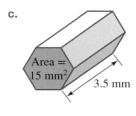

8 cm

10 cm

c.

Area = 15 mm²

3.5 mm

16. Calculate the capacity of the object shown (to the nearest litre).

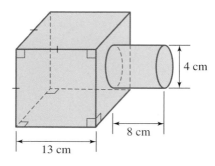

4 cm

8 cm

13 cm

Extended response

17. A car's engine has 4 cylinders, each of which has a diameter of 8.5 cm and a height of 11.2 cm. Calculate the capacity of the engine in cubic centimetres. Give your answer correct to 1 decimal place.

18. A copper pipe has an internal diameter of 35.66 mm and a thickness of copper of 4.34 mm. Determine the volume of copper used per metre in making this pipe. Give your answer in cm³, correct to the nearest whole number.

19. A cylindrical can holds 4 tennis balls stacked one on top of another, packed tightly. If the diameter of a tennis ball is 8.8 cm, determine the volume of the can correct to 1 decimal place.

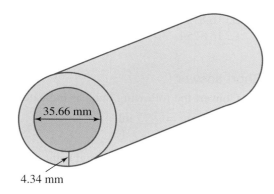

35.66 mm

4.34 mm

20. A lift has a maximum capacity of 1500 kg. If the average weight of each of the 8 people in the lift is 78.45 kg, then determine how far below maximum capacity is the lift with these 8 people in it.

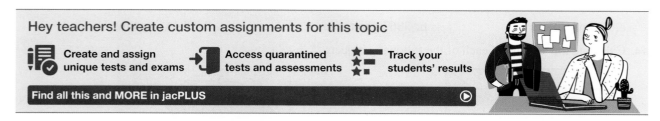

Answers

Topic 13 Volume, capacity and mass

13.2 Volume of prisms and capacity

13.2 Exercise

1. $36.288\,\text{cm}^3$
2. a. $12\,\text{m}^3$ b. $6144\,\text{cm}^3$
   c. $192\,\text{m}^3$ d. $101.25\,\text{cm}^3$
3. a. $30\,624\,\text{mm}^3$ b. $66.528\,\text{m}^3$ c. $1695\dfrac{3}{32}\,\text{cm}^3$
4. a. $97.02\,\text{cm}^3$ b. $0.96\,\text{m}^3$ (or $960\,000\,\text{cm}^3$)
5. a. $382.88\,\text{mm}^3$ b. $4497.64\,\text{cm}^3$
6. $768\,\text{m}^3$
7. a. $3000\,\text{mL}$ b. $524\,\text{mL}$ c. $0.425\,\text{mL}$
   d. $40\,\text{mL}$ e. $2.5\,\text{L}$ f. $1300\,\text{mL}$
8. a. $165\,\text{mL} = 0.165\,\text{L}$
   b. $6\,350\,000\,\text{mL} = 6\,350\,000\,\text{cm}^3$
   c. $4200\,\text{mL} = 4200\,\text{cm}^3$
9. a. $112\,500\,\text{cm}^3$ b. $112.5\,\text{L}$
10. B
11. B
12. a. $48\,000\,\text{cm}^3$ b. $40\,800\,\text{cm}^3$
13. a. 4 teaspoons
    b. 12.5 tablespoons
    c. 6 tablespoons and 1 teaspoon
    d. 3 tablespoons and $\dfrac{1}{2}$ teaspoon
14. a. $72\,\text{cm}^3$
    b. i. $576\,\text{cm}^3$ ii. $1944\,\text{cm}^3$
    c. i. $1:8$ ii. $1:27$
    d. n^3
15. Sample responses can be found in the worked solutions in the online resources.
16. $9.15
17. The tank has a height of 68.3 cm and a diameter of 136.6 cm.
18. a. $402.12\,\text{cm}^3$ b. 21.46%

13.3 Volume of spheres and composite shapes

13.3 Exercise

1. a. $20\,580\,\text{mm}^3$ b. $2310\,\text{mm}^3$
   c. $28\,731\,\text{mm}^3$ d. $8579\,\text{cm}^3$
2. $5\,\text{cm}$
3. $1.1 \times 10^{12}\,\text{km}^3$
4. A
5. $486\,\text{cm}^3$
6. a. $654.12\,\text{cm}^3$ b. $906.39\,\text{mm}^3$
7. $91.63\,\text{cm}^3$
8. $67.56\,\text{L}$
9. a. $9.425\,\text{m}^3$ b. $9425\,\text{L}$
10. a. $99\,\text{m}^2$ b. $14.85\,\text{m}^3$ c. $2673
11. Sample responses can be found in the worked solutions in the online resources.
12. $176.7\,\text{s}$
13. a. i. $2187\,\text{cm}^3$
    ii. $1717.67\,\text{cm}^3$
    iii. Empty space in the rectangular prism: $1041.89\,\text{cm}^3$
    iv. Empty space in the cylinder: $572.56\,\text{cm}^3$
    b. Cylinder
14. 52.4%

13.4 Mass

13.4 Exercise

1. a. kilogram   b. milligram   c. tonne
   d. gram   e. kilogram   f. kilogram
2. a. $200\,\text{g}$ b. $300\,\text{g}$ c. $300\,\text{g}$
3. a. $10\,\text{g}$ b. $2000\,\text{g}$ c. $5\,\text{g}$
4. a. $18\,900\,\text{g}$ b. $0.076\,490\,\text{kg}$ c. $83\,\text{kg}$
5. a. $0.717\,\text{g}$ b. $0.003\,867\,\text{t}$ c. $8400\,\text{g}$
6. a. $2\,\text{kg}$ b. $250\,\text{g}$
   c. $750\,\text{g}$ d. $380\,\text{g}$
7. a. $2400\,\text{mg}$ b. $46\,700\,\text{kg}$
   c. $2.510\,\text{g}$ d. $0.082\,\text{kg}$
   e. $30\,\text{g}$ f. $7\,893\,000\,\text{mg}$
8. a. $0.025\,781\,\text{kg}$ b. $\dfrac{7\,000\,000}{8}\,\text{mg}$
   c. $0.3842\,\text{t}$ d. $75\,000\,\text{mg}$
   e. $4.5\,\text{g}$ f. $0.033\,456\,\text{kg}$
9. a. $2000\,\text{g}$
   b. *Net* means the weight of the chocolate not including the packaging.
10. $43\,\text{kg}$
11.

Item	Cost per kg	Cost
1.5 kg of bananas	$3.50	$5.25
3.4 kg of apples	$3.95	$13.43
1.2 kg of carrots	$2.80	$3.36
2.3 kg of grapes	$7.90	$18.17
Total cost		$40.21

12. $26.6\,\text{kg}$

13. See the table at the bottom of the page.*
    a. Jack lost 32.112 kg.
    b. Paul lost 13.9 kg.
    c. Jack, Matthew, Paul, Kelly, Sarah, Jane
    d. 114.752 kg
14. a. 7350 N   b. 0.91 N
    c. 124.54 N   d. 0.541 N

13.5 Review

13.5 Exercise

Multiple choice

1. B   2. D   3. C   4. B   5. D
6. A   7. B   8. D   9. B   10. A

Short answer

11. a. 300 mL   b. 0.75 L   c. 3.3 L
    d. 25.2 L
12. Yes, as the total weight of the items is 2.651 kg.
13. a. 0.717 g   b. 0.003 867 t   c. 8400 g
14. a. 340 m^2 b. 1425.6 cm^2
15. a. 792 mm^3 b. 502.65 cm^3 c. 52.5 mm^3
16. 2 L

Extended response

17. 2542 cm^3 18. 0.5 cm^3
19. 2140.9 cm^3 20. 872.4 kg

---

*13

Name	Original mass (kg)	Loss	New mass (kg)
Matthew	112.3	16.7 kg	95.6 kg
Jack	133.8	24% of the original mass	101.688 kg
Sarah	93.6	$\frac{1}{6}$ of the original mass	78 kg
Jane	88.3	Jane now weighs $\frac{7}{9}$ of the original mass.	68.68 kg
Paul	105.6	13.9 kg	91.7 kg
Kelly	96.1	17.5% of the original mass	79.28 kg

14 Presenting and interpreting data

Fully worked solutions for this topic are available online.

14.1 Overview

14.1.1 Introduction

Data collection and analysis has become a major focus of governments and businesses in the 21st century. How to effectively handle the data collected is one of the biggest challenges to overcome in the modern world. You may have seen graphs or charts of housing prices in the news or online. One house price report stated that in December 2021 the median house price was $1.6 million in Sydney, $1.12 million in Melbourne and $1.4 million in Brisbane. What is the median house price? Why are house prices discussed in terms of the median and not the mean? This topic will help you to understand the answers to these questions.

When analysed properly, data can be an invaluable aid to good decision-making. However, deliberate distortion of the data or meaningless pictures can be used to support almost any claim or point of view. Whenever you read an advertisement, hear a news report or are given some data by a friend, you need to have a healthy degree of skepticism about the reliability of the source and nature of the data presented.

In 2020, when the COVID-19 pandemic started, news and all forms of media were flooded with data. These data were used to inform governments worldwide about infection rates, recovery rates and all sorts of other important information. These data guided the decision-making process in determining the restrictions

that were imposed or relaxed to maintain a safe community. A solid understanding of data analysis is crucially important, as it is very easy to fall prey to statistics that are designed to confuse and mislead.

KEY CONCEPTS

This topic covers the following key concepts from the VCE Mathematics Study Design:
- creation of a range of charts, tables and graphs to represent and compare data
- measures of central tendency and simple measures of spread (such as range and interquartile range) to summarise and interpret data and compare sets of related data
- interpretation, summary and comparison of related data sets to report findings and draw possible conclusions.

Source: VCE Mathematics Study Design (2023–2027) extracts © VCAA; reproduced by permission.

14.2 Measures of central tendency and the outlier effect

LEARNING INTENTION

At the end of this subtopic you should be able to:
- calculate common measures of central tendency (mean, median, mode)
- determine the mean, median and mode from a frequency distribution table
- determine outliers and their effect on data.

14.2.1 Measures of central tendency — the mean

Two key measures are used in statistics to examine a set of data: **measures of central tendency** and **measures of spread**. The focus of this section is on the measures of central tendency.

Central tendency is the centre or the middle of a frequency distribution.

Mean

The **mean**, also called the **arithmetic mean**, is the average value of all the data.

Calculating the mean

To calculate the mean of a set of data, calculate the sum of all data values and then divide it by the total number of data.

$$\text{mean} = \frac{\text{sum of all data values}}{\text{total number of data values}}$$

This is also written as $\bar{x} = \frac{\sum x_i}{n}$, where:
- x_i = data values (x_1 is the first data score, x_2 is the second data score, ..., x_n is the nth data score)
- $\bar{x}$ = mean
- $\sum x_i$ = sum of all data values. Σ (sigma) is the capital letter S in the Greek alphabet. In mathematics, Σ stands for the 'sum of ...'
- n = total number of data values.

WORKED EXAMPLE 1 Calculating the mean for a data set

The data below represents the results obtained by 10 students in a statistics test. The maximum number of marks for the test was 20. Calculate the mean for this set of data.

$$15, 20, 20, 17, 19, 11, 6, 18, 18, 16$$

THINK	WRITE
1. Write the formula for the mean.	$\bar{x} = \frac{\sum x_i}{n}$, where: $x = 15, 20, 20, 17, 19, 11, 6, 18, 18, 16$ $n = 10.$

2. Substitute the data values into the formula for the mean.

$$\bar{x} = \frac{15 + 20 + 20 + 17 + 19 + 11 + 6 + 18 + 18 + 16}{10}$$

3. Determine the answer.

$$= \frac{160}{10}$$

$$= 16$$

Limitations of the mean as a measure of centre

An important property of the mean is that all the data are included in its calculation. As such, it has genuine credibility as a representative value for the distribution. On the other hand, this property also makes it susceptible to being adversely affected by the presence of extreme values when compared to the majority of the distribution.

Consider the data set 3, 4, 5, 6, 7, 8, 9. It has a mean of:

$$\frac{3 + 4 + 5 + 6 + 7 + 8 + 9}{7} = 6$$

Compare this to a data set with the same values with the exception of the largest one: 3, 4, 5, 6, 7, 8, 90.

It has a mean of:

$$\frac{3 + 4 + 5 + 6 + 7 + 8 + 90}{7} = 17.6$$

As we can see, the mean has been significantly influenced by the one extreme value. When the data is skewed or contains extreme values, the mean becomes less reliable as a measure of centre.

14.2.2 Measures of central tendency — the median

The **median** represents the *middle* score when the data values are in ascending or descending order, such that an equal number of data values will lie below the median and above it.

Calculating the median

When calculating the median:
1. **arrange the data values in order (usually in ascending order)**
2. **the *position* of the median is the $\left(\dfrac{n+1}{2}\right)$th data value, where n is the total number of data values.**

Note: If there is an even number of data values, then there will be two middle values. In this case, the median is the average of those data values.

When there is an odd number of data values, the median is the middle value.

Median = 4

When there is an even number of data values, the median is the average of the two middle values.

$$2 \quad 3 \quad 3 \quad \boxed{5} \quad \boxed{6} \quad 6 \quad 7 \quad 9$$

$$\text{Median} = \frac{5+6}{2} = 5.5$$

WORKED EXAMPLE 2 Calculating the median of a data set

tlvd-4953

Calculate the median pulse rate for the two sets of data given.
a. Individual pulse rates of a group of 10 adults:

$$69, 62, 83, 75, 71, 66, 64, 67, 70, 62$$

b. Individual pulse rates of a group of 11 newborns (0−3 months):

$$129, 121, 116, 132, 124, 136, 124, 129, 133, 128, 129$$

THINK	WRITE
a. There is an even number of data values.	**a.** $62, 62, 64, 66, 67, 69, 70, 71, 75, 83$
1. Arrange the data in increasing order.	
2. Determine the position of the median.	$\text{Position} = \dfrac{n+1}{2}$ $= \dfrac{10+1}{2}$ $= 5.5$ (between 5th and 6th scores)
3. Determine the median.	For an even number of data values, the position of the median is going to be halfway between the two middle scores. The median is the average or the mean of these two scores. 62, 62, 64, 66, 67, 69, 70, 71, 75, 83 Median $\text{Median} = \dfrac{67+69}{2}$ $= \dfrac{136}{2}$ $= 68$ The median of the set of data given is 68.
b. There is an odd number of data values.	**b.** $116, 121, 124, 124, 128, 129, 129, 129, 132, 133, 136$
1. Arrange the data in increasing order.	

2. Determine the position of the median.	$$\text{Position} = \frac{n+1}{2}$$ $$= \frac{11+1}{2}$$ $$= \text{6th score}$$ For an odd number of data, the position of the median is exactly in the middle.
3. Determine the median.	The median is the score in the middle. 116, 121, 124, 124, 128, (129,) 129, 129, 132, 133, 136 The median of the set of data given is 129.

Is the median as a measure of centre affected by an extreme value in the data set?

Previously we saw that the mean is affected by an extreme value in the data set; however, the same cannot be said for the median. In both data sets the median will be the value in the fourth position.

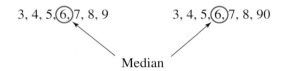

The median is therefore more reliable than the mean when the data is skewed or contains extreme values.

14.2.3 Measures of central tendency — the mode

The **mode** is the score that occurs most often.

The data set can have no modes, one mode, two modes (bimodal) or multiple modes (multimodal).

Calculating the mode

When determining the mode:
1. **Arrange the data values in ascending order (smallest to largest). This step is optional but helpful.**
2. **Look for the number that occurs most often (has the highest frequency).**

If no value in a data set appears more than once, then there is no mode.

If a data set has multiple values that appear the most, then it has multiple modes. All values that appear the most are modes.

For example, the set 1, 2, 2, 4, 5, 5, 7 has two modes, 2 and 5.

Michelle is a real estate agent and has listed ten properties for sale in the last fortnight. She has listed the following numbers of bedrooms per house.

$$3, 2, 4, 5, 2, 2, 1, 2, 3, 2$$

Determine the mode for this data.

THINK	WRITE
1. Count the number of each score.	Score 1 (one bedroom) has a frequency of one. Score 2 (two bedrooms) has a frequency of five. Score 3 (three bedrooms) has a frequency of two. Score 4 (four bedrooms) has a frequency of one. Score 5 (five bedrooms) has a frequency of one.
2. Determine the highest frequency and the mode.	The highest frequency is 5, belonging to score 2 (two bedrooms). Therefore, mode = 2.

14.2.4 Determining the mean, median and mode from a frequency distribution table

When working with large amounts of data (e.g. 100 data scores), it is quite tedious to calculate the three measures of central tendency. In this case, it helps to work with the data using a frequency distribution table.

tlvd-4954

Using the random sampler from the website Census@School (Australian Bureau of Statistics), Alexandra downloaded the responses of 200 students across Australia who participated in a questionnaire. She displayed the data values of the students' year levels in a frequency table.

a. Calculate the mean of this data.
b. Determine the median of this data.
c. Determine the mode of this data.

Year level	Frequency
4	8
5	19
6	40
7	23
8	36
9	29
10	27
11	11
12	7
Total	200

Source: ABS; http://www.cas.abs.gov.au/cgi-local/cassampler.pl

Note: Year level 4 represents '4 or below'. For the purpose of this example, we are going to use 4 instead of '4 or below'.

THINK

a. 1. Add one more column to the frequency table and label it $x \times f$, where x is the variable and f is the frequency.

WRITE

a. *Note:* The frequency table shows 8 students who were in Year 4.
$$4 + 4 + 4 + 4 + 4 + 4 + 4 + 4 = 4 \times 8$$
$$= 32$$

Year level x	Frequency f	$x \times f$
4	8	32
5	19	95
6	40	240
7	23	161
8	36	288
9	29	261
10	27	270
11	11	121
12	7	84
Total	**200**	**1552**

2. Calculate the mean.
Note: Quite often, the mean is not one of the scores of the data set. It does not need to be. It can even be an impossible score, like in this example.

$\bar{x} = \dfrac{\sum x_i}{n}$, where $\sum x = 1552$ and $n = 200$

$\bar{x} = \dfrac{1552}{200}$

$= 7.76$

Mean $= 7.76$

b. 1. Determine the position of the median.

b. Position $= \dfrac{n+1}{2}$

$= \dfrac{200+1}{2}$

$= 100.5$ (between 100th and 101st scores)

2. Determine the median.

Add up the frequencies until the sum is greater than 100.
8 (Year 4 students) + 19 (Year 5 students) = 27
27 + 40 (Year 6 students) = 67
67 + 23 (Year 7 students) = 90
From 90 to 100 we only need 10 more scores. These scores fall into the Year 8 frequency. Therefore, the median is Year 8.

c. Determine the highest frequency and hence the mode.

c. The highest frequency occurs for Year 6. Therefore, the mode is Year 6.

14.2.5 Outliers and their effect on data

A score that is considerably lower or considerably higher than the rest of the data is called an **outlier**. Outliers can influence some of the measures of central tendency. For example, the median house price paid at auction on one particular Saturday in a city could be influenced by a very large price paid for one house. However, for a large number of data, i.e. 2000 scores, this influence could be quite insignificant.

WORKED EXAMPLE 5 Determining the outliers and their effect on the data

The marks of 10 students on a test out of a total of 50 marks are:

$$26, 50, 49, 36, 18, 27, 43, 43, 33, 35$$

Discuss the effect of adding a very small test mark to this set of data.

THINK

1. Calculate the mean, median and mode for this set of data.

2. Calculate the mean, median and mode for this set of data after adding a score of 1 mark to the set of data.

3. Describe the differences between the mean, median and mode.

WRITE

$\bar{x} = \dfrac{\sum x_i}{n}$, where $\sum x_i = 360$ and $n = 10$

$= \dfrac{360}{10}$

$= 36$

Median: $18, 26, 27, 33, 35, 36, 43, 43, 49, 50$

Position $= \dfrac{n+1}{2}$

$= \dfrac{10+1}{2}$

$= 5.5$ (between the 5th and 6th scores)

Median $= \dfrac{35+36}{2}$

$= 35.5$

Mode $= 43$

$1, 18, 26, 27, 33, 35, 36, 43, 43, 49, 50$

$\bar{x} = \dfrac{\sum x_i}{n}$ where $\sum x_i = 361$ and $n = 11$

$= \dfrac{361}{11}$

$= 32.8$

Median position $= \dfrac{11+1}{2}$

$= 6$ (6th score)

Median $= 35$

Mode $= 43$

Both the mean and the median have changed, while the mode has stayed the same. The score 1 is an outlier for this data and has changed the mean from 34.5 to 32.8.

4. Discuss the impact of adding a very small test score to the data set.

Adding a very small score or a very large score to a set of data always changes the mean. It sometimes changes the median and rarely changes the mode.

The most suitable measure for this type of data is the median, as it shows that half of the students performed higher than 35.5 while half performed lower than 35.5.

The mean is less appropriate as it is influenced by outliers.

 Resources

 Interactivity Mean, median, mode and quantities (int-6496)

14.2 Exercise

Students, these questions are even better in jacPLUS

 Receive immediate feedback and access sample responses

 Access additional questions

 Track your results and progress

Find all this and MORE in jacPLUS ▶

1. **WE1** A postman delivered the following number of letters to 15 households:

$$2, 5, 3, 1, 3, 4, 6, 2, 2, 2, 4, 2, 4, 3, 2$$

Calculate the mean number of letters per household delivered by the postman.

2. The following data set represents the heights in centimetres of 10 students. Calculate the mean height of the students.

$$171.5, 162.9, 158.4, 156.8, 142.3, 170.3, 163.7, 158.1, 160.6, 162.9$$

3. Calculate the mean of each of the following sets of data.
   a. 13, 20, 17, 16, 19, 11, 24, 17, 13
   b. 163.2, 219.6, 157.0, 206.4, 211.3, 234.3, 195.8, 179.1

4. Car battery manufacturers offer a certain lifetime guarantee to their customers. This means that if something goes wrong with the car battery within its lifetime, the manufacturer will replace it for free. To determine the lifetime guarantee of their car batteries, CarBat Company checked the lifetime of 200 batteries.

The data is displayed in the frequency distribution table shown.

Lifetime (months)	Frequency	fx
21	9	
22	18	
23	44	
24	67	
25	29	
26	20	
27	13	
Total	200	

a. Calculate the entries in the column labelled fx.

b. Hence calculate the average lifetime guarantee for this set of car batteries.

5. The principal of Achieve Secondary College wanted to determine the average wage of her teachers. She collected the data and displayed it in the frequency table shown.

Wage ($/annum)	Frequency	fx
57 797	37	
62 880	24	
66 517	16	
70 771	21	
75 248	8	
85 458	5	
90 609	2	
Total	113	

a. Calculate the entries in the column labelled fx.

b. Calculate the average teacher wage of the teachers at Achieve Secondary College.

6. **WE2** Amir is a market researcher. He used a random numbers program to collect a set of random numbers from 1 to 20.

12	15	3	19	17	6	11	4	8
10	14	12	7	13	18	20	9	19
7	16	1	5	8	10	15	3	2

a. Determine the position of the median.   b. Determine the median of this data set.

7. Richard is a fencer and has to calculate eight quotes for his clients. The following data represents the lengths of the fences he has to build.

142.3, 96.2, 57.3, 101.8, 53.4, 81.4, 114.3, 72.6, 81.7, 92.5

All measurements are given in metres. Calculate the position and the value of the median length for this data set.

8. Determine the median of each of the following sets of data.
   a. 4.5, 3.6, 5.7, 8.2, 2.9, 6.5, 1.8, 7.4, 9.3
   b. 142, 187, 167, 128, 115, 132, 155, 194, 173, 168

9. A group of 20 students were tested on their knowledge of measures of central tendency by answering a test with 10 multiple choice questions. The numbers of correct answers per student are displayed in the list below. Determine the mean number of multiple choice questions correctly answered by this group of students.

7	3	4	5	6	6	7	7	8	8
9	9	8	8	9	7	9	8	10	7

10. **WE3** Determine the mode for the frequency distribution shown below.

14	14	15	13	17	13	13	14	14	14
16	16	17	17	16	16	16	13	14	14
16	17	17	14	14	15	16	16	16	16

11. A group of people aged 20–29 were asked to state their age. Determine the mode of this set of data.

21	25	23	21	23	24	26	22	22	20
24	27	24	23	29	24	28	24	25	29

12. A shoe store collected data from 15 people about their shoe size. Calculate the average shoe size for this group of people.

$$8, 8\frac{1}{2}, 10\frac{1}{2}, 9, 9\frac{1}{2}, 10\frac{1}{2}, 10, 9, 8, 8, 10\frac{1}{2}, 8\frac{1}{2}, 9\frac{1}{2}, 9\frac{1}{2}, 9$$

13. Determine the mode of the following sets of data.
   a. 18, 21, 16, 19, 22, 21, 23, 21, 18, 18, 21, 19
   b. 144.1, 144.8, 144.5, 144.8, 144.9, 144.2, 144.9, 144.8, 144.7, 144.6, 144.3

14. **WE4** The frequency distribution table shown displays the data collected in an airport on the number of return trips that 120 passengers booked in the last month.

Number of bookings	Frequency
1	27
2	32
3	19
4	26
5	12
6	3
7	1
Total	120

   a. Add a third column to the frequency distribution table, label it *fx* and fill it in.
   b. Calculate the mean of this data.
   c. Determine the median of this data.
   d. Determine the mode of this data.

15. Jenny started a new online store. She recorded the numbers of times her customers logged in over the first month of business. The data is displayed in the frequency distribution table below.

a. Add a third column to the frequency distribution table, label it fx and fill it in.
b. Calculate the mean of this data.
c. Determine the median of this data.
d. Determine the mode of this data.

Number of logins	Frequency
1	23
2	45
3	69
4	51
5	38
6	17
Total	243

16. **WE5** Ling owns a school uniform shop and has to order new stock for the year. She collects the dress sizes of a sample of 20 girls from the nearby school.

12	14	8	8	8	10	12	10	8	8
10	14	8	10	8	8	8	8	24	8

a. Identify the outlier and discuss the effect of this outlier on the set of data by calculating the mean, median and mode of the data:

i. without the outlier
ii. with the outlier.

b. Determine which measure of central tendency (mean, median or mode) Ling is more interested in. Explain your answer.

14.3 Measures of spread

LEARNING INTENTION

At the end of this subtopic you should be able to:
• calculate the range, quantiles and interquartile range (IQR)
• calculate the standard deviation
• calculate outliers.

14.3.1 The quantiles

The **spread** of a set of data indicates how far the data values are spread from the centre or from each other. This is also known as the distribution.

The **range** is the difference between the highest and lowest values of the data set.

Data sets can be split up into any given number of equal parts called **quantiles**.

Quantiles are named after the number of parts that the data is divided into.

Deciles divide the data into 10 equal-sized parts.

Percentiles divide the data into 100 equal-sized parts.

Quartiles divide the data into 4 equal-sized parts. For example, 25% of the data values lie at or below the first quartile.

Percentile	Quartile and symbol	Common name
25th percentile	First quartile, Q_1	Lower quartile
50th percentile	Second quartile, Q_2	Median
75th percentile	Third quartile, Q_3	Upper quartile
100th percentile	Fourth quartile, Q_4	Maximum

A percentile is named after the percentage of data that lie at or below that value. For example, 60% of the data values lie at or below the 60th percentile.

Percentiles can be read off a percentage cumulative frequency curve.

tlvd-4955

WORKED EXAMPLE 6 Calculating the quantiles from a cumulative frequency graph

The following cumulative frequency graph shows the cumulative frequency versus the number of skips with a skipping rope per minute for Year 11 students at a school.

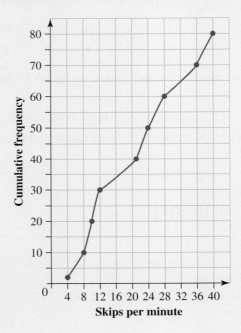

Calculate:

a. the median

b. Q_1

c. Q_3

d. the number of skips per minute for the 40th percentile.

THINK

a. 1. Determine the total number of subjects. This is the highest vertical value on the cumulative frequency curve.

WRITE

a. 80 students

2. The median occurs at the middle of the total cumulative frequency, so divide the total frequency by 2.

3. The median occurs at the $\frac{80}{2} = 40$th person.

4. Draw a line from 40 on the vertical axis until it hits the curve. Read down to the corresponding skips per minute on the horizontal axis.

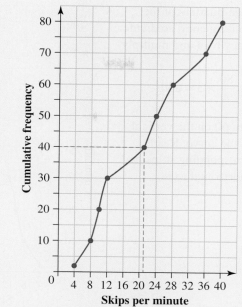

The median number of skips per minute is 21.

b. • Q_1 occurs at a quarter of the total cumulative frequency, so divide the total frequency by 4.

• Q_1 occurs at the $\frac{80}{4} = 20$th person.

• Draw a line from 20 on the vertical axis until it hits the curve. Read down to the corresponding skips per minute on the horizontal axis.

b.

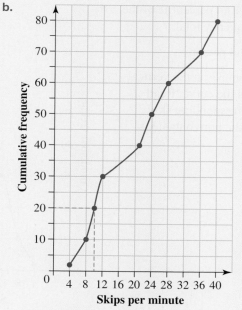

Q_1 is at 10 skips per minute.

c. • Q_3 occurs at three-quarters of the total cumulative frequency, so divide the total frequency by 4 and multiply the result by 3.

• Q_3 occurs at the $\dfrac{80}{4} \times 3 = 60$th person.

• Draw a line from 60 on the vertical axis until it hits the curve. Read down to the corresponding skips per minute on the horizontal axis.

c.

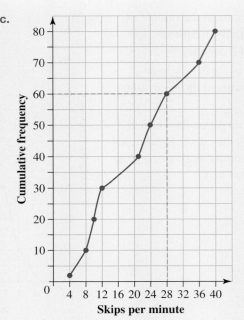

Q_3 is at 28 skips per minute.

d. • Determine the 40th percentile of the cumulative frequency total.

• $\dfrac{40}{100} \times 80 = 32$nd person

• Draw a line from 32 on the vertical axis until it hits the curve. Read down to the corresponding skips per minute on the horizontal axis.

d.

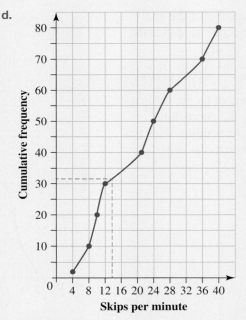

The 40th percentile is at 14 skips per minute.

14.3.2 The range

The most basic measure of spread is the range.

The range is defined as the difference between the highest and the lowest values in the set of data.

Calculating the range of a data set

Range = highest score − lowest score

$$= X_{max} - X_{min}$$

WORKED EXAMPLE 7 Calculating the range of a data set

Calculate the range of the data set 2.1, 3.5, 3.9, 4.0, 4.7, 4.8, 5.2.

THINK

1. Identify the lowest score (X_{min}) of the data set.
2. Identify the highest score (X_{max}) of the data set.
3. Write the rule for the range.
4. Substitute the known values into the rule.
5. Evaluate and write the answer.

WRITE

Lowest score $= 2.1$

Highest score $= 5.2$

$$Range = X_{max} - X_{min}$$
$$= 5.2 - 2.1$$
$$= 3.1$$

14.3.3 The interquartile range (IQR)

The **interquartile range (IQR)** is the range of the middle 50% of the data set. It measures the spread of the middle 50% of data about the median. The IQR is found by subtracting the lower quartile from the upper quartile.

The lower quartile (Q_1) is the median of the bottom half of the data and the upper quartile (Q_3) is the median of the top half of the data.

If the median is one of the actual data values, it is not considered to be in either the upper or lower half of the data. For example:

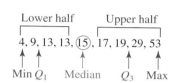

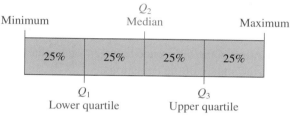

Calculating the IQR

Interquartile range (IQR) = upper quartile − lower quartile

$$= Q_{upper} - Q_{lower}$$
$$= Q_3 - Q_1$$

tlvd-4956

WORKED EXAMPLE 8 Calculating the IQR of a data set

Calculate the IQR for the following data.

26, 32, 15, 12, 35, 27, 22, 31, 38, 20, 41, 26, 17, 29

THINK

1. Put the data in ascending order.

2. The median value will split the data into the top and bottom half.
There are 14 values, so the median is at the $\dfrac{14+1}{2} = 7.5$th value.

WRITE

$12, 15, 17, 20, 22, 26, 26, 27, 29, 31, 32, 35, 38, 41$

$12, 15, 17, 20, 22, 26, 26, |27, 29, 31, 32, 35, 38, 41$

$$Median = \frac{26+27}{2}$$
$$= 26.5$$

3. List the bottom half of data. $12, 15, 17, 20, 22, 26, 26$
   Find the median of the bottom half of the data. $12, 15, 17, 20, 22, 26, 26$
   This is Q_1. $Q_1 = 20$
   There are 7 values, so the median is at the
   $\dfrac{7+1}{2} = $ 4th value.

4. List the top half of data. $27, 29, 31, 32, 35, 38, 41$
   Find the median of the top half of the data. $27, 29, 31, 32, 35, 38, 41$
   This is Q_3. $Q_3 = 32$
   There are 7 values, so the median is at the
   $\dfrac{7+1}{2} = $ 4th value.

5. Calculate the IQR.
$$\begin{aligned} IQR &= Q_3 - Q_1 \\ &= 32 - 20 \\ &= 12 \end{aligned}$$

14.3.4 The standard deviation

The **standard deviation** for a set of data is the measure of how far the data values are spread out (deviate) from the mean. The value of the standard deviation tells you the average deviation of the data from the mean.

Deviation is the difference between each data value and the mean $(x - \overline{x})$. The standard deviation is calculated from the square of the deviations.

Standard deviation formula

- **Standard deviation is denoted by the lowercase Greek letter sigma, σ, and can be calculated using the following formula.**

$$\sigma = \sqrt{\dfrac{\sum (x - \overline{x})^2}{n}}$$

Where $\overline{x}$ is the mean of the data values and n is the number of data values.

A low standard deviation indicates that the data values tend to be close to the mean.

A high standard deviation indicates that the data values tend to be spread out over a large range, away from the mean.

Standard deviation can be calculated using a scientific calculator, a spreadsheet or using a frequency table by following the steps below.

Step 1: Calculate the mean.

Step 2: Calculate the deviations.

Step 3: Square each deviation.

Step 4: Sum the squares.

Step 5: Divide the sum of the squares by the number of data values.

Step 6: Take the square root of the result.

The number of chocolate drops in different bags was recorded. These values were:

$$271, \ 211, \ 221, \ 288, \ 209, \ 285, \ 230, \ 220, \ 296, \ 216$$

Calculate the standard deviation, correct to 1 decimal place, of the number of chocolate drops in a bag.

THINK

1. Enter the data into the digital technology of your choice. This example is done in Excel.

WRITE

A12	⇕ ⊗ ✓	*fx*
A	**B**	**C**
1 **Chocolate drops**		
2 271		
3 211		
4 221		
5 288		
6 209		
7 285		
8 230		
9 220		
10 296		
11 216		
12		

2. To calculate the sample standard deviation, enter the formula '= STDEV()' and select the values to be included in the calculation.

SUM	⇕ ⊗ ✓
A	**B**
1 **Chocolate drops**	
2 271	
3 211	
4 221	
5 288	
6 209	
7 285	
8 230	
9 220	
10 296	
11 216	
12	
13	
14 =STDEV(A2:A11)	
15	

3. Press ENTER to calculate the sample standard deviation.

	A	B
	A15	
	A	**B**
1	**Chocolate drops**	
2	271	
3	211	
4	221	
5	288	
6	209	
7	285	
8	230	
9	220	
10	296	
11	216	
12		
13		
14	35.65903345	
15		

4. Write the answer correct to 1 decimal place.   The standard deviation is 35.7.

14.3.5 Calculating outliers

The outliers are extreme values on either end of a data set that appear very different from the rest of the data.

Outliers can be calculated by considering the distance a data point is from the mean or median compared to the rest of the data. If the value of a data point is clearly far away from the mean or median, as well as from other data points, it may be considered an outlier. For example, in the data set $2, 2, 4, 6, 7, 7, 100$, it is reasonable to conclude that 100 is an outlier, as it is much larger than the rest of the data and is far away from the median of 5.

It is not always obvious if a data point is an outlier.

The IQR can be used in a calculation to determine if a data point is far away enough from the median to be considered an outlier.

Calculating outliers

To determine if a data point is an outlier, first calculate the upper and lower fences:

$$\text{lower fence} = Q_1 - 1.5 \times \text{IQR}$$
$$\text{upper fence} = Q_3 + 1.5 \times \text{IQR}$$

where Q_1 is the lower quartile, Q_3 is the upper quartile and IQR is the interquartile range.

Any data point below the lower fence is an outlier.

Any data point above the upper fence is an outlier.

WORKED EXAMPLE 10 Calculating the IQR and using it to determine any outliers

Consider the following stem-and-leaf plot.

Key: 1|4 = 14

Stem	***Leaf***
0	**13**
1	**00134446899**
2	**2255678889**
3	**4577**
4	**01129**
5	
6	
7	**1**

Calculate the IQR and use this to determine if there are any outliers.

THINK	WRITE
1. To calculate the IQR, you must first calculate Q_1 and Q_3. Q_1 is at a quarter of the data. There are 33 data values, so Q_1 occurs at the $\frac{33+1}{4} = 8.5$th value.	*Stem* | *Leaf* 0 | 13 1 | 001344 (Q_1) | 46899 2 | 2255678889 3 | 4577 4 | 01129 5 | 6 | 7 | 1 $Q_1 = 14$
Q_3 is at three-quarters of the data. There are 33 data values, so Q_3 occurs at the $\frac{33+1}{4} \times 3 = 25.5$th value.	*Stem* | *Leaf* 0 | 13 1 | 00134446899 2 | 2255678889 3 | 45 (Q_3) | 77 4 | 01129 5 | 6 | 7 | 1 $Q_3 = 36$
Calculate the IQR.	$\text{IQR} = Q_3 - Q_1$ $= 36 - 14$ $= 22$

2. Calculate the lower fence.

$$\begin{aligned}\text{Lower fence} &= Q_1 - 1.5 \times \text{IQR}\\ &= 14 - 1.5 \times 22\\ &= -19\end{aligned}$$

Determine if there are any values below the lower fence.

No values are below this point.

Calculate the upper fence.

$$\begin{aligned}\text{Upper fence} &= Q_3 + 1.5 \times \text{IQR}\\ &= 36 + 1.5 \times 22\\ &= 69\end{aligned}$$

Determine if there are any values below the upper fence.

There is a value above this point, so 71 is an outlier.

 Resources

 Interactivities Mean, median, mode and quartiles (int-6496)
The median, the interquartile range, the range and the mode (int-6244)
The mean and the standard deviation (int-6246)

14.3 Exercise

Students, these questions are even better in jacPLUS

Receive immediate feedback and access sample responses

Access additional questions

Track your results and progress

Find all this and MORE in jacPLUS

1. Calculate the range of each of the following sets of data.

a. 4, 6, 8, 11, 15
b. 1.7, 1.9, 2.5, 0.5, 3.1, 1.9, 1.7, 1.6, 1.2
c.

Score (x)	Frequency (f)
110	12
111	9
112	18
113	27
114	5

2. a. Calculate the maximum value of a data set if its minimum value is 23 and its range is 134.
   b. Calculate the minimum value of a data set if its range is 32 and its maximum value is 101.

3. Put these expressions in order from smallest value to largest value:
   upper quartile; minimum; median; maximum; lower quartile.

4. Aptitude tests are often used by companies to help them decide who to employ. An employer gave 30 potential employees an aptitude test with a total of 90 marks. The scores achieved are shown below.

   | 67 | 67 | 68 | 68 | 68 | 69 | 69 | 72 | 72 | 73 | 73 | 74 | 74 | 75 | 75 |
   | 77 | 78 | 78 | 78 | 79 | 79 | 79 | 81 | 81 | 81 | 82 | 83 | 83 | 83 | 86 |

   Only applicants who score above the 80th percentile receive an interview. Use your knowledge of percentiles to work out how many interviews the employer will have to run.

5. **WE6** The number of steps per minute taken by runners at a local park run event was recorded. The cumulative frequency graph is as shown.

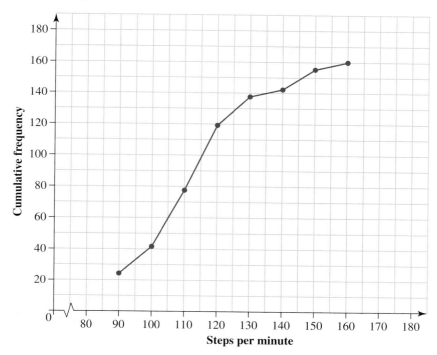

a. Determine the number of people who participated in the park run.
b. Calculate:

   i. the median   ii. Q_1 iii. Q_3.
c. A local running group invited all of the people over the 65th percentile to join the group. Determine the number of steps per minute that a runner would need to take to be invited to join the running group.

Use the following data set to answer questions 6–8.

171	122	182	153	167	184	171	177
189	175	128	190	135	147	171	

6. **WE7** **MC** The range of the data is:
   A. 67   B. 68   C. 69   D. 70   E. 71

7. **WE8** **MC** The IQR of the data is:
   A. 32   B. 33   C. 34   D. 35   E. 36

8. **WE9** **MC** The standard deviation of the data (correct to 1 decimal place) is:
   A. 19.5   B. 21.3   C. 22.0   D. 24.3   E. 25.0

9. The speed of 20 cars (in km/h) is monitored along a stretch of road that is a designated 80 km/h zone.

80	82	77	75	80	80	81	78	79	78
80	80	85	70	79	81	81	80	80	80

Calculate the range and IQR of the data.

10. 30 pens are randomly selected off the conveyor belt at the factory and are tested to see how long they will last (values given in hours). Calculate the range and IQR of the data.

20	32	38	22	25	34	47	31	26	29	30	36	28	40	31
26	37	38	32	36	35	25	29	30	40	35	38	39	37	30

11. The ages of patients (in years) that came into a hospital emergency room during an afternoon were recorded. Using a digital technology of your choice, calculate the standard deviation of the data.

14	1	3	87	27	42	19	91
17	73	68	83	62	29	32	2

12. **WE10** Consider the following stem plot. Calculate:

a. the range
b. the standard deviation
c. the IQR, and use this to determine if there are any outliers.

Key: $1\,|\,6 = 16$

Stem	Leaf
0	1
1	5
2	
3	
4	2
5	6 7
6	0 1 4 5 6
7	1 1 2 3 5 7 9
8	2 2 4 4 4 8 8
9	3 5

13. The number of swimmers at a pool were recorded over a fortnight. The numbers are as follows.

56	69	59	113	9	100	80
111	94	77	57	166	101	96

Determine if any outliers exist and justify your answer with a calculation.

14. Create a data set that fits the following descriptions.

a. 8 data values with a range of 43
b. 6 data values with a lower quartile of 5 and an upper quartile of 12
c. 9 data values with a lower quartile of 7 and an upper quartile of 13

15. Explain how the standard deviation can be calculated for grouped data.

16. A Mathematics teacher wanted to design a question for her class. She wanted the question to have 10 data points, a lower quartile of 14 and an IQR of 20. State a data set that could be used in her question.

17. The number of trees in several different parks was recorded, and the mean number was calculated to be 15. The council wanted to determine whether the majority of parks contained close to 15 trees, or whether there was a large difference in the number of trees. Describe how the council could use a measure of spread to investigate this problem.

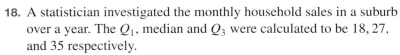

18. A statistician investigated the monthly household sales in a suburb over a year. The Q_1, median and Q_3 were calculated to be 18, 27, and 35 respectively.

a. Prove that 64 houses per month is an outlier.
b. The statistician realised he made a mistake and the outlier of 64 was recorded incorrectly. The actual value was 46 houses per month. Using your answer from **a** to help you, explain what would happen to the values of Q_1, Q_3 and the IQR if the data point was changed from 64 to 46 houses per month.

14.4 Describing graphical distributions

14.4.1 Modality

To describe a distribution, there are several key features like modality and shape that should be included. The **modality** of a graph refers to the mode or most common value. A graph may have one mode, two modes (called bimodal), or many modes (called multimodal).

Data that is **unimodal** will have one obvious peak.

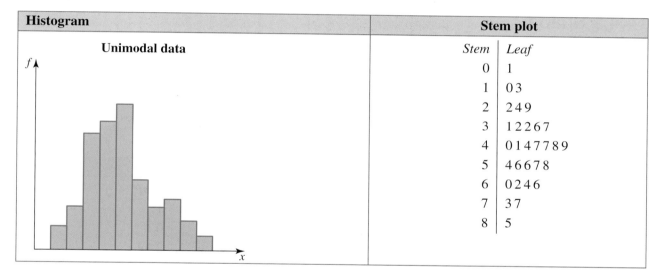

Histogram	Stem plot	
Unimodal data	*Stem*	*Leaf*
	0	1
	1	0 3
	2	2 4 9
	3	1 2 2 6 7
	4	0 1 4 7 7 8 9
	5	4 6 6 7 8
	6	0 2 4 6
	7	3 7
	8	5

Data that is **bimodal** will have two obvious peaks.

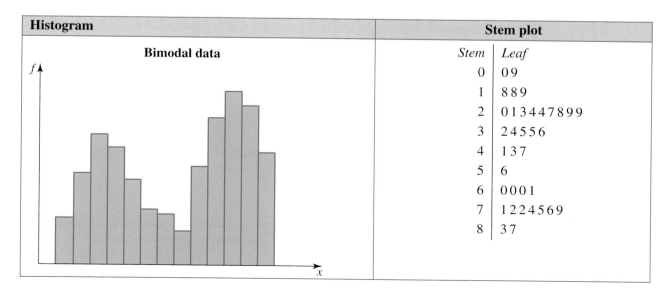

Histogram	Stem plot	
Bimodal data	*Stem*	*Leaf*
	0	0 9
	1	8 8 9
	2	0 1 3 4 4 7 8 9 9
	3	2 4 5 5 6
	4	1 3 7
	5	6
	6	0 0 0 1
	7	1 2 2 4 5 6 9
	8	3 7

Data that is **multimodal** will have three or more obvious peaks.

Histogram	Stem plot
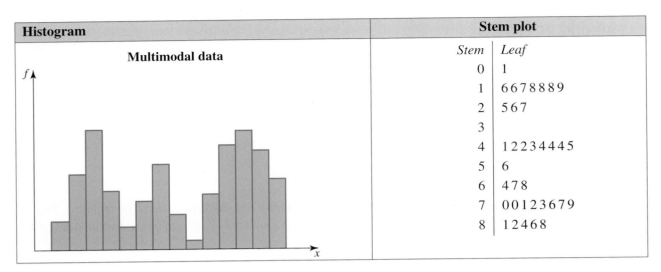	Stem | Leaf 0 | 1 1 | 6 6 7 8 8 8 9 2 | 5 6 7 3 | 4 | 1 2 2 3 4 4 4 5 5 | 6 6 | 4 7 8 7 | 0 0 1 2 3 6 7 9 8 | 1 2 4 6 8

14.4.2 Shape

The shape of a data distribution can be described as symmetric or asymmetric (skewed).

Symmetric distribution

Distributions are **symmetric** if the two halves of the distribution are roughly equal about the mean. The distributions will tail off evenly to the left and right of the peak, with the bulk of the data generally close to the centre of the distribution.

Histogram	Stem plot
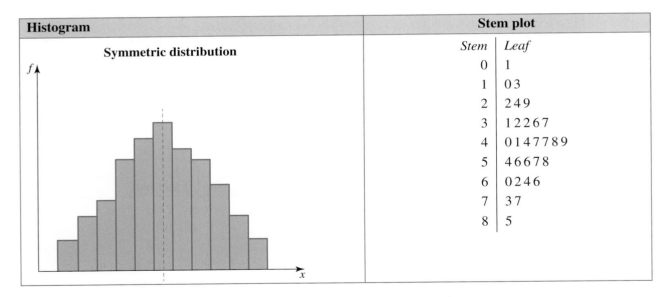	Stem | Leaf 0 | 1 1 | 0 3 2 | 2 4 9 3 | 1 2 2 6 7 4 | 0 1 4 7 7 8 9 5 | 4 6 6 7 8 6 | 0 2 4 6 7 | 3 7 8 | 5

If symmetry exists, the median and mean will occur at approximately the same position. An example of a symmetric distribution might be the examination results for Maths students.

Asymmetric distribution

Asymmetric distributions are sets of data that display different shapes and different amounts of data above and below the centre. An asymmetric (skewed) distribution can either be **positively** or **negatively skewed**.

Positively skewed

A distribution is **positively skewed** if it tapers off in a long tail to the right-hand side of the peak of the distribution. The bulk of the data is usually around the peak of the distribution, which is on the left-hand side of the distribution.

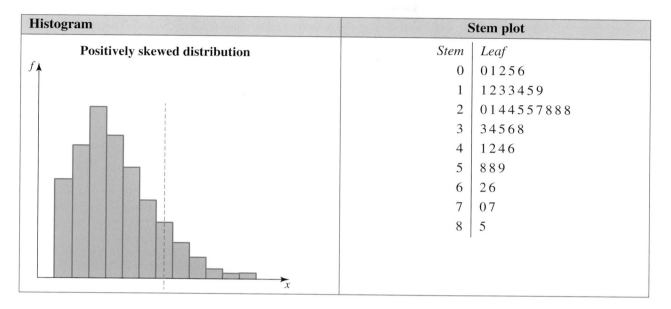

Histogram	Stem plot	
	Stem	Leaf
	0	0 1 2 5 6
	1	1 2 3 3 4 5 9
	2	0 1 4 4 5 5 7 8 8 8
	3	3 4 5 6 8
	4	1 2 4 6
	5	8 8 9
	6	2 6
	7	0 7
	8	5

For positively skewed distributions, the median is usually smaller than the mean (the mean will be on the same side of the median as the tail or skew). An example of a positively skewed distribution would be the number of hours spent playing sports versus age, as the number of hours spent playing sport is likely to reduce with age.

Negatively skewed

A distribution is **negatively skewed** if it tapers off in a long tail to the left-hand side of the peak of the distribution. The bulk of the data is usually around the peak of the distribution, which is on the right-hand side of the distribution.

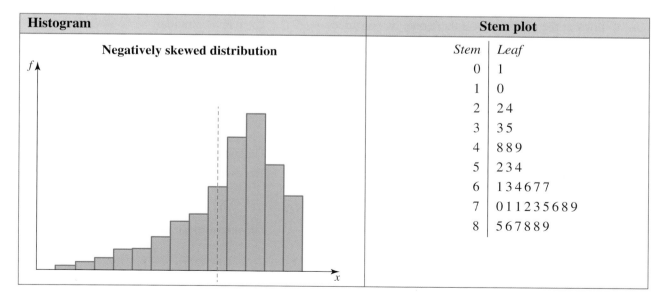

Histogram	Stem plot	
	Stem	Leaf
	0	1
	1	0
	2	2 4
	3	3 5
	4	8 8 9
	5	2 3 4
	6	1 3 4 6 7 7
	7	0 1 1 2 3 5 6 8 9
	8	5 6 7 8 8 9

For negatively skewed distributions, the median is usually greater than the mean. An example of a negatively skewed distribution would be the distribution of the number of heart attacks with age, as the number of people having heart attacks would increase with age.

Identify the modality and shape of each of the following diagrams.

a.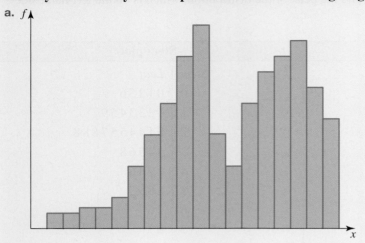

b.

Stem	Leaf
0	13
1	22467
2	1679
3	08
4	
5	13
6	4578
7	013
8	89
9	

c.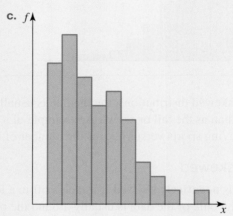

THINK

a. 1. Look at the modality. Are there one, two or three or more peaks?

2. Does the data appear asymmetric or symmetric?

3. If the data is asymmetric, does the distribution tail off to the left or right?

b. 1. Look at the modality. Are there one, two, or three or more peaks?

2. Does the data appear asymmetric or symmetric?

c. 1. Look at the modality. Are there one, two, or three or more peaks?

2. Does the data appear asymmetric or symmetric?

3. If the data is asymmetric, does the distribution tail off to the left or right?

WRITE

The data is bimodal.

The data is negatively skewed.

The data is bimodal.

The data is symmetric.

The data is unimodal.

The data is positively skewed.

14.4.3 Central tendency, spread and outliers

Measures of central tendency

A **histogram** is a graphical representation of group data. Measures of central tendency (mean, median and mode) can be calculated from a histogram.

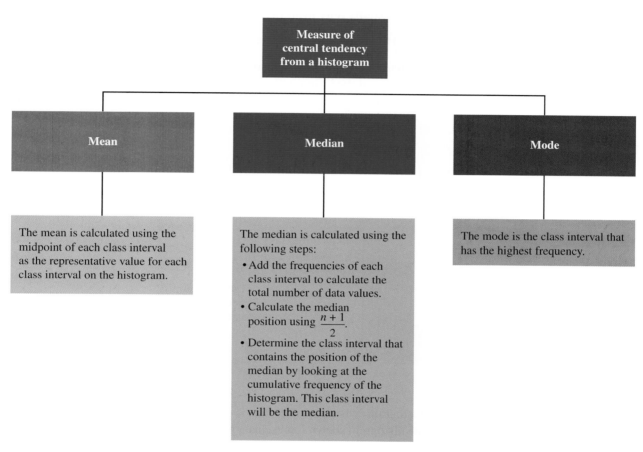

Measure of central tendency from a histogram

Mean

The mean is calculated using the midpoint of each class interval as the representative value for each class interval on the histogram.

Median

The median is calculated using the following steps:
- Add the frequencies of each class interval to calculate the total number of data values.
- Calculate the median position using $\frac{n+1}{2}$.
- Determine the class interval that contains the position of the median by looking at the cumulative frequency of the histogram. This class interval will be the median.

Mode

The mode is the class interval that has the highest frequency.

Spread

The spread of a histogram examines how closely packed together the data is.

Bulk of data close together

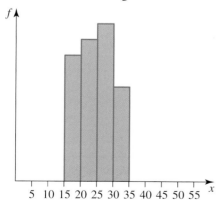

The histogram has a small spread as the bulk of the data is close together.

Bulk of data far apart

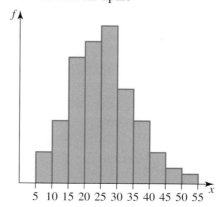

The histogram has a large spread as the bulk of the data is far apart.

To discuss the spread of a histogram

The measure most commonly used to discuss the spread of a histogram is the range.

- Range = maximum value − minimum value
- The lower and upper quartiles of a histogram can be written as class intervals. The IQR is more complicated to calculate, but it will usually only need to be determined if a list of data values is given. Q_1 will occur at the $\left(\dfrac{n+1}{4}\right)$th value and Q_3 will occur at the $\left(\dfrac{n+1}{4}\times 3\right)$th value.
- The standard deviation can be calculated from a histogram using digital technologies that allow the input of class intervals. The standard deviation will usually only need to be calculated if the data values of a histogram are provided.

Outliers

Spaces between bars on a histogram represent class intervals that have a frequency of 0.

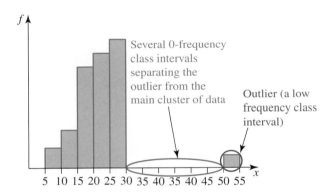

In this course, all outliers read from a histogram (where no other data is provided) must be written as 'possible outliers'. The class interval 50–55 on the histogram shown represents a possible outlier.

---

tlvd-4957

WORKED EXAMPLE 12 Determining measures of central tendency from a histogram

The histogram shows the heights of students in a class.
a. Using the histogram, calculate the following statistics for heights of students.
  i. Mean
  ii. Median
  iii. Mode
  iv. Range
  v. Q_1 and Q_3
b. Determine any possible outliers.

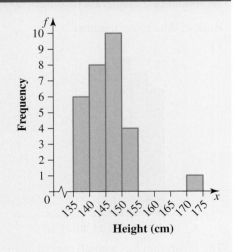

THINK

a. i. 1. The mean can be calculated using the formula for the mean of grouped data. Use the midpoint to calculate the mean.

WRITE

Class interval	Frequency (*f*)	Midpoint (*x*)	*xf*
135–140	6	137.5	825
140–145	8	142.5	1140
145–150	10	147.5	1475
150–155	4	152.5	610
155–160	0	157.5	0
160–165	0	162.5	0
165–170	0	167.5	0
170–175	1	172.5	172.5
Totals	29		4222.5

$$\bar{x} = \frac{\sum xf}{n}$$
$$= \frac{4222.5}{29}$$
$$= 145.6$$

ii. To calculate the median, count the total number of data values by adding the frequency of each class interval on the histogram, then divide the total by 2.

- The median occurs at the $\frac{29+1}{2} = 15$th value.

- Determine which class interval contains the 15th value by calculating the cumulative frequency. The 15th value is contained in the 145–150 cm class interval, which starts at the 15th value and finishes at the 24th value.

Total number of data values $= 6 + 8 + 10 + 4 + 1$
$$= 29$$
The median occurs at the $\frac{29+1}{2} = 15$th value.
The median occurs at the 15th value.
The median occurs in the 145–150 cm class interval.

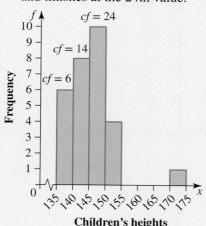

Children's heights

iii. The mode is the class interval that has the highest frequency.

The mode is 145–150 cm.

| iv. Calculate the range by subtracting the minimum class interval value from the maximum class interval value. | Range $= 175 - 135$ $= 40\,\text{cm}$ |

v. 1. Determine at which value Q_1 occurs. Determine which class interval contains the 7.5th value by calculating the cumulative frequency.

Q_1 occurs at the $\dfrac{29+1}{4} = 7.5$th value.

Q_1 occurs in the 140–145 cm class interval.

2. Determine at which value Q_3 occurs. Determine which class interval contains the 22.5th value by calculating the cumulative frequency.

Q_3 occurs at the $\dfrac{29+1}{4} \times 3 = 22.5$th value.

Q_3 occurs in the 145–150 cm class interval.

b. Look at the graph to see if any extreme values occur. Make sure to state that it is a possible outlier as no calculations have been made to prove whether it is or isn't an outlier.

A possible outlier occurs at the height of 170–175 cm.

14.4 Exercise

Note: Where necessary, give answers correct to 1 decimal place.

1. **MC** The histogram shown is:

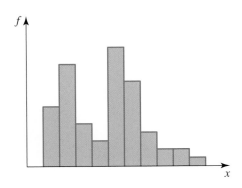

A. unimodal and positively skewed.
B. bimodal and positively skewed.
C. unimodal and symmetric.
D. unimodal and negatively skewed.
E. bimodal and negatively skewed.

2. **WE11** Identify the modality and shape of the following stem plot, and state whether there are any possible outliers.

Key: 4 | 3 = 43

Stem	Leaf
0	7
1	1
2	0 1
3	4 8
4	5 6 7 9
5	2 3 4 5 9
6	0 1 2 2 2 3 4 7 8
7	4 5 5 6 7 8
8	
9	0

3. Answer the following questions.
   i. Identify the modality of each of the histograms.
   ii. State the shape of each of the histograms.

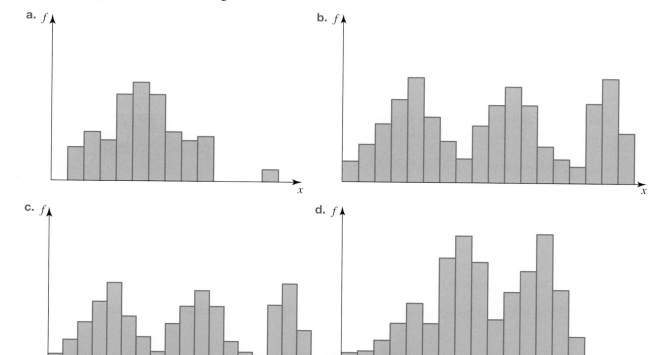

a. f

b. f

c. f

d. f

Use the histogram shown to answer questions 4–6.

4. **MC** The range of student weights is:

   A. 20 kg   **B.** 18 kg
   C. 36 kg   **D.** 38 kg
   E. 80 kg

5. **MC** The mean weight of students is closest to:

   A. 53.0 kg   **B.** 54.6 kg
   C. 54.7 kg   **D.** 52.8 kg
   E. 52.9 kg

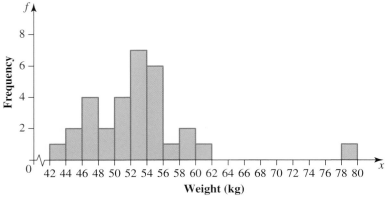

6. **MC** The median weight falls into class interval:

   A. 50–52 kg   **B.** 52–54 kg   **C.** 54–56 kg   **D.** 56–58 kg   **E.** 58–60 kg

7. Calculate the mean, median and mode for the data represented by the histogram shown.
   Compare these values.

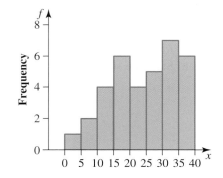

8. A florist recorded the number of bouquets of flowers sold daily over a three-week period. The results are recorded below.

   $$2, 14, 15, 1, 14, 17, 8, 19, 8, 5, 6, 16, 12, 9, 4, 14, 2, 3, 9, 19, 16$$

   a. Calculate:

      i. the mean   **iii.** the IQR
      ii. the median   **iv.** the standard deviation.

   b. Draw a histogram showing the number of bouquets of flowers sold.

   c. Mark the median, mean and IQR on the histogram.

9. **WE12** The following histogram shows the number of 50-m laps completed by swimmers at a local pool.

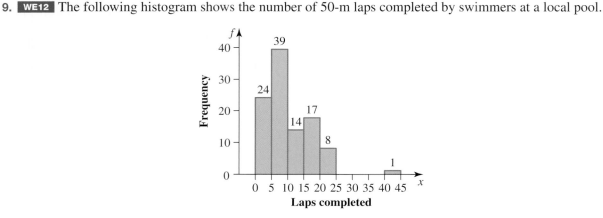

a. Using the histogram, calculate the following measures of central tendency for laps completed.

   i. Mean   **ii.** Median   **iii.** Mode   **iv.** Range   **v.** Q_1 and Q_3

b. Determine any possible outliers.

10. Describe how an outlier can be identified from a histogram.

11. The mean of a histogram is 22.5 cm and the median is in the class interval 30–35 cm. Explain the shape of the histogram.

12. Give a possible real-life example of:

   a. bimodal distribution

   b. negatively skewed distribution.

13. Darwyn and Morgan were discussing the stem plot shown. Darwyn said that the plot was symmetric but Morgan thought it was negatively skewed. Explain who is correct and why.

Key: $1\,|\,2 = 12$

Stem	Leaf
0	1
1	0
2	
3	
4	
5	
6	1 2 4
7	2 5 7 7
8	0 3 3 5 8 9
9	0 1 2 2 2 6 7 8 8
10	4 4 5 6 7 8 9
11	1 2 3 4 5
12	1 8 9

14. The following histograms show the Mathematics examination results for two different classes. Write a report comparing the results of the two classes in terms of central tendency and spread (range only).

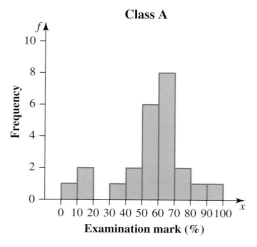

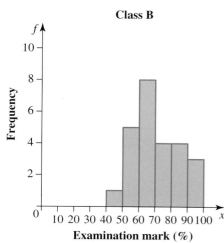

15. The median of the histogram shown occurs at the 8th data value. State a possible exact value of the median and justify your answer.

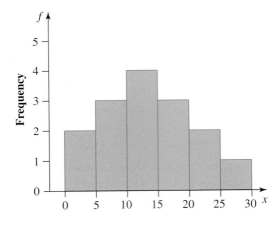

14.5 Characteristics of data sets

LEARNING INTENTION

At the end of this subtopic you should be able to:
- read and interpret two-way tables
- interpret line, conversion and step graphs
- compare line graphs.

14.5.1 Reading and interpreting two-way tables

In this section, we consider in more detail how to read and interpret two-way tables.

The table shown is a two-way table that relates two variables: the ability to swim and age group.

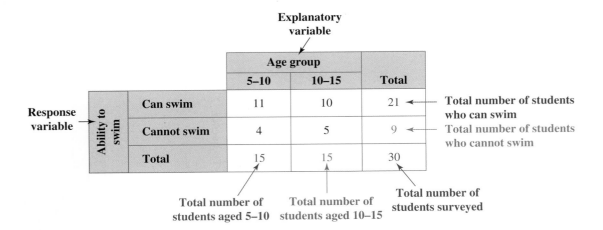

How do we interpret this information? If we want to find the number of students aged 5–10 who can swim, draw an imaginary line through the row 'Can swim' and an imaginary line through the column '5–10'. The cell where the two imaginary lines meet has the answer: 11 students aged 5–10 can swim.

		Age group		Total
		5–10	10–15	
Ability to swim	Can swim	11	10	21
	Cannot swim	4	5	9
	Total	15	15	30

The other cells of the table represent:

		Age group		Total
		5–10	10–15	
Ability to swim	Can swim	11 students 5–10 years old who can swim	10 students 10–15 years old who can swim	21 students who can swim
	Cannot swim	4 students 5–10 years old who cannot swim	5 students 10–15 years old who cannot swim	9 students who cannot swim
	Total	15 students 5–10 years old surveyed	15 students 10–15 years old surveyed	30 students surveyed

tlvd-4958

WORKED EXAMPLE 13 Interpreting a two-way frequency table

Two new school policies were introduced at Burumin High School.
School policy 1: The school uniform has to be worn on the way to school as well as on the way home from school.
School policy 2: Mobile phones are not to be used at school between 8:30 am and 3:30 pm.
The two-way table shown displays the responses to a survey conducted on 200 students.

		School policy 1		Total
		Agree	Disagree	
School policy 2	Agree	68	14	82
	Disagree	103	15	118
	Total	171	29	200

a. State the number of students who agree with both new school policies.
b. State the number of students who agree with school policy 1 but disagree with school policy 2.
c. State of the total number of students who disagree with school policy 2.
d. State the number of students who disagree with school policy 1 but agree with school policy 2.

a. 1. Draw a horizontal imaginary line through the row 'Agree' with school policy 2 and a vertical imaginary line through through the column 'Agree' with school policy 1.

		School policy 1		
		Agree	Disagree	Total
School policy 2	Agree	68	14	82
	Disagree	103	15	118
	Total	171	29	200

2. Read the number of students in the box where the two lines meet.

68 students agree with both new school policies.

b. 1. Draw a horizontal imaginary line through the row 'Disagree' with school policy 2 and a vertical imaginary line through the column 'Agree' with school policy 1.

		School policy 1		
		Agree	Disagree	Total
School policy 2	Agree	68	14	82
	Disagree	103	15	118
	Total	171	29	200

2. Read the number of students in the box where the two lines meet.

103 students agree with school policy 1 but disagree with school policy 2.

c. 1. Draw a horizontal imaginary line through the row 'Disagree' with school policy 2 and a vertical imaginary line through the column 'Total'.

		School policy 1		
		Agree	Disagree	Total
School policy 2	Agree	68	14	82
	Disagree	103	15	118
	Total	171	29	200

2. Read the number of students in the box where the two lines meet.

118 students disagree with school policy 2.

d. 1. Draw a horizontal imaginary line through the row 'Agree' with school policy 2 and a vertical imaginary line through the column 'Disagree' with school policy 1.

		School policy 1		
		Agree	Disagree	Total
School policy 2	Agree	68	14	82
	Disagree	103	15	118
	Total	171	29	200

2. Read the number of students in the box where the two lines meet.

14 students disagree with school policy 1 but agree with school policy 2.

14.5.2 Graphical representations of data

Data is often presented in the form of graphs. Graphical representations of data are easier to read and interpret. In this section we are going to discuss line graphs, conversion graphs and step graphs.

Line graphs

Line graphs are used to represent the relationship between two numerical continuous data sets.

These graphs consist of individually plotted points joined with a straight line.

The graph shown represents the relationship between two numerical continuous data sets: the temperature readings (°C) in Canberra recorded hourly on one day in August.

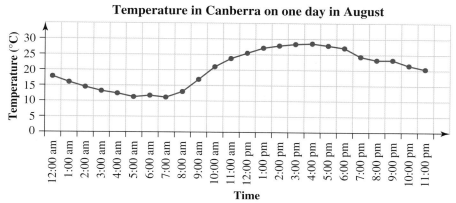

Temperature in Canberra on one day in August

Source: Adapted from Bureau of Meteorology

The information we can gather from this graph is:
- the minimum (lowest) temperature of the day and the time when it occurred (10.9 °C at 7:00 am)
- the maximum (highest) temperature of the day and the time when it occurred (28.5 °C at 4:00 pm)
- the temperature at any hour of the day
- the time of the day when a certain temperature occurred
- the times between which the temperature was increasing (from 7:00 am to 4:00 pm)
- the times between which the temperature was decreasing (from 4:00 pm to 8:00 pm).

The graph shown is a representation of the temperature readings (°C) in Sydney on one day in March, recorded hourly.

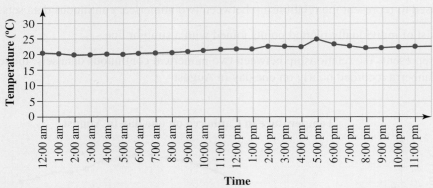

Temperature in Sydney on one day in March

Source: Adapted from Bureau of Meteorology, http://www.bom.gov.au/products/IDN6090 1/IDN60901.94768.shtml

a. **State the maximum temperature in Sydney on this day.**
b. **State the minimum temperature in Sydney on this day.**
c. **State the times between which the temperature in Sydney was increasing.**
d. **State the temperature in Sydney at 12:00 pm.**
e. **State the times when the temperature in Sydney was 24 °C.**

THINK

a. 1. The maximum temperature occurs at the highest point on the graph.

WRITE

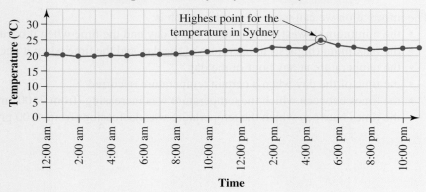

Temperature in Sydney on one day in March

Highest point for the temperature in Sydney

2. State the temperature shown on the graph.

Approximately 25 °C at 5:00 pm

b. 1. The minimum temperature occurs at the lowest point on the graph.

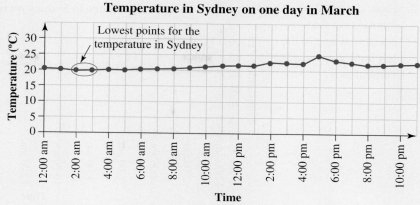

2. State the temperature shown on the graph.

Approximately 20 °C between 2:00 am and 3:00 am

c. 1. The increase in temperature is shown by a line that slopes upwards from left to right.

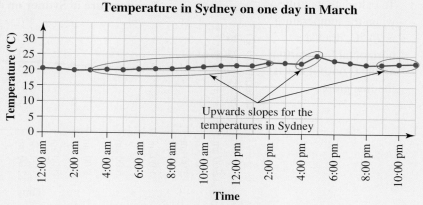

2. State the times required.

The temperature increase in Sydney on this day occurred between 2:00 am and 2:00 pm, between 4:00 pm and 5:00 pm and between 8:00 pm and 11:00 pm.

d. 1. To calculate the temperature at 12:00 pm, draw a vertical line from 12:00 pm until it meets the line graph.

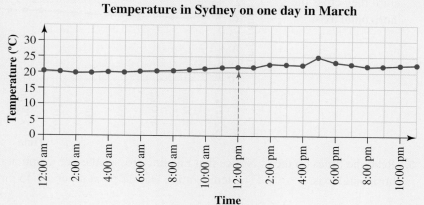

2. From the point of intersection draw a horizontal line to the left until it crosses the vertical axis.

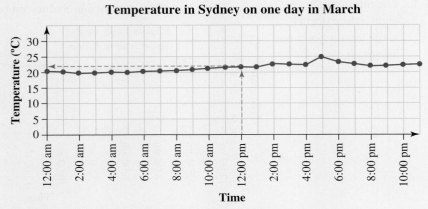

3. Read the temperature on the vertical axis.

The temperature at 12:00 pm was approximately 22 °C.

e. 1. To calculate the time when the temperature is 24 °C, draw a horizontal line starting at 24 °C until it meets the line graph.

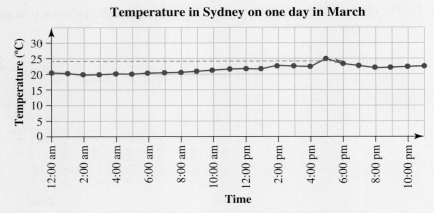

2. From the points of intersection draw vertical lines down until they cross the horizontal axis.

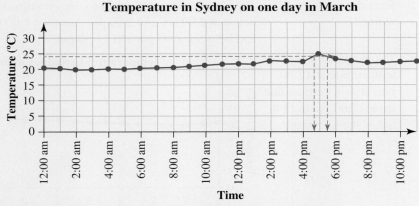

3. Read the times on the horizontal axis.

Notice that this line crosses the line graph in two points.
The temperature was 24 °C at around 4:45 pm and 5:30 pm.

14.5.3 Conversion graphs

As the name explicitly states, **conversion graphs** are just that: line graphs that convert one quantity into another.

Conversion graphs are often used to convert between foreign currencies or between measurements. These graphs are always represented by a straight line because the relationship between the two variables does not change.

Temperature can be measured in either degrees Celsius (°C) or degrees Fahrenheit (°F). In Australia we measure temperature using degrees Celsius. Other countries, such as the USA, measure temperature using degrees Fahrenheit. The graph shown represents this relationship.

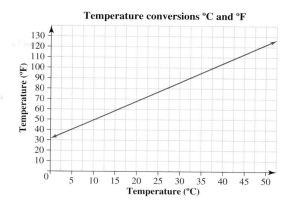

Temperature conversions °C and °F

WORKED EXAMPLE 15 Intrepreting a conversion graph

The conversion table shown displays the relationship between speed given in km/h and speed given in m/s. Use the conversion graph to convert

a. 180 km/h into m/s

b. 100 m/s into km/h.

Conversions km/h and m/s

THINK

a. 1. Draw a dashed horizontal line starting at 180 km/h until it touches the line describing the relationship between the two variables.

WRITE

a.

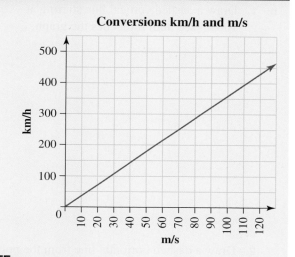

Conversions km/h and m/s

2. Draw a dashed vertical line from the point where the horizontal line intersects with the graph to the horizontal axis.

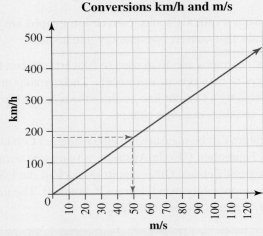

Conversions km/h and m/s

3. Read the value found and write the answer.

A speed of $180 \text{ km/h} = 50 \text{ m/s}$

b. 1. Draw a dashed vertical line starting at 100 m/sec until it touches the graph.

b.

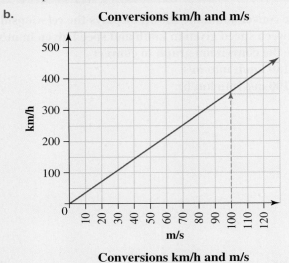

Conversions km/h and m/s

2. Draw a dashed horiontal line from the point on the graph that intersects with the vertical line to the vertical axis.

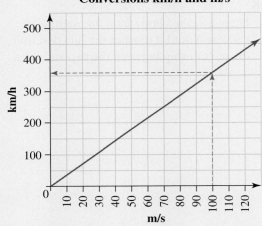

Conversions km/h and m/s

3. Read the value found and write the answer.

A speed of $100 \text{ m/s} = 360 \text{ km/h}$

14.5.4 Step graphs

Step graphs are made up of horizontal straight lines. These graphs are used when the value remains constant over intervals. Postage and parking costs are examples of step graphs.

The first step represents a cost of $0.60 for envelopes with a weight of less than 20 g. This means that regardless of whether the envelope weighs 10 g or 19 g, the postage cost is $0.60.

The open circle at the end of the interval means that the postage cost for 20 g is not $0.60, it jumps to the next step which is $1.20. This is why the next step has a closed dot at the beginning of the interval.

- The closed dot • means that the interval includes that value.
- The empty circle ○ means that the interval does not include that value.

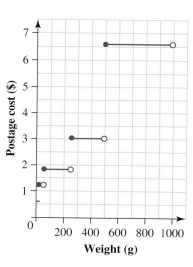

WORKED EXAMPLE 16 Interpreting a step graph

The step graph shown displays the parking cost in a car park.

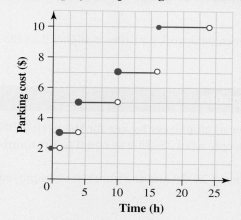

a. **Calculate the parking cost for 4 hours.**
b. **Calculate the parking cost for 15 hours.**

THINK	WRITE
a. 1. Describe the first step of the graph.	a. The cost for parking for one hour or less is $2.
2. Describe the second step of the graph.	As soon as an hour has passed, the cost for between 1 hour and up to 4 hours of parking is $3.
3. State the parking cost.	The four-hour mark is on the second step so the cost will be $3.
b. 1. Describe the third step of the graph.	b. Step three shows a cost of $5 for between 4 hours and up to 10 hours.
2. Describe the fourth step of the graph.	Step four shows a cost of $7 for parking between 10 hours and up to 16 hours.
3. State the parking cost.	The fifteen-hour mark is on the fourth step so the cost will be $7.

14.5.5 Comparing line graphs

Line graphs are very useful to help compare similar types of data. The line graph shown displays the proportion of ongoing employees working part-time by gender for a 15-year period.

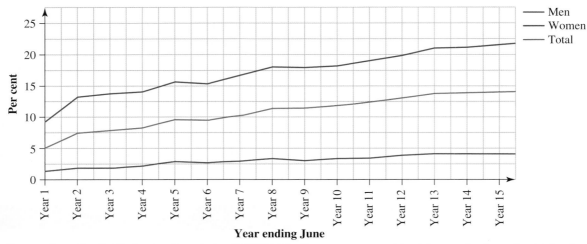

Source: Australian Public Service Commission (APSCD), http://www.apsc.gov.au/about-theapsc/parliamentary/state-of-the-service/new-sosr/06-diversity

This graph clearly shows that the percentage of women who work part-time is higher than the percentage of men working part-time for the 15-year period.

tlvd-4959

WORKED EXAMPLE 17 Comparing line graphs

The line graph shown displays the total number of employees by age for the period 2008 to 2022.
a. Determine when the number of employed people aged 25–35 was equal to the number of employed people aged 35 and above.
b. Determine the number of employed people aged 25–35 and the number of employed people aged 35 and above in 2018.

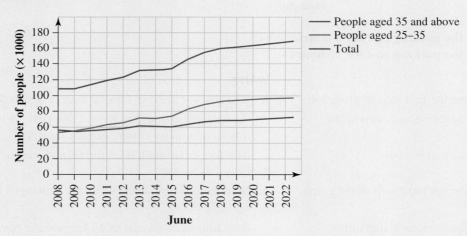

a. **1.** Mark the point where the two line graphs intersect.

a.

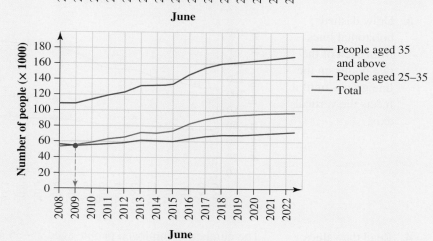

2. Draw a dotted vertical line from this point until it meets the horizontal axis.

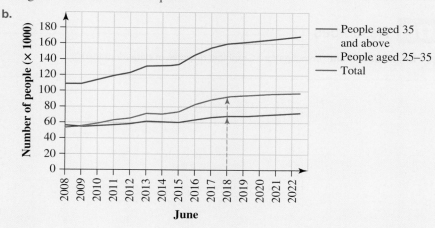

3. Write the answer.

In 2009 the numbers of employed people aged 25–35 and employed people aged 35 and above are equal.

b. **1.** Draw a dashed vertical line starting at 2018 until it touches the two line graphs.

b.

2. Mark the two points.

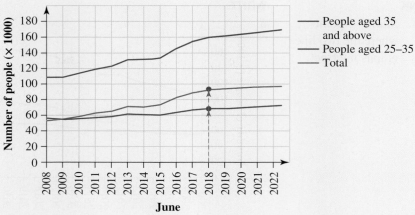

June

3. Draw dashed horizontal lines from each of the two points on the line graphs until it meets the vertical axis.

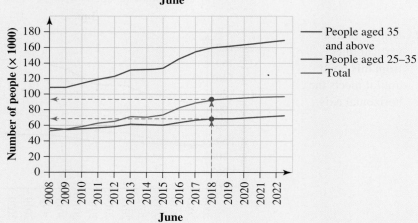

June

4. Read the values required.

The number of employed people aged 25–35 in 2018 was 95 000 and the number of employed people aged 35 and above was 70 000.

 Resources

 Digital document Investigation: Graphical displays of data (doc-15910)

1. **WE13** The students in a school were surveyed on whether they wear glasses for reading or not. The data is displayed in the two-way table shown.

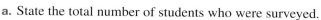

		Year level		
		Year 11	Year 12	Total
Reading glasses	Wear glasses for reading	26	23	49
	Do not wear glasses for reading	291	313	603
	Total	317	335	652

a. State the total number of students who were surveyed.
b. State the number of students in Year 11 who wear glasses for reading.
c. State the number of students in both years who do not wear glasses for reading.
d. State the number of students in Year 12 who were surveyed.

2. The two-way table shown displays the way a group of 150 students check the weather.

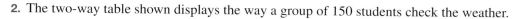

		Weather app		
		Weather app	TV news	Total
Students	Junior (Years 7–10)	21	47	68
	Senior (Years 11 and 12)	75	7	82
	Total	96	54	150

a. State the number of students who check the weather using a weather app.
b. State the number of senior students who check the weather by watching the TV news.
c. State the number of junior students who check the weather using a weather app.
d. State the number of junior students who were surveyed.

3. The contingency table shown below displays the information gained from a medical test screening for a virus. A positive test indicates that the patient has the virus.

		Test results		
		Accurate	**Not accurate**	**Total**
Virus	**With virus**	45	3	48
	Without virus	922	30	952
	Total	967	33	1000

a. State the number of patients who were screened for the virus.
b. State the number of positive tests that were recorded (that is, the number of tests in which the virus was detected).
c. Determine the percentage of test results that were accurate.
d. Based on the medical results, if a positive test is recorded, determine the percentage chance that a patient actually has the virus.

4. The two-way table below indicates the results of a radar surveillance system. If the system detects an intruder, an alarm is activated.

		Test results		
		Alarm activated	**Not activated**	**Total**
Detection	**Intruders**	40	8	48
	No intruders	4	148	152
	Total	44	156	200

a. State over how many nights the system was tested.
b. Determine the number of occasions when the alarm was activated.
c. If the alarm is activated, determine the percentage chance that there actually is an intruder.
d. If the alarm is not activated, determine the percentage chance that there is an intruder.
e. Calculate the percentage of accurate results over the test period.
f. Comment on the overall performance of the radar detection system.

5. **MC** For the two-way table shown, the number of nurses in regional area is:

		Location		
		Metro	**Regional**	**Total**
Career	**Teachers**	58	26	84
	Nurses	77	39	116
	Total	135	65	200

A. 77   B. 65   C. 39   D. 116   E. 55

6. The two-way table shown displays the relationship between two categorical variables: year level and pierced ears.

		Year level		
		Year 12	**Year 11**	**Total**
Pierced ears	**Yes**	2	35	37
	No	41	7	48
	Total	43	42	85

a. State the number of students in Year 11 who have had their ears pierced.
b. Calculate the total number of students surveyed.
c. State the number of students in Year 12 who have had their ears pierced.
d. Calculate the total number of students who have not had their ears pierced.

The information below is to be used in questions 7–9.

A test for a medical disease does not always produce the correct result. A positive test indicates that the patient has the condition. The table indicates the results of a trial on a number of patients who were known to either have the disease or known not to have the disease.

		Test results		Total
		Accurate	Not accurate	
Disease status	With disease	57	3	60
	Without disease	486	54	540
	Total	543	57	600

7. **MC** The overall accuracy of the test is:

   A. 9.5%   B. 90%   C. 90.5%   D. 92.5%   E. 95%

8. **MC** Based on the table, select the percentage of patients with the disease who have had it detected by the test.

   A. 9.5%   B. 90%   C. 90.5%   D. 95%   E. 92.5%

9. **MC** Select which of the following statements is correct.

   A. The test has a greater accuracy with positive tests than with negative tests.
   B. The test has a greater accuracy with negative tests than with positive tests.
   C. The test is equally accurate with positive and negative test results.
   D. The test is equally inaccurate with positive and negative test results.
   E. None of these.

10. Airport scanning equipment was tested by scanning 200 pieces of luggage. Prohibited items were placed in 50 bags and the scanning equipment detected 48 of them. The equipment detected prohibited items in five bags that did not have any forbidden items in them.

   a. Use the above information to complete the contingency (or two-way frequency) table below.

		Test results		Total
		Accurate	Not accurate	
Bags	With prohibited items			
	With no prohibited items			
	Total			

   b. Use the table to answer the following.

      i. Determine the percentage of bags with prohibited items that were detected.
      ii. Determine the percentage of false positives among the bags that had no prohibited items.
      iii. Determine the percentage of prohibited items that passed through the scanning equipment undetected.
      iv. Determine the overall percentage accuracy of the scanning equipment.

11. **WE14** The graph shows the temperature readings (°C) in Melbourne on a day in March, recorded hourly.

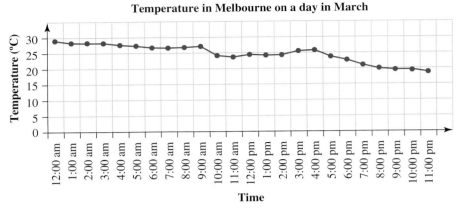

Temperature in Melbourne on a day in March

Source: Exchange-Rates.com http://www.exchange-rates.org/history/AUD/USD/G/30.

a. State the maximum temperature in Melbourne on this day.
b. State the minimum temperature in Melbourne on this day.
c. State the times between which the temperature in Melbourne was decreasing.
d. State the temperature in Melbourne at 9:00 am on this day.
e. State the times at which the temperature in Melbourne was 21 °C.

12. The line graph shown displays the average rainfall levels (mm) in Australia for a 13-year period.

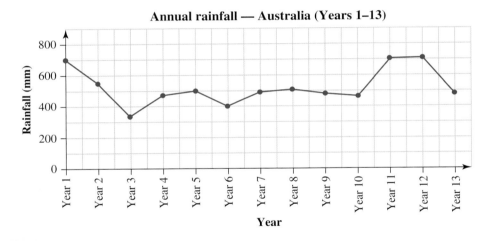

Annual rainfall — Australia (Years 1–13)

a. State when the maximum average annual rainfall occurred.
b. State the average annual rainfall in Australia for the sixth year.
c. State the years between which the average annual rainfall in Australia was increasing.
d. State the years in which the average annual rainfall in Australia was 500 mm.
e. State when the minimum average annual rainfall in Australia occurred and what it was.

13. The line graph shown displays the daily maximum temperature in Launceston, Tasmania, for a month in winter.

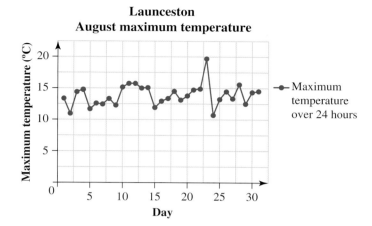

Launceston
August maximum temperature

—•— Maximum temperature over 24 hours

Use the line given to answer the following questions.

a. State which day had the lowest maximum temperature for the month and state this temperature.
b. State which day had the highest maximum temperature for the month and state this temperature.
c. Determine the difference between the maximum temperatures on days 14 and 19.
d. State the days on which the maximum temperature was 15.5 °C.
e. Determine for how many days the maximum temperature was less than 12 °C.

14. **WE15** The conversion graph shown displays the relationship between the Australian dollar (A$) and the American dollar (US$).

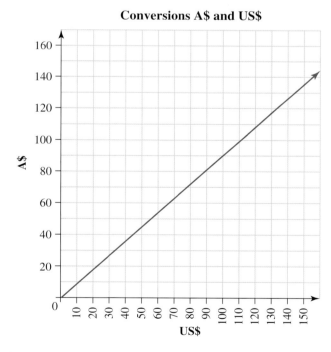

Conversions A$ and US$

Use the conversion graph to convert:

a. US$80 into A$
b. A$100 into US$.

15. The line graph shown displays the mean maximum temperature in Antarctica at the station Mawson for the years 1954–2018.

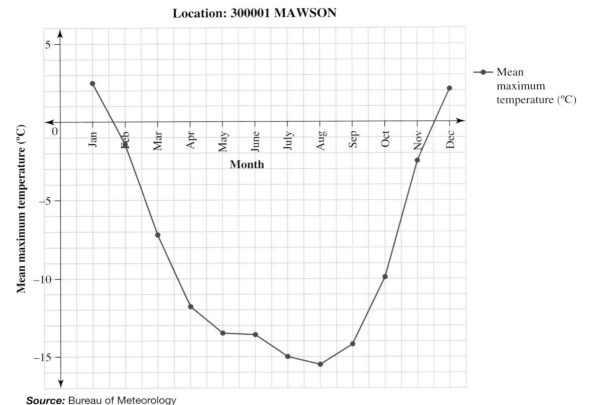

Location: 300001 MAWSON

Source: Bureau of Meteorology

Use the line graph given to answer the following questions.

a. State which month had the lowest mean maximum temperature in Antarctica and state this temperature.
b. State which month had the highest mean maximum temperature in Antarctica and state this temperature.
c. Determine the difference between the two maximum temperatures found in parts **a** and **b**.
d. State the month in which the mean maximum temperature in Antarctica was $-2.1\,°C$.
e. State for how many months the mean maximum temperature in Antarctica was greater than $-8\,°C$.

16. The graph shown represents a conversion graph between miles and kilometres.

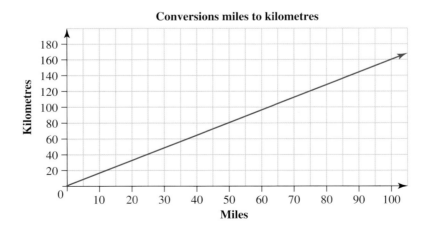

Conversions miles to kilometres

a. Calculate the approximate number of kilometres equal to 80 miles.

b. Calculate the approximate number of kilometres equal to 50 miles.

c. Calculate the approximate number of miles equal to 30 kilometres.

d. Calculate the approximate number of miles equal to 140 kilometres.

17. The conversion graph shown displays the relationship between centimetres and inches.
Use the conversion graph to convert:

a. 10 cm, 15 cm, 46 cm and 51 cm into inches

b. 20 inches and 35 inches, 44 inches and 51 inches into centimetres.

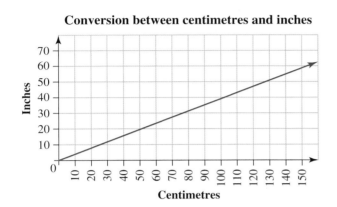

Conversion between centimetres and inches

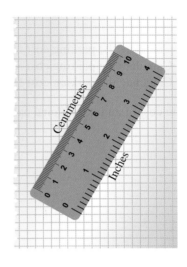

18. Use the temperature conversion graph given to answer the following questions.

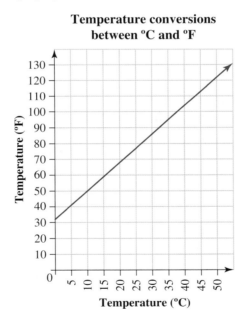

Temperature conversions between °C and °F

a. State the temperature in °F equivalent to 35 °C.

b. State the temperature in °F equivalent to 10 °C.

c. State the temperature in °C equivalent to 41 °F.

d. State the temperature in °C equivalent to 104 °F.

19. **WE16** The step graph shown represents the cost of parking at the domestic terminal at the Brisbane Airport.

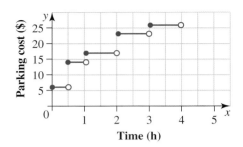

a. Calculate the parking cost for 45 minutes.
b. Calculate the parking cost for 2 hours.
c. Calculate the parking cost for 3.5 hours.

20. The cost of an international call is displayed in the step graph shown.

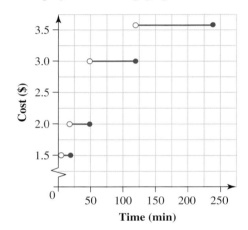

a. Determine the cost of a 5-minute call.
b. Determine the cost of a 50-minute call.
c. Determine the length of a call if the cost of the call is $3.50.

21. **WE17** The line graph shown displays the percentage of full-time regional secondary school students by gender in over a 12-month period.

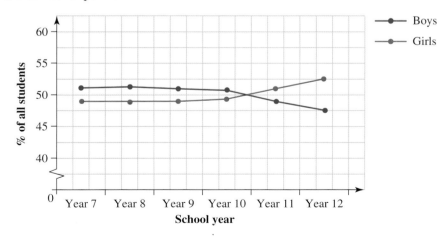

a. Determine the percentage of boys and the percentage of girls enrolled in Year 7 during the year.
b. Determine the percentage of boys and the percentage of girls enrolled in Year 12 during the year.
c. Determine in which year level the percentage of boys was equal to the percentage of girls.

22. The line graph shown displays the gender wage gap in Australia over a 20-year period.

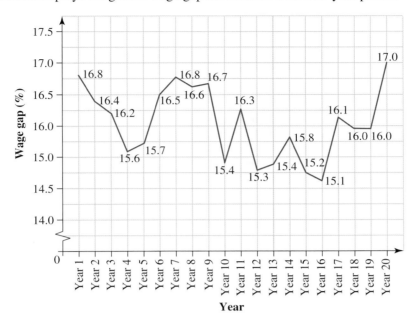

a. State in which year the wage gap was at its lowest percentage and state its value.
b. State in which year the wage gap was at its highest percentage and state its value.
c. Calculate the percentage increase in the wage gap in years 1 to 20.
d. Determine between which years the highest decrease in the wage gap occurred.
e. State in which year the wage gap was 16.5%.

14.6 Drawing and using column and pie graphs

LEARNING INTENTION

At the end of this subtopic you should be able to:
• draw and interpret column and pie graphs.

14.6.1 Drawing column graphs

In this subtopic you will learn to draw both **column graphs** and pie graphs. Recall that column graphs are used to represent categorical data. Pie graphs are circular graphical representations of data.

Drawing column graphs

Column graphs consist of vertical or horizontal bars of equal width. The frequency is measured by the height of the column. Both axes have to be clearly labelled, and appropriate and accurate scales are required. The title should explicitly state what the column graph represents.

Horizontal column graphs

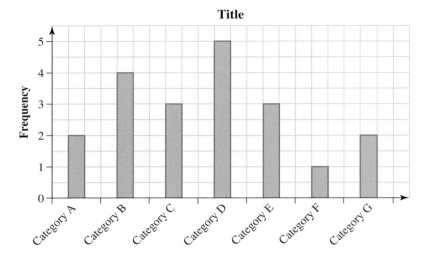

The frequency is always shown on the horizontal axis and the categories of the data are shown on the vertical axis. Horizontal column graphs are usually drawn when the category names are long.

Vertical column graphs

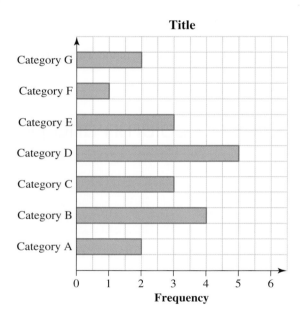

The frequency is always shown on the vertical axis and the categories of the data are shown on the horizontal axis.

Features of a column graph

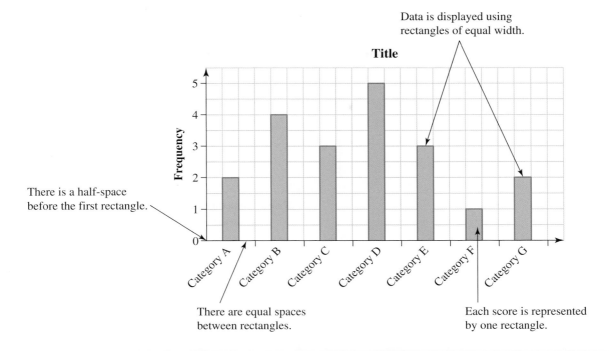

Data is displayed using rectangles of equal width.

Title

There is a half-space before the first rectangle.

There are equal spaces between rectangles.

Each score is represented by one rectangle.

tlvd-4960

WORKED EXAMPLE 18 Drawing a column graph

The data shown is part of a random sample of 25 Year 11 students. The answers were related to how often the students use the Internet to do research for school work. Draw a column graph displaying the data for Year 11 only.

<table>
<thead>
<tr><th colspan="2"></th><th colspan="2">Year level</th><th rowspan="2">Total</th></tr>
<tr><th colspan="2"></th><th>Year 11</th><th>Year 12</th></tr>
</thead>
<tbody>
<tr><td rowspan="4">Use of internet for school work</td><td>Rarely</td><td>2</td><td>8</td><td>10</td></tr>
<tr><td>Sometimes</td><td>4</td><td>4</td><td>8</td></tr>
<tr><td>Often</td><td>6</td><td>1</td><td>7</td></tr>
<tr><td>Total</td><td>12</td><td>13</td><td>25</td></tr>
</tbody>
</table>

THINK

1. Draw the two axes and label them with the two variables.

WRITE/DRAW

The categorical variable, internet usage, is displayed on the horizontal axis and the numerical discrete variable, number of Year 11 students, on the vertical axis.

Use of internet for school work

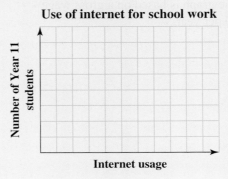

Number of Year 11 students

Internet usage

2. Choose the scales of the two axes.

The categorical variables are equally spaced on the horizontal axis. The scale of the vertical axis will have 6 ticks from 0 to 6.

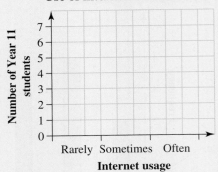

Use of internet for school work

3. Draw a rectangle for each category.

The category 'rarely' is 2 units high because there are 2 Year 11 students using the internet for school work rarely.

The category 'sometimes' is 4 units high, and the category 'often' is 6 units high.

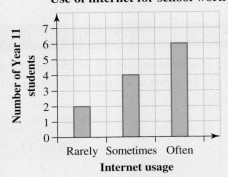

Use of internet for school work

14.6.2 Drawing pie graphs

Pie graphs are circles divided into sectors. Each sector represents one item of the data. The pie graph shown has four sectors: net wage, tax, Medicare levy and superannuation contribution.

Each sector represents the percentage of each item in the gross wage. This pie graph is based on a gross wage (before tax) of $65 000 per year.

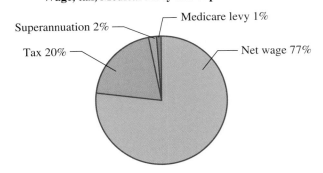

Wage, tax, Medicare levy and superannuation

Percentage	Angle
1% is the Medicare levy: $1\% \text{ of } \$65\,000 = \dfrac{1}{100} \times 6500$ $\qquad\qquad\qquad = \$650$	The corresponding angle on the circle is: $1\% \text{ of } 360° = \dfrac{1}{100} \times 360°$ $\qquad\qquad\quad = 3.6°$
2% is the superannuation contribution: $2\% = \dfrac{2}{100} \times 65\,000$ $\quad = \$1300$	The corresponding angle on the circle is: $2\% \text{ of } 360° = \dfrac{2}{100} \times 360°$ $\qquad\qquad\quad = 7.2°$
20% is the tax: $20\% = \dfrac{20}{100} \times 65\,000$ $\qquad = \$13\,000$	The corresponding angle on the circle is: $20\% \text{ of } 360° = \dfrac{20}{100} \times 360°$ $\qquad\qquad\qquad = 72°$
77% is the net wage (after tax): $77\% = \dfrac{77}{100} \times 65\,000$ $\qquad = \$50\,050$	The corresponding angle on the circle is: $77\% \text{ of } 360° = \dfrac{77}{100} \times 360°$ $\qquad\qquad\qquad = 277.2°$

Features of pie graphs

Each sector has to be coloured differently. All labels and other writing have to be horizontal and equally distanced from the circle.

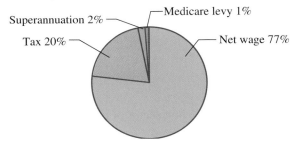

Wage, tax, Medicare levy and superannuation

Superannuation 2% — Medicare levy 1%

Tax 20% — Net wage 77%

---

tlvd-4961

WORKED EXAMPLE 19 Constructing and interpreting pie graphs

A recipe for baked beans contains the following ingredients:
300 g dry beans, 50 g onion, 750 g diced tomatoes, 30 g mustard and 870 g smoked hock.
a. Construct a pie graph to represent these ingredients.
b. Determine how many grams of dry beans are required to make 100 g of this dish.
c. Determine how many grams of dry beans are required for a serve of 250 g of this dish.

THINK	WRITE/DRAW
a. 1. Establish the number of sectors required for the pie graph.	a. The pie graph requires 5 sectors, one for each ingredient: dry beans, onion, diced tomatoes, mustard and smoked hock.
2. Calculate the total quantity.	$300 + 50 + 750 + 30 + 870 = 2000$ g

3. Calculate the percentage for each quantity.

Dry beans: $\dfrac{300}{2000} \times 100 = 15\%$

Onion: $\dfrac{50}{2000} \times 100 = 2.5\%$

Diced tomatoes: $\dfrac{750}{2000} \times 100 = 37.5\%$

Mustard: $\dfrac{30}{2000} \times 100 = 1.5\%$

Smoked hock: $\dfrac{870}{2000} \times 100 = 43.5\%$

4. Check that the percentage sum is 100%.

$15\% + 2.5\% + 37.5\% + 1.5\% + 43.5\% = 100\%$

5. Calculate the angles for each sector out of 360°.
To calculate the angle for each sector, use the formula:
$$\text{angle} = x\% \times 360°$$
$$= \dfrac{x}{100} \times 360°$$

Dry beans: $\dfrac{15}{100} \times 360° = 54°$

Onion: $\dfrac{2.5}{100} \times 360° = 9°$

Diced tomatoes: $\dfrac{37.5}{100} \times 360° = 135°$

Mustard: $\dfrac{1.5}{100} \times 360° = 5.4°$

Smoked hock: $\dfrac{43.5}{100} \times 360° = 156.6°$

6. Check that the sum of the angles is 360°.

$54 + 9 + 135 + 5.4 + 156.6 = 360°$

7. Draw a circle for the pie graph and a vertical radius from the centre of the circle to the top of the circle.

8. Draw the first sector of the pie graph.

The first ingredient is dry beans with an angle of 54°. Measure 54° from the vertical line in a clockwise direction.

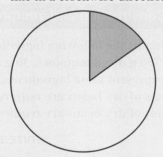

9. Draw the second sector of the pie graph.

The second ingredient is onion with an angle of 9°. Measure 9° from the right-hand side of the sector in a clockwise direction.

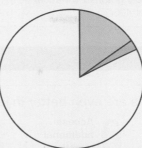

10. Draw the rest of the sectors of the pie graph.

Continue to draw the remaining sectors going in a clockwise direction around the circle.

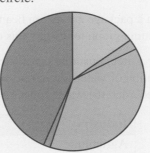

11. Label all sectors.

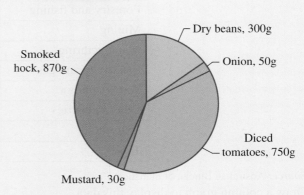

Smoked hock, 870g
Dry beans, 300g
Onion, 50g
Diced tomatoes, 750g
Mustard, 30g

b. 1. Calculate the quantity required. Use the formula:
$$\frac{\text{quantity of the item}}{\text{total quantity}} \times \text{new quantity}$$

b. Dry beans:
$$\frac{300}{2000} \times 100 = 15 \text{ g}$$

c. 1. Calculate the quantity required. Use the formula:
$$\frac{\text{quantity of the item}}{\text{total quantity}} \times \text{new quantity}$$

c. Dry beans:
$$\frac{300}{2000} \times 250 = 37.5 \text{ g}$$

14.6 Exercise

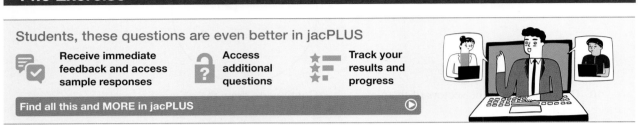

1. **WE18** The ingredients on a package are: green beans 70 g, baby corn spears 63 g, asparagus 67 g, butter 4 g and water 6 g. Draw a column graph to display this data.

2. Vlad was interested in the water consumption in Victoria. He researched the Australian Bureau of Statistics to determine the consumption of water for various industries. He downloaded the Water Account released on 27 November 2018 and recorded the water consumption for Victoria in the table below.

Industry	ML (million litres)
Agriculture	1234
Forestry and fishing	13
Mining	10
Manufacturing	144
Electricity and gas	117
Water supply	310
Other industries	220
Household	3 11
Total	**2359**

Source: Australian Bureau of Statistics

Draw a column graph to display this data.

3. A group of students was surveyed about their favourite type of chocolate. The results have been recorded in the table shown.

Type of chocolate	Frequency
Dark chocolate	35
Milk chocolate	47
White chocolate	18

Draw a column graph for this data set.

4. A new program to improve school attendance was introduced at Diramu College. A survey was conducted at the end of the program to seek teachers' opinion about the program's effect on attendance. These opinions were recorded in the table shown.

Opinion	Frequency
No improvement at all	9
Some improvement	15
Moderate improvement	56
High improvement	28
Total	**108**

   a. Using this data, draw a vertical column graph.
   b. Calculate the percentages for each category.
   c. Construct a pie graph to represent this data.

5. Fiona has been the manager of a toy shop for six months. She recorded the sales over this period of time in the table shown.

	Jan	Feb	Mar	Apr	May	Jun
Sales ($)	24 000	26 000	30 000	35 000	32 000	27 000

   Construct a column graph to represent this data.

6. The top 12 countries based on the gold medal tally at the 2012 London Olympics are recorded in the table shown.

Country	Gold medals
United States of America	46
People's Republic of China	38
Great Britain	29
Russian Federation	24
Republic of Korea	13
Germany	11
France	11
Italy	8
Hungary	8
Australia	7
Japan	7
Kazakhstan	7
Total	**209**

   Construct a column graph to represent this data.

7. **WE19** The ingredients on a package are 70 g green beans, 63 g baby corn spears, 67 g asparagus, 4 g butter and 6 g water.

   a. Draw a pie graph to display this data.
   b. Determine how many grams of asparagus there are in a 1-kg package.

8. Nicole has a gross wage of $95 000. She pays $40 000 in tax and $1425 for the Medicare levy, and deposits $1575 in her superannuation fund.

   a. Determine how much money she actually receives.
   b. Draw a pie graph to display this data, including the amount of money she actually receives.

9. An apple cake recipe has the following ingredients.

300 g apples

250 g flour

200 g caster sugar

250 g butter

4 eggs (240 g)

10 g ground cinnamon

a. Calculate the individual percentages of each ingredient.

b. Draw a pie graph to represent this data.

c. Determine the amount of flour that would be needed if the total weight of the ingredients was 1.5 kg.

10. The table shown displays a typical family budget.

Area	Cost ($)
Housing	420
Food	180
Clothing	72
Transport	96
Entertainment	144
Health	60
Savings	120
Miscellaneous	108

Construct a pie graph to display this data.

14.7 Misleading graphs

LEARNING INTENTION

At the end of this subtopic you should be able to:
- explore methods of misrepresenting data
- identify whether a graph is misleading.

14.7.1 Methods of misrepresenting data

Many people have reasons for misrepresenting data. For example, politicians may wish to magnify the progress achieved during their term, or business people may wish to accentuate their reported profits. There are numerous ways of misrepresenting data. In this section, only graphical methods of misrepresentation are considered.

The vertical and horizontal axes

It is a truism that the steeper the graph, the better the growth appears. A 'rule of thumb' for statisticians is that for the sake of appearances, the vertical axis should be two-thirds to three-quarters the length of the horizontal axis. This rule was established in order to have some comparability between graphs.

The figure shown illustrates how distorted a graph appears when the vertical axis is disproportionately large.

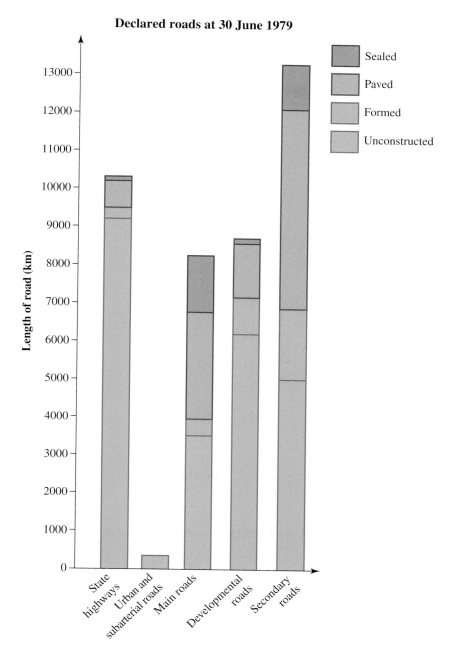

Declared roads at 30 June 1979

Legend:
- Sealed
- Paved
- Formed
- Unconstructed

Length of road (km) (vertical axis: 0 to 13000)

Horizontal axis categories: State highways, Urban and subarterial roads, Main roads, Developmental roads, Secondary roads

Changing the scale on the vertical axis

The following table gives the holdings of a corporation during a particular year.

Quarter	Holdings in $1 000 000
Jan–Mar	200
Apr–June	200
July–Sept	201
Oct–Dec	202

Here is one way of representing these data.

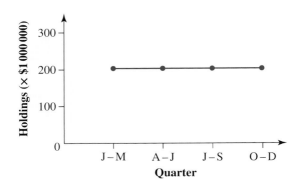

But it is not very spectacular, is it? Now look at the following graph showing the same information. The shareholders would be happier with this one.

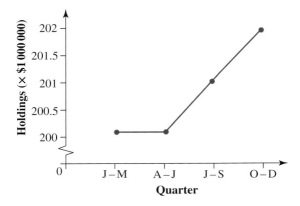

Omitting certain values

If one chose to ignore the second quarter's value, which shows no increase, then the graph would look even better.

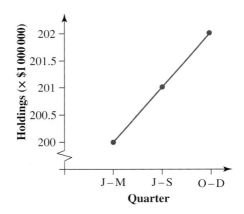

Foreshortening the vertical axis

Look at the figures below. Notice in graph (a) that the numbers from 0 to 4000 have been omitted. In graph (b) these numbers have been inserted. The rate of growth of the company looks far less spectacular in graph (b) than in graph (a). This is known as foreshortening the vertical axis.

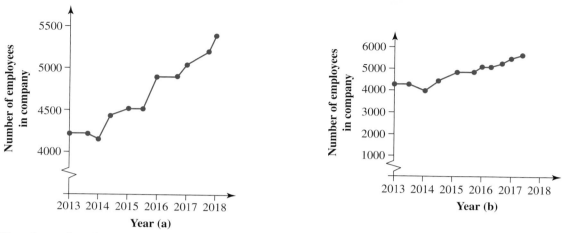

Foreshortening the vertical axis is a very common procedure. It does have the advantage of giving extra detail, but it can give the wrong impression about growth rates.

Visual impression

In this graph, height is the property that gives the true relation, yet the impression of a much greater increase is given by the volume of each money bag.

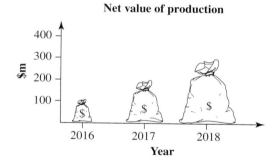

A non-linear scale on an axis or on both axes

Consider the following two graphs.

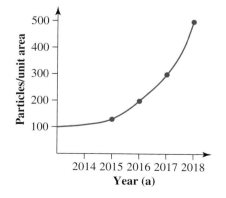

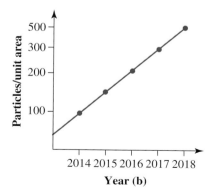

Both of these graphs show the same numerical information, but graph (a) has a linear scale on the vertical axis and graph (b) does not. Graph (a) emphasises the ever-increasing rate of growth of pollutants, whereas graph (b) suggests a slower, linear growth.

WORKED EXAMPLE 20 Determining whether a graph is misleading

The following data give wages and profits for a certain company. All figures are in millions of dollars.

Year	2015	2016	2017	2018
Wages	6	9	13	20
% increase in wages	25	50	44	54
Profits	1	1.5	2.5	5
% increase in profits	20	50	66	100

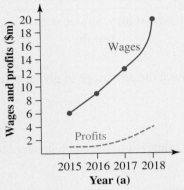

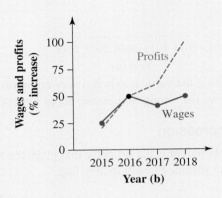

a. State whether the graphs accurately reflect the data.
b. Suggest which graph would you rather have published if you were:
   i. an employer dealing with employees requesting pay increases
   ii. an employee negotiating with an employer for a pay increase.

THINK	WRITE
a. 1. Look at the scales on both axes. Both scales are linear.	**a.** The graphs do represent data accurately. However, quite a different picture of wage and profit increases is painted by graphing with different units on the *y*-axis.
2. Look at the units on both axes. Graph (a) has its *y*-axis in $, whereas graph (b) has its *y*-axis in %.	
b. i. 1. Compare wage increases with profit increases.	**b. i.** The employer would prefer graph (a) because he/she could argue that employees' wages were increasing at a greater rate than profits.
2. The employer wants a graph that shows wages are already increasing exponentially to argue against pay increases.	
ii. 1. Consider again the increases in wages and profits.	**ii.** The employee would choose graph (b), arguing that profits were increasing at a great rate while wage increases clearly lagged behind.
2. The employee doesn't like to see profits increasing at a much greater rate than wages.	

14.7 Exercise

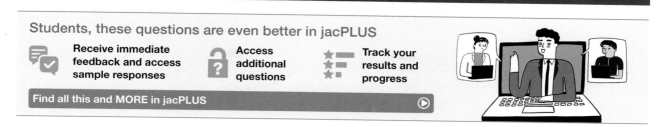
1. **WE20** The graph shows the dollars spent on research in a company for 2010, 2014 and 2018.

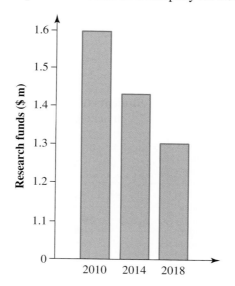

Draw another bar graph that minimises the appearance of the fall in research funds.

2. Examine this graph of employment growth in a company.

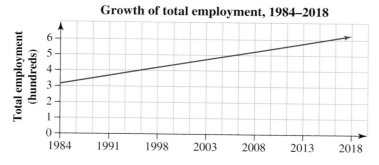

Explain why this graph is misleading.

3. Examine this graph.

International visitor arrivals, by month of visit — 2018

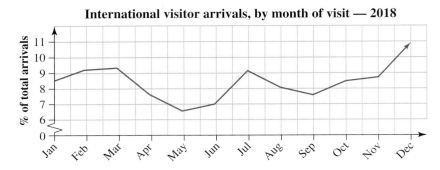

a. Redraw this graph with the vertical axis showing percentage of total arrivals starting at 0.
b. Explain whether the change in visitor arrivals appears to be as significant as the original graph suggests.

4. This graph shows the student-to-teacher ratio in Australia for the years 2008 and 2018.

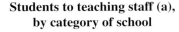

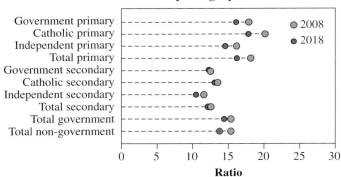

Note: Ratio = number of full-time equivalent students divided by the number of full-time equivalent teaching staff.

a. Describe what has generally happened to the ratio of students to teaching staff over the 10-year period.
b. A note says that the graph should not be used as a measure of class size. Explain why.

5. You run a company that is listed on the Stock Exchange. During the past year you have given substantial rises in salary to all your staff. However, profits have not been as spectacular as in the year before. The following table gives the figures for the mean salary and profits for each quarter.

Quarter	1st quarter	2nd quarter	3rd quarter	4th quarter
Profits (× $1 000 000)	6	5.9	6	6.5
Salaries (× $1 000 000)	4	5	6	7

Draw two graphs, one showing profits, the other showing salaries, that will show you in the best possible light to your shareholders.

6. You are a manufacturer and your plant is discharging heavy metals into a waterway. Your own chemists do tests every 3 months, and the following table gives the results for a period of 2 years.

Year	2017				2018			
Month	Jan.	Apr.	Jul.	Oct.	Jan.	Apr.	Jul.	Oct.
Concentration (parts per million)	7	9	18	25	30	40	49	57

Draw a graph that will show your company in the best light.

7. This pie graph shows the break-up of national health expenditure in 2017–18 from three sources: the Australian Government, state and local government, and non-government. (This expenditure relates to private health insurance, injury compensation insurers and individuals.)

Source	Expenditure ($m)	%
Australian Government	37 229	45
State and local government	21 646	25
Non-government	28 004	30

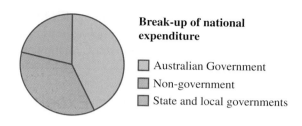

Break-up of national expenditure

- ☐ Australian Government
- ☐ Non-government
- ☐ State and local governments

a. Comment on the claim that $87 000 m was spent on health from these three sources.
b. State which area contributes least to national health expenditure. Comment on its quoted percentage.
c. State which area contributes the next greatest amount to national health expenditure. Comment on its quoted percentage.
d. The Australian Government contributes the greatest amount. Comment on its quoted percentage.
e. Consider the pie graph.

   i. Based on the percentages shown in the table, determine what the angles should be.
   ii. Based on the actual expenditures, determine what the angles should be.
   iii. Measure the angles in the pie graph and comment on their values.

8. This graph shows how the $27 that a buyer pays for a new album by their favourite singer-songwriter is distributed among the departments of a major recording company involved in its production and marketing.

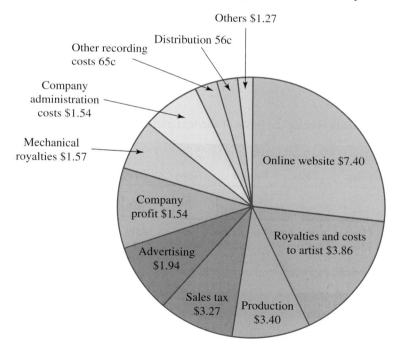

You are required to find out whether or not the graph is misleading. You must fully explain your reasoning and support any statements that you make.

a. Comment on the shape of the graph and how it could be obtained.
b. State whether your visual impression of the graph supports the figures.

14.8 Graphing and technology

14.8.1 Drawing lines in Excel worksheets

Excel worksheets allow us to draw various graphs accurately and easily from the data sets entered, as well as providing a convenient way of constructing line graphs.

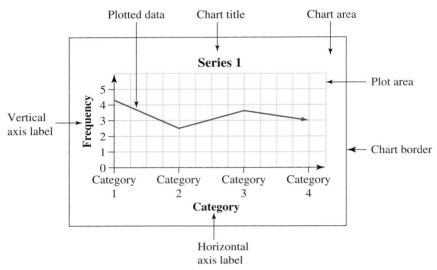

Chart tools

The Chart tools allow us to easily format the fonts of the text and the colours of various parts of the graph.

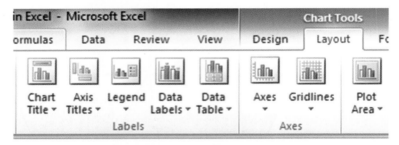

WORKED EXAMPLE 21 Using an Excel worksheet to draw a line graph

Using an Excel worksheet, draw a line graph for the data recorded in the table of values shown.

Time, min	0	1	2	3	4	5	6	7	8
Temperature, °C	21	25	29	31	33	30	31	32	33

THINK

1. Open a new Excel worksheet.

WRITE/DRAW

	A	B	C	D	E	F	G	H	I	J
1										

---

2. Label the two columns for the two variables. The explanatory variable will be written in column A.

Click in cell A1 and type 'Time, min'.

	A	B	C	D	E
1	Time, min				
2					

The response variable will be written in column B.

Click in cell B1 and type 'Temperature,°C'.

	A	B	C	D	E
1	Time, min	Temperature, °C			
2					

3. Insert the data.

Click in cell A2 and start typing in the data given for the explanatory variable.

	A	B	C
1	Time, min	Temperature, °C	
2	0		
3	1		
4	2		
5	3		
6	4		
7	5		
8	6		
9	7		
10	8		
11			

Click in cell B2 and start typing in the data given for the response variable.

	A	B	C
1	Time, min	Temperature, °C	
2	0	21	
3	1	25	
4	2	29	
5	3	31	
6	4	33	
7	5	30	
8	6	31	
9	7	32	
10	8	33	
11			

4. Draw the line graph.

Select the two columns starting from cell A1 to cell B10. To select these cells, click in cell A1, hold, move to cell B10 and release.

	A	B	C
1	Time, min	Temperature °C	
2	0	21	
3	1	25	
4	2	29	
5	3	31	
6	4	33	
7	5	30	
8	6	31	
9	7	32	
10	8	33	
11			

Go to Insert and click the 'Lines' icon on the 'Charts' menu.

Click any of the templates for a line. The graph shown is the default graph Excel draws. It draws the two data sets as two separate sets of data rather than as related to each other.

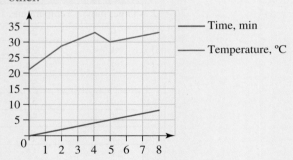

5. Format the line graph.

This graph can be formatted the way we want. Click 'Select data' on the menu bar. The screen shown will open.

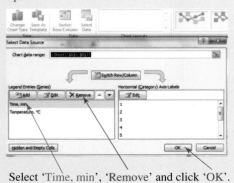

Select 'Time, min', 'Remove' and click 'OK'.

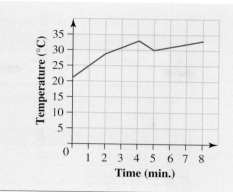

14.8.2 Drawing a column graph in an Excel worksheet

Excel worksheets provide a convenient way of constructing column graphs.

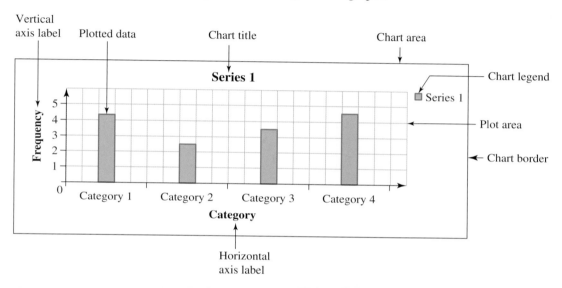

WORKED EXAMPLE 22 Using an Excel worksheet to draw a column graph

Using an Excel worksheet, draw a column graph for the data recorded in the table of values shown.

Opinion	Very poor	Poor	Average	Good	Very good	Excellent
Frequency	8	15	52	44	27	16

THINK

1. Open a new Excel worksheet.

2. Label the two columns for the two variables. The explanatory variable will be written in column A.

WRITE/DRAW

◢	A	B	C	D	E	F	G	H	I	J
1										

Click in cell A1 and type 'Opinion'.

◢	A	B	C	D
1	Opinion			
2				

The response variable will be written in column B.

Click in cell B1 and type 'Frequency'.

	A	B	C	D
1	Opinion	Frequency		
2				

3. Insert the data.

Click in cell A2 and start typing in the data given for the explanatory variable.

	A	B
1	Opinion	Frequency
2	Very poor	
3	Poor	
4	Average	
5	Good	
6	Very good	
7	Excellent	

Click in cell B2 and start typing in the data given for the response variable.

	A	B
1	Opinion	Frequency
2	Very poor	8
3	Poor	15
4	Average	52
5	Good	44
6	Very good	27
7	Excellent	16

4. Draw the column graph.

Select the two columns starting from cell A1 to cell B7. To select these cells, click in cell A1, hold, move to cell B7 and release.

	A	B
1	Opinion	Frequency
2	Very poor	8
3	Poor	15
4	Average	52
5	Good	44
6	Very good	27
7	Excellent	16

Go to Insert and click the 'Column' icon on the 'Charts' menu.

Click any of the templates for a column. The graph shown is the default graph Excel draws.

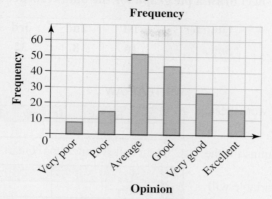

5. Format the column graph.

This graph can be formatted the way we want. Click 'Chart Tools' on the menu bar and format the graph as desired.

14.8.3 Drawing a pie graph in an Excel worksheet

Pie graphs are easy to draw in Excel worksheets because the program calculates the percentages and angles required automatically.

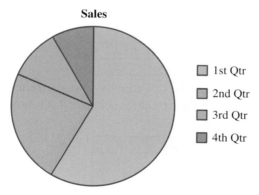

Using an Excel worksheet draw a pie graph for the data recorded in the table of values shown.

Quarter	1st	2nd	3rd	4th
Sales	16	8	32	40

THINK

1. Open a new Excel worksheet.

2. Label the two columns for the two variables.
The explanatory variable will be written in column A.
The response variable will be written in column B.

3. Insert the data.

4. Draw the pie graph.

WRITE/DRAW

	A	B	C	D	E	F
1						

Click in cell A1 and type 'Quarter'.

	A	B	C	D
1	Quarter			

Click in cell B1 and type 'Sales'.

	A	B	C	D
1	Quarter	Sales		

Click in cell A2 and start typing in the data given for the explanatory variable.

	A	B
1	Quarter	Sales
2	1st Qtr	
3	2nd Qtr	
4	3rd Qtr	
5	4th Qtr	

Click in cell B2 and start typing in the data given for the response variable.

	A	B
1	Quarter	Sales
2	1st Qtr	8.2
3	2nd Qtr	3.2
4	3rd Qtr	1.4
5	4th Qtr	1.2

Select the two columns starting from cell A1 to cell B5. To select these cells, click in cell A1, hold, move to cell B5 and release.

	A	B
1	Quarter	Sales
2	1st Qtr	8.2
3	2nd Qtr	3.2
4	3rd Qtr	1.4
5	4th Qtr	1.2

Go to Insert and click on the 'Pie' icon on the 'Charts' menu.

Click on any of the templates for a pie graph. The graph shown is the default graph Excel draws.

Sales

- 1st Qtr
- 2nd Qtr
- 3rd Qtr
- 4th Qtr

5. Format the pie graph.

This graph can be formatted the way we want. Click 'Chart Tools' on the menu bar and format the graph as desired.

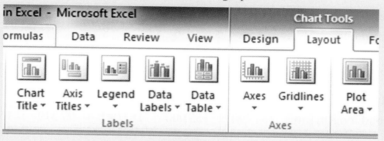

14.8 Exercise

Students, these questions are even better in jacPLUS

Receive immediate feedback and access sample responses

Access additional questions

Track your results and progress

Find all this and MORE in jacPLUS

1. **WE21** Using an Excel worksheet, draw a line graph for the data recorded in the table of values shown.

Day of the week	Mon	Tue	Wed	Thu	Fri	Sat	Sun
Temperature, °C	32	35	34	29	21	26	34

2. Using an Excel worksheet draw a line graph for the data recorded in the table of values shown.

Time, days	0	1	2	3	4	5	6	7	8	9	10
Height, cm	3.2	3.5	4.1	4.7	5.9	6.0	6.2	6.5	7.0	7.2	7.3

3. Answer the following questions.
   a. State the symbols used to label rows and columns in Excel worksheets.
   b. State the row and the column of cell A4.
   c. Name the cell shown.

	A	B
1		
2		X

4. Using an Excel worksheet, construct a line graph for the data recorded in the table shown.

Day	Mon	Tue	Wed	Thu	Fri	Sat	Sun
Temperature, °C	25	21	24	29	18	21	22

5. **WE22** Using an Excel worksheet, draw a column graph for the data recorded in the table of values shown.

Month	Jan	Feb	Mar	Apr	May	Jun	Jul
Electricity, kWh	4.1	3.7	3.9	3.8	4.3	5.6	5.2

6. Using an Excel worksheet, draw a line graph for the data recorded in the table of values shown.

Month	Jan	Feb	Mar	Apr	May	Jun	Jul	Aug	Sep	Oct	Nov	Dec
Gas, MJ	70	60	65	75	82	120	490	530	380	420	120	45

7. Using an Excel worksheet, construct a column graph for the data recorded in the table shown.

Pet	Cat	Dog	Rabbit	Fish
Frequency	15	36	8	17

8. The table shown displays the total road length in Australia by state and territory.

State	NSW	VIC	QLD	SA	WA	TAS	NT	ACT
Road length, km	184 794	153 000	183 041	97 319	154 263	25 599	22 239	2 963

Draw a column graph for the data set given using an Excel worksheet.

9. The table shown displays the total road length in Australia for a 5-year period.

Year	Total road length, km
1	815 588
2	816 949
3	822 649
4	825 592
5	823 217

Draw a line graph for the data set given using an Excel worksheet.

10. **WE23** Using an Excel worksheet, draw a pie graph for the data recorded in the table of values shown.

Ingredients	Flour	Sugar	Butter	Cinnamon	Cocoa
Quantity, g	500	250	100	20	60

Label the sectors of the pie graph with the names of the ingredients and their quantities in grams.

11. Using an Excel worksheet, draw a pie graph for the data recorded in the table of values shown.

Type of payment	Net wage	Tax	Superannuation	Medicare levy
Amount, $	57 000	10 230	1200	855

Label the sectors of the pie graph with the names of the categories and their percentages.

12. The total road accident costs in one year amounted to $6.1 billion. These costs are estimated for the following categories.
Lost earnings of victims: $829.1 million
Family and community losses: $587.8 million
Vehicle damage: $1868.2 million
Pain and suffering: $1463.3 million

Insurance administration: $571.1 million

Other: $816.4 million

a. Calculate the percentages for each category.

b. Construct a pie graph for this data set. Label each sector with the category and the corresponding percentage.

13. The table shows the percentage of international passengers carried by major airlines for the year ended December 2018 to/from Australia.

Airline	Percentage of international passengers
Qantas Airways	17.7%
Singapore Airlines	9.2%
Emirates	8.4%
Virgin Australia	8.3%
Jetstar	8.3%
Air New Zealand	8.0%
Cathay Pacific Airways	4.9%
Malaysia Airlines	3.7%
Thai Airways International	3.5%
AirAsia	2.8%
Others	25.3%

Construct a pie graph for the data given using an Excel worksheet.

14. The table shows the numbers of international passengers carried (in thousands) to and from Australia for the year ended December 2018.

Month	Inbound passengers (1000s)	Outbound passengers (1000s)
Dec 2017	1220	1451
Jan 2018	1523	1281
Feb 2018	1202	1019
Mar 2018	1141	1184
Apr 2018	1177	1203
May 2018	1001	1098
Jun 2018	1094	1267
Jul 2018	1411	1186
Aug 2018	1191	1229
Sep 2018	1241	1294
Oct 2018	1410	1164
Nov 2018	1207	1247
Dec 2018	1314	1539

Construct a line graph for the two data sets given using an Excel worksheet.

14.9 Review

14.9.1 Summary

doc-38055

14.9 Exercise

Multiple choice

1. **MC** An outlier is a score:
   A. with an extreme value far less than the minimum score or far greater than the maximum score.
   B. bigger or smaller than the middle half of the data values.
   C. different to all other data values.
   D. that lies between the smallest and largest data values.
   E. with the lowest value.

2. **MC** The mean, median and mode respectively for the data set displayed in the frequency table shown are:

Outcome	Frequency	fx
1	12	12
2	31	62
3	15	45
4	24	96
5	19	95
6	8	48

   A. 5.19, 3 and 2.
   B. 2, 3.28, and 3.
   C. 3, 3.28 and 4.
   D. 3.28, 3 and 4.
   E. 3, 3.28 and 2.

3. **MC** The interquartile range of the data distribution shown in the stem plot is:

Key: $2\,|\,6 = 26$

Stem	Leaf
0	2
1	1 5
2	6 6 7 8
3	8 8 9
4	3 4
5	2

   A. 41
   B. 50
   C. 28
   D. 20.5
   E. 26

4. **MC** The mean of the data distribution shown in the table is:

Interval	Frequency (f)
0– <15	5
15– <30	7
30– <45	6
45– <60	2

A. 22.4 **B.** 26.25 **C.** 24.35 **D.** 25.65 **E.** 27.45

5. **MC** As part of his preparation for the Foundation Mathematics exam, Andrew completes a weekly test. The results, in percentages, for his last 9 tests are:

$$67, \ 61, \ 69, \ 66, \ 68, \ 66, \ 71, \ 69, \ 69$$

Choose the statement that explains how the mean, median and mode of this set of data will be affected if he scored 100% in the 10th test.

A. The mean increases, the median increases and the mode does not change.
B. The mean increases, the median does not change and the mode does not change.
C. The mean does not change, the median increases and the mode does not change.
D. The mean does not change, median does not change and the mode increases.
E. The mean decrease, the median does not change and the mode increases.

6. **MC** The literacy and numeracy skills of 142 primary school students are recorded in the table shown.

		Literacy skills		
		Good	Poor	Total
Numeracy skills	Good	101	12	113
	Poor	26	3	29
	Total	127	15	142

The number of students who have good numeracy skills but poor literacy skills is:

A. 3 **B.** 15 **C.** 29 **D.** 12 **E.** 26

7. **MC** The daily minimum temperatures for a month recorded at Mount Wellington station are displayed in the graph shown.

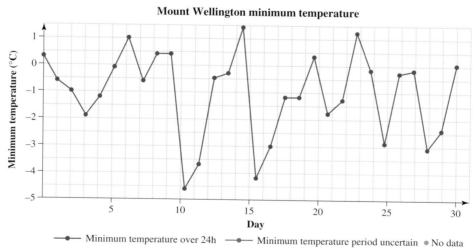

Mount Wellington minimum temperature

—•— Minimum temperature over 24h —•— Minimum temperature period uncertain •No data

The highest minimum temperature for the month was recorded on:

A. 10 June **B.** 11 June **C.** 14.5 June **D.** 25 June **E.** 28 June

8. **MC** A skip can be hired for \$120 for the first 2 days, \$160 for more than 2 days up to and including 4 days, \$210 for more than 4 days up to and including 7 days, and \$280 for more than 7 days and up to and including 14 days.

The graph that represents the relationship between the cost of hiring the skip and the number of days hired is:

A.

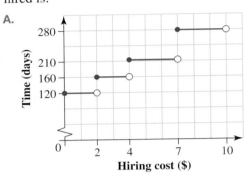

B.

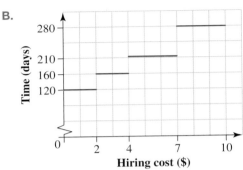

C.

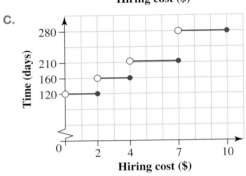

D.

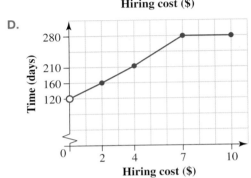

E. None of the above.

9. **MC** Select which of the following graphs displays the data recorded in the table shown.

		Time (days)					
		0	**1**	**2**	**3**	**4**	**5**
Height (cm)	**Plant A**	10	15.6	21.5	29.3	34.7	35.1
	Plant B	10	13.2	17.4	20.8	25.6	29.2

A.

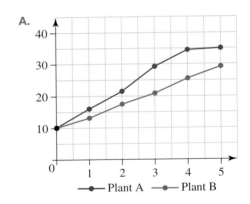

B.

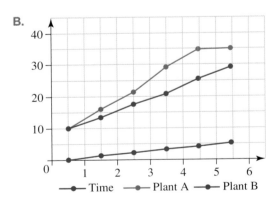

C.

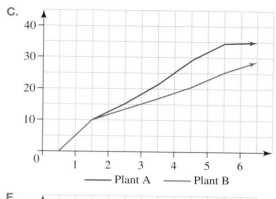

Plant A — Plant B

D.

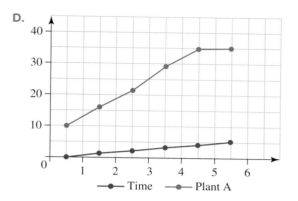

Time — Plant A

E.

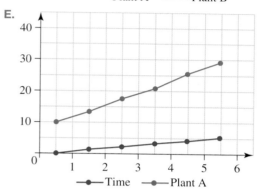

Time — Plant A

10. **MC** The graph shown displays superannuation contributions, in billions of dollars, for a 10-year period.

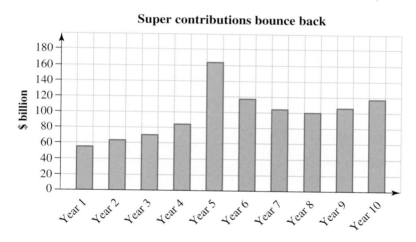

Super contributions bounce back

The highest increase in the superannuation contributions was:

A. from year 4 to year 5.
B. from year 5 to year 6.
C. from year 7 to year 8.
D. from year 8 to year 9.
E. from year 9 to year 10.

Short answer

11. A class participated in a read-a-thon. The number of books read in a week by each student was recorded. The cumulative frequency graph of the data is as shown.

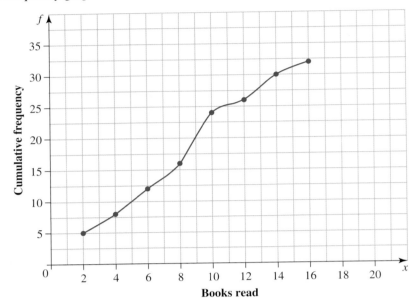

a. State the number of students who were in the class.

b. Calculate:
     i. the median   ii. Q_1 iii. Q_3.

c. The top 20% of children in the competition were given a prize. Calculate number of books that would a child need to read in order to be given a prize.

12. The following data shows the number of ice creams sold nightly over two weeks.

$$4, \ 6, \ 48, \ 29, \ 39, \ 48, \ 44, \ 45, \ 39, \ 47, \ 48, \ 32, \ 31, \ 51$$

Determine if there are any outliers.

13. Consider the following stem plot.
    Calculate:
    a. the mean
    b. the median
    c. the IQR
    d. the standard deviation.
    e. Using the data from parts **a–d**, construct a histogram using a class interval of 10. Show the median, mean and IQR on the histogram.
    f. Describe the modality and shape of the histogram.

Key: $11|2 = 112$

Stem	Leaf
5	0 1 2
6	4 5 8 9
7	0 7
8	0 3 3
9	5 5 5 7 8
10	1 1 2 3 4 6 6
11	0 1 3 3 4 5 5 5 7 9
12	1 1 2 2 3 3 4 7
13	2 6 9

14. The table shown displays the number of words that users from Australia typed in the search engines during March.

   a. Construct a column graph to represent this data set.
   b. From a total of 5000 searches, determine how many would be two-word searches.

Words:	🚗 Au
1	47.14%
2	21.98%
3	18.06%
4	6.51%
5	3.37%
6	1.43%
7	0.80%

15. The graph shows monthly car sales for a local car yard over the past year.

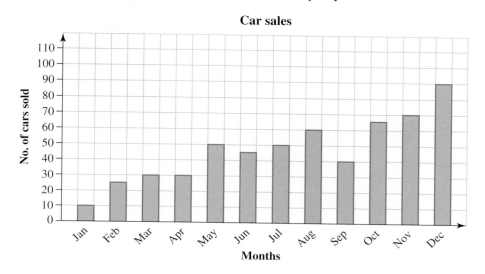

Car sales

Use the graph to answer the following questions.

   a. State the month in which the lowest sales figures were recorded.
   b. State the months in which the number of cars sold was equal.
   c. State the highest number of cars sold and the month in which this occurred.
   d. State by how much the December sales exceeded the January sales.
   e. Calculate the difference between the highest and the second highest sales figures over the last year.
   f. Determine the total sales for the year.

16. A vegetable shop surveyed 60 of its customers on their favourite vegetable. The data was collected and recorded in the table shown.

Favourite vegetable	Frequency
Carrots	11
Tomatoes	27
Cabbage	8
Broccoli	5
Cauliflower	9

   a. Choose five pictures to represent each vegetable.
   b. Construct a picture graph to represent this data set using a picture to represent:
      i. 3 items   **ii.** 5 items   **iii.** 10 items.
   c. State which graph from part **b** is clearest to read: **i, ii,** or **iii.** Explain your response.

17. The firing temperature in a kiln was collected every 5 minutes and recorded in the table shown.

Time, min	5	10	15	20	25	30	35	40	45
Temperature (°C)	204	288	338	382	410	438	450	466	471

Construct a line graph to display this data set using an Excel worksheet.

Extended reponse

18. A car company surveyed a group of 250 people on the number of cars per household. The data is displayed in the frequency distribution table below.

a. Calculate the entries in the column labelled fx.
b. Calculate the mean of this data set correct to 2 decimal places.
c. Determine the median of this data set.
d. Explain if this data set has a mode, and if so, state what it is.

Number of cars	Frequency	fx
1	36	
2	107	
3	89	
4	13	
5	5	
Total	250	

19. A small family business has 6 fulltime employees with the following salaries:

$$\$54\ 283.10, \ \$55\ 681.70, \ \$64\ 709.30, \ \$69\ 281.50, \ \$63\ 458.30, \ \$54\ 283.10$$

As the business is growing, there is a need to employ a new manager on a salary of $93 591.80. Explain how this salary is going to affect the mean and the median of the given data set.

20. Shirly recorded the approximate times she spends on various task during a 24-hour period in the table shown.

Activity	Time (hours)
School	7.0
Going to and from school	0.5
Homework	3.5
Sleep	8.5
Exercise	1.0
Relaxing	1.5
Other	2.0

a. Display this data in a column graph.
b. Calculate the percentage that each activity represents out of the 24-hour period.
c. Display this data as a pie graph.
d. Calculate the number of hours Shirly spends on her homework over five days.

21. The daily maximum temperatures in May in Perth are displayed in the table shown.

Day	Maximum temperature (°C)	Day	Maximum temperature (°C)
1st	25.7	17th	20.3
2nd	22.9	18th	20.5
3rd	22.6	19th	21.7
4th	23.6	20th	19.2
5th	25.4	21st	21.2
6th	22.3	22nd	21.6
7th	24.1	23rd	20.8
8th	20.8	24th	22.0
9th	20.8	25th	23.2
10th	20.1	26th	24.9
11th	19.7	27th	23.0
12th	20.7	28th	20.0
13th	21.1	29th	19.6
14th	21.8	30th	14.6
15th	22.6	31st	16.2
16th	23.4		

a. Construct a line graph for this data set.
b. State the day in May that recorded the highest maximum temperature for the month.
c. State the day in May that recorded the lowest maximum temperature for the month.
d. Calculate the temperature difference between 7 and 8 May. Explain whether this was a decrease or an increase in temperature.
e. Calculate the temperature difference between 25 and 26 May. Explain whether this was a decrease or an increase in temperature.

22. Use the data on the incidence of communicable diseases in Australia to answer the following questions.

Incidence of communicable diseases in Australia over two consecutive years

Disease	Year 1	Year 2
Hepatitis C	11 089	7286
Typhoid fever	116	96
Legionellosis	302	298
Meningococcal disease	259	230
Tuberculosis	1324	1327
Influenza (laboratory confirmed)	59 090	13 419
Measles	104	70
Mumps	165	95
Chickenpox	1753	1743
Shingles	2716	2978
Dengue virus infection	1406	1201
Malaria	508	399
Ross River virus infection	4796	5147

a. Calculate the mean (correct to 1 decimal place) and median number of cases of communicable diseases of the sample for each year.
b. Comment on the differences between the mean and median values calculated in part **a**.

23. The price of a barrel of oil in US dollars over a particular 18-month time period is shown in the following table.

Month	Price (US$)	Month	Price (US$)
Jan	102.96	Oct	92.44
Feb	97.63	Nov	87.05
Mar	108.76	Dec	88.69
Apr	105.25	Jan	93.14
May	106.17	Feb	97.46
June	83.17	Mar	90.71
July	83.72	Apr	97.1
Aug	88.99	May	90.74
Sept	95.34	June	93.41

a. Calculate the mean and median for this data set. Give your answer correct to 1 decimal place.
b. Calculate the interquartile range for this data set. Give your answers correct to 2 decimal places.

24. A transport company charges different fees depending on the weight of the luggage. Luggage less than or equal to 5 kg is free of charge. Luggage over 5 kg and less than or equal to 10 kg has a fee of $15, and luggage over 10 kg and less than or equal to 20 kg has a fee of $25. Luggage heavier than 20 kg and less than or equal to 30 kg incurs a charge of $30.
Construct a step graph to represent this data.

25. The two-way table shown displays the extracurricular activities that 90 Year 11 students undertake outside school hours.

		Other		
		Singing	Playing an instrument	Total
Sport played	Soccer	4	12	16
	Netball	23	17	40
	Basketball	8	26	34
	Total	35	55	90

a. State the number of students who play soccer and learn to sing.
b. State the number of students who play basketball.
c. State the number of students who learn to play an instrument.

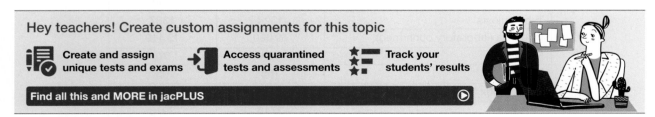

Hey teachers! Create custom assignments for this topic

Create and assign unique tests and exams

Access quarantined tests and assessments

Track your students' results

Find all this and MORE in jacPLUS

Answers

Topic 14 Presenting and interpreting data

14.2 Measures of central tendency and the outlier effect

14.2 Exercise

1. Mean = 3 letters
2. Mean = 160.75 cm
3. a. Mean = 15
   b. Mean = 195.8
4. a.

Lifetime (months)	Frequency	fx
21	9	189
22	18	396
23	44	1012
24	67	1608
25	29	725
26	20	520
27	13	351
Total	200	4801

   b. Mean = 24.0 months

5. a.

Wage ($/annum)	Frequency	fx
57 797	37	2 138 489
62 880	24	1 509 120
66 517	16	1 064 272
70 771	21	1 486 191
75 248	8	601 984
85 458	5	427 290
90 609	2	181 218
Total	113	7 408 564

   b. Average = $65 562.5
6. a. Position 14
   b. Median = 10
7. Median = 87.1 m
8. a. Median = 5.7 m
   b. Median = 161
   c. Mean = 7.25
10. Mode = 16
11. Mode = 24 years
12. Mean = 9.2
13. a. 21
    b. 144.8

14. a.

Number of bookings	Frequency	fx
1	27	27
2	32	64
3	19	57
4	26	104
5	12	60
6	3	18
7	1	7
Total	120	337

   b. Mean = 2.81 trips per given month
   c. Median = 3 trips per given month
   d. Mode = 2 trips per given month

15. a.

Number of logins	Frequency	fx
1	23	23
2	45	90
3	69	207
4	51	204
5	38	190
6	17	102
Total	243	816

   b. Mean = 3.36 logins
   c. Median = 3 logins
   d. Mode = 3 logins
16. a. i. Mean = 9.5, median = 8, mode = 8
       ii. Mean = 10.2, median = 8, mode = 8. When size 24 is included, this large score increases the mean from 9.5 to 10.2. The median and the mode are not affected by this score.
    b. It is more appropriate to use the median or the mode for this data set as extreme sizes highly affect the value of the mean.

14.3 Measures of spread

14.3 Exercise

1. a. 11   b. 2.6   c. 4
2. a. 157   b. 69
3. Minimum, lower quartile, median, upper quartile, maximum
4. 5
5. a. 160
   b. i. 110   ii. 99   iii. 120
   c. 117 steps/minute
6. B
7. D
8. C
9. 15 km/h, 2 km/h
10. Range = 27, IQR = 8 hours
11. 32.0 years

12. a. 94   **b.** 21.9

  c. 23; outliers are 1 and 15.

13. 166 is an outlier.

14. Answers will vary. Example answers are shown.

  a. 2, 5, 17, 21, 29, 35, 39, 45. (Several different data sets are possible.)

  b. 1, **5**, 6, 8, **12**, 14. (Several different data sets are possible, but the numbers in bold must be present in the same position.)

  c. 5, 6, 8, 9, 10, 11, 12, 14, 15. (Several different data sets are possible.)

15. The midpoint of the class interval could be used as the data value.

16. 5, 10, **14**, 21, 22, 23, 31, **34**, 35, 40. (Several different data sets are possible, but the numbers in bold must be present in the same position.)

17. Calculate the standard deviation. If the standard deviation is small, then the majority of parks will contain close to 15 trees.

18. a. $IQR = 35 - 18 = 17$

  Upper fence $= Q_3 + 1.5 \times IQR = 35 + 1.5 \times 17 = 60.5$

  Any value above 60.5 will be an outlier, so 64 is an outlier.

  b. Q_1, Q_3 and IQR would remain unchanged.

14.4 Describing graphical distributions

14.4 Exercise

1. B

2. Unimodal, negatively skewed with no outliers

3. i. a. Unimodal   **b.** Multimodal

    c. Multimodal   **d.** Bimodal

  ii. a. Symmetric   **b.** Symmetric

    c. Positively skewed   **d.** Negatively skewed

4. D

5. E

6. B

7. Mean: 24.4; median: 25–30; mode: 30–35

8. a. i. 10.1   **ii.** 9   **iii.** 11   **iv.** 5.9

  b. The histogram may vary due to the class intervals used.

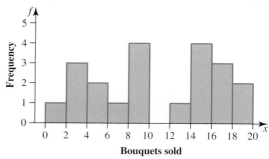

  c. See the figure at the bottom of the page.*

9. a. i. 10.2 laps

    ii. 5–10 laps

    iii. 5–10 laps

    iv. 45

    v. Q_1 is in the 5–10 laps class interval; Q_3 is in the 15–20 laps class interval.

  b. The data point in the 40–45 class interval is a possible outlier.

10. Outliers, on a histogram, will usually have a low frequency, and there will be several class intervals with zero data separating the outlier from the main cluster of data.

11. The histogram is likely to be negatively skewed.

12. Answers will vary. Sample response are shown.

  a. Peak hours in a coffee shop

  b. The age at which people retire

13. Darwyn is correct. The stem plot is roughly symmetric about the 90–99 class interval with two possible outliers of 1 and 10.

---

*8. c.

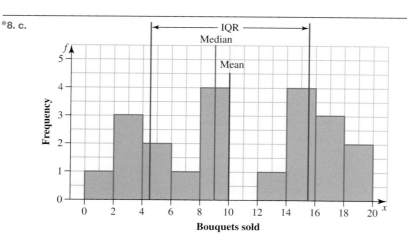

14. The two histograms show that Class A had a larger range of examination marks (100%) than Class B (60%). The median and modal examination mark for both classes was in the 60–70% class interval. However, the mean examination mark for Class A (56%) was lower than for Class B (71%). Overall, Class B achieved better examination results than Class A.

15. The median could be 13.75. The 8th value lies in the 10–15 class interval, which contains 4 data values; the 8th value will be the 3rd value in that group of 4 data values.

14.5 Characteristics of data sets

14.5 Exercise

1. a. 652   b. 26   c. 603   d. 335
2. a. 96   b. 7   c. 21   d. 68
3. a. 1000   b. 75   c. 96.7%   d. 60%
4. a. 200
   b. 44
   c. 90.9%
   d. 5.1%
   e. 94%
   f. Sample responses are available in the worked solutions in the online resources.
5. C
6. a. 35   b. 85   c. 2   d. 48
7. C
8. D
9. A
10. a.

		Test results		
		Accurate	**Not accurate**	**Total**
Bags	**With prohibited items**	48	2	50
	With no prohibited items	145	5	150
	Total	193	7	200

   b. i. 96%   ii. 3.3%   iii. 4%   iv. 96.5%
11. a. 29 °C
    b. 19 °C
    c. The temperature is decreasing during 12:00–7:00 am, 9:00–11:00 am, 12:00–2:00 pm and 4:00–11:00 pm.

    d. 27 °C
    e. 7 pm
12. a. Year 12
    b. 400 mm
    c. Years 3–5, years 6–8 and years 10–12
    d. Years 5 and 8
    e. Year 3, 330 mL
13. a. 24th, 10.6 °C   b. 23rd, 19.6 °C
    c. 2 °C   d. 11th, 12th and 28th
    e. 4 days
14. a. A$72   b. US$110
15. a. August, −15.5 °C   b. January, 2.5 °C
    c. 18 °C   d. In February and November
    e. 5
16. a. 128 km   b. 80 km
    c. 19 miles   d. 87 miles
17. a. 4 inches, 6 inches, 18 inches, 20 inches
    b. 51 cm, 89 cm, 112 cm, 130 cm
18. a. 95 °F   b. 50 °F   c. 5 °C   d. 40 °C
19. a. $14   b. $17   c. $26
20. a. $1.50
    b. $2.00
    c. Between 2 hours and 4 hours, not including 2 hours exactly
21. a. 51% Year 7 boys and 49% Year 7 girls
    b. 47.5% Year 12 boys and 52.5% Year 12 girls
    c. Year 10
22. a. Year 16, 15.1%
    b. Year 20, 17.0%
    c. 0.9%
    d. From Year 9 to Year 10, 1.3%
    e. Year 6

14.6 Drawing and using column and pie graphs

14.6 Exercise

1. See the figure at the bottom of the page.*

---

*1.

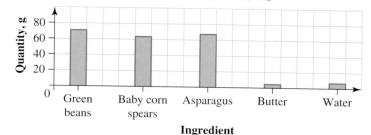

Ingredients on a package, grams

2. See the figure at the bottom of the page.*

3.

Favourite type of chocolate

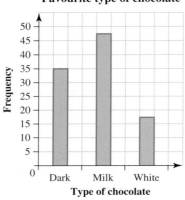

c.

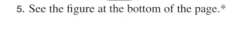

- Moderate improvement
- High improvement
- Some improvement
- No improvement at all

5. See the figure at the bottom of the page.*

4. a. See the figure at the bottom of the page.*

   b. 8% no improvement at all, 14% some improvement, 52% moderate improvement, 26% high improvement

---

*2.

Victoria, million litres

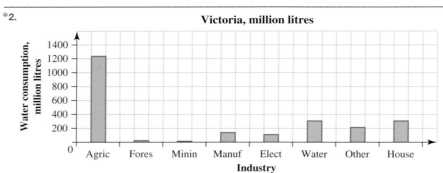

*4. a.

School attendance program

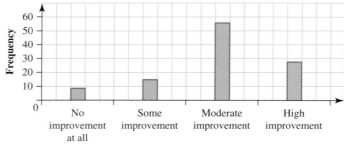

*5.

Toy shop sales

6. See the figure at the bottom of the page.*

7. a.

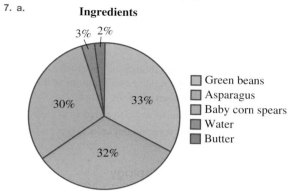

Ingredients

3% 2%
33% Green beans
30% Asparagus
32% Baby corn spears
Water
Butter

Legend:
- Green beans
- Asparagus
- Baby corn spears
- Water
- Butter

b. 320 g asparagus

8. a. $52 000

b.

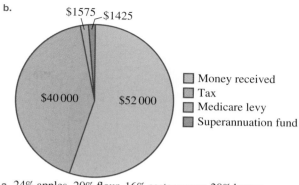

$1575 $1425

$40 000 $52 000

Legend:
- Money received
- Tax
- Medicare levy
- Superannuation fund

9. a. 24% apples, 20% flour, 16% caster sugar, 20% butter, 19% eggs, 1% ground cinnamon

*6.

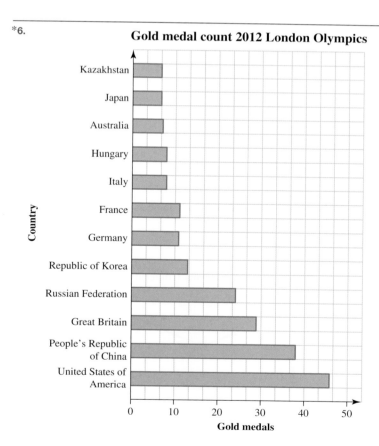

Gold medal count 2012 London Olympics

Country (y-axis), Gold medals (x-axis):
- Kazakhstan
- Japan
- Australia
- Hungary
- Italy
- France
- Germany
- Republic of Korea
- Russian Federation
- Great Britain
- People's Republic of China
- United States of America

x-axis: 0, 10, 20, 30, 40, 50 — Gold medals

b.

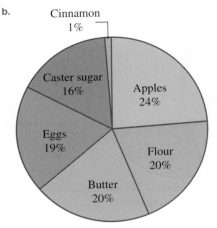

Cinnamon 1%

Caster sugar 16%

Apples 24%

Eggs 19%

Flour 20%

Butter 20%

c. 300 g

10.

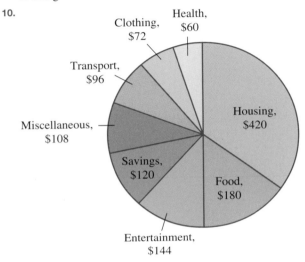

Clothing, $72

Health, $60

Transport, $96

Miscellaneous, $108

Savings, $120

Housing, $420

Food, $180

Entertainment, $144

14.7 Misleading graphs

14.7 Exercise

1. Sample responses are available in the worked solutions in the online resources.

2. The horizontal axis uses the same division for 5- and 7-year periods.

3. **a.** Sample responses are available in the worked solutions in the online resources.

   b. No

4. **a.** Student-to-teacher ratios have improved slightly.

   b. Country schools have smaller class sizes.

5. Sample responses are available in the worked solutions in the online resources.

6. Sample responses are available in the worked solutions in the online resources.

7. **a.** The claim is accurate enough in the context ($86 879 m actually).

   b. State and local governments. The stated 25% is correct (rounded up from 24.9%).

   c. Non-government organisations. The stated 30% is rounded down from 32.2%. The percentages being quoted seem to be rounded to the nearest 5%.

d. The quoted percentage (45%) has been rounded up from 42.9%. This could be considered misleading in some contexts.

e. i. $162°, 90°, 108°$

   ii. $154°, 90°, 116°$

   iii. $154°, 78°, 128°$. Even though the pie graph gives a rough picture of the relative contributions of the three sectors, it has not been carefully drawn.

8. **a.** It is a circle viewed on an angle to produce an ellipse.

   b. No, because it causes some angles to be larger and others to be smaller.

14.8 Graphing and technology

14.8 Exercise

1.

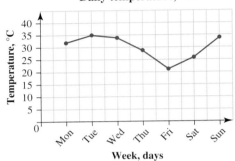

Daily temperature, °C

2.

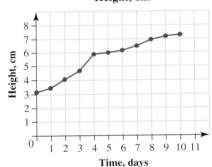

Height, cm

3. **a.** Rows are labelled with numbers and columns are labelled with capital letters.

   b. Row 4, column A

   c. Cell B2

4.

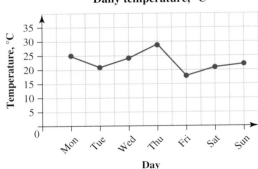

Daily temperature, °C

5.

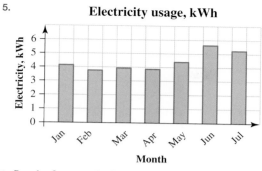

Electricity usage, kWh

6. See the figure at the bottom of the page.*

7.

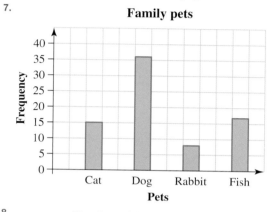

Family pets

8.

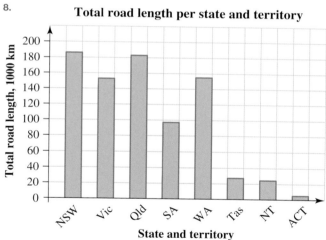

Total road length per state and territory

9.

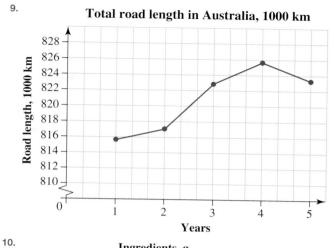

Total road length in Australia, 1000 km

10.

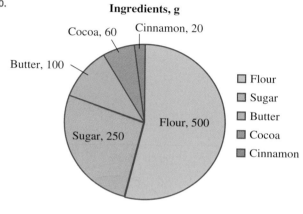

Ingredients, g

---

*6.

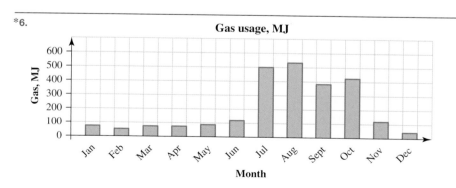

Gas usage, MJ

11.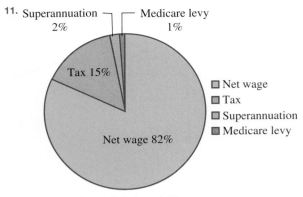

b.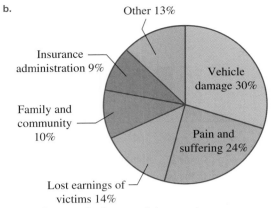

12. a. Lost earnings of victims: 14%
Family and community losses: 10%
Vehicle damage: 30%
Pain and suffering: 24%
Insurance administration: 9%
Other: 13%

13. See the figure at the bottom of the page.*
14. See the figure at the bottom of the page.*

14.9 Review

14.9 Exercise
Multiple choice

1. A	**2.** D	**3.** D	**4.** B	**5.** A
6. D	**7.** C	**8.** C	**9.** A	**10.** A

---

*13.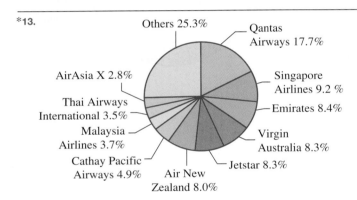

*14.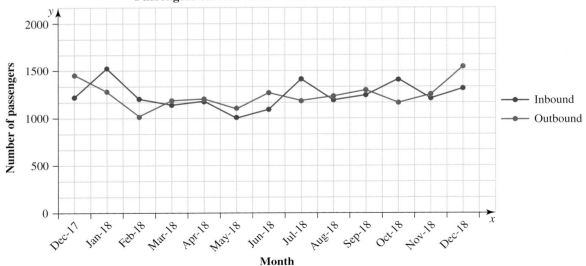

Short answer

11. a. 32

   b. i. 8   **ii.** 4   **iii.** 10

   c. 12

12. 4 is an outlier.

13. a. 101.0

   b. 106

   c. 37

   d. 23.5

   e. See the figure at the bottom of the page.*

   f. Unimodal, negatively skewed

14. a.

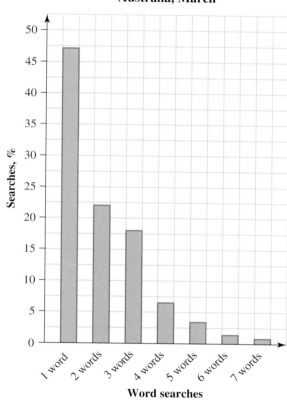

Type of word searches on the Internet in Australia, March

   b. 1099

15. a. The lowest sales figures were recorded in January.

   b. Equal sales figures were recorded in March and April, and May and July.

   c. The highest sales figures were recorded in December (90 cars).

   d. 80

   e. 20

   f. 565

16. a. Answers will vary.

   b. Answers should be similar to the graphs shown.

   i.

Favourite vegetable

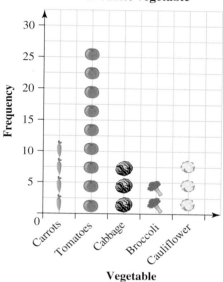

Vegetable

Key: Each picture equals 3 items.

---

*13. e.

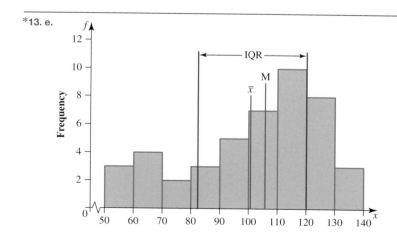

ii.

Favourite vegetable

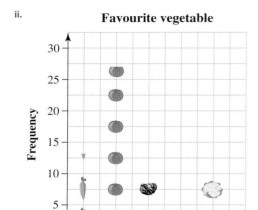

Key: Each picture equals 5 items.

iii.

Favourite vegetable

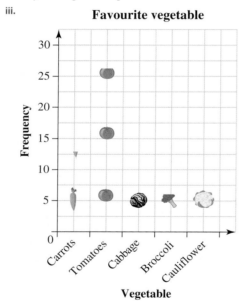

Key: Each picture equals 10 items.

c. Graph ii is the easiest to read, because the fewer items a picture represents, the easier the graph is to read.

17. See the figure at the bottom of the page.*

Extended response

18. a.

Number of cars	Frequency	fx
1	36	36
2	107	214
3	89	267
4	13	52
5	5	25
Total	250	594

b. Mean $= 2.38$

c. Median $= 2$

d. The data set has a mode of 2.

19. The mean increases from \$60 282.80 to \$65 041.30. The median increases from \$59 570 to \$63 458.30.

20. a.

Activities per 24-hour day

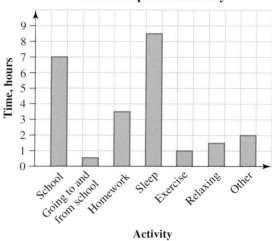

*17.

Firing temperature in a kiln, °C

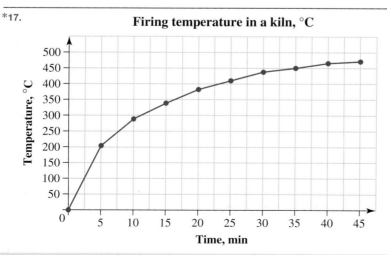

b. School = 29.2%
Going to and from school = 2%
Homework = 14.6%
Sleep = 35.4%
Exercise = 4.2%
Relaxing = 6.3%
Other = 8.3%

c. School = 105°
Going to and from school = 7.5°
Homework = 52.5°
Sleep = 127.5°
Exercise = 15°
Relaxing = 22.5°
Other = 30°
See the figure at the bottom of the page.*

d. 17.5 hours

21. a. See the figure at the bottom of the page.*

b. 1 May

c. 30 May

d. 3.3 °C

e. 1.7 °C

22. a. Year 1: mean = 6432.9, median = 1324
Year 2: mean = 2637.6, median = 1201

b. The mean for Year 1 is significantly greater than for Year 2, while the values of the medians are quite similar. In year 1 the number of incidences of influenza was very large compared to year 2 (59 090 compared to 13 419). Similarly, the number of incidences of hepatitis C was much larger in Year 1 (11 089 compared to 7286). This resulted in mean value being much larger for year 1. Since medians are not affected by extreme values, the medians for both years are similar.

23. a. Mean = 94.6, median = 93.3

b. IQR = 8.64

24.

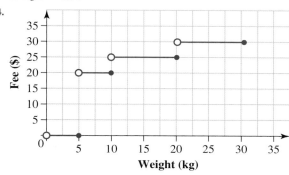

25. a. 4   **b.** 34   **c.** 55

---

*20. c.

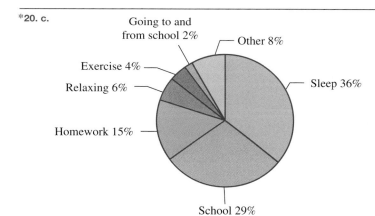

*21. a.

Perth Metro May maximum temperature

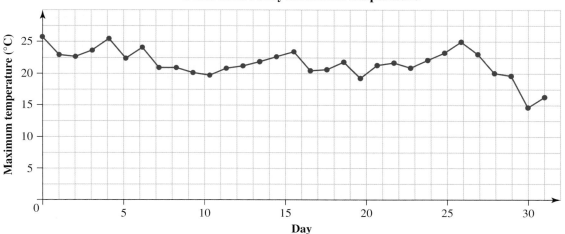

15 Managing finances

LEARNING SEQUENCE

Fully worked solutions for this topic are available online.

15.1 Overview

15.1.1 Introduction

It is important to develop a strong understanding of how to manage financial matters, as money is needed to pay for basic personal essentials: food, shelter and clothing. Informed spending choices will help you make smart financial decisions and not only provide the essentials but allow you to successfully manage your personal finances and give you the ability to engage in community financial matters. This topic may help you understand some of these concepts.

Before you take the first steps to becoming an independent adult, it's a good idea to be aware of some of the costs you'll be expected to pay.

Basic expenses, such as internet services, utilities bills and car insurance, can't be avoided so should be included in your budget. Many companies offer discounts for their services, so research your options. By managing your finances, you can have an understanding of your income and expenses and can make good decisions now and in the future.

KEY CONCEPTS

This topic covers the following key concepts from the VCE Mathematics Study Design:
- products and services such as comparison of health products, informed spending choices, decision-making according to criteria
- managing money: earning and spending, life-stage financial planning, servicing of current and future commitments such as HECS-HELP debt, childcare support and other benefits
- local, community and national financial and economic data and trends over time (national/community/local) such as CPI, interest rates, wages and house prices.

Source: VCE Mathematics Study Design (2023–2027) extracts © VCAA; reproduced by permission.

15.2 Products and services

15.2.1 Internet services

Access to the internet has become part of our daily life, enabling us to use email, social media, online games, file sharing and media streaming services. When considering an internet provider, it is important to consider what type of connection is available and how often and what you use the internet for. Finding a plan that suits you will depend on your individual needs and circumstances.

These days most plans offer unlimited data with a fixed cost structure, but there are still some that have a monthly data limit. This would have variable costs each month if the data limit is exceeded, and can get expensive.

tlvd-5080

WORKED EXAMPLE 1 Determining the most cost-effective option for products

Byron is trying to decide between the two internet plans as shown.

Option 1: Unlimited data for $54.99 per month

Option 2: 200 GB of data for $49.95 per month + $10 for every extra 50 GB of data

Determine the most cost-effective option if Byron uses:

a. 100 GB of data per month

b. 250 GB of data per month.

OPTION 01
Unlimited data for $54.99 per month

OPTION 02
200 GB of data for $49.95 per month + $10 for every extra 50 GB of data

THINK	WRITE
a. 1. Option 1 has unlimited data, and option 2 will not exceed the 200 GB in the plan.	**a.** Option 1 = $54.99 Option 2 = $49.95
2. Write the answer.	Option 1 costs $54.99 and option 2 costs $49.95, so option 2 is more cost-effective.
b. 1. Option 2 will exceed the data limit by 50 GB, which is an additional $10.	**b.** Option 1 = $54.99 Option 2 = $49.95 + $10 $\qquad\qquad$ = $59.95
2. Write the answer.	Option 1 costs $54.99 and option 2 costs $59.95, so option 1 is more cost-effective

15.2.2 Health insurance

Health insurance is a type of insurance that helps to cover the costs of medical services not already covered by Medicare. To be covered by health insurance, a regular fee is paid to an insurer weekly, fortnightly, monthly or annually. The amount of the fee will depend on the type of cover you require (hospital or extras), and your age. All private health insurers can be compared at the Australian government website www.privatehealth.gov.au.

In Australia, having appropriate private hospital health insurance may mean that you do not have to pay the Medicare levy surcharge, which can be up to an extra 1.5% of your taxable income, on top of the Medicare levy that is paid by everyone.

It is also important to know about the lifetime health cover loading, which encourages younger people to get private health insurance by charging an extra 2% of your insurance fee for every year after you turn 30 if you have not joined a private health fund prior to turning 30.

WORKED EXAMPLE 2 Calculating costs associated with health insurance

HealthCo Hospital offers Diya private health insurance to cover hospital costs for $79 per month.
a. Calculate the annual cost for Diya.
b. Diya's taxable income is $70 000 per annum. Determine if it is more cost-effective for Diya to join HealthCo or pay the Medicare levy surcharge.
c. Diya doesn't join HealthCo until she turns 32. Determine the cost per month for Diya, taking into account the lifetime health cover loading.

THINK	WRITE
a. 1. Multiply the monthly cost by 12 to find the annual cost.	**a.** $79 \times 12 = 948$
2. Write the answer.	The annual cost for Diya is $948.
b. 1. Calculate the 1.5% Medicare levy surcharge on Diya's taxable income of $70 000.	**b.** $1.5\% \text{ of } 70\,000 = \dfrac{1.5}{100} \times 70\,000$ $= 1050$
2. Write the answer.	The Medicare levy surcharge on Diya's taxable income is $1050, which is more than the private health insurance cost. Therefore, it is more cost-effective for Diya to join HealthCo.
c. 1. Diya will be charged an extra 2% for every year after she turned 30.	**c.** $2 \times 2\% = 4\%$
2. Calculate 4% of 79.	$\dfrac{4}{100} \times 79 = 3.16$
3. Add the lifetime health cover loading to the monthly cost.	$79 + 3.16 = 82.16$
4. Write the answer.	Diya will pay $82.16 per month.

15.2.3 Car insurance

Car insurance covers the costs associated with vehicle accidents, such as repairs, replacement vehicles and medical expenses.

Types of vehicle insurance			
Compulsory third-party (CTP)	**Comprehensive**	**Third-party property**	**Third-party fire and theft**
Covers injuries to you and others in motor vehicle accidents and is mandatory insurance in all states and territories of Australia	Covers damage to your own vehicle and other people's property, as well as theft and some other risks, plus legal costs	Covers damage to other people's property and legal costs, but not damage to your own vehicle	Covers everything included in third-party property insurance, as well as your vehicle in the event of fire, theft or both

The cost of vehicle insurance depends on the driver's experience, the type of the vehicle, the age of the car, the address where the vehicle is kept and the type of insurance required. For comprehensive insurance, vehicles can be insured at the market value (the vehicle's current value) or the agreed value (a value that is agreed on by the owner and insurer).

There is an excess cost for insurance claims. This is the cost the insured driver needs to pay before the claim is processed. Drivers under a certain age usually have to pay higher excess costs for insurance claims.

tlvd-5081

WORKED EXAMPLE 3 Determining the most cost-effective car insurance

A 19-year-old driver has three options for comprehensive car insurance for a small car that will be parked in a locked garage in a suburban property. All options include a replacement vehicle, provision of transportation if the car is unable to be driven, and costs for injuries for all passengers.
Option A: $1143, 10% online discount, $850 excess
Option B: $1149, $850 excess
Option C: $1155, 15% online discount, $800 excess
The excess fee is the amount that is paid to the insurer if a claim is made.
If the driver had an accident in the first year of their policy, determine which option is the most cost-effective.

THINK	WRITE
1. Calculate the policy amount with the 10% discount from option A.	Option A with 10% discount: $10\% \times 1143 = \$114.30$ Policy amount: $1143 - 114.30 = \$1028.70$
2. Add the excess to the policy amount for option A to calculate the total yearly cost.	Option A with excess: $1028.70 + 850 = \$1878.70$
3. Calculate the total policy amount for option B.	Option B with excess: $1149 + 850 = \$1999$

4. Calculate the policy amount with the 15% discount for option C.

Option C with 15% discount: $15\% \times 1155 = \$173.25$
Policy amount: $1155 - 173.25 = \$981.75$

5. Add the excess to the policy amount for option C to calculate the total yearly cost.

Option C with excess: $981.75 + 800 = \$1781.75$

6. Write the answer.

Option C is the most cost-effective.

15.2 Exercise

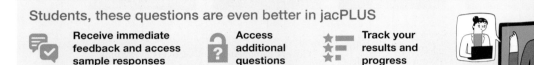

1. **WE1** Connie is trying to decide between the two internet plans shown.

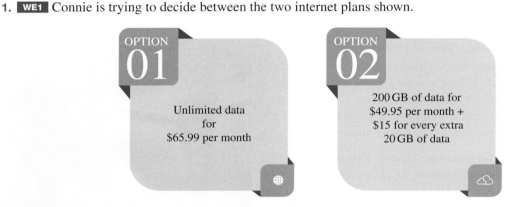

OPTION 01
Unlimited data for $65.99 per month

OPTION 02
200 GB of data for $49.95 per month + $15 for every extra 20 GB of data

Determine the most cost-effective option if Connie uses:

a. 150 GB of data per month

b. 220 GB of data per month

c. 260 GB of data per month.

2. Aaron is trying to decide between the two internet plans shown.

a. Determine the most cost-effective option if Aaron uses:

  i. 3 GB of data per month

  ii. 5 GB of data per month

  iii. 7 GB of data per month.

b. HFC offers Aaron a deal whereby he pays $1 for the first month and then normal rates for the remainder of the year. If Aaron regularly uses 7 GB of data per month, calculate the yearly cost for each plan and determine the best plan for Aaron.

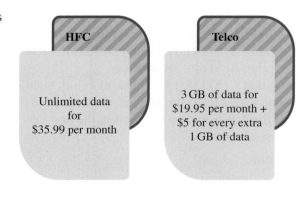

HFC
Unlimited data for $35.99 per month

Telco
3 GB of data for $19.95 per month + $5 for every extra 1 GB of data

3. **WE2** HIB Hospital + Extras offers John private health insurance to cover hospital costs for $125 per month.

   a. Calculate the annual cost for John.
   b. John's taxable income is $90 000 per annum. Determine if it is more cost-effective for John to join HIB or pay the Medicare levy surcharge.
   c. John doesn't join HIB until he turns 37. Determine the cost per month for John, taking into account the lifetime health cover loading.

4. Visit the Australian government website www.privatehealth.gov.au and determine the most appropriate private health insurance for yourself.

5. Private health insurance costs for Marco are provided below:

Type	Cost per month ($)
Hospital	83
Hospital + extras	131

   a. Calculate the difference in cost per year for hospital cover and hospital + extras cover.
   b. Determine the cost of each type of insurance per month if Marco waits to join until he is 43 years old.

6. **WE3** A 21-year-old driver has three options to purchase comprehensive car insurance for their car. All options include a replacement vehicle and costs for injuries for all passengers.

OPTION A

$1850,
15% online discount,
$850 excess

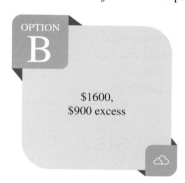

OPTION B

$1600,
$900 excess

OPTION C

$1650,
5% online discount,
$900 excess

The excess fee is the amount that is to be paid to the insurer if a claim is made.
If the driver had an accident in the first year of their policy, determine which option is the most cost-effective.

7. Burke is 19 years old and has just had a car accident in which she was at fault. The following excesses apply to claims for at-fault motor vehicle accidents for his comprehensive car insurance:

   • basic excess of $750 for each claim
   • an additional age excess of $1500 for drivers under 25 years of age
   • an additional age excess of $300 for drivers 25 years of age or over with no more than 2 years of driving experience.

Determine how much excess Burke is required to pay her insurance company.

8. An insurance company offers customers the following discounts on the annual cost for car insurance.

Type of discount	Discount	Conditions
Multi-policy discount	15%	Owner has 2 or more insurance policies with the company.
No claim bonus	20%	Owner has had at least 5 years without an insurance claim.

Calculate the discounted annual cost for car insurance for:

a. a driver whose basic cost is $870 per year, who has three insurance policies and has never made a claim in their 10 years with this insurance company
b. a driver whose basic cost is $1100 per year, who has no other insurance policies and has never made a claim in their 6 years with this insurance company
c. a driver whose basic cost is $85 per month, who has two insurance policies but made a claim last year after a small accident.

9. Josie is 22 years old and has been with the same car insurance company since obtaining her driving licence at 18 years old. She has 3 other types of insurance with this company. Her car insurance costs $1035 per year prior to any discounts being applied.

Using the information from questions **7** and **8**, calculate the annual amount that Josie is required to pay after having her first accident.

10. Use an online car insurance calculator to determine the cost of annual insurance for your family car. Use two other different companies and calculate the most cost-effective insurance option for your family car.

15.3 Rent and utilities costs

LEARNING INTENTION

At the end of this subtopic you should be able to:
- calculate rental costs
- calculate household electricity and gas bills
- calculate household water bills and council rates.

15.3.1 Rent

When renting a property, a rental agreement or lease must state how much rent is to be paid, how frequently and for how long. All rent payments need to be paid in advance.

A deposit, known as a rental bond, is also usually part of a rental agreement. It is often equivalent to four weeks' worth of rent and is kept as a form of insurance in case of damage to the property.

It is important to understand the terms of a rental agreement before signing a contract.

tlvd-5082

A rental agreement is shown below.

Rental agreement

By this agreement made at _____, on the
_____ day of _____, 20 ___, the Landlord
_____ and the
Tenant _____ agree as follows:

1. Rent
   Tenant agrees to pay rent in the amount of ($) ⊡380⊡ (payable in advance).
   To be paid per ⊡x⊡ week ☐ fortnight ☐ calendar month
   Day rent is to be ⊡Every Friday⊡ paid (e.g. each Thursday or the 11ᵗʰ of each month).
   Date first rent ⊡14ᵗʰ Dec 2021⊡

2. Bond
   The renter has been asked to pay the bond specified below.
   The bond is 4 weeks rent.
   Bond amount [   ]

   Date of bond ⊡14ᵗʰ Dec 2021⊡

Calculate:
a. the bond amount payable
b. the total amount payable on 14 December 2021
c. the amount of rent to be paid each month, to the nearest dollar.

THINK	WRITE
a. 1. Calculate the amount of bond that is equivalent to four weeks' rent.	**a.** Bond amount $= 4 \times 380$ $\qquad\qquad\quad = 1520$
2. Write the answer.	The bond amount payable is $1520.
b. 1. Calculate the sum of the bond and the weekly rent.	**b.** $1520 + 380 = 1900$
2. Write the answer.	The total amount payable on 14 December is $1900.
c. 1. Calculate the amount of rent to be paid per year.	**c.** $380 \times 52 = 19\,760$
2. Divide the yearly amount of rent by 12.	$\dfrac{19760}{12} = 1647$
3. Write the answer.	To the nearest dollar, the amount of rent to be paid each month is $1647.

15.3.2 Electricity and gas bills

Electricity and gas are supplied to our homes to help us keep warm and cool, to allow us to cook, and to provide lighting.

There are three main parts to our electricity and gas supply systems: wholesalers (or generators), distributors and retailers.
- Wholesalers produce electricity and extract gas.
- Distributors own and maintain infrastructure such as power poles, wiring and gas lines. When there are disruptions to the electricity or gas supply, the distributor is contacted to fix the problem.

- Retailers purchase electricity and gas from wholesalers and then sell them to customers. There are many electricity and gas options available to customers, and customers can select whichever deal best suits their needs. Therefore, it is important to know how much you will be charged per unit of energy used and what discounts are available before you sign any contracts.

The standard measurement unit of electricity consumption

The standard measurement unit of electricity consumption is the kilowatt hour (kWh). One kWh is the amount of energy produced by an appliance outputting 1 kW (1000 W) of power for an hour. Gas consumption is measured in megajoules (MJ).

To understand electricity and gas bills, it is important to know some of the terms that are used.

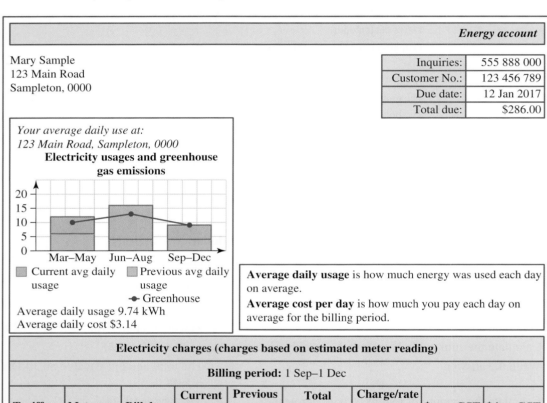

		Energy account

Mary Sample
123 Main Road
Sampleton, 0000

Inquiries:	555 888 000
Customer No.:	123 456 789
Due date:	12 Jan 2017
Total due:	$286.00

Your average daily use at:
123 Main Road, Sampleton, 0000

Electricity usages and greenhouse gas emissions

Current avg daily usage
Previous avg daily usage
Greenhouse

Average daily usage 9.74 kWh
Average daily cost $3.14

Average daily usage is how much energy was used each day on average.
Average cost per day is how much you pay each day on average for the billing period.

Electricity charges (charges based on estimated meter reading)

Billing period: 1 Sep–1 Dec

Tariff	Meter no.	Bill days	Current reading	Previous reading	Total usage (kWh)	Charge/rate (c/kWh)	$ exc. GST	$ inc. GST
Peak	111111	91	793	459	334	28.9	96.53	106.18
Off-peak	111111	91	4861	4309	552	12.8	70.66	77.72
Service to property		91		**$1.020/day**			92.82	102.10

Bill days is the number of days for the billing period.

Charge/kWh: how much you are being charged, in cents, per kilowatt hour.

Peak and off-peak are the different time periods for electricity usage. They will have different charges per kWh depending on the energy plan.

Service to property is a fixed charge that is also called the 'daily supply charge'.

tlvd-5083

WORKED EXAMPLE 5 Calculating the total electricity bill amount

The meter readings and charge/rates for a household are shown.

Electricity charges (charges based on meter reading)								
Billing period: 1 Jan–31 March								
Tariff	Meter no.	Bill days	Current reading	Previous reading	Total usage (kWh)	Charge/rate (c/kWh)	$ exc. GST	$ inc. GST
Peak		90	12 791	12 353	438	29.85		
Off-peak		90	2 587	1 883	704	13.37		
Service to property	90				$1.254/day			

Calculate the total bill amount.

THINK

1. State the electricity usage by reading from the table for both peak and off-peak.

2. Calculate the cost for each time period by multiplying the usage by the charge/rate.

3. Add GST by adding 10% of each amount.

4. Calculate the costs of service to property by multiplying the cost per day by the number of billing days.

5. Add 10% GST.

6. Calculate the total cost by adding the usage cost to the service cost.

7. Write the answer as a sentence.

WRITE

Peak: 438
Off-peak: 704

Peak: $438 \times 0.2985 \approx \$130.74$
Off-peak: $704 \times 0.1337 \approx \$94.12$

Peak: $10\% \times \$130.74 = \$13.074$
$\$130.74 + \$13.074 = \$143.81$
Off-peak: $10\% \times \$94.12 = \$9.412$
$\$94.12 + \$9.412 = \$103.53$
Total for usage $= \$143.81 + 103.53$
$\qquad = \$247.34$

Service costs $= 1.254 \times 90$
$\qquad = \$112.86$

$10\% \times \$112.86 = \$11.286$
Service costs $= \$112.86 + \$11.286$
$\qquad = \$124.15$

Total cost $= \$247.34 + 124.15$
$\qquad = \$371.49$

The bill total is $371.49.

15.3.3 Water bills

Water is provided to homes and businesses through a series of water pipes that are managed and serviced by providers.

Home owners are charged a fixed fee to provide water and remove wastewater in a system known as sewerage. Households are also charged a fee for the management of stormwater services and some households have to pay a recycled-water usage fee, depending on their property.

The amount of water used in the home over a fixed period, usually 3 months, is measured through a water meter fixed to the main water supply in the home.

Water usage

Home owners are charged for their water usage at a rate, in dollars, per kilolitre (kL).

$$1000 \text{ litres} = 1 \text{ kilolitre}$$

Note: GST does not apply to water bills.

WORKED EXAMPLE 6 Calculating water rates with usage and fixed charges

Calculate the water rates for a home with the following usage and fixed charges.

Bill details as of 31 March		Value	Price
Water service charge	1 Jan to 31 Mar	22.51	
Sewerage service charge	1 Jan to 31 Mar	145.90	
Stormwater	1 Jan to 31 Mar	18.70	
	Volume	Charge/kL	
Water usage	55	$2.00	
Recycled water usage	35	$1.79	

THINK

1. Calculate the cost of the water usage by multiplying the volume by the cost.

2. Calculate the cost of the recycled water usage by multiplying the volume by the cost.

3. Calculate the fixed service cost.

4. Add the fixed service cost to the water usage cost.

5. Write the answer as a sentence.

WRITE

$$\text{Water usage} = 55 \times 2$$
$$= \$110$$

$$\text{Recycled water} = 35 \times 1.79$$
$$= \$62.65$$

$$22.51 + 145.90 + 18.70 = 187.11$$
$$187.11 + 110 + 62.65 = \$359.76$$

The water rates are $359.76.

15.3.4 Council rates

Councils provide services and infrastructure to people living in communities, such as parklands, libraries, rubbish collection and road maintenance. To support councils in providing these services, home owners pay **council rates**.

Council rates are calculated as a percentage on the capital improved value of the land and buildings within the council municipality.

The rate at which council rates are charged — the rate in the dollar — is determined by dividing the revenue the council plans to raise by the amount of capital improvement of land and buildings within the council municipality.

Calculating council rates

$$\text{rate in the dollar} = \frac{\text{revenue}}{\text{council capital improved value}}$$

Then multiply the rate in the dollar by the capital improved value of the property.

annual council rates = rate in the dollar × property capital improved value

For example, if the council plans to raise $15 million for the annual budget and there is $2 billion in capital improved value within the municipality, then the rate in the dollar is $\frac{15\,000\,000}{2\,000\,000\,000} = 0.0075$.

If the capital improvement on a ratepayer's house and land is $750\,000$, then the payable council rates for that property are $0.0075 \times 750\,000 = \$5625$.

Note: GST does not apply to council rates.

WORKED EXAMPLE 7 Calculate the payable council rates for a property

A council municipality plans to raise $15 million. There is $3.25 billion in capital improved value in the municipality. Calculate the payable council rates for a property with $275 000 capital improved value. Write your answer correct to the nearest cent.

THINK	WRITE
1. Calculate the rate in the dollar. Divide the revenue to be raised by the capital improved value.	$\frac{15\,000\,000}{3\,250\,000\,000} = \frac{15}{3250}$ $= 0.004\,62...$
2. Multiply the rate in the dollar by the capital improved value.	$0.004\,62... \times 275\,000 = \$1269.23$
3. Write the answer as a sentence.	The council rates payable are $1269.23.

on Resources

🔷 **Interactivity** Electricity bills (int-6912)

15.3 Exercise

1. **WE4** A rental agreement is shown.

   Calculate:

   a. the bond amount payable
   b. the total amount payable on 21 August 2021
   c. the amount of rent to be paid each month, to the nearest dollar.

Rental agreement

By this agreement made at _____, on the _____ day of _____, 20 ___, the Landlord _____ and the

Tenant _____ agree as follows:

1. Rent

Tenant agrees to pay rent in the amount of ($) 550 (payable in advance).

To be paid per ☒ week ☐ fortnight ☐ calendar month

Day rent is to be Every Wednesday paid (e.g. each Thursday or the 11th of each month).

Date first rent 21st August 2021

2. Bond

The renter has been asked to pay the bond specified below.

The bond is 4 weeks rent.

Bond amount [   ]

Date of bond 21st August 2021

---

2. A rental agreement is shown.

   a. Calculate:

      i the amount of rent to be paid each week, to the nearest dollar
      ii the total amount payable on 16 January 2022.

   b. A family has the choice between this rental property and the rental property given in question 1. Discuss which of the two options is the most cost-effective for the family.

Rental agreement

By this agreement made at _____, on the _____ day of _____, 20 ___, the Landlord _____ and the

Tenant _____ agree as follows:

1. Rent

Tenant agrees to pay rent in the amount of ($) 2250 (payable in advance).

To be paid per ☐ week ☐ fortnight ☒ calendar month

Day rent is to be 16th of each month paid (e.g. each Thursday or the 11th of each month).

Date first rent 16th Jan 2021

2. Bond

The renter has been asked to pay the bond specified below.

The bond is 4 weeks rent.

Bond amount [   ]

Date of bond 16th Jan 2021

---

3. **WE5** The meter readings and charges/rates for a household are as shown.

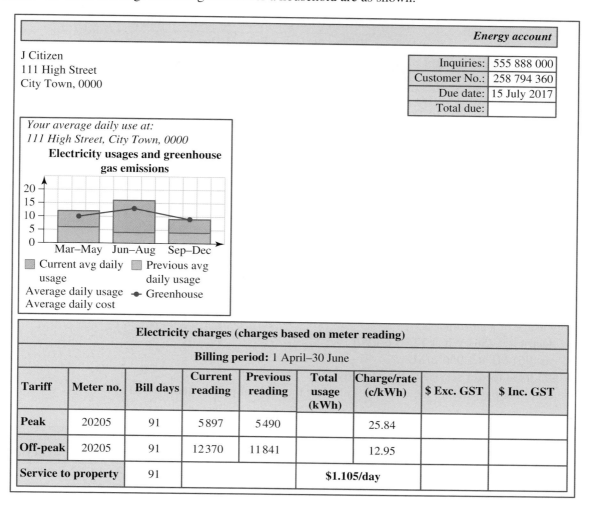

Energy account		
	Inquiries:	555 888 000
	Customer No.:	258 794 360
	Due date:	15 July 2017
	Total due:	

J Citizen
111 High Street
City Town, 0000

Your average daily use at:
111 High Street, City Town, 0000

Electricity charges (charges based on meter reading)

Billing period: 1 April–30 June

Tariff	Meter no.	Bill days	Current reading	Previous reading	Total usage (kWh)	Charge/rate (c/kWh)	$ Exc. GST	$ Inc. GST
Peak	20205	91	5897	5490		25.84		
Off-peak	20205	91	12370	11841		12.95		
Service to property		91			$1.105/day			

Calculate the total bill amount. (Remember to include GST.)

4. Calculate the energy charges (excluding GST) for the following, giving your answers correct to the nearest cent.

	Energy usage	Charge/rate
a.	563 kWh	10.56 c/kWh
b.	1070 MJ	2.629 c/MJ
c.	895 kWh	14.64 c/kWh
d.	6208 MJ	1.827 c/MJ

5. The usage component of a family's energy bill for a quarter is $545 excluding GST. The charge/rate is 11.43 c/kWh. Calculate the family's energy usage for the quarter. Give your answer to the nearest whole number.

6. Part of an electricity bill is shown below.

Electricity charges (charges based on meter reading)								
Billing period: 1 Sep–31 Dec								
Tariff	**Meter no.**	**Bill days**	**Current reading**	**Previous reading**	**Total usage (kWh)**	**Charge/rate (c/kWh)**	**$ exc. GST**	**$ inc. GST**
All day	28 710	91	47 890	46 995		30.54		
Service to property		91			**$1.017/day**			

a. Write:

   i. the charge/rate
   ii. the daily service-to-property fee
   iii. the usage amount in kWh.

b. Calculate:

   i. the amount charged for electricity usage excluding GST
   ii. the total amount for the bill, excluding GST
   iii. the total amount for the bill, including GST.

7. There is a 15% discount off the total bill for customers who pay their energy bill before the due date. The energy usage of a customer is:
   - electricity: 456 kWh at 12.46 c/kWh
   - gas: 967 MJ at 1.956 c/MJ
   - $95 fixed fee for service to the property.
   If the customer pays before the due date, calculate the amount paid including GST.

8. **MC** A family's electricity usage is shown in the following table.

Tariff	**Bill days**	**Current reading**	**Previous reading**	**Total usage**	**Charge/rate c/kWh**
Peak	90	58 973	58 615		29.05
Off-peak	90	8 569	8 028		15.59
Service to property	90			**$1.08/day**	

The total cost of the family's electricity usage, including GST and correct to the nearest 5 cents, is:

A. $188.35   **B.** $269.00   **C.** $285.55   **D.** $314.10   **E.** $214.10

9. **WE6** Calculate the water rates for a home with the following usage and fixed charges.

Bill details as of 31 March		**Value**	**Price**
Water service charge	1 Jan to 31 Mar	22.51	
Sewerage service charge	1 Jan to 31 Mar	145.90	
Stormwater	1 Jan to 31 Mar	18.70	
	Volume (L)	**Charge/kL**	
Water usage	58 000	$2.00/kL	
Recycled water usage	39 000	$1.79/kL	

10. A customer pays $245.85 for their water bill for 3 months. They are charged a fixed service fee of $180.30. The charge/kL is $1.95. The customer does not have recycled-water service. Calculate the amount of water in kilolitres, used by the customer for the 3 months. Give your answer correct to 2 decimal places.

11. **WE7** A council municipality plans to raise $12 million. There is $1.8 billion in capital improved value in the municipality. Calculate the payable council rates for a property with $275 000 capital improved value.

12. The council rates for a home that has a capital improved value of $325 000 are $1950.
    a. Calculate the rate in the dollar charged by the council.
    b. Calculate the amount of revenue the council intends to collect if the total amount of capital improved value for the municipality is $2.67 billion.

13. A customer has a choice between two electricity plans, A and B. The two plans are shown below.

| Plan A | | Plan B | |
Charge/rate		Charge/rate	
All day	12.03 c/kWh	Off-peak (11 pm – 7 am)	9.08 c/kWh
		Peak	13.01 c/kWh
Service to property	$1.01/day	Service to property	$0.99/day
Pay-on-time discount	15%	Pay-on-time discount	12%

The customer's energy usage over 90 billing days is 534 (peak time) and 757 (off-peak time).

Which plan is more cost-effective for the customer? Justify your answer by calculating the electricity costs for each plan if the customer pays on time and if they don't pay on time.

14. Here is an incomplete water bill received by a household.

Bill details as of 30 June		Value	Price
Water service charge	1 Apr to 30 Jun	23.80	
Sewerage service charge	1 Apr to 30 Jun	150.20	
Stormwater	1 Apr to 30 Jun	17.95	
	Volume (L)	Charge/kL	
Water usage	64 000		130.56
Recycled water usage	38 000		
		Total:	$393.57

a. Complete the water bill by filling in the missing values.

As of 1 July, the charges per kL will increase by 4%, the water service charge will increase by $2.50 and the sewerage service charge will increase by $1.50.

b. Calculate the new monthly bill.
c. Determine the percentage by which the average water bill will increase (assuming water usage remains the same). Give your answer correct to 2 decimal places.

15. A family has both electricity and gas connected to their house. Their average electricity usage for 90 billing days is 987 kWh and their average gas usage is 1650 MJ per month (30 days). The family decides to change providers. They have the following choices.

Choice 1: Separate electricity and gas providers
- Electricity: all-day rate 10.03 c/kWh
- Pay-on-time discount (electricity only): 12.5%
- Gas: all-day rate 1.89 c/MJ

Choice 2: Bundle both electricity and gas with one provider and receive the following discounts
- 17% discount on gas
- 10% discount on electricity
- Electricity daily rate: 12.01 c/kWh
- Gas daily rate: 2.03 c/MJ

Service-to-property charges are the same for both providers.

Determine which choice will be more cost-effective for the family, assuming they always pay on time.

16. A family household's bills for 3 months (1 quarter) are shown.

Electricity bill

Tariff	Bill days	Current reading	Previous reading	Total usage	Charge/rate (c/kWh)
Peak	90	12 589	11 902		31.08
Off-peak	90	2 058	1 511		14.05
Service to property	90				$1.075/day

Water bill

Bill details as of 31 March		Value	Price
Water service charge	1 Jan to 31 Mar	22.51	
Sewerage service charge	1 Jan to 31 Mar	145.90	
Stormwater	1 Jan to 31 Mar	18.70	
	Volume (kL)	**Charge/kL**	
Water usage	55	$2.00/kL	
Recycled water usage	35	$1.79/kL	

Council rates

Capital improved value $280 000, rate in the dollar 0.008

a. Show that the quarterly council rate instalment is $560.
b. Calculate the amount that the family spends on electricity, water and council rates for the quarter. (Remember to include GST where relevant.)
c. The family's monthly net income is $6019.35 (after tax). Calculate the percentage of their income that goes towards paying the household bills. Give your answer correct to 2 decimal places.

15.4 Credit cards

15.4.1 Credit card fees and surcharges

In Topic 11, we learnt about credit cards and how they can be used to pay for transactions, and also that interest is charged for such an agreement with a financial institution. Now we will look further at the fees and surcharges associated with credit cards.

In Australia, financial institutions partner with credit card companies including Visa, Mastercard and American Express to issue credit cards. A credit card statement, usually issued monthly, outlines transactions and information about how much needs to be paid and how that can be done.

WORKED EXAMPLE 8 Reading and interpreting a credit card statement

A monthly credit card statement is shown.

COMMUNITY BANK

Ms S Sample
111 Main Street
Cityville NSW 2222

Your Statement | Gold VISA card

Account number	121213589000
Statement period	8 June – 8 July
Credit limit	$2000.00
Available credit	$1129.50
Annual percentage rate	15.95%
Daily percentage rate	0.04370%
Next statement end date	8 August

Your account balance

Opening balance 8 June	$705.85
New transaction charges	$870.50
Payment/refunds	$705.85
Closing balance at 8 July	$870.50

Your payment summary

Minimum payment due	$25.00
Payment due by	2 August
Total amount owing	$870.50

Transactions

Date	Transaction details	Amount (A$)
8 Jun	Toll road	31.50
8 Jun	Supermarket	155.85
9 Jun	Hairstyling	205.00
10 Jun	Bookstore	25.95
12 Jun	Petrol	65.85
15 Jun	Mobile phone bill	85.00
22 Jun	Electricity bill	220.25
23 Jun	Petrol	55.65
24 Jun	Payment – thank you	−705.85
30 Jun	Cinema tickets	25.00

How to pay

Online: Pay with your credit card using combanknet

BPAY: Biller code: 0000
Reference:#########

Autopay: Set up direct debit to pay your credit card each month.

Phone: Call 10 00 10 to transfer funds.

Mail: This slip with your cheque to:
Locked bag CB101, Sydney NSW 2000.

We're here to help:
Call 10 00 00, 24 hours a day
www.communitybank.com.au

a. **State the opening balance of the statement.**
b. **State the closing balance of the statement.**
c. **Show how the daily percentage rate is calculated from the annual percentage rate. Assume 365 days = 1 year.**
d. **Explain how the amount of available credit was calculated.**

COMMUNITY BANK

Ms S Sample
111 Main Street
Cityville NSW 2222

Available credit: $1129.50,
Credit limit: $2000

Closing balance:
$870.50

Annual percentage rate:
15.95%

Opening balance:
$705.85

Your Statement | Gold VISA card

Account number	121213589000
Statement period	8 June – 8 July
Credit limit	$2000.00
Available credit	$1129.50
Annual percentage rate	15.95%
Daily percentage rate	0.04370%
Next statement end date	8 August

Your account balance

Opening balance 8 June	$705.85
New transaction charges	$870.50
Payment/refunds	$705.85
Closing balance at 8 July	$870.50

Your payment summary

Minimum payment due	$25.00
Payment due by	2 August
Total amount owing	$870.50

Start date 8 June; payment due 2 August

Transactions

Date	Transaction details	Amount (A$)
8 Jun	Toll road	31.50
8 Jun	Supermarket	155.85
9 Jun	Hairstyling	205.00
10 Jun	Bookstore	25.95
12 Jun	Petrol	65.85
15 Jun	Mobile phone bill	85.00
22 Jun	Electricity bill	220.25
23 Jun	Petrol	55.65
24 Jun	Payment – thank you	–705.85
30 Jun	Cinema tickets	25.00

How to pay

Online: Pay with your credit card using combanknet

BPAY: Biller code: 0000
Reference:##########

Autopay: Set up direct debit to pay your credit card each month.

Phone: Call 10 00 10 to transfer funds.

Mail: This slip with your cheque to:
Locked bag CB101, Sydney NSW 2000.

We're here to help:
Call 10 00 00, 24 hours a day
www.communitybank.com.au

---

a. 1. Look at the opening balance on the statement.

   2. Write the answer as a sentence.

b. 1. Look at the closing balance on the statement.

   2. Write the answer as a sentence.

c. 1. Look at the annual percentage on the statement.

   2. Calculate the daily percentage rate.

d. 1. Look at the credit limit.

   2. Subtract the closing balance from the credit limit.

   3. Give an explanation.

a. $705.85

The opening balance was $705.85.

b. $870.50

The closing balance was $870.50.

c. Annual % rate = 15.95%

$$\frac{15.95}{365} = 0.0437\%$$

d. The credit limit is $2000.

Available credit = 2000 − 870.50
                 = $1129.50

The available credit is the difference between the credit limit and the closing balance.

15.4.2 Interest-free period

For some credit cards, there is an **interest-free period**. No interest is charged on the debt if the debt is paid back in full before the end of the given interest-free period. For example, an up-to-55-days-interest-free period means that the card holder has a maximum of 55 days to pay the debt without accruing interest.

If the balance is not paid before the due date, then the interest-free period is forfeited and interest can be charged from the date of purchase. Some credit cards don't have an interest-free period. This means that the card holder is charged interest on the outstanding amount of each purchase from the date of purchase until the purchase is repaid in full.

WORKED EXAMPLE 9 Calculating the amount of interest charged on an outstanding balance for an interest-free credit card

For an up-to-55-days-interest-free credit card with an interest rate of 16.9% per annum, calculate the interest payable on an outstanding balance of $450 if it were repaid:
a. after 40 days
b. after 65 days.

THINK	WRITE
a. 1. 40 days is less than the interest-free period.	a. $0
2. Write the answer as a sentence.	No interest is charged as 40 days is within the 55-days-interest-free period.
b. 1. 65 days is longer than the interest-free period, so interest is charged on the entire amount.	b.
2. Calculate the daily interest rate.	Daily interest rate $= \dfrac{16.9\%}{365}$ $= 0.046\,301\%$
3. Calculate the interest charged on $450.	Interest $= \dfrac{0.046\,301}{100} \times 450 \times 65$ $= 13.54$
4. Write the answer as a sentence.	The amount of interest charged on a balance of $450 after 65 days is $13.54.

---

tlvd-5085

WORKED EXAMPLE 10 Calculating the amount of interest charged on an outstanding balance for a no-interest-free credit card

For a no-interest-free-period credit card, calculate the interest charged on the average outstanding daily balance of $220 with an interest rate of 18.24% p.a. if the statement covers a 30-day period.

THINK	WRITE
1. Calculate the daily percentage rate.	$\text{Daily interest rate} = \dfrac{\text{annual interest rate}}{365}$ $= \dfrac{18.24}{365}$ $= 0.049\,97\%$
2. Calculate the daily interest charged on the outstanding balance.	$\text{Interest} = 220 \times \dfrac{0.049\,97}{100}$ $= 0.109\,93$
3. Calculate the interest charged over the 30-day period and round to the nearest cent.	$\text{Total interest (30 days)} = 0.109\,93 \times 30$ $= 3.2979$ $= \$3.30 \text{ (to the nearest cent)}$
4. Write the answer as a sentence.	The amount of interest charged was $3.30.

15.4.3 Annual fees and cash advances

Credit card issuers often charge card holders an **annual fee** for the benefit of having a credit card. Some card issuers don't charge an annual fee, but others can charge hundreds of dollars, depending on the benefits and rewards being offered with the card. More benefits usually means a higher annual fee.

Credit cards can also be used to get cash or a cash equivalent, known as a **cash advance**. Withdrawing money from an ATM, transferring funds from your credit card to other bank accounts, and buying foreign currency are some examples of cash advances. Cash advances attract a higher interest fee and are not eligible for interest-free periods or contributions towards the rewards associated with some credit cards. A cash advance is similar to taking out a short-term loan; however, the transaction is considered a higher risk to the lender, so a higher interest is charged to offset the risk.

WORKED EXAMPLE 11 Calculating the total interest charged on a cash advance

A credit card holder uses their card on 20 March to obtain a cash advance of $450. They repay the cash advance on 31 March. Calculate the amount of interest charged on the cash advance at a rate of 21.24% p.a. Write your answer correct to the nearest cent.

THINK	WRITE
1. Calculate the number of days that interest is charged on the cash advance.	The card holder is charged interest for 11 days.
2. Calculate the daily interest rate.	$\dfrac{21.24}{365} = 0.058\,19$
3. Calculate the interest, in dollars, charged per day.	$\text{Interest} = \dfrac{0.058\,19}{100} \times 450$ $= 0.26$
4. Calculate the total interest charged.	$\text{Total interest} = 0.26 \times 11 \text{ days}$ $= \$2.88$

15.4.4 Surcharges

Some businesses may charge the customers a **surcharge** to make purchases using a credit card. The surcharge covers the costs charged to the business to process the credit card payments.

Businesses are permitted to cover their costs for credit card payments, but there are Australian laws to prohibit businesses from charging excessive credit card surcharges to customers. The costs incurred to businesses for processing credit card payments are 1–1.5% for Visa and Mastercard and 1.5–3% for American Express credit cards. A flat surcharge fee, such as charging an additional 50c for using a credit card, is not permitted.

WORKED EXAMPLE 12 Calculating the total interest charged on a cash advance

A local restaurant charges customers a credit card surcharge. The cost incurred by the restaurant to process a payment on Visa or Mastercard is 1.25% of the payment; the cost for them to process a payment on American Express is 2.5%.

A group of diners ate at the restaurant on the weekend. They paid for their bill using a credit card.

The dinner bill, the total cost charged and the type of credit card used are shown.

Dinner bill	Total cost including surcharge	Credit card used
$195.30	$200	Visa

By calculating the cost incurred by the restaurant to process the credit card payment, determine whether the diners have been charged an excessive credit card surcharge.

THINK	WRITE
1. Determine the cost incurred by the restaurant to process the Visa payment.	$1.25\% \times \$195.30 = 0.0125 \times 195.30$ $\qquad\qquad\qquad = 2.44$
2. Calculate the actual total cost for the diners.	Total cost = bill + surcharge $\qquad\quad = 195.30 + 2.44$ $\qquad\quad = 197.74$
3. Write the answer as a sentence.	The diners paid $200, which is more than the actual cost ($197.74). Yes, the diners were charged an excessive surcharge.

15.4.5 Comparing credit cards and personal loans

Another type of loan that can be considered instead of using a credit card is a personal loan. A personal loan is made by a lending institution to an individual. Interest rates usually range from 7% to 20%.

Whether using a credit card or a personal loan to make purchases, you need to be well informed about the associated costs. Regardless of the type of loan, all money borrowed must be paid back with interest.

This table makes some comparisons between credit cards and personal loans.

Credit card	Personal loan
The borrower pays interest only on the amount spent.	The borrower pays interest on the entire amount of the loan.
Interest fees vary between 15% and 30%.	Interest charged varies between 7% and 20%, depending on the borrower's credit rating.
There are no regular monthly repayments. However, paying off overdue amounts requires the card holder to be in control of their finances.	Paying the loan off in regular monthly instalments makes it easier to budget.
There may be an interest-free period to pay off the amount owing without being charged interest.	No interest-free period. Interest is charged on the entire amount.
Rewards are offered for using the credit card. These can be used for flights, accommodation and other purchases.	No rewards are offered to spend the money.

WORKED EXAMPLE 13 Calculating repayment and comparing cost-effective options for personal loans

A family are renovating their kitchen and will need $4500 to cover the cost of purchasing new appliances. They use their credit card to pay for the appliances. They have a credit limit of $5000 and interest of 16.95% p.a. Their minimum payments are $25 or 2% of the balance, whichever is greater.
a. Calculate the minimum repayment.
b. Use technology to show that if they make $300 payments each month, it will take them 17 months to pay off their credit card balance. Assume 30 days in each month.
c. They also have the option of taking out a personal loan for $4500, with flat-rate interest of 14.69% p.a. for 2 years and regular monthly repayments of $242.59. Determine which option is more cost-effective for them. Justify your answer using calculations.

THINK

a. 1. Calculate 2% of the balance.

    2. Write the answer as a sentence.

b. 1. Set up a spreadsheet with formulas as shown.

WRITE

a. $2\% \times \$4500 = \$90$

    2% of balance $> \$25$

    The minimum payment is $90.

b.

	A	B	C	D
1	Month	Balance ($)	Balance + interest ($)	Payment ($)
2	1	4500.00	= 16.95/36 500 *b2*30 + b2	300
3	2	= c2 − d2	= 16.95/36 500 *b3*30 + b3	300

2. Complete the table using the fill-down option.

	A	B	C	D
			Balance +	
1	Month	Balance ($)	interest ($)	Payment ($)
2	1	4500.00	4562.69	300
3	2	4262.69	4322.08	300
4	3	4022.08	4078.11	300
5	4	3778.11	3830.75	300
6	5	3530.75	3579.93	300
7	6	3279.93	3325.63	300
8	7	3025.63	3067.78	300
9	8	2767.78	2806.34	300
10	9	2506.34	2541.26	300
11	10	2241.26	2272.48	300
12	11	1972.48	1999.96	300
13	12	1699.96	1723.64	300
14	13	1423.64	1443.48	300
15	14	1143.48	1159.41	300
16	15	859.41	871.38	300
17	16	571.38	579.34	300
18	17	279.34	283.23	283.23
19			Total	5083.23

3. Write the answer as a sentence.

As shown in the spreadsheet, it will take 17 months for the family to pay off their credit card. After the 16th month, they have a balance of $283.23, which is paid off at the end of the 17th month.

c. 1. Calculate the total amount the family will pay using their credit card.

c. $300 per month for 16 months + $283.23
$= 16 \times 300 + 283.23$
$= 5083.23$

2. Calculate the total amount they will pay for the personal loan.

2 years at $242.59 per month $= 2 \times 12 \times 242.59$
$= 5822.16$

3. Write the answer as a sentence.

If the family pay $300 on their credit card each month, they will take less time to pay back $4500. They will have an overall repayment of $5083.23 on the credit card, compared to $5822.16 on the personal loan. Therefore, using the credit card is the more cost-effective option for them.

1. **WE8** A credit card statement is shown.

COMMUNITY BANK

Mr A Smith
258 City Road
Cityville NSW 2222

Your Statement | Silver Mastercard

Account number	44780369
Statement period	15 Sep – 15 Oct
Credit limit	$5000.00
Available credit	$2950.00
Annual percentage rate	16.95%
Daily percentage rate	0.04644%
Next statement end date	15 Nov

Your account balance

Opening balance 15 Sep	$1158.00
New transaction charges	$2050.00
Payment/refunds	$1158.00
Closing balance at 15 Oct	$2050.00

Your payment summary

Minimum payment due	$59.00
Payment due by	9 Nov
Total amount owing	$2050.00

Transactions

Date	Transaction details	Amount (A$)
15 Sep	Electricity bill	458.00
16 Sep	Payment – thank you	−1158.00
20 Sep	Registration fees	75.00
20 Sep	Groceries	97.00
25 Sep	Fuel	125.00
28 Sep	Entertainment	110.00
30 Sep	Dinner	58.00
3 Oct	Groceries	195.00
5 Oct	Fuel	105.00
10 Oct	Council rates	512.00
12 Oct	Clothing	225.00
14 Oct	Fuel	90.00

- -

How to pay

Online: Pay with your credit card using combanknet

BPAY: Biller code: 0000
Reference:#########

Autopay: Set up direct debit to pay your credit card each month.

Phone: Call 10 00 10 to transfer funds.

Mail: This slip with your cheque to:
Locked bag CB101, Sydney NSW 2000.

We're here to help:
Call 10 00 00, 24 hours a day
www.communitybank.com.au

---

a. State the opening balance of this statement.
b. State the closing balance of this statement.
c. Show how the daily percentage rate is calculated from the annual percentage rate.
d. Explain how the amount of available credit was calculated.

2. A credit card statement is shown. The credit limit on the card is $3000 and the annual interest is 17.50% p.a.

   a. State the opening balance of this statement.
   b. Determine the values of A and B.

COMMUNITY BANK

Your Statement | Gold VISA card

Ms S Sample
111 Main Street
Cityville NSW 2222

Account number	121213589000
Statement period	15 May – 15 June
Credit limit	$3000.00
Available credit	$1129.50
Annual percentage rate	17.50%
Daily percentage rate	0.04370%
Next statement end date	15 July

Your account balance

Opening balance 15 May	$985.00
New transaction charges	$A
Payment/refunds	$1135.00
Closing balance at 15 Jun	$B

Your payment summary

Minimum payment due	
Payment due by	9 July
Total amount owing	$B

Transactions

Date	Transaction details	Amount (A$)
15 May	Fuel	55.85
16 May	Energy bill	358.40
18 May	Groceries	65.75
20 May	Public transport fare top up	50.00
25 May	Opera House tickets	600.00
28 May	Cash advance	150.00
31 May	Car service	125.00
1 Jun	Mobile phone bill	60.45
3 Jun	Direct debit – car loan	58.65
6 Jun	Fuel	65.40
7 Jun	Payment – thank you	−1135.00
8 Jun	Road	31.05
10 Jun	Interest – 0.04795%	0.79
11 Jun	Annual card fee	87.00
13 Jun	Council rates	475.25

- -

How to pay

Online: Pay with your credit card using combanknet

BPAY: Biller code: 0000
Reference:#########

Autopay: Set up direct debit to pay your credit card each month.

Phone: Call 10 00 10 to transfer funds.

Mail: This slip with your cheque to:
Locked bag CB101, Sydney NSW 2000.

We're here to help:
Call 10 00 00, 24 hours a day
www.communitybank.com.au

---

3. **WE9** For an up-to-55-days-interest-free credit card with an interest rate of 16.9% per annum, calculate the interest payable on an outstanding balance of $625 if it were repaid:

   a. after 20 days
   b. after 70 days.

4. **WE10** For a no-interest-free-period credit card, calculate the interest charged on an average outstanding daily balance of $430 with a percentage interest rate of 18.24% p.a. if the statement covers a 30-day period.

5. **WE11** An up-to-55-days-interest-free credit card holder used his card on 15 March to obtain a cash advance of $365, which he repaid on 20 March. Determine the amount of interest charged on the cash advance at a rate of 20.74% p.a.

   Write your answer correct to the nearest cent.

6. **MC** The calculation that determines the amount of interest charged on a cash advance of $200 obtained on 3 October and paid back on 20 October at a rate of 19.75% p.a. is:

   A. $\dfrac{19.75}{100} \times 200 \times 17$

   B. $\dfrac{19.75}{100} \times 200$

   C. $\dfrac{19.75}{36\,500} \times 200 \times 17$

   D. $\dfrac{19.75}{36\,500} \times 200$

   E. $\dfrac{19.75}{100} \times 200$

7. **WE12** A local restaurant charges customers a credit card surcharge. The cost incurred by the restaurant to process a Visa or Mastercard payment is 1.15% of the payment; the cost for them to process an American Express payment is 2.2%.

   The following groups of diners ate at the restaurant over the weekend. They paid for their bills using credit cards. The dinner bill, total cost charged and the type of credit card used are shown for each group. All money shown is correct to the nearest cent.

Group	Dinner bill	Total cost (including surcharge)	Credit card used
A	$35.80	$36.21	Visa
B	$112.50	$115.00	American Express
C	$78.95	$82.10	Mastercard

   By calculating the cost incurred by the restaurant to process each credit card payment, determine whether any of the groups of diners have been charged an excessive credit card surcharge.

8. **WE 13** Bailey uses his credit card to purchase accessories for his classic car. The only purchase he will make on his credit card is for the $3000 worth of accessories. He has a credit limit of $5000, with a 55-day interest-free period and interest of 17.65% p.a. His minimum payment is $25 or 3% of the balance, whichever is greater.

   a. Calculate the minimum payment if Bailey does not pay back the amount in full by the end of the 55-day interest-free period. Write your answer correct to the nearest cent.

   b. Bailey makes a regular payment of $250 at the end of each month, with interest calculated daily and added to his balance at the end of month. Use technology to determine how many months it will take him to pay off his credit card balance.

   c. Bailey also has the option of taking out a personal loan for $3000 with an interest rate of 15.40% p.a. for 2 years and regular monthly repayments of $163.50. Determine which option is the most cost-effective for Bailey.

   Justify your answer using calculations.

15.5 Personal income–related expenses

15.5.1 Youth Allowance

Centrelink is an Australian government agency that provides income support and other payments to Australians. Various government allowances and benefits are available for students, families, carers, unemployed, retirees and disabled people.

Youth Allowance is financial help for young people who fall into one of the following categories:
- 16 to 21 and looking for full-time work
- 18 to 24 and studying full-time
- 16 to 24 and doing a full-time Australian Apprenticeship
- 16 to 17 and independent or needing to live away from home to study
- 16 to 17, studying full time and have completed Year 12 or equivalent.

To be eligible, people must also meet Australian residence rules, complete an approved course or apprenticeship, and satisfy income and assets tests or the parental means test.

If eligible for Youth Allowance, a claim must be submitted to Centrelink and, if approved, maximum fortnightly payments can be received at the following rates.

Your circumstances	Your maximum fortnightly payment
Single, no children, younger than 18, and living at your parents' home	$313.80
Single, no children, younger than 18, living away from your parents' home to study, train or look for work	$530.40
Single, no children, 18 or older and living at your parents' home	$367.00
Single, no children, 18 or older and need to live away from your parents' home	$530.40
Single, with children	$679.00
A couple, no children	$530.40
A couple, with children	$577.40

Note: The payment table is subject to change and was correct at the time of writing.

tlvd-5086

WORKED EXAMPLE 14 Calculate the maximum fortnightly payment for Youth Allowance

Determine the maximum fortnightly payment for Youth Allowance for each of the following young people using the payment table provided. Assume all eligibility criteria have been met.

a. Jeroen, aged 19, is completing an approved Australian apprenticeship and living at home with his parents.

b. Chrissie, aged 24, is studying an approved course at university and lives with her partner and young child.

c. Xiao, aged 17, had to move out of his family's rural home to study at university full-time.

THINK	WRITE
a. 1. Look in the table to find Jeroen's circumstances, which are: *single, no children, 18 or older and living at your parents' home.*	a. $367.00
2. Write the answer as a sentence.	Jeroen's maximum fortnightly payment could be $367.00.
b. 1. Look in the table to find Chrissie's circumstances which are: *a couple, with children.*	b. $577.40
2. Write the answer as a sentence.	Chrissie's maximum fortnightly payment could be $577.40.
c. 1. Look in the table to find Xiao's circumstances, which are: *single, no children, younger than 18, living away from your parents' home to study, train or look for work.*	c. $530.40
2. Write the answer as a sentence.	Xiao's maximum fortnightly payment could be $530.40.

15.5.2 Child Care Subsidy

The government can provide monetary assistance to help with the cost of childcare expenses. The percentage of the subsidy you may get depends on your annual family income.

To be eligible for Child Care Subsidy, families must meet the following criteria:

- care for a child 13 or younger who's not attending secondary school, unless an exemption applies
- use an approved childcare service
- be responsible for paying the childcare fees
- meet residency and immunisation requirements.

If eligible for the Child Care Subsidy, a claim must be submitted to Centrelink and, if approved, the Child Care Subsidy can be applied at the following rate.

Your family income	Child Care Subsidy percentage
$0 to 70 015	85%
More than $70 015 to below $175 015	Between 85% and 50% The percentage goes down by 1% for every $3000 of income your family earns above $70 015
$175 015 to below $254 305	50%
$254 305 to below $344 305	Between 50% and 20% The percentage goes down by 1% for every $3000 of income your family earns above $254 305
$344 305 to below $354 305	20%
$354 305 or more	0%

Note: The payment table is subject to change and was correct at the time of writing.

WORKED EXAMPLE 15 Calculate the Child Care Subsidy percentage for families

Determine the Child Care Subsidy percentage for each of the following families using the payment table provided. Assume all eligibility criteria have been met.
a. A family earning a combined income of **$68 000** per annum
b. A family earning a combined income of **$190 000** per annum
c. A family earning a combined income of **$120 000** per annum

THINK

a. 1. Look in the table for the percentage for a family income of $68 000.

2. Write the answer as a sentence.

b. 1. Look in the table for the percentage for a family income of $190 000.

2. Write the answer as a sentence.

c. 1. Look in the table for the percentage for a family income of $120 000.

2. Calculate the income amount above $70 015 that this family earns.

3. Calculate the percentage decrease for this family if for every $3000 of income the percentage decreases by 1%.

4. Decrease the 85% subsidy by the percentage decrease amount.

5. Write the answer as a sentence.

WRITE

a. 85%

The family's Child Care Subsidy percentage is 85%.

b. 50%

The family's Child Care Subsidy percentage is 50%.

c. Between 85% and 50%
The percentage goes down by 1% for every $3000 of income your family earns above $70 015.

$120 000 - 70 015 = 49 985$

$49 985 \div 3000 = 16.66\%$ decrease

$85 - 16.66 = 68.34\%$

The family's Child Care Subsidy percentage is 68.34%.

15.5.3 Age Pension

Senior Australian residents aged 66 or over, whose assets and income come under the government test limits, can be eligible for the Age Pension to assist with their living requirements during retirement.

If eligible for the Age Pension, a claim must be submitted to Centrelink and, if approved, maximum fortnightly payments can be received at the following rates:

Per fortnight	Single	Couple each	Couple combined	Couple apart due to ill health
Maximum basic rate	$882.20	$665	$1,330	$882.20

Note: The payment table is subject to change and was correct at the time of writing.

WORKED EXAMPLE 16 Calculating the Age Pension fortnightly payment for retirees

Determine the Age Pension fortnightly payment for each of the following retirees using the payment table provided. Assume all eligibility criteria have been met.
a. Elsie, aged 68, who lives at home with her husband
b. John, aged 82, who lives at home on his own as his wife is unwell and in an aged care home
c. Rupert, aged 67, who is still working full-time

THINK	WRITE
a. 1. Look in the table to find Elsie's circumstances, which are: *couple each*.	a. $665
2. Write the answer as a sentence.	The amount of Age Pension Elise can receive per fortnight is $665.
b. 1. Look in the table to find John's circumstances, which are: *couple apart due to ill health*.	b. $882.20
2. Write the answer as a sentence.	The amount of Age Pension John can receive per fortnight is $882.20.
c. 1. Rupert is not eligible to receive the Age Pension as he is still working full-time.	c.
2. Write the answer as a sentence.	Rupert will not be able to receive the Age Pension.

15.5.4 Student loans

To assist students to cover the costs of university study and/or vocational education and training courses, the Australian government offers Commonwealth assistance loans known as HECS-HELP or VET Student Loans.

HECS-HELP

Australian universities offer Commonwealth Supported Places (CSP) to Australian residents, which are university enrolments subsided by the Australian government. If enrolled in a CSP the Australian government will pay some of your fees and it is up to you to pay the rest, known as your *student contribution amount*, which can be quite costly. To cover the student contribution amount, a government loan is offered: HECS-HELP.

HECS-HELP only covers the student contribution amount and cannot be used to pay for books, accommodation or other costs association with university study. The limit to a HECS-HELP loan is usually $109 206 and it needs to be repaid through the Australian tax system once you earn above the compulsory repayment threshold, which is $47 014 in 2021–22. The table below shows the repayment interest rates for taxable annual incomes in the 2021–22 financial year.

2021–2022 repayment threshold	Repayment % rate
$47 014–$54 282	1.0%
$54 283–$57 538	2.0%
$57 539–$60 991	2.5%
$60 992–$64 651	3.0%
$64 652–$68 529	3.5%
$68 530–$72 641	4.0%
$72 642–$77 001	4.5%
$77 002–$81 620	5.0%
$81 621–$86 518	5.5%
$86 519–$91 709	6.0%
$91 710–$97 212	6.5%
$97 213–$103 045	7.0%
$103 046–$109 227	7.5%
$109 228–$115 781	8.0%
$115 782–$122 728	8.5%
$122 729–$130 092	9.0%
$130 093–$137 897	9.5%
Over $137 898	10.0%

VET Student Loans

If studying an approved higher level (diploma and above) vocational education and training (VET) course, you can access the Australian government VET Student Loans program to assist in paying tuition fees.

The limit, or loan cap, for a VET Student Loan differs based on the course being studied. If you undertake a course that costs more than the loan cap, you will need to pay the difference up-front to your course provider.

WORKED EXAMPLE 17 Calculate the monthly HECS-HELP repayment

Courtney studied a Bachelor of Psychological Science at university, accruing a total HECS-HELP debt of $35 124. She is now a practising psychologist and is expected to repay her HECS-HELP loan from her monthly gross income.

a. If Courtney earns $120 000 per annum, calculate her monthly HECS-HELP repayment.

b. Determine how long, to the nearest month, it will take Courtney to pay off her HECS-HELP debt if she remains on this annual salary.

THINK	WRITE
a. 1. Calculate the repayment percentage rate for an annual income of $120 000.	**a.** 8.5%
2. Calculate the annual amount of HECS-HELP owing by finding 8.5% of 120 000.	$\dfrac{8.5}{100} \times 120\,000 = 10\,200$
3. Divide by 12 to calculate the monthly HECS-HELP to be repaid.	$10\,200 \div 12 = 850$
4. Write the answer.	Courtney's monthly HECS-HELP repayment is $850.
b. Divide the total HECS-HELP debt by the monthly payment amount. Round up to calculate the total number of months and write the answer.	**b.** $35\,124 \div 850 = 41.32$ It will take Courtney 42 months to repay her HECS-HELP debt.

15.5 Exercise

Students, these questions are even better in jacPLUS

 Receive immediate feedback and access sample responses

 Access additional questions

 Track your results and progress

Find all this and MORE in jacPLUS ▶

1. **WE14** Determine the maximum fortnightly Youth Allowance payment for each of the following people. Assume all eligibility criteria have been met.

   a. Steffan, aged 20, is currently looking for work and lives at home with his mum.
   b. Jenna, aged 22, is studying an approved course at university and lives away from home.
   c. Ken, aged 23, is a single dad studying full-time and supporting his one-year-old daughter.

2. Consider the pathway you would like to follow after secondary school. Use the Centrelink Youth Allowance calculator to determine how much Youth Allowance you might receive per fortnight.

3. **WE15** Determine the Child Care Subsidy percentage for each of the following families using the payment table provided. Assume all eligibility criteria have been met.

   a. A family earning a combined income of $360 000 per annum.
   b. A family earning a combined income of $35 000 per annum.
   c. A family earning a combined income of $300 000 per annum.

4. **WE16** Determine the Age Pension fortnightly payment for each of the following retirees using the payment table provided. Assume all eligibility criteria have been met.

   a. Sandra, aged 50, who lives at home with her partner
   b. Phillip, aged 72, a single man
   c. Bonnie and Peter, both in their 90s, a married couple

5. Research the details of the Age Pension income and assets tests.

6. **WE17** Anoush studied a Bachelor of Arts at university, accruing a total HECS-HELP debt of $42 400. She is now working and expected to repay her HECS-HELP loan from her monthly gross income.

   a. If Anoush earns $70 000 per annum, calculate her monthly HECS-HELP repayment.
   b. Determine how long, to the nearest month, it will take Anoush to pay off her HECS-HELP debt if she remains on this annual salary.

7. A computer technician's annual salary is $67 374.

   a. The technician is paid monthly. Calculate their gross monthly salary.
   b. Determine the amount of HECS-HELP that will be withheld from the technician's monthly salary.

8. Nathan earns $90 000 per annum. Based on the table shown in section 15.5.4, calculate:

   a. the amount of tax that Nathan is expected to pay annually
   b. the amount of HECS-HELP Nathan is expected to pay annually
   c. the monthly income Nathan will receive after tax and HECS-HELP have been taken out.

15.6 Car expenses

LEARNING INTENTION

At the end of this subtopic you should be able to:

- calculate different types of costs involved with buying vehicles
- calculate on-road costs for new and used vehicles
- calculate the fuel consumption rate
- calculate the cost of servicing vehicles
- calculate and compare the cost of purchasing different vehicles using a spreadsheet.

15.6.1 Planning for the purchase of a vehicle

Buying a vehicle for the first time can be very exciting. However, it pays to do research and know exactly which type of vehicle will best suit your needs. For example, do you need a small, medium or large vehicle? How many seats and how big an engine do you need?

Many costs are associated with buying a vehicle, but the primary cost is the purchase price or sale price. The advertised price for the vehicle is not always the sale price. Many vehicle dealerships are willing to negotiate on price or match other competitive prices for a similar or same-model vehicle. It is important to know how much the vehicle is worth and then shop around for the best price.

Paying for the vehicle

Many financial institutions offer loans to purchase vehicles. It is also possible to obtain finance through vehicle dealerships. As with all loans, it is important to know exactly how much interest is charged and what additional costs apply, such as administration fees and loan servicing fees.

A **comparison rate** gives a more accurate indication of the interest rate for the loan. It includes all the additional fees in the calculation.

At times, 0% finance is advertised for the purchase of a vehicle. This means no interest is charged on the loan. However, this type of finance may apply only to certain models and the sale price may not be negotiable.

tlvd-5087

WORKED EXAMPLE 18 Calculating costs involved with buying a vehicle

A customer takes out a loan to purchase a car for $15 090. The monthly repayments are $294.26 and the loan is for 5 years. The customer pays a 10% deposit. Calculate:
a. the deposit paid
b. the total amount paid for the car
c. the interest rate charged for the loan.

THINK	WRITE
a. 1. Calculate 10% of $15 090.	**a.** $\dfrac{10}{100} \times 15\,090 = 1509$
2. Write the answer.	The deposit paid is $1509.
b. 1. Calculate the total amount paid in repayments.	**b.** Monthly repayments $= \$294.26$ for 5 years $\$294.26 \times 12 \times 5 = \$17\,655.60$
2. Calculate the amount paid for the car by adding the deposit paid to the total amount in monthly repayments.	Amount paid for the car $= \$1509 + \$17\,655.60$ $= \$19\,164.60$
3. Write the answer.	The total amount paid for the car is $19 164.60.
c. 1. Calculate the principal by subtracting the deposit from the cost of the car.	**c.** $\$15\,090 - \$1509 = \$13\,581$
2. Calculate the interest paid for the car by subtracting the principal from the total amount in monthly repayments.	Interest $= \$17\,655.60 - \$13\,581$ $= \$4074.60$
3. Use the formula $I = \dfrac{Prn}{100}$ to calculate the rate. $P = \$13\,581$, $I = \$4074.60$, $n = 5$	$4074.60 = \dfrac{13\,581 \times r \times 5}{100}$ $407\,460 = 13\,581 \times r \times 5$ $\dfrac{407\,460}{13\,581 \times 5} = r$ $r = 6\%$
4. Write the answer as a sentence.	The interest rate charged is 6% per year.

15.6.2 On-road costs for new and used vehicles

As well as car insurance, which we looked at in subtopic 15.2, there are other on-road costs such as registration and stamp duty. A **registration fee** is a combination of administration fees, taxes and charges paid to legally drive a vehicle on the roads. Registered vehicles have number plates that help emergency services identify to whom the vehicle belongs and help transport authorities keep track of vehicle ownership. Registration costs depend on the type of vehicle and the location.

Stamp duty is a government tax on certain transactions, including the purchase of a motor vehicle. It is charged based on:
- the date the vehicle was registered or the registration transferred
- whether it is a passenger car or a non-passenger car
- the dutiable value of the vehicle.

Calculate the amount of stamp duty to be paid for the following vehicle purchases.

Type of vehicle	Rate
• hybrid — any number of cylinders • electric	Up to $100 000: $2 for each $100, or part of $100 More than $100 000: $4 for each $100, or part of $100
• 1 to 4 cylinders • 2 rotors • steam powered	Up to $100 000: $3 for each $100, or part of $100 More than $100 000: $5 for each $100, or part of $100

a. A 4-cylinder car purchased for $14 560
b. A 4-cylinder car purchased for $104 900

THINK	WRITE
a. 1. Determine the percentage to be paid by reading the table.	a. The vehicle costs less than $100 000 and has 4 cylinders. Therefore, the percentage is 3%.
2. Calculate 3% of the vehicle purchase price.	$\dfrac{3}{100} \times 14\,560 = \$436.80$
3. Answer the question.	The stamp duty to be paid on a 4-cylinder vehicle purchased for $14 560 is $436.80.
b. 1. Determine the percentage to be paid.	b. The vehicle costs more than $100 000 and has 4 cylinders. Therefore, the percentage is 5%.
2. Calculate 5% of the vehicle purchase price.	$\dfrac{5}{100} \times 104\,900 = \$5245$
3. Write the answer as a sentence.	The stamp duty to be paid on a 4-cylinder vehicle purchased for $104 900 is $5245.

15.6.3 Sustainability, fuel consumption rates, servicing and tyres

The running costs of a vehicle include the fuel consumption and maintenance. The latter includes the cost to service the vehicle and replace the tyres.

Fuel consumption rates show the amount of fuel (petrol or diesel) used per 100 km. For example, 7.8 L/100 km means that the vehicle uses an average of 7.8 litres of fuel for every 100 km travelled.

Fuel consumption rates will vary depending on where you drive. Driving in city areas usually uses more fuel because of the constant stopping and starting. Driving on freeways, where the speed travelled is more constant, generally has a lower fuel consumption rate.

WORKED EXAMPLE 20 Calculating the fuel consumption rate

A vehicle travels 580 km and uses 62 litres of fuel. Calculate the fuel consumption rate in L/100 km. Give your answer correct to 2 decimal places.

THINK	WRITE
1. Divide the total distance travelled by 100 km (this gives the distance in 100-km lots).	$\dfrac{580}{100} = 5.8$
2. Divide the amount of fuel in litres by the answer in step 1. Round to 2 decimal places.	$\dfrac{62}{5.8} = 10.6896$ ≈ 10.69
3. Write the answer as a sentence.	The fuel consumption rate is 10.69 L/100 km.

Vehicles require servicing by experienced mechanics to maintain performance. The cost of servicing a vehicle varies according to the distance travelled and general wear on moving parts, such as brakes. Services are generally required every 10 000–15 000 km, depending on the manufacturer's guidelines. The service cost will include labour, plus the cost to replace parts.

WORKED EXAMPLE 21 Calculating the cost of servicing a vehicle

A driver travels on average of 450 km every week.
a. If the vehicle requires servicing every 15 000 km, determine how often this vehicle will need servicing over 3 years.
b. At the 30 000-km service, the vehicle needs the following parts. (The prices for the parts include GST.)
 • Oil filter $30
 • Fuel filter $25
 • Oil $35
 • Windscreen wiper blades $15
The service will take 3 hours of labour at $105 per hour excluding GST.
Calculate the cost for the 30 000-km service.

THINK	WRITE
a. 1. Calculate the number of weeks the driver takes to travel 15 000 km.	a. $\dfrac{15\,000}{450} = 33.33$
2. Calculate the number of weeks in 3 years.	$52 \times 3 = 156$
3. Divide the number in step 2 by the number in step 1.	$\dfrac{156}{33.33} = 4.68$
4. Write the answer as a sentence.	The number of 15 000-km services in 3 years is 4.
b. 1. Calculate the labour costs.	b. $3 \times \$105 = \$315$ Add 10%. GST: 10% of $315 = \$31.50$ Labour costs $= \$315 + \$31.50$ $= \$346.50$

2. Calculate the cost of parts.	$30 + 25 + 35 + 15 = $105
3. Add the labour costs to the cost of the parts.	$105 + $346.50 = $451.50
4. Write the answer as a sentence.	The total cost for the 30 000-km service is $451.50.

Tyre tread helps keep vehicles stable on the road and is subject to roadworthy inspection by police. Vehicles without the regulated amount of tyre tread are deemed unsafe and the owner is required to have the tyres replaced. The wear on tyres depends on driving conditions such as the road surface, vehicle performance and style of driving.

The tyres are usually checked during routine vehicle servicing to ensure that they are safe to drive on and wearing evenly. If there is an issue with the vehicle's steering alignment, the tyres may wear more on one side.

WORKED EXAMPLE 22 Determining how many times tyres need to be replaced

For a particular brand of tyres, the manufacturer recommends replacement after 40 000 km. A driver using these tyres travels an average of 550 km each week. Determine how many times they should expect to replace the tyres in 5 years.

THINK	WRITE
1. Calculate the number of weeks it takes to travel 40 000 km. Divide 40 000 km by the average weekly distance.	$\dfrac{40\,000}{550} = 72.73$
2. Calculate the number of years by dividing the answer by 52.	$\dfrac{72.73}{52} = 1.3986$
3. Calculate the number of times in 5 years by dividing 5 by the answer from step 2.	$\dfrac{5}{1.399} = 3.574$
4. Write the answer as a sentence.	In 5 years the driver should expect to replace the tyres 3 times.

15.6.4 Calculate and compare the cost of purchasing different vehicles using a spreadsheet

A spreadsheet can be used to calculate and compare the cost of purchasing different vehicles. Many resources that compare vehicle performance and costs are available.

To compare vehicles it is important to compare similar vehicles and common elements, such as fuel consumption and servicing costs. Registration and insurance costs will vary due to other factors, such as location of the vehicle and state of registration.

WORKED EXAMPLE 23 Determining the most effective purchase using a spreadsheet

Using a spreadsheet, determine which one of the following three cars is the most cost-effective to buy. Assume that the average distance travelled each year is 15 000 km, the average price of fuel per litre is $1.20, and all of the cars have the same safety rating.

Four-door sedan 1.6-litre engine				
Model	Sale price	Fuel consumption (L/100 km)	Servicing (km)	Average service costs
Car A	$15 480	8.2	15 000	$450
Car B	$16 250	7.5	20 000	$350
Car C	$14 999	9.8	10 000	$475

THINK

1. Create a spreadsheet.

2. Calculate the number of litres used (on average) for one year of driving.

3. Calculate the cost to fuel the car for one year at $1.20/L.

4. Calculate the servicing costs. To compare the costs, we need to calculate the proportional service costs by dividing the average distance travelled (15 000 km) by the service distance for each car, then multiply by the average service costs.

WRITE

	A	B	C	D	E
1	Vehicle	Sales price	Fuel consumption (L/100 km)	Servicing (km)	Average service costs
2	Car A	$15 480.00	8.2	15 000	$450.00
3	Car B	$16 250.00	7.5	20 000	$350.00
4	Car C	$14 999.00	9.8	10 000	$475.00

Average distance travelled = 15 000 km

Number of 100 km units: $\dfrac{15\,000}{100} = 150$

Insert the formula '= 150*C2' into the cell F2.
Copy the formula into cells F3 and F4.

Enter the formula '= F2*1.2' into cell G2.
Copy the formula into cells G3 and G4.

	A	B	C	D	E	F	G
1	Vehicle	Sales price	Fuel consumption (L/100 km)	Servicing (km)	Average service costs	Litres/ years	Fuel cost
2	Car A	$15 480.00	8.2	15 000	$450.00	1230	$1476.00
3	Car B	$16 250.00	7.5	20 000	$350.00	1125	$1350.00
4	Car C	$14 999.00	9.8	10 000	$475.00	1470	$1764.00

Enter the formula '= 15 000/D2*E2' into cell H2.
Copy the formula into cells H3 and H4.

	A	B	C	D	E	F	G	H
1	Vehicle	Sales price	Fuel consumption (L/100 km)	Servicing (km)	Average service costs	Litres/ years	Fuel cost	Service cost
2	Car A	$15 480.00	8.2	15 000	$450.00	1230	$1476.00	$450.00
3	Car B	$16 250.00	7.5	20 000	$350.00	1125	$1350.00	$262.50
4	Car C	$14 999.00	9.8	10 000	$475.00	1470	$1764.00	$712.50

5. Calculate the total cost for each car over the year.

Insert the formula '= B2 + G2 + H2' into cell I2.
Copy the formula into cells I3 and I4.

	A	B	C	D	E	F	G	H	I
1	Vehicle	Sales price	Fuel consumption (L/100 km)	Servicing (km)	Average service costs	Litres/ years	Fuel cost	Service cost	Total cost
2	Car A	$15 480.00	8.2	15 000	$450.00	1230	$1476.00	$450.00	$17 406.00
3	Car B	$16 250.00	7.5	20 000	$350.00	1125	$1350.00	$262.00	$17 862.50
4	Car C	$14 999.00	9.8	10 000	$475.00	1470	$1764.00	$712.00	$17 475.50

6. Write the answer as a sentence.

The most cost-effective car to buy is car A, with an overall annual cost in the first year of $17 406.

 Resources

 Interactivity   Car loans (int-6913)

15.6 Exercise

Students, these questions are even better in jacPLUS

 Receive immediate feedback and access sample responses

 Access additional questions

 Track your results and progress

Find all this and MORE in jacPLUS

1. **WE18** A customer takes out a loan to purchase a car for $17 995. The monthly repayments are $313.38 and the loan is for 5 years. The customer pays a 10% deposit. Calculate:

   a. the deposit paid
   b. the total amount paid for the car
   c. the interest rate charged for the loan, correct to 2 decimal places.

2. A $24 995 car is purchased through finance of 5.99% per annum. A 10% deposit is paid, with monthly repayments for 5 years. Calculate:

   a. the amount borrowed for the car (principal)
   b. the monthly repayments for the 5 years
   c. the total amount paid for the car.

3. **WE19** Stamp duty payable on the purchase of 4-cylinder vehicles is:
   $100 000 or less — 3 per cent ($3 per $100 or part thereof)
   more than $100 000 — 5 per cent ($5 per $100 or part thereof).
   Calculate the amount of stamp duty to be paid for the following vehicle purchases.

   a. $22 995
   b. $125 000

4. The stamp duty paid for a 4-cylinder vehicle was $250. Determine the price paid for the vehicle.

5. **WE20** A vehicle travels 640 km and uses 58 litres of fuel. Calculate the fuel consumption rate in L/100 km. Give your answer correct to 2 decimal places.

6. A vehicle's fuel consumption rate for a journey was calculated at 8.9 L/100 km. If the vehicle used a total of 71 L of fuel for the journey, determine how far in km it travelled. Give your answer to the nearest whole number.

7. A family is on a road trip. The vehicle they are travelling in averages 765 km for 99 L of fuel.

   a. Calculate the fuel consumption rate correct to 2 decimal places.
   b. If the average fuel price is $1.34/L, calculate the cost per 100 km.

8. **WE21** A driver travels on average 350 km every week. Assume 52 weeks in a year.

   a. If the vehicle requires servicing every 10 000 km, calculate how often his vehicle will need servicing over 2 years.
   b. At the 30 000 km service, the vehicle needs the following parts and labour. (The prices for the parts include GST.)
   • Oil filter $20
   • Fuel filter $35
   • Oil $40
   • Windscreen wiper blades $10
   • 2.5 hours labour at $85 per hour excluding GST.
   Calculate the cost for the 30 000 km service.

9. **WE22** For a particular brand of tyres, the manufacturer recommends replacement after 35 000 km. A driver using these tyres travels an average of 2050 km each month. Determine how many times she should expect to replace her tyres in 10 years.

10. **MC** A vehicle's average fuel consumption is 9.8 L/100 km. The driver fills the tank with 65 litres of fuel. The average number of kilometres she can travel before the tank is empty is:

   A. 6.63   B. 65   C. 151   D. 663   E. 15.1

11. **WE23** Using a spreadsheet, work out which one of the following three cars is the most cost-effective to buy. Assume that the average distance travelled each year is 15 000 km and the cost of fuel is $1.15.

Four-door sedan (medium-sized car)				
Model	Sale price	Fuel consumption (L/100 km)	Servicing (km)	Average service costs
Car A	$37 990	12.7	15 000	$255
Car B	$36 748	8.6	20 000	$595
Car C	$38 490	10.5	10 000	$235

12. The fuel consumption for a vehicle travelling in city traffic is recorded as 9.8 L/100 km. When travelling along highways, the fuel consumption is recorded as 7.5 L/100 km.

   a. Show that the average fuel consumption is 8.65 L/100 km.
   b. The vehicle travels 126 km along city roads and 458 km along the highway. Calculate the total number of litres of fuel used, correct to 1 decimal place.
   c. The cost of fuel is $1.25 per litre. Calculate the cost of the fuel for this trip, correct to the nearest cent.
   d. When calculating the fuel costs using the average fuel consumption, the value is different to the actual cost. Explain why.

13. A vehicle is purchased for $24 980 through finance that attracts 6% interest for 5 years. The owner pays a 10% deposit and then pays $496.28 monthly. This value for the monthly repayments includes a $250 one-off administration fee to set up the loan and a monthly $5 loan-servicing fee. Show that the comparison rate for the loan is 6.489% per annum.

14. A person is looking to buy a new car. They drive to work in the city, a return distance of 45 km, Monday to Friday. On the weekend, they drive along a major highway to visit a relative who lives in a regional town 112.5 km from the city.

Car	Fuel consumption for city driving (L/100 km)	Fuel consumption for highway driving (L/100 km)	Average fuel consumption (L/100 km)
A	12.5	7.8	
B	11.2	8.5	

a. Calculate the average fuel consumption for both cars.
b. Based on the weekly distances driven by the person, determine which car would be most cost-effective to purchase. Justify your answer using calculations.

15.7 Managing money

LEARNING INTENTION

At the end of this subtopic you should be able to:
• calculate monthly income
• calculate total and monthly expenses
• use a spreadsheet to prepare a budget to estimate income and expenditure.

15.7.1 Monthly income

When managing money, it can be helpful to create a monthly budget.

A **budget** is a list of all planned income and costs. A budget can be prepared for long periods of time, but for individuals it is usually a monthly plan.

In order to create a monthly budget, monthly income and expenses need to be known. **Monthly income** is $\frac{1}{12}$ of the total annual income.

WORKED EXAMPLE 24 Calculating monthly income

Calculate the monthly income of a worker who has a net annual salary of \$71 400 and earns \$1590 each year in bank interest.

THINK	WRITE
1. Calculate the total annual income as the sum of salary and bank interest.	Total annual income after tax $= 71\,400 + 1590$ $= 72\,990$
2. Calculate the total monthly income after tax by dividing the answer in step **1** by 12.	Total monthly income after tax $= \dfrac{72\,990}{12}$ $= 6082.50$
3. Write the answer as a sentence.	The worker's monthly income after tax is \$6082.50.

15.7.2 Fixed and variable expenses

Fixed expenses are expenses that occur every month (or week, fortnight, year or other time period) and are always the same amount. They may include rent or mortgage, school fees, health and car insurance. Variable expenses occur from time to time and are not a fixed amount. They may include food, entertainment, clothing and other items that can be controlled or varied.

To reduce your expenses in order to save more money, you would look at reducing your variable expenses.

WORKED EXAMPLE 25 Calculating total monthly expenses

Determine the total monthly expenses from the following list of fixed and variable expenses. The variable expenses represent the amounts spent in a month.

Item	Fixed expenses		Variable expenses	
	Frequency	Amount	Item	Amount
Rent	Weekly	$230	Food	$423
Health insurance	Monthly	$78	Clothing	$107
Vehicle registration	Yearly	$620	Entertainment	$85
Vehicle insurance	Yearly	$389	Car repairs	$325

THINK

1. Convert any weekly expenses to monthly expenses by multiplying by $\frac{52}{12}$. The only weekly cost is the rent of $230.

2. Find the total of all of the annual expenses and divide by 12 to convert it to monthly expenses.

3. Calculate the total fixed expenses by adding the numbers from steps **1** and **2** to the monthly fixed cost of health insurance ($78).

4. Calculate the total variable expenses by adding the numbers in the last column of the table.

5. To calculate the total monthly expenses, add the total fixed expenses to the total variable expenses obtained in step **4**.

6. Write the answer as a sentence.

WRITE

Monthly rent $= 230 \times \dfrac{52}{12}$
$= 996.67$ (correct to 2 decimal places)

Annual expenses $= 620 + 389$
$= 1009$
Monthly cost $= \dfrac{1009}{12}$
$= 84.08$ (correct to 2 decimal places)

Total monthly fixed expenses $= 996.67 + 84.08 + 78$
$= 1158.75$

Total monthly variable expenses $= 423 + 107$
$+ 85 + 325$
$= 940$

Total monthly expenses $=$ fixed expenses
$+$ variable expenses
$= 940 + 1158.75$
$= 2098.75$

The total expenses are $2098.75 per month.

15.7.3 Financial planning

To help in planning your finances, you can prepare a budget to estimate income and expenditure. A personal or family budget can be prepared using a spreadsheet to calculate the total fixed and variable expenses for each month and then subtracting this from the net monthly income.

Profit and loss

A *profit* is made when the net monthly income is more than the total monthly expenses.

A *loss* is made when the net monthly income is less than the total monthly expenses.

A personal or family budget can help to:
- ensure that you do not spend more money that you earn
- decide what you can and can't afford
- estimate the amount of money that you can save
- control and decrease your expenses in order to save more money.

It should be noted that a budget gives only an approximation of the real-life situation as it is based on estimates and does not include unexpected expenses.

If help is needed to create a budget or to develop strategies to manage financial affairs and reach long-term financial goals, a financial planner or financial advisor can be hired.

WORKED EXAMPLE 26 Using a spreadsheet to determine the profit or loss at the end of a month

A nurse's net annual salary is $59 500. Her expenses are shown in the following table.

Spending description	Amount
Rent (monthly)	$658
Electricity (quarterly)	$295
Water usage (quarterly)	$150
Health insurance (monthly)	$205
Car finance (fortnightly)	$172.50
Vehicle registration (yearly)	$595
Vehicle insurance (half-yearly)	$680
Food (weekly)	$75
Entertainment (monthly)	$200
Fuel (weekly)	$68

Using a spreadsheet, determine whether the nurse has made a profit or loss at the end of each month.

THINK

1. Calculate the monthly costs. For yearly, divide by 12.
For half-yearly, divide by 6.
For weekly, multiply by 52, then divide by 12.
For quarterly, divide by 3.
For fortnightly, multiply by 26, then divide by 12.

WRITE

Insert the needed formulas into the relevant cells in column C:
= (cell in column B) /12
= (cell in column B) /6
= (cell in column B)*52/12
= (cell in column B) /3
= (cell in column B)*26/12

	A	B	C
1	**Spending description**	**Amount**	**Monthly expenses**
2	Rent (monthly)	$658.00	$658.00
3	Electricity (quarterly)	$295.00	$98.33
4	Water usage (quarterly)	$150.00	$50.00
5	Health insurance (monthly)	$205.00	$205.00
6	Car finance (fortnightly)	$172.50	$373.75
7	Vehicle registration (yearly)	$595.00	$49.58
8	Vehicle insurance (half yearly)	$680.00	$113.33
9	Food (weekly)	$75.00	$325.00
10	Entertainment (monthly)	$200.00	$200.00
11	Fuel (weekly)	$68.00	$294.67

2. Calculate the total monthly spending.

Create a new row: Monthly expenses
Insert '= sum(C2 : C11)' into cell C12.

3. Calculate the net annual income.

Create a new row: Net monthly income
Insert '59500' into B13.
Insert '= B13/12' into C13.

4. Calculate the profit/loss for the month.

Create new row: Profit/loss
Insert '= C13 − C12' into cell C14.

	A	B	C
1	**Spending description**	**Amount**	**Monthly expenses**
2	Rent (monthly)	$658.00	$658.00
3	Electricity (quarterly)	$295.00	$98.33
4	Water usage (quarterly)	$150.00	$50.00
5	Health insurance (monthly)	$205.00	$205.00
6	Car finance (fortnightly)	$172.00	$373.75
7	Vehicle registration (yearly)	$595.00	$49.58
8	Vehicle insurance (half yearly)	$680.00	$113.33
9	Food (weekly)	$75.00	$325.00
10	Entertainment (monthly)	$200.00	$200.00
11	Fuel (weekly)	$68.00	$294.67
12	**Monthly expenses**		**$2 367.66**
13	**Net monthly income**		**$4 958.33**
14	**Profit/loss**		**$2 590.67**

5. Write the answer as a sentence.

The nurse makes a profit of $2590.67.

 Resources

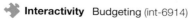 **Interactivity** Budgeting (int-6914)

1. **WE24** Calculate the net monthly income of a worker who has a net annual salary of $63 200 and earns $3745 each year in bank interest.

2. A bank teller earns $1100 per week (after tax) plus $2200 per year in bank interest (after tax). Calculate their net monthly income.

3. Complete the following table.

	Net annual salary	Total net annual bank interest	Total net monthly income
a.	$70 500	$4 812	
b.	$81 234	$10 387	
c.	$90 349	$9 885	
d.	$24 500	$42 500	

4. **WE25** Determine the total monthly expenses from the following list of fixed and variable expenses. The variable expenses represent the amounts spent in a month.

| Item | Fixed expenses | | Variable expenses | |
	Frequency	Amount	Item	Amount
Rent	Weekly	$205	Food	$494
Health insurance	Monthly	$89	Clothing	$205
Vehicle registration	Yearly	$540	Entertainment	$123
Vehicle insurance	Yearly	$499	Car repairs	$72

5. A baker earns $18.50 an hour after tax. She works an average of 38 hours a week plus 7 hours overtime, paid at time and a half. Her monthly expenses is 47% of her net monthly income. Calculate how much she spends.

6. Determine the total monthly cost from the following list of fixed and variable expenses. Variable expenses represent the amounts spent in a month.

| Item | Fixed expenses | | Variable expenses | |
	Frequency	Amount	Item	Amount
Home mortgage	Monthly	$1254	Food	$563
Health insurance	Monthly	$124	Clothing	$321
Vehicle registration	Yearly	$702	Entertainment	$389
Vehicle insurance	Yearly	$899	Car repairs	$396
School fees	Half-yearly	$5300	Travel	$1379

7. Complete this table of income and costs for a family of five over 4 months.

	Net annual salary	Net monthly bank interest	Monthly fixed expenses	Monthly variable expenses	Monthly profit/loss
a.	$73 700	$784	$3623	$2563	
b.	$73 700	$792	$3623	$639	
c.	$73 700	$804	$3623	$4456	
d.	$73 700	$823	$3623	$1065	

8. **WE26** The net annual salary for a childcare worker is $43 470. Their expenses are shown in the following table.

Expenses	Amount
Rent (weekly)	$225
Electricity (quarterly)	$150
Water usage (quarterly)	$75
Travel to work (weekly)	$60
Food (weekly)	$45
Entertainment (monthly)	$250

Using a spreadsheet, determine the amount of profit or loss the childcare worker makes at the end of each month.

9. A university student who lives with her parents has the following expenses: she pays her parents $70 per week for board and food; a monthly ticket for public transport costs her $80; she spends on average $45 a month on books and stationery; her single health insurance premium is $68.55 a month; entertainment and snacks cost her about $90 a month; the university enrolment fee takes $900 a year; and she also needs clothes and accessories that cost approximately $80 per month.

Prepare a monthly budget if the student's income consists of Youth Allowance ($355.40 per fortnight) and birthday and Christmas presents ($250 a year), and calculate the amount of money that she can save per month.

10. The following information shows the expenses and income for a couple.

Expenses		
	Mortgage	$672 per fortnight
	Food	$123 per week
	Clothing	$78 per week
	Transport	$35 per day
	Entertainment	$120 per week
	Health insurance	$200 per month
	Car (insurance, registration, fuel, servicing)	$600 per month
Net income	Partner A	$1680 per fortnight
	Partner B	$1100 per week

a. Using a spreadsheet, calculate the total monthly spending.
b. Show that the couple make a monthly profit.

The couple would like to save to buy a block of land for $350 000. To buy the block of land, they will need to make a 10% deposit.

c. If the couple's spending does not change, determine how many months it will take to save the 10% deposit. Give your answer correct to 1 decimal place.

d. The couple's financial adviser recommends they reduce their spending so that it's less than 50% of their net monthly income. Suggest where they should adjust their spending and justify your answer by using calculations.

11. The following shows a personal budget prepared by a graduate IT technician.

Item	Frequency	Amount	Monthly
Rent	Weekly	$253.85	$1100
Bills	Monthly	$145	$145
Travel	Monthly	$350	$350
Entertainment	Monthly	$330	$330
Clothes	Fortnightly	$203.08	$440
Insurance	Yearly	$1150	$95.83
Monthly spending			$2460.83
Net monthly income			
Profit			**$2355.92**

a. Determine the technician's net monthly income and hence calculate their net annual income.
b. Calculate the percentage amount they spend of their net income. Give your answer correct to 2 decimal places.
c. The technician would like to save up for an overseas holiday and has budgeted $5500 for the trip. If they can save 50% of their monthly net income, determine how many months it will take to have saved enough money for the trip.

12. A family's net annual income is $69 480. Their spending is shown in the following table.

Item	Frequency	Amount
Holiday/entertainment	Yearly	$4500
Mortgage	Weekly	$245
Insurance	Yearly	$1980
Car costs	Monthly	$950
Food	Weekly	$145
Electricity, water	Quarterly	$580
School fees	Yearly	$8500
Health insurance	Monthly	$368

a. Prepare a monthly budget and show whether the family are living within their means (i.e. making a profit).
b. The family need to adjust their spending. By identifying the fixed and variable spending items, explain which items the family should consider reducing.

13. A family of four are going on an overseas holiday for 6 weeks. They have saved $20 000 for the trip and recorded the costs for the essentials but don't know how much to spend on food.

Item	Cost
Passports	$832
Flights	$5668
Travel insurance	$658
Accommodation	$245 per night
Food	

a. They estimate food costs to be about $45 per day. Calculate the amount of money remaining from the $20 000 that they can spend on their trip.

b. The family reassess and decide to adjust their spending to save more money for their trip. Their monthly spending is $2585 and net monthly income is $4580. If they can save half of their monthly profit for 4 months, calculate the amount of money they will have for their overseas trip.

14. Prepare your personal monthly budget (or your family budget if you do not have any income). Determine the possibilities for cutting some of the expenses.

15.8 Financial and economic data trends

LEARNING INTENTION

At the end of this subtopic you should be able to:
- interpret financial and economic data trends and make future predictions
- calculate the effect of inflation on prices.

15.8.1 Data trends

To maximise our own financial wellbeing, and benefit from monetary opportunities, it is important to understand local, community and national financial and economic data trends. By examining data trends and making careful choices, we may be able to make future predictions that will build wealth over time. We will look at share performance and inflation in detail, but other important trends to consider include interest rates, house prices and wages.

Shares

When it comes to investing funds, there are three broad areas to be considered. These are shares, interest-bearing deposits and property. Although there have been short-term fluctuations in these areas, over the long term, domestic shares have been shown to perform well, as seen in the graph on the next page.

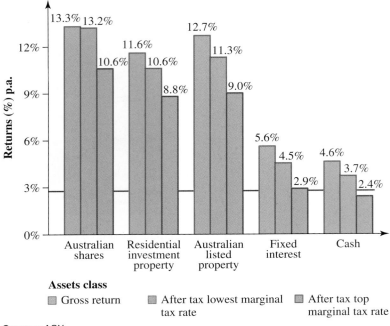

Investment performance comparison over 10 years

There is an element of risk when investing funds by purchasing **shares**. The shares have the potential to return more money to the investor than depositing money in a bank; however, there is also a chance that the shares may fall in value.

Because shares offer no guaranteed returns, we can only use the past performance of a share to try to predict its future performance. One simple way to do this is to graph the value of the share at regular intervals and then draw a line of best fit to try to model the trend.

By continuing the **line of best fit**, you can make a prediction for future share prices. This is called **extrapolating** information from the graph. **Interpolation** is the opposite of extrapolation and occurs when making a prediction using data found between the end points.

Many share traders use sophisticated analysers of graphs (charts) to determine both short- and long-term movements of share prices, but this 'technical analysis' is beyond the scope of this book.

WORKED EXAMPLE 27 Predicting the future share price by using a line of best fit

Below is the share price of a company taken on the first day of the month for one year.

Month	Share price	Month	Share price
January	$10.34	July	$10.98
February	$10.54	August	$11.56
March	$10.65	September	$11.34
April	$10.89	October	$11.23
May	$10.72	November	$11.48
June	$11.10	December	$11.72

a. Plot the share price for each month and draw a line of best fit.
b. Predict the share price in June of the following year.

a. 1. Draw up a set of axes and plot the data.

a.

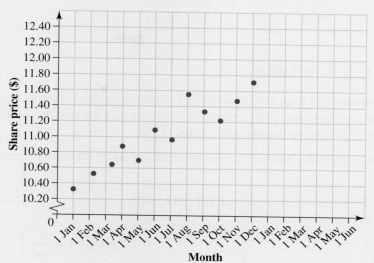

2. Draw a straight line on the graph that best fits in with the marked points.

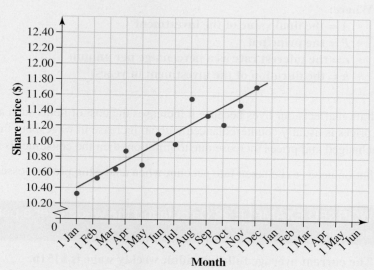

b. 1. Extend the line of best fit for six months.

b.

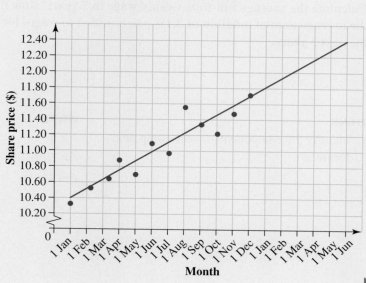

2.	Predict the share price by reading from the line of best fit.	The predicted share price is approximately $12.40.

15.8.2 Inflation

Inflation is a term used to describe a general increase in prices over time that effectively decreases the purchasing power of a currency. Inflation can be measured by the inflation rate, which is an annual percentage change of the **consumer price index (CPI)**.

The compound interest formula is used to calculate the price of items after inflation is added.

Formula for compound interest

$$A = p\left(1 + \frac{r}{100}\right)^n$$

Where:
- A = the final value of the investment
- P = the principal
- r = the interest rate as a percentage per annum
- n = the duration of the investment in years.

Inflation affects our lives because it increases the cost of items such as food, utilities and housing. The increase in prices is usually counterbalanced with an increase in wages, but this is not always the case. If wages remain unchanged and prices increase due to inflation, then the consumer is paying more in relative terms for the items.

Inflation needs to be taken into account when analysing profits, losses and wage increases over a period of time.

WORKED EXAMPLE 28 Calculating the wage over a period of time, taking inflation into account

The current average full-time adult weekly wage is $1516. Calculate the average full-time weekly wage in 5 years' time if the average annual inflation of 2.1% p.a. is added to wages for the next 5 years.

THINK	WRITE
1. Identify the components for the compound formula.	$n = 5$
	$r = 2.1\%$
	$P = \$1516$

2. Substitute the values into the compound
   interest formula.

$$A = P\left(1 + \frac{r}{100}\right)^n$$

$$= 1516\left(1 + \frac{2.1}{100}\right)^5$$

$$= 1516(1 + 0.021)^n$$

$$= 1516(1.021)^5$$

$$= \$1682.01$$

3. Write the answer.

In 5 years, the average weekly wage will be $1682.01.

tlvd-5088

WORKED EXAMPLE 29 Determining the effect of inflation on the selling price of a property

An investment property is purchased for $300 000 and is sold 3 years later for $320 000. If the average annual inflation is 2.5% p.a., determine whether the investment has been profitable.

THINK

1. Recall that inflation is an application of compound interest and identify the components of the formula.

2. Substitute the values into the formula and evaluate the amount.

3. Compare the inflated amount to the selling price.

4. Write the answer.

WRITE

$P = 300\,000$, $r = 2.5\%$, $n = 3$

$$A = P\left(1 + \frac{r}{100}\right)^n$$

$$= 300\,000\left(1 + \frac{2.5}{100}^n\right)$$

$$= 300\,000(1 + 0.025)^3$$

$$= 323\,067.19 \text{ (to 2 decimal places)}$$

Inflated amount: $323 067.19
Selling price: $320 000

This has not been a profitable investment, as the selling price is less than the inflated purchase price.

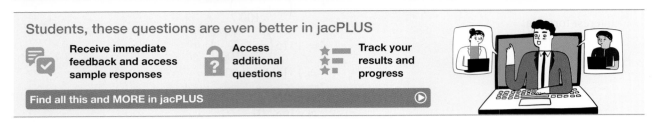

1. The following graph shows the share performance of a large telecommunications company from January 2022. The line shows a prediction of the share value in the next 12 months.

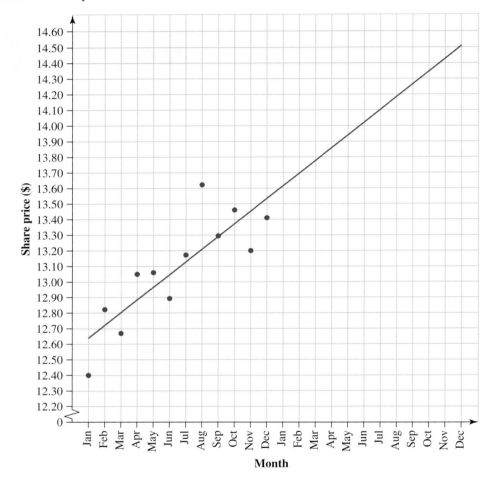

a. Use the graph to determine when the shares will first be worth more than $14.00.
b. Use the graph to predict the value of the shares at the end of 2023.

2. **WE27** The table shows the share price of a large multinational company over a 12-month period.

Month	Share price	Month	Share price
January	$12.86	July	$13.45
February	$13.43	August	$13.86
March	$11.98	September	$14.40
April	$12.10	October	$13.65
May	$12.11	November	$13.20
June	$12.98	December	$12.86

a. Plot the share prices on a set of axes and draw a line of best fit.
b. Use your graph to predict the value of the share after a further 6 months.

3. **WE28** The current average casual junior weekly wage is $750. Calculate the average weekly wage in 3 years if the average annual inflation of 1.9% p.a. is applied for the next 3 years.

4. The current prices for 4 household items are shown below. If the average annual inflation rate remains steady at 2.3% p.a., determine how much a customer could expect to pay for these items in 2 years' time. Write your answers correct to the nearest cent.

a. Orange juice: $4.50
b. Packet of 24 toilet rolls: $8.65
c. 1 kg of apples: $4.95
d. Margarine: $3.20

5. **WE29** An investment property is purchased for $325 000 and is sold 5 years later for $370 000. If the average annual inflation is 2.73% p.a., determine whether the investment has been profitable.

6. The graph shows the performance of a company's shares over the last 6 months and the line of best fit.

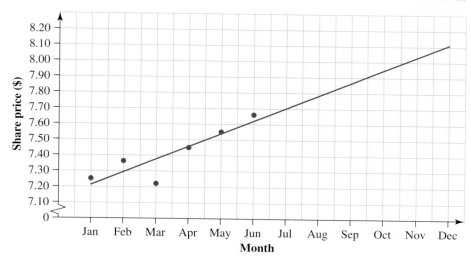

a. Use the graph to write down the predicted share prices for the following months.

i. September
ii. December

b. Explain how you would predict the share market price in 12 months.

7. A newspaper article stated: 'Our current wages are 10 times what they were 40 years ago.' Forty years ago, the average annual salary was $7618; today it is $72 000. If the annual average inflation rate over the last 40 years was 2.3% p.a., determine whether this statement is accurate.
Justify your answer using calculations.

8. Research online data trends and write about one of the following topics. The Australian Bureau of Statistics and the Reserve Bank of Australia are good places to find data.
   • Average weekly earnings
   • Wage price index
   • Median house prices in Victoria
   • Average monthly exchange rate
   • Lender's variable interest rates

15.9 Review

📄 **15.9.1 Summary**
doc-38056

Hey students! Now that it's time to revise this topic, go online to:

📄 **Access the topic summary**

✅ **Review your results**

▶ **Watch teacher-led videos**

A⁺ **Practise questions with immediate feedback**

Find all this and MORE in jacPLUS ▶

15.9 Exercise

Multiple choice

1. **MC** An insurance company offers customers the following discounts on the annual cost for car insurance.

Type of discount	Discount	Conditions
Multi-policy discount	7%	Owner has 2 or more insurance policies with the company.
No claim bonus	15%	Owner has had at least 5 years without an insurance claim.

The following excesses apply to claims for at-fault motor vehicle accidents for the comprehensive car insurance:
- basic excess of $550 for each claim
- an additional age excess of $1100 for drivers under 25 years of age.

Fred has been with the same insurance company since obtaining his licence at the age of 18, has never had an accident, and he has 4 other insurance policies with this company. His car insurance costs $1450 per year prior to any discounts being applied. The annual amount that Fred is required to pay after having his first accident at the age of 24 years old is:

   A. $1450
   B. $2000
   C. $3100
   D. $2728
   E. $2781

2. **MC** In its annual budget, the local council plans to raise $25 million in revenue to repair local infrastructure. The capital improvement for the area is valued at $1.5 billion. The council rate quarterly instalment for a rate payer whose house has a capital improved value of $475 000 is:

   A. $712.50
   B. $7916.76
   C. $1979.17
   D. $2850.00
   E. $1970.00

3. **MC** A person spends 65% of their net annual income. If their net annual income is $58 480, the monthly profit they make is closest to:

   A. $3167.67
   B. $1705.67
   C. $20 468
   D. $4873.33
   E. $38 012.00

4. **MC** A person earns $74 600 net annual salary and also receives a bonus payment from their employer of $3500. They have a fortnightly salary of:

   A. $1501.92
   B. $3003.85
   C. $3124.00
   D. $6369.23
   E. 6508.33

5. **MC** In determining the budget for a small business, select which of the following can be considered the variable expense.

   A. Rent
   B. Employee wages
   C. Public liability insurance
   D. Electricity bills

6. **MC** The energy usage and rates for a household are as shown.

Electricity bill

Tariff	Bill days	Current reading	Previous reading	Total usage (kWh)	Charge/rate (c/kWh)
All day	90	78 250			25.08
Service to property	90			$1.105/day	

Gas

Tariff	Bill days	Current reading	Previous reading	Total usage (kWh)	Charge/rate (c/MJ)
All day	90	8920	8138		2.452
Service to property	90			$0.985/day	

There is a 25% discount for paying the bill before the due date. If the bill was paid before the due date, select the amount paid from the following options. (Remember to include GST.)

A. $107.82   B. $241.90   C. $262.29   D. $288.52   E. $384.69

7. **MC** A manager's annual salary is $65 850. Calculate the amount of HECS-HELP payable per month.

A. $2305   B. $2287   C. $191   D. $192   E. $44

8. **MC** The Child Care Subsidy percentage for a family with a combined income of $275 000 per annum is closest to:

A. 50%   B. 20%   C. 6.89%   D. 26.89%   E. 43.11%

9. **MC** The following sign was on display at a local café. If the charge to the café to process credit card payments is 2% for all credit cards, determine which one of the following statements is False.

A. There is an excessive surcharge on the credit card purchase of a espresso and a babyccino.

B. A customer purchased an espresso, a cappuccino and 2 babyccinos and was charged the correct surcharge using his credit card.

C. A student purchased 2 hot chocolates on her credit card and was charged the correct credit card surcharge.

D. A group of four people individually purchased hot chocolates using their credit cards. They would have each saved 7.5 cents if they purchased all the hot chocolates on one credit card.

E. There is no excessive surcharge on the credit card purchase of a cappuccino and a babyccino.

COFFEE

ESPRESSO	$3.50
CAPPUCCINO	$4.50
LATTE	$4.20
HOT CHOCOLATE	$2.50
BABYCCINO	$1.00

CREDIT CARD SURCHARGE 10c

10. **MC** For a no-interest-free credit card, the interest charged on the average outstanding daily balance of $380 with a percentage interest rate of 16.80% p.a. over a 30-day period, correct to the nearest cent, is:

A. $2.13   B. $5.25   C. $5.32   D. $63.84   E. $5.50

Short answer

11. A retailer charges customers a 1.75% credit card surcharge. Calculate the surcharge, correct to the nearest cent, for the following purchases.

    a. $35.95

    b. $105.75

12. Calculate the monthly net income for each of the following.

    a. A net annual income of $63 480
    b. A weekly net income of $945.35
    c. A fortnightly net income of $1450

13. A household's daily water usage is 1700 litres. The total water bill for 60 billing days is $392.12, which includes $187.10 for water services. Calculate the charge per kilolitre.

14. The following table shows the net annual incomes and monthly spending of 4 people. Calculate the profit or loss for each person.

	Net annual income	Monthly spending	Profit/loss
a.	$51 890	$3950	
b.	$84 755	$8640	
c.	$75 120	$6580	
d.	$62 308	$4980	

15. There is a 12% discount for customers who pay their energy bill before the due date. The energy usage of a customer is:

    • electricity: 875 kWh at 19.02 c/kWh
    • gas: 720 MJ at 2.034 c/MJ.

    The customer is also charged a $112 fixed fee for service to the property.

    If the customer pays the bill before the due date, calculate the amount they paid, including GST.

16. The quarterly council rates for a property with capital improved value of $385 000 are $721.88.

    a. Calculate the amount in the dollar charged annually by the council.
    b. If the total capital improved value for all properties within the council municipality is $1.5 billion, calculate the amount of revenue the council plans to raise this year. Give your answer correct to the nearest thousand dollars.

Extended response

17. The cost of bread has increased each year due to inflation. Over the last 2 years the average annual inflation rates were 1.5% and 2% respectively. If the cost of bread today is $3.04, explain how you would determine the cost of bread 2 years ago.

    Justify your answer using calculations.

18. Candy uses her credit card to pay for $1500 in university fees at the start of March. Her credit card's annual interest rate is 12.99% p.a. Her card has no interest-free period and no annual fees.

 a. i. Calculate the daily percentage interest, correct to 4 decimal places.
    ii. Show that Candy was charged $0.53, correct to the nearest cent, each day for the first month.
 b. Candy works part-time and pays $400 off her credit card balance each month at the end of the statement period after interest has been calculated and added.
    Complete the table that shows her credit card balance over 3 months.

Month	Balance ($)	Interest ($)	Payment ($)	Balance owing ($)
1	1500	16.02	400	1116.02
2	1116.02	11.92	400	
3			400	

 c. Instead of using her credit card to pay for her university fees, Candy has the option to take out a student reducing-balance loan with interest of 7.75% and monthly payments of $130.30 for 1 year. If she pays $400 in monthly payments, it will take her 3.8 months to pay off the loan. In the 4th month the balance owing is $321.
    Explain why the personal loan would be the more cost-effective choice for Candy.

19. Calco and Dynico are two companies listed on the stock exchange. The table shows Calco's market share price over the first 6 months of the new financial year. The graph shows the market share price for Dynico over the same period as Calco.

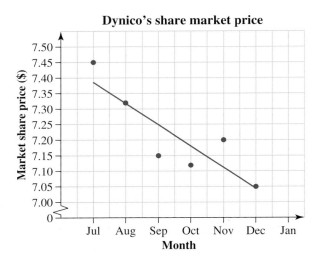

Dynico's share market price

Calco's market share price	
Month	**Market share price ($)**
July	3.25
August	3.19
September	3.35
October	3.58
November	3.67
December	3.75

 a. Explain the trend in the market share price for Dynico.
 b. By recording the market share price on a graph and finding the line of best fit, describe the expected market share price trend for Calco. Support your answer by providing a predicted market share price at the end of June.

20. A family of 6 lives in a 5-bedroom house in the outer suburbs of a major city. They own two cars. Partner A uses one car to drive to the local train station; partner B uses the other car to drive their children to school, sports training and holidays.

They spend $165 on food each week. Their health insurance costs $368 per month, their house insurance costs $1580 a year, and their annual council rates are $3580. The children's school fees total $2850 a year. Partner A's net annual salary is $74 805.

Partner B works part-time Monday–Friday as an administrator at a local law firm. Their hourly rate is $27.50 and their hours are 9 am to 3 pm with a 45-minute unpaid lunch break. Their employer withholds $85 in PAYG tax each week.

Their bills and other expenses are shown below.

Electricity bill

Billing period: 1 Jul–30 Sep					
Tariff	Bill days	Current reading	Previous reading	Total usage	Charge/rate (c/kWh)
Peak	91	23 890	23 150		33.58
Off-peak	91	48 901	48 334		15.60
Service to property	91				$1.148/day

Water bill

Bill details as of 30 September		Value	Price
Water service charge	1 Jul to 30 Sep	25.75	
Sewerage service charge	1 Jul to 30 Sep	149.80	
Stormwater	1 Jul to 30 Sep	22.05	
	Volume (kL)	Charge/kL	
Water usage	65	$2.05/kL	
Recycled water usage	45	$1.82/kL	

Cars

Average fuel price: $1.31/litre					
Vehicle	Fuel consumption (L/100 km)	Registration (annually)	Insurance (6-monthly)	Servicing (every 15 000 km)	Average distance travelled each year
Partner A	8.8	$568	$302	$250	12 500
Partner B	10.2	$568	$257	$320	17 500

a. Calculate their joint net monthly income.
b. Prepare a budget for the family.
c. To be solvent means to be able to pay one's debts (expenses, bills). State whether the family is solvent. Justify your answer using your calculations.

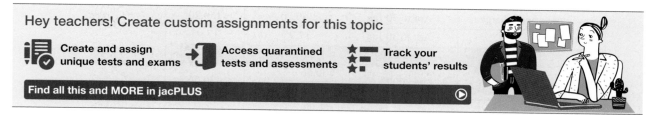

Hey teachers! Create custom assignments for this topic

Create and assign unique tests and exams

Access quarantined tests and assessments

Track your students' results

Find all this and MORE in jacPLUS

Answers

Topic 15 Managing finances

15.2 Products and services

15.2 Exercise

1. a. Option 2   b. Option 2
   c. Option 1

2. a. i. Telco   ii. Telco   iii. HFC
   b. HFC

3. a. $1500
   b. Medicare levy surcharge
   c. $142.50

4. Sample responses can be found in the worked solutions in the online resources.

5. a. $576
   b. Hospital = $104.58
      Extras = $165.06

6. Option A

7. $2250

8. a. $565.50   b. $880   c. $867

9. $3129.75

10. Sample responses can be found in the worked solutions in the online resources.

15.3 Rent and utilities costs

15.3 Exercise

1. a. $2200   b. $2750   c. $2383

2. a. i. $519   ii. $4326
   b. The rental property in question 2 is more cost-effective as it costs $519 per week, whereas the rental property in question 1 costs $550 per week.

3. $301.67

4. a. $59.45   b. $28.13
   c. $131.03   d. $113.42

5. 4768 kWh

6. a. i. 30.54c/kWh
      ii. $1.017
      iii. 895 kWh
   b. i. $273.33   ii. $365.88   iii. $402.47

7. $159.63

8. D

9. $372.92

10. 33.62 kL

11. $1833.33

12. a. 0.006   b. $16 020 000

13. Plan B is more cost-effective for both paying on time and not paying on time.

14. a. See the table at the bottom of the page.*
    b. $405.63
    c. 3.06%

15. Choice 1 is more cost-effective for the family as it costs $60.06 and choice 2 costs $63.36.

16. a. $\dfrac{0.008 \times 280\,000}{4} = \$560$
    b. $1345.60
    c. 7.45%

15.4 Credit cards

15.4 Exercise

1. a. $1158   b. $2050
   c. 0.046 64%   d. $2950

2. a. $985
   b. A = $2183.59
      B = $2033.59

3. a. $5.79
   b. $20.26

4. $6.45

5. $1.04

6. C

7. The $35.80 and $112.50 bills were not charged an excessive credit card surcharge. The $78.95 bill was charged an excessive credit card surcharge.

8. a. $90
   b. 15 months
   c. Using the credit card to purchase his accessories and paying more than the minimum monthly payment required is the better option, because he will pay $354.76 less in interest.

*14. a.

Bill details as of 30 June		Value	Price
Water service charge	1 Apr to 30 Jun	23.80	23.80
Sewerage service charge	1 Apr to 30 Jun	150.20	150.20
Storm water	1 Apr to 30 Jun	17.95	17.95
	Volume (L)	Charge/kL	
Water usage	64 000	$\dfrac{130.56}{64} = \$2.04$	130.56
Recycled water usage	38 000	$\dfrac{71.06}{38} = \$1.87$	71.06
		Total	$393.57

15.5 Personal income–related expenses

15.5 Exercise

1. a. $367 **b.** $530.40 **c.** $679

2. Sample responses can be found in the worked solutions in the online resources.

3. a. 0% **b.** 85% **c.** 34.768%

4. a. $0 **b.** $882.20 **c.** $1330

5. Sample responses can be found in the worked solutions in the online resources.

6. a. $233.33 **b.** 182 months

7. a. $5614.50 **b.** $196.51

8. a. $19 717 **b.** $5400 **c.** $5406.92

15.6 Car expenses

15.6 Exercise

1. a. $1799.50 **b.** $20 602.30 **c.** 3.22% p.a.

2. a. $22 495.50 **b.** $487.22 **c.** $31 732.40

3. a. $689.85 **b.** $6250

4. $8333.33

5. 9.06 L/100 km

6. 798 km

7. a. 12.94 L/100 km

  b. $17.34

8. a. 3 times **b.** $338.75

9. 7 times

10. D

11. Car B

12. a. $\dfrac{9.8 + 7.5}{2} = 8.65$ L/100 km

  b. 46.7 L

  c. $58.38

  d. The average fuel consumption is a value that assumes equal distance (1 : 1) of driving under the two conditions (city and highway). In this case, the ratio of city to highway driving was 126 : 458 (1 : 3.6). This means the overall fuel consumption is lower than the average value would predict, because 3.6 times more driving took place at the lower fuel consumption.

13. Total monthly payments: $496.28 \times 60 = \$29\ 776.80$
  Deposit: $10\% \times 24\ 980 = \$2498$
  Principal borrowed: $24\ 980 - 2498 = \$22\ 482$
  Interest paid: $29\ 776.80 - 22\ 482 = \$7294.80$

$$I = \frac{Prn}{100}$$

$$7294.80 = \frac{22\ 482 \times r \times 5}{100}$$

$$\frac{729\ 480}{22\ 482 \times 5} = r$$

$$r \approx 6.489\%$$

14. a. Car A: 10.15 L/100 km
  Car B: 9.85 L/100 km

b. Car B has a lower average fuel consumption (9.85 L/100 km) and will therefore be more cost-effective. (The average is valid here because the distances of city and highway driving come to the same value, 225 km.)

15.7 Managing money

15.7 Exercise

1. $5578.75

2. $4950

3. a. $6276 **b.** $7635.08
  c. $8352.83 **d.** $5583.33

4. $1957.91

5. $1827.40

6. $5392.75

7. a. $739.67 profit **b.** $2671.67 profit
  c. $1133.33 loss **d.** $2276.67 profit

8. $1867.50 profit

9. $48.98

10. a. $4711.58

  b. The net monthly income of $8406.67 exceeds the monthly spending, for a monthly profit of $3695.09.

  c. 9.5 months

  d. Areas in which spending can be reduced (adjusted) include clothing, entertainment, transport and food. Mortgage, health insurance and car registration/insurance are fixed spending.

11. a. Net monthly income: $4816.75
  Net annual income: $57 801

  b. 51.09%

  c. 2.28 months

12. a. See the table at the bottom of the next page.*

  b.

Fixed	Variable
Mortgage	Holiday/entertainment
Insurance	Car costs
Health insurance	Food
School fees	Electricity, water

Variable items can be reduced, particularly non-essential items such as holidays/entertainment. These items do not have fixed costs, which means that their costs can be adjusted based on usage or need.

13. a. $662 **b.** $3990

14. Sample responses can be found in the worked solutions in the online resources.

15. 8 Financial and economic data trends

15.8 Exercise

1. a. Between May and June in 2023
   b. $14.50

2. a.

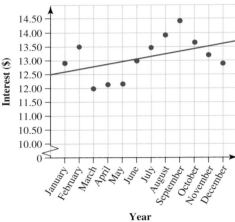

Share price

b. Approximately $14.15

3. $793.57

4. a. $4.71   b. $9.05   c. $5.18   d. $3.35

5. Inflated price: $371 851.73; selling price: $370 000
   Selling price < inflated price, so this has not been a profitable investment.

6. a. i. $7.85   ii. $8.10
   b. Extrapolating the line of best fit for another 12 months. Marking on the horizontal the 12th month and extending this to the line of best fit and then reading off the market share price value.

7. Current wages have increased over the last 40 years with inflation, so comparing current wages with wages from 40 years ago is not an accurate comparison. Inflating the wages from 40 years ago gives a wage of $18 917.61, which makes current wages approximately 4 times greater than 40 years ago, not 10 times greater.

8. Please see the worked solution in the online resources.

15.9 Review

15.9 Exercise
Multiple choice

1. E
2. C
3. B
4. B
5. D
6. D
7. D
8. E
9. B
10. B

Short answer

11. a. 0.63   b. $1.85

12. a. $5290   b. $4096.52   c. $3141.67

13. $2.01/kL

14. a. $374.17 profit
    b. $1577.08 loss
    c. $320 loss
    d. $212.33 profit

15. $283.69

*12. a.

Item	Frequency	Amount	Monthly cost
Holiday/entertainment	Yearly	$4500	$\frac{4500}{12} = \$375$
Mortgage	Weekly	$245	$245 \times \frac{52}{12} = \$1061.67$
Insurance	Yearly	$1980	$\frac{\$1980}{12} = \$165$
Car costs	Monthly	$950	$950.00
Food	Weekly	$145	$145 \times \frac{52}{12} = \$628.33$
Electricity, water	Quarterly	$580	$\frac{580}{3} = \$193.33$
School fees	Yearly	$8500	$\frac{8500}{12} = \$708.33$
Health insurance	Monthly	$368	$368.00
		Total	$375 + 1061.67 + 165 + 950 + 628.33 + 193.33 + 708.33 + 368 = \textbf{\$4449.66}$
Income	Yearly	**$69 480**	$\frac{69\,480}{12} = \textbf{\$5790.00}$
		Profit	$5790 - 4449.66 = \textbf{\$1340.34}$

16. a. 0.0075   **b.** $11 250 000

Extended response

17. Divide the current price of bread by 1.015×1.02.

18. a. i. 0.0356%

  ii. $\dfrac{0.0356}{100} \times 1500 = \$0.53$

b.

Month	Balance ($)	Interest ($)	Payment ($)	Balance owing ($)
1	1500	16.02	400	1116.02
2	1116.02	11.92	400	727.94
3	727.94	7.77	400	335.71

c. Taking out a personal loan and paying it off in 4 months rather than using a credit card saves Candy $18.30 in interest. Therefore, the personal loan is the more cost-effective option.

19. a. Decreasing trend

b.

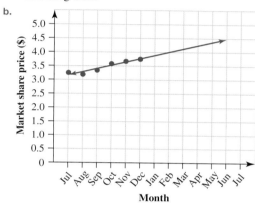

The market share price is increasing over the 6 months (July to December). If this increasing trend continues, then the predicted market share price at the end of the following June will be $4.50.

20. a. $8993.56

b.

Item	Cost	Frequency	Monthly cost
Food	$165.00	Weekly	$715.00
Health insurance	$368.00	Monthly	$368.00
House insurance	$1580.00	Yearly	$131.67
School fees	$2850.00	Yearly	$237.50
Council rates	$3580.00	Yearly	$298.33
Electricity	$485.56	Quarterly	$161.85
Water	$412.75	Quarterly	$137.58
Car insurance	$559.00	Half-yearly	$93.17
Car registration	$1136.00	Yearly	$94.67
Fuel	$3779.35	Yearly	$314.95
Servicing	$581.67	Yearly	$48.47
		Total	**$2601.19**
		Net monthly income	**$8993.56**
		Profit	**$6392.37**

c. Yes, the family are solvent. Their net monthly income is greater than their spending.

16 Space, time and measurement

LEARNING SEQUENCE

Fully worked solutions for this topic are available online.

16.1 Overview

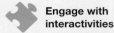

16.1.1 Introduction

We live in a world surrounded by shapes and objects. Often, we ask questions such as 'how long?', 'how far?' or 'how big?'. These questions are all answered using measurement.

From backyard sheds and home extensions to city buildings, architecture is founded on congruence. Many structures, from the modern to the very old, were created by people who had the opportunity to read a chapter similar to this one. Grand structures such as Flinders Street Station, and even the very house in which you live, are based on the principles of geometry. Builders, surveyors and engineers all use their knowledge of geometry to ensure buildings are stable and visually pleasing.

Designers, interior decorators and architects use measurement in their drawings and calculations.
Chefs measure ingredients in their cooking. Nurses and health professionals follow instructions regarding the amount of a drug to administer to a patient. To maximise profits, manufacturers need to minimise the amount of raw materials used in production. This means knowing the measurements of various parts. Understanding the basic concepts involved in measurement is beneficial in many real-world situations.

KEY CONCEPTS

This topic covers the following key concepts from the VCE Mathematics Study Design:
- description, representation and properties of simple and composite shapes and objects
- two-dimensional plans, models and diagrams of objects
- reading and interpretation of scales on digital and analogue instruments
- time and duration including time and date specifications, conventions, schedules and timetables
- location, maps, directories and digital maps including bird's-eye and street views
- routes and itineraries, including location and direction, speeds, distances and estimated travel times, for example daily work route and diversions, and itinerary for travel.

Source: VCE Mathematics Study Design (2023–2027) extracts © VCAA; reproduced by permission.

16.2 Conventions and geometric shapes

LEARNING INTENTION

At the end of this subtopic you should be able to:
- identify terms and names of common geometric shapes and objects
- identify properties of triangles
- identify properties of quadrilaterals
- calculate values of pronumerals by applying the properties of geometric shapes.

16.2.1 Geometric terms

Convention	Definition
Point	A point has a precise position with no size.
Line	A line is straight and doesn't end in either direction.
Ray	A ray is part of a line that has a start but no end.
Curve	A curve is a line that is not straight. A closed curve joins up from where it started. Curve Closed curve
Vertex	A vertex is a point where two or more line segments meet. Vertex Vertex Vertex
Edge	An edge is a line segment on the boundary joining one vertex to another. Edge Edge Edge
Diagonal	A diagonal is a line segment that goes from one corner (vertex) to another, but not to an edge. Diagonal

Boundary and perimeter	The boundary is a line or border around the outside of a shape. The distance around the boundary is the perimeter.
	Boundary
Face	A face is any of the flat surfaces of a solid object.
	Face, Face, Face
Surface	A surface is the outside of a solid object. Total surface area is the total area of the outside of the surface.
	Vertex, Edge, Face, Vertex, Edge, Diagonal, Face, Diagonal, Vertex, Edge

16.2.2 Polyhedron solids

A **polygon** is a closed shape where each edge is a straight line and all edges meet at vertices. A 3-dimensional solid where each of the faces is a polygon is called a **polyhedron**. If all the faces are congruent, the solid is called a regular polyhedron or a platonic solid.

A **cube** has 6 faces, each of which is a square.

A **tetrahedron** has 4 faces, each of which is an equilateral triangle.

An **octahedron** has 8 faces, each of which is an equilateral triangle.

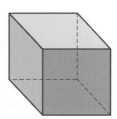

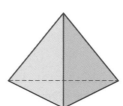

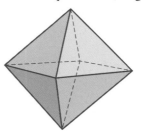

A **dodecahedron** has 12 faces, each of which is a regular pentagon.

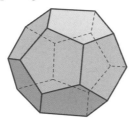

An **icosahedron** has 20 faces, each of which is an equilateral triangle.

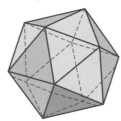

Euler (1701–1783) was a Swiss mathematician who is regarded as one of the greatest mathematicians in history for his work in graph and number theory, complex analysis and calculus. He was credited with popularising the Greek letter π. **Euler's formula** shows the relationship between the number of **edges**, the number of **faces** and the number of vertices in any polyhedron. Note that a **vertex** (the singular of vertices) is a point or a corner of a shape where the straight edges meet.

> ### Euler's formula
>
> **Euler's formula states that for any polyhedron:**
>
> **number of faces (F) + number of vertices (V) − 2 = number of edges (E)**
>
> $$F + V - 2 = E$$

WORKED EXAMPLE 1 Determining faces, edges and vertices

Consider the solid shown.
a. Determine:
   i. the number of faces
   ii. the number of edges
   iii. the number of vertices.
b. Use your values to prove Euler's formula.

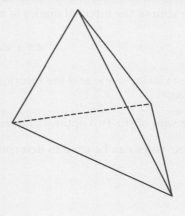

THINK

a.   i. A face is the flat surface of the solid.
    ii. An edge is the line joining the vertices.
    iii. A vertex is where the lines meet.

b. Prove the formula $F + V - 2 = E$.

WRITE

a. There are 4 faces.
   There are 6 edges.
   There are 4 vertices.

b. $F + V - 2 = E$
   $4 + 4 - 2 = 6$
        $6 = 6$
   Euler's formula is proven.

16.2.3 Angles and triangles

Types of triangles

Types of triangles	Description	Shape
Equilateral triangle	All edges are the same length and all angles are the same size. This is shown by the identical marks on each edge and each angle.	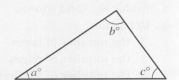
Isosceles triangle	Two edges are the same length. The third edge is a different length and is often called the base of the triangle. The two base angles are the same size.	
Scalene triangle	All edges are different lengths and all angles are different sizes. This is shown by the unique markings on each edge and each angle.	

Properties of triangles

The sum of the internal angles of a triangle equals 180°.

$$\angle a + \angle b + \angle c = 180°$$

The exterior angle and the interior angle adjacent (next) to it add up to 180°.

The sum of the two opposite interior angles is equal to an exterior angle.

These rules can be used to determine the size of unknown angles.

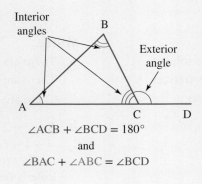

Interior angles

Exterior angle

$\angle ACB + \angle BCD = 180°$

and

$\angle BAC + \angle ABC = \angle BCD$

tlvd-6022

WORKED EXAMPLE 2 Calculating the values of pronumerals in a triangle

Calculate the value of each angle in the triangle shown.

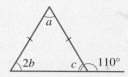

THINK

1. Angle c and the exterior angle (110°) are supplementary.

2. Solve for c by subtracting 110° from both sides of the equation.

3. • The triangle is isosceles, so the opposite angles of the base are equal.
   • Solve for b by dividing both sides of the equation by 2.
   • Substitute the value of c (70°) and complete the division.

4. • The angle sum of a triangle is 180°. Substitute the values of the angles where possible ($b = 35°$, $c = 70°$).
   • Simplify and solve for a.

5. Write the value of each pronumeral in degrees.

WRITE

$c + 110° = 180°$

$c = 180° - 110°$
$c = 70°$

$2b = c$
$\dfrac{2b}{2} = \dfrac{c}{2}$
$b = \dfrac{70°}{2°}$
$\quad = 35°$

$a + 2b + c = 180°$
$a + 2(35°) + 70° = 180°$
$a + 70° + 70° = 180°$
$a + 140° = 180°$
$\quad\quad a = 180° - 140°$
$\quad\quad\quad = 40°$

The values of the pronumerals are $a = 40°$, $b = 35°$, $c = 70°$.

WORKED EXAMPLE 3 Calculating values of all the angles in a triangle

Calculate the value of each angle in the following triangle.

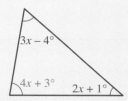

THINK

1. All angles in a triangle add up to 180°, so we have to add all these angles together.

WRITE

$2x + 1 + 3x - 4 + 4x + 3 = 180°$

2.	• Collect like terms. • Simplify by dividing both sides by 9 to find x.	$2x + 3x + 4x + 1 - 4 + 3 = 180°$ $9x = 180°$ $x = 20°$
3.	Substitute the value of x into each of the three angles to calculate the value of each angle.	Angle 1: $2x + 1$ $2 \times 20° + 1 = 41°$ Angle 2: $3x - 4$ $3 \times 20° - 4 = 56°$ Angle 3: $4x + 3$ $4 \times 20° + 3 = 83°$
4.	Write the value of each angle in degrees.	The three angles inside the triangle are 41°, 56° and 83°.

16.2.4 Quadrilaterals

A **quadrilateral** is a 2-dimensional closed shape with four straight sides.

Types of quadrilaterals	Description	Shape
Regular	All four edges are of equal length and their four internal angles are identical (they are squares).	
Irregular	Some (or all) of the sides and angles are different (they are not squares).	

Angles and quadrilaterals

A quadrilateral is made up of two triangles. The sum of the internal angles of a quadrilateral is equal to twice the sum of the internal angles of a triangle.

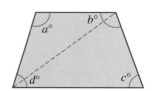

Sum of angles in a quadrilateral

The four internal angles *of* any quadrilateral add up to 360°.

$$\angle a + \angle b + \angle c + \angle d = 360°$$

WORKED EXAMPLE 4 Calculating the value of the pronumeral in a quadrilateral

Calculate the value of the pronumeral in the diagram.

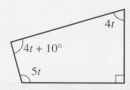

THINK	WRITE
1. The sum of the four angles in a quadrilateral is 360°. Write this as an equation.	$4t + 90° + 5t + (4t + 10°) = 360°$
2. Simplify the equation by adding like terms.	$4t + 5t + 4t + 90° + 10° = 360°$ $13t + 100° = 360°$
3. Solve the equation for *t*. • Subtract 100° from both sides. • Divide both sides by 13.	$13t = 360° - 100°$ $= 260°$ $\dfrac{\cancel{13}t}{\cancel{13}} = \dfrac{260°}{13°}$ $t = 20°$
4. Write the value of the pronumeral in degrees.	The value of the pronumeral *t* is 20°.

WORKED EXAMPLE 5 Determining the values of the pronumerals

Calculate the values of the pronumerals *m* and *n* in the following diagram.

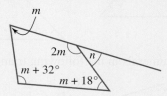

THINK	WRITE
1. The sum of the four interior angles in a quadrilateral is 360°. Write this as an equation.	$(m) + (2m) + (m + 18°) + (m + 32°) = 360°$
2. Simplify the equation by adding like terms.	$m + 2m + m + m + 18° + 32° = 360°$ $5m + 50° = 360°$

3. Solve the equation for *m*; that is, subtract 50° from both sides and then divide both sides by 12.

$$5m = 360° - 50°$$
$$= 310°$$
$$5 = 310°$$
$$\frac{5m}{5} = \frac{310°}{5}$$
$$m = 62°$$

4. Since the angles *n* and 2*m* are on a straight line, they sum to 180°. Use the value *m* = 62°, and solve.

$$n + 2m = 180°$$
$$m = 62, \text{ so } n + 2 \times 62° = 180°$$
$$n + 124° = 180°$$
$$n = 56°$$

Write the values of the pronumerals in degrees.

The values of the pronumerals are *m* = 62° and *n* = 56°.

16.2 Exercise

1. **WE1** A 3-dimensional solid has 6 faces and 8 vertices. State the number of edges of the 3-dimensional solid.

2. A 3-dimensional solid has 8 faces and 12 edges. State the number of vertices of the 3-dimensional solid.

3. Explain the difference between an interior angle and an exterior angle of a triangle. Support your explanation with a diagram.

4. **WE2** Calculate the values of the pronumerals in the following triangles.

a.

b.

c.

d.

5. **WE3** Calculate the values of the pronumerals in the following triangles.

a.
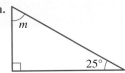

b.
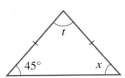

c.
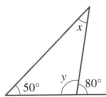

d.
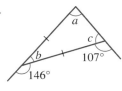

6. a. Name the type of triangle that best describes the road sign shown.
   b. Use an appropriate method to determine the size of the three interior angles of the sign.
   c. Explain how you would go about determining the value of the pronumeral *m*. Calculate the value of *m*.

7. a. Calculate the values of the pronumerals in each of the diagrams below.

   i. 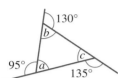   ii. 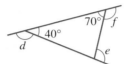   iii. 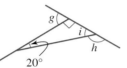   iv.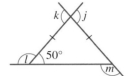

   b. Calculate the sum of the exterior angles in each of the triangles.
   c. Write a statement about the sum of the exterior angles of a triangle.

8. Explain why the angles in a quadrilateral add to 360°.

9. Calculate the value of the pronumeral in each of the following quadrilaterals.

   a. 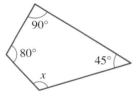   b. 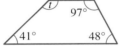   c.    d.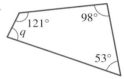

10. The frame of the swing shown in the photograph is a trapezium. The diagonal sides make an angle of 73° with the ground. Evaluate the angle, *x*, that the horizontal crossbar makes with the sides.

11. **WE4** Calculate the value of the pronumeral in each of the following quadrilaterals.

   a.    b. 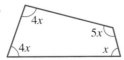   c.    d.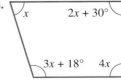

12. **WE5** For each of the quadrilaterals in question **11**, use the value of the pronumeral to calculate the size of each interior angle.

13. The pattern for a storm sail for a sailboard is shown in the photograph. Calculate the size of each interior angle.

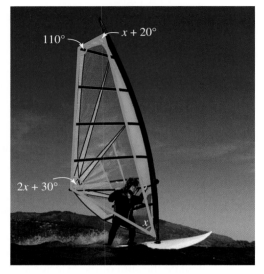

14. Use illustrations and the rule for the sum of the interior angles of quadrilaterals for the following question.

   a. Prove that a quadrilateral can have no more than one interior angle that is greater than 180°.
   b. If all interior angles are integers, determine the largest possible interior angle that a quadrilateral can have. Use your protractor to draw it.

16.3 Angles and polygons

LEARNING INTENTION

At the end of this subtopic you should be able to:
- identify properties of polygons
- calculate the sum of interior angles of polygons
- calculate the values of pronumerals using properties of polygons.

16.3.1 Polygons

A polygon is a closed shape where each edge is a straight line and all edges meet at vertices.

Sum of the angles of a polygon

The sum of the interior angles of any polygon can be found using the rule:

$$\text{angle sum} = 180° \times (n - 2)$$

where n = the number of sides of the polygon.

Type of polygon	Description	Triangle 3 sides	Quadrilateral 4 sides	Pentagon 5 sides	Hexagon 6 sides	Heptagon 7 sides	Octagon 8 sides
Regular	All sides have equal length and all internal angles are equal.	Triangle	Quadrilateral	Pentagon	Hexagon	Heptagon	Octagon
Irregular	The edges are of different lengths and their internal angles are not all the same.	Scalene triangle	Rectangle	Irregular pentagon	Irregular hexagon	Irregular heptagon	Irregular octagon

Shapes that are not polygons	Circles	Any shape that includes a curve	Any shape that isn't 'closed'	Three-dimensional objects

WORKED EXAMPLE 6 Calculating the sum of the interior angles of a regular polygon

Answer the following questions about the stop sign shown.
a. Calculate the sum of the interior angles of the stop sign.
b. Calculate the size of each interior angle.

THINK

a. 1. Write the formula for the sum of the interior angles of a polygon.

   2. n is equal to the number of sides of the polygon; the stop sign has eight sides.

   3. Substitute the value of n (8) into the formula and solve.

   4. Write the answer as a sentence.

WRITE

a. Angle sum $= 180° \times (n - 2)$

n = number of sides of polygon
$n = 8$

Angle sum $= 180° \times (8 - 2)$
$= 180° \times 6$
$= 1080°$

The angle sum of a regular octagon is $1080°$.

b. 1. Each angle size can be calculated by dividing the angle sum by the number of interior angles. There are eight interior angles in the polygon.

b. $1080° × 8 = 135°$

2. Write the answer as a sentence.

Each interior angle is 135°.

---

tlvd-6024

WORKED EXAMPLE 7 Calculating the sum of the interior angles of an irregular polygon

The road arrow shown has been redrawn with angles marked.

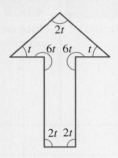

a. Calculate the value of the pronumeral.
b. Calculate the size of each interior angle.

THINK

a. 1. The polygon has seven sides. A polygon with seven sides has an angle sum of 900° (angle sum = $180° × (7 − 2)$). Write an equation for the sum of the angles in the polygon using the information given.

WRITE

a. Angle sum $= 180° × (7 − 2)$
$\qquad\qquad = 900°$
$2t + t + t + 6t + 6t + 2t + 2t = 900°$

2. Simplify the equation by adding like terms and solving for t.

$20t = 900°$
$t = \dfrac{900°}{20}$
$t = 45°$

b. Use the value of t to calculate the value of each interior angle.

b. The angles marked as t equal 45°.
The angles marked as $2t$ equal $2 × 45° = 90°$.
The angles marked as $6t$ equal $6 × 45° = 270°$.

16.3 Exercise

1. Explain the difference between a regular polygon and an irregular polygon.

2. **WE6** For each of the polygons shown, calculate the sum of the interior angles.

a.

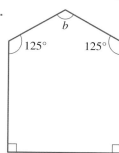

b.

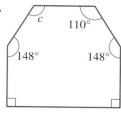

c.

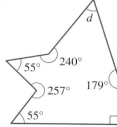

d.

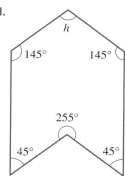

e.

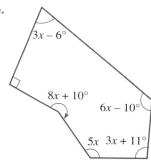

f.

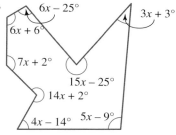

3. **WE7** For each of the polygons in question 2:
   i. calculate the value of the pronumeral
   ii. calculate the size of each unknown angle.

4. A basketball key has been drawn with angles represented as shown in the figure.

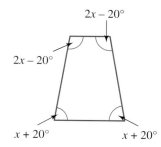

Calculate the value of the pronumeral and, hence, the size of each interior angle.

5. Calculate the value of the pronumeral in the diagram of the beach house shown.

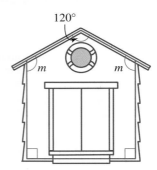

120°

m   *m*

6. State whether the following polygons are regular. Give a reason for each answer.

a.

b.

c.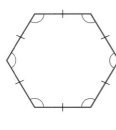

d.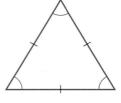

e.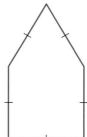

f.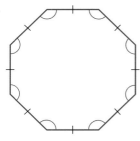

7. Explain how you calculate the size of the internal angles for any regular polygon.

8. For each of the regular polygons in question **6**, calculate the value of each interior angle.

9. Two angles of a particular hexagon are right angles. A second pair of angles are equal to each other. The third pair of angles are equal to each other and are triple the size of the second equal pair of angles.

   a. Draw a diagram and label it using pronumerals to show each of the conditions outlined above.

   b. Calculate the size of each of the angles.

10. Platonic solids are three-dimensional solids with faces that are regular polygons. There are five platonic solids. Three have faces composed of regular triangles, and the other two have faces that are squares and regular pentagons. Research each of these solids and then sketch and label them.

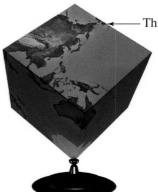

This is a platonic solid.

16.4 Angles, parallel lines and transversals

LEARNING INTENTION

At the end of this subtopic you should be able to:
- identify different types of angles formed at the line intersections
- identify properties of parallel lines and transversals
- calculate the sizes of the angles by applying the properties of parallel lines and transversals.

16.4.1 Parallel lines and transversals

Parallel lines are two or more lines that are simple translations of each other.

The distance between parallel lines is the same across their entire lengths; they never intersect each other. The train tracks in the picture are parallel to each other and only *appear* to intersect at the horizon.

The term *transverse* means *crossways*. A line that intersects with a pair of parallel lines is called a **transversal**. The road crossing the two parallel lines in the diagram shown represents a transversal.

When a transversal cuts a set of parallel lines, a number of angles are created. The table below shows how these pairs of angles are related.

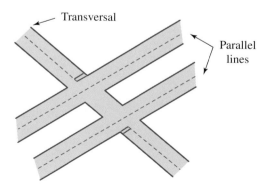

Angle identification	Relationship	Example
Corresponding angles are positioned on the same side of the transversal and are either both above or both below the parallel lines; think of them as **F**-shaped.	$\angle a = \angle b$	
Co-interior angles are positioned 'inside' the parallel lines, on the same side of the transversal; think of them as **C**-shaped.	Co-interior angles are **supplementary**: $\angle a + \angle b = 180°$	
Alternate angles are positioned 'inside' the parallel lines on alternate sides of the transversal; think of them as **Z**-shaped.	$\angle a = \angle b$	
Vertically opposite angles are created when two lines intersect. The angles opposite each other are equal in size; think of them as **X**-shaped.	$\angle a = \angle a$ $\angle b = \angle b$	

WORKED EXAMPLE 8 Identifying angles

Use the diagram shown to answer the questions below.
a. State which angle is vertically opposite c.
b. State which angle is alternate to d.
c. State which angle is a corresponding angle to g.
d. State which angle is co-interior with f.
e. State which of the angle pairs are equal and which are supplementary.

THINK	WRITE
a. *Vertically opposite* means that the angle is opposite the intersection from c. They are X-shaped. 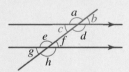	**a.** b is vertically opposite c.
b. Alternate angles are in a Z shape. 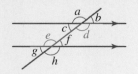	**b.** e is alternate to d.

c. Corresponding angles are in an F shape.

d. Co-interior angles are in a C shape.

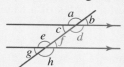

e. Colour all the angles that are equal to *a* in blue:
- *d* is vertically opposite to *a*.
- *e* is a corresponding angle to *a*.
- *h* is vertically opposite to *e*.

Colour all the angles that are equal to *b* in red:
- *c* is vertically opposite to *b*.
- *f* is a corresponding angle to *b*.
- *h* is vertically opposite to *f*.

c. *c* is a corresponding angle to *g*.

d. *d* is co-interior with *f*.

e.

Evaluate the sizes of unknown angles

Knowing angle relationships formed between parallel lines and a transversal can help you evaluate the sizes of unknown angles:
- **Identify the relationship between a known angle and an unknown angle.**
- **Use your knowledge of whether angles are equal or supplementary.**

---

tlvd-6025

WORKED EXAMPLE 9 Calculating the value of the pronumerals

Use the diagram to calculate the value of each pronumeral. Provide a reason for each answer.

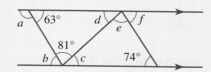

THINK

1. Angles *a* and 63° are supplementary so they sum to 180°. Write an equation and solve for *a*.

2. Angles *b* and 63° are alternate, and alternate angles are equal.

WRITE

$$a + 63° = 180° \quad \text{(supplementary)}$$
$$a = 180° - 63°$$
$$a = 117°$$

$$b = 63° \quad \text{(alternate)}$$

3. Angles b, 81° and c are supplementary so they sum to 180°. Write an equation and solve for c.

$$63° + 81° + c = 180° \quad \text{(supplementary)}$$
$$144° + c = 180°$$
$$c = 180° - 144°$$
$$c = 36°$$

4. Angles c and d are alternate, so they are equal.

$$d = c = 36° \quad \text{(alternate)}$$

5. The triangle contains angles c, e and 74°. The sum of the interior angles in a triangle is 180°.

$$c + e + 74° = 180° \quad \text{(interior angles of a triangle)}$$
$$36° + e + 74° = 180°$$
$$110° + e = 180°$$
$$e = 180° - 110°$$
$$e = 70°$$

6. Angles f and 74° are alternate, so they are equal.

$$f = 74° \quad \text{(alternate)}$$

16.4 Exercise

Students, these questions are even better in jacPLUS

Receive immediate feedback and access sample responses

Access additional questions

Track your results and progress

Find all this and MORE in jacPLUS

1. **MC** Describe the marked angles in each of the following diagrams using the appropriate name from these four options.

   A. Vertically opposite angles   B. Co-interior angles   C. Corresponding angles
   D. Alternate angles   E. Exterior angles

   a. 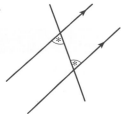   b. 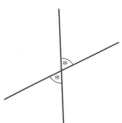   c. 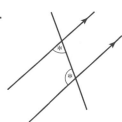   d.

2. Draw a series of parallel lines, each with a transversal. Label a set of:

   a. co-interior angles   b. vertically opposite angles
   c. corresponding angles   d. alternate angles.

3. Examine the picture of the intersection shown. State which pairs of objects are:

   a. co-interior
   b. corresponding
   c. vertically opposite
   d. alternate.

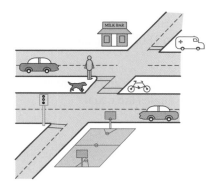

4. **WE 4** Use the diagram shown to answer the following questions.
   a. State which angles are vertically opposite to *c*.
   b. State which angles are alternate to *d*.
   c. State which angles are corresponding angles to *g*.
   d. State which angles are co-interior with *f*.

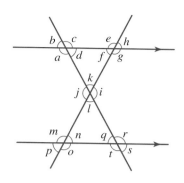

5. Carefully examine the diagram shown.
   a. List all the sets of angles that sum to 180°. In each case, explain why you think the sum is 180°.
   b. List all the pairs of vertically opposite angles.
   c. List all the pairs of alternate angles.

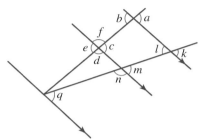

6. Make up your own design, which must include at least one pair of parallel lines. Indicate some angles by using pronumerals.

7. For each of the following diagrams:
   i. state the type of angle relationship between the given angle and the pronumeral
   ii. calculate the value of the pronumeral.

   a.

   b.

   c.

   d.

8. **WE9** In each of the following diagrams, calculate the values of the pronumerals, stating the types of angle relationships.

   a.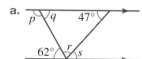

   b.

   c.

   d.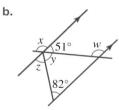

9. Consider the following diagram of jet trails. The vertical line represents the direction to the North Pole. Angle e represents the north-east direction of the single-engine jet aeroplane. Evaluate e.

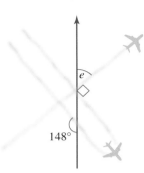

10. The pitch of a house roof is 43°. The owner renovates and installs a balcony. Evaluate the angle p that the balcony makes with the roof.

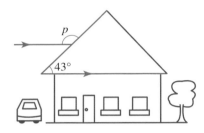

11. A satellite dish needs to be tilted at an angle of 70° from vertical. The pitch of the house roof is 38°. Evaluate the angle x that the dish makes with the roof.

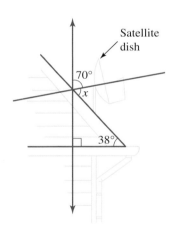

12. In each of the following diagrams, calculate the value of the pronumeral x and hence the sizes of the angles marked.

a.

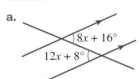

b.

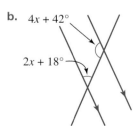

13. The angle relationships discussed in this section hold only if the transversal cuts through *parallel* lines. Study the following figures and use the angle measurements provided to decide whether the two lines cut by the transversal are parallel.

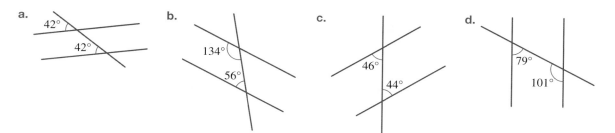

14. Consider the figure shown and answer the following questions.

    a. Calculate the size of the angle *a*. Explain how you found it.
    b. Determine whether the line KL is parallel to line MN. Justify your answer.
    c. Calculate all remaining internal and external angles of the triangle in the diagram.

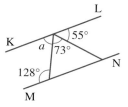

15. Calculate the angles indicated in the diagram shown. Provide reasons for
each answer.

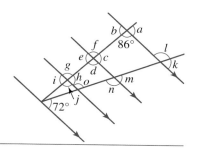

16.5 Constructing triangles

16.5.1 Construct a triangle

To construct a triangle, one of the following conditions is required.

Three edge lengths	
	3.5 cm 2.9 cm 4 cm
Two adjacent angles and the edge length between them	35° 55° 2 cm
Two adjacent edges and the angle between them	6 cm 50° 10 cm

Use a ruler and a pair of compasses to construct a triangle with edge lengths 4 cm, 3.5 cm and 2.9 cm.

THINK	WRITE

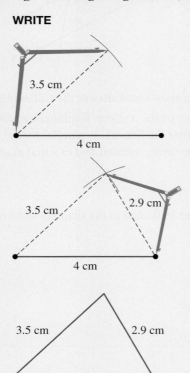

1. Rule a horizontal line representing the longest edge (4 cm). Open the pair of compasses to 3.5 cm and draw an arc from the left end of the 4-cm baseline.

2. Open the pair of compasses to 2.9 cm and draw an arc from the other end of the baseline so that it intersects the first arc.

3. The point of intersection between the two arcs is the position of the third vertex of the triangle. Rule a line from this vertex to each end of the baseline.

Use a ruler and protractor to construct a triangle with angles 35° and 55° and an edge between them of length 2 cm.

THINK	WRITE

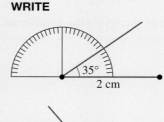
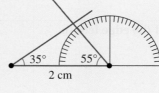
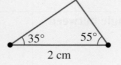

1. Rule a horizontal line of length 2 cm. Place the centre point of the protractor on the left end of the baseline. Mark a dot at 35° on the outer margin of the protractor and rule a line between this dot and the left end of the baseline.

2. Place the centre point of the protractor on the right end of the baseline. Mark a dot at 55° on the outer margin of the protractor, and rule a line between this dot and the right end of the baseline. (If these lines do not intersect, simply extend them.)

3. The point of intersection between the two lines is the third vertex of the triangle. Erase any excess length.

WORKED EXAMPLE 12 Constructing a triangle given two adjacent edge lengths and the angle between them

Use a ruler and protractor to construct a triangle with sides measuring 6 cm and 10 cm, with an angle between them of 50°.

THINK

1. Rule a horizontal line 10 cm long. Place the centre of the protractor on the left end of the baseline. Mark a dot at 50° on the outer margin of the protractor, and rule a 6-cm line from the left end of the 10-cm baseline through this dot.

2. Complete the triangle by using the ruler to join the end points of the two lines.

WRITE

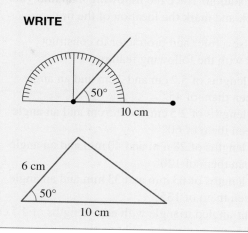

16.5 Exercise

1. State the minimum requirements needed to draw a triangle.

2. **WE10** Use a ruler and a pair of compasses to construct triangles with the following edge lengths.
   a. 5 cm, 4 cm, 3 cm
   b. 4.8 cm, 2.5 cm, 3.3 cm
   c. 2 cm, 2.5 cm, 1.5 cm
   d. 3 cm, 2.5 cm, 2.5 cm

3. A bushwalker is injured and cannot walk. Before the battery of his mobile phone died, he reported that he was equidistant from the peak of Mount Buller and Sheepyard Flat where he parked his car. He estimated that he had walked 6 km and had crossed Howqua River before his fall. At rescue headquarters, the SES captain looked at a map. Sheepyard Flat is approximately 10 km from Mount Buller. Trace the map shown into your workbook and draw the point where the SES should send the rescue helicopter.

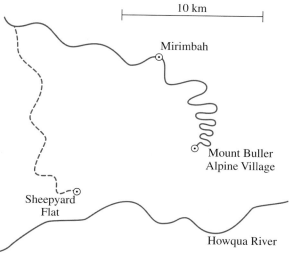

4. **WE11** Use a ruler and protractor to construct triangles with the following features.
   a. Angles of 30° and 90° with a 3-cm edge between them
   b. Angles of 45° and 45° with a 2.5-cm edge between them
   c. Angles of 60° and 100° with a 40-mm edge between them
   d. Angles of 22° and 33° with a 33-mm edge between them
   e. An isosceles triangle with adjacent angles of 57° and a 4-cm baseline

5. A forestry ranger informs the CFA that he sees smoke rising from behind a mountain range at 40° east of due north from his park office. The CFA chief sees the same smoke at an angle of 30° west of due north from her station. Trace the following map into your workbook and mark the location of the fire.

6. **WE12** Use a ruler and protractor to construct triangles with the following features.

   a. Edge lengths of 5.2 cm and 3 cm and an angle between them of 45°
   b. Edge lengths of 2.5 cm and 2.5 cm and an angle between them of 60°
   c. Edge lengths of 28 mm and 40 mm and an angle between them of 120°
   d. Edge lengths of 63 mm and 33 mm and an angle between them of 135°
   e. A right-angled triangle with edge lengths of 4.5 cm and 2.5 cm creating the right angle

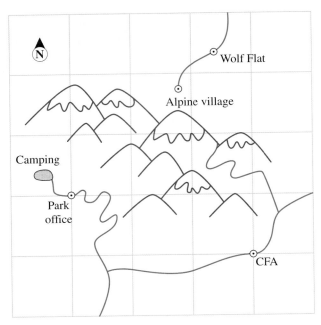

16.6 Using scales

LEARNING INTENTION

At the end of this subtopic you should be able to:
- use scales to calculate the actual length of an object
- determine the new dimensions of an object using a scale ratio.

16.6.1 Scales

A **scale** describes how much larger or smaller the image is compared to the original object. The architect writes the numerical scale on the plans, such as 1 : 100. The first number is the size of the image (drawing) and the second number is the size of the real object. The builder simply measures the figure on the plan and multiplies this by the **scale factor**.

For example, if the scale is 1 : 100, a wall that is 2.8 centimetres long on the plan is 280 centimetres long in the actual building.

Representing scales

A scale may be written as:
- **a ratio of length on a plan to the actual length, such as 1 : 100 000**
- **a statement, such as '1 cm on the plan represents a length of 100 000 cm or 1 km in real life'**
- **a statement, such as '1 cm to 1 km'.**

A scale ratio is often written as:

$$\text{representation of the original object : original object}$$

The representation of the original object may be a map, a plan, a diagram or a scale model.

WORKED EXAMPLE 13 Representing the size of the original object on a scale

a. **Rewrite the scale ratio 1 : 10 000 using appropriate units.**
b. **Rewrite the scale 4 cm to 5 km as a ratio.**

THINK	WRITE
a. 1. 1 : 10 000 means that 1 cm on the plan is 10 000 cm in actual length. Convert 10 000 cm into m by dividing by 100.	a. 10 000 cm = 100 m
2. Rewrite the scale ratio using cm and m. *Note:* It would also be appropriate to use mm and m (1 mm to 10 m).	1 cm to 100 m
b. 1. 4 cm to 5 km means that 4 cm on the plan is 5 km in actual length. Make the units the same by converting 5 km to cm, by multiplying by 100 000.	b. 5 km = 500 000 cm
2. Now that the units are the same, 4 cm and 500 000 cm can be written as a ratio.	4 : 500 000
3. Rewrite as a ratio in simplest form by dividing each part by 4.	1 : 125 000

tlvd-6026

WORKED EXAMPLE 14 Calculating the original length of an object using a scale ratio

A scale ratio of 1 : 10 is used to make a toy horse.

195 mm

Calculate the actual length of the horse if the toy is 195 millimetres long.

THINK	WRITE
1. The length of the toy horse is measured in mm. A scale ratio of 1 : 10 means that 1 mm on the toy horse is equal to 10 mm on the actual horse. To find the length of the actual horse, multiply the length of the toy horse by 10.	Actual length = length of the toy horse × 10 = 195 × 10 = 1950 mm
2. Convert 1950 mm into m, which is a more appropriate measurement for the length of a horse. To do this, divide by 1000.	1950 mm = 1.95 m
3. Write the answer.	The actual horse is 1.95 m long.

WORKED EXAMPLE 15 Determining the new dimensions of an object using a scale ratio

Determine the new dimensions of a football (30 cm long by 21 cm wide) if it is reduced to make a toddler's toy, using a scale ratio of toy : football = 1 : 3.

THINK	WRITE
A scale ratio of 1 : 3 means that 1 cm on the new toy represents 3 cm on the original football. To get the length and width of the new toy, divide each dimension by 3.	New length $= \dfrac{30}{3}$ $= 10$ cm New width $= \dfrac{21}{3}$ $= 7$ cm
Write the answer.	The toy football is 10 cm long and 7 cm wide.

16.6 Exercise

Students, these questions are even better in jacPLUS

 Receive immediate feedback and access sample responses

 Access additional questions

 Track your results and progress

Find all this and MORE in jacPLUS ⊙

1. Explain how a scale is written.

2. **WE13a** Rewrite the following scale ratios using appropriate units.
   a. 1 : 10   **b.** 1 : 100   **c.** 1 : 1000   **d.** 1 : 100 000   **e.** 1 : 1500   **f.** 1 : 3 300 000

3. **WE13b** Rewrite the following scales as ratios.
   a. 1 cm to 50 000 cm
   b. 1 cm to 200 m
   c. 1 cm to 50 km
   d. 2 cm to 2 km
   e. 3 cm to 6 km

4. State whether 6 : 1 is the same as 1 : 6. Explain.

5. Determine whether the following scales are the same and explain your answer.
   a. A dilation factor of $\dfrac{1}{50}$ and a scale of 1 : 50
   b. A scale of 1 : 150 and a scale of 3 : 450

6. State the new dimensions $(l \times w \times h)$ of a car $(360 \times 180 \times 150$ cm$)$ reduced by a scale factor of 60.

7. **WE14** Evaluate the new dimensions of a photograph $(12 \times 17$ cm$)$ enlarged by a factor of 5.

8. **WE15** A doll's house is made to a scale of 1 : 150.

   a. Calculate the height and width of the windows of the doll's house if real windows are 1500 millimetres by 1800 millimetres.
   b. Normal ceiling height is 2.7 metres. Calculate the height of the staircase in the doll's house.
   c. The seat of a real kitchen chair is 45 centimetres high. The seat itself is 34 cm deep and 38 cm wide. The top of the backrest of the chair is 80 cm high. Determine the dimensions of a kitchen chair in the doll's house.

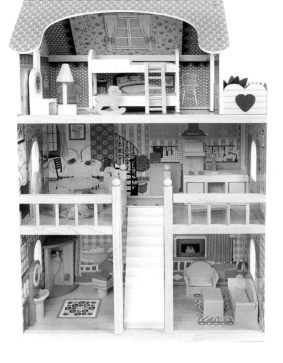

9. A photograph is taken of a local skate ramp and published in the paper. It is found that the photograph represents the ramp on a scale of 1 : 75.

   a. If the ramp is 3 m high in real life, determine its height in the photograph.
   b. If a separate ramp measured 5 cm in the photograph, determine how high it would be in real life.
   A skater is pictured in the photograph and is 3 mm in the air above one of the ramps.
   c. Calculate how far above the ramp the skater is in real life.
   d. If the ramp the skater is on is 1.5 m high, determine how far the skater is above the ground in both real life and in the photograph.

10. Determine the scale to represent a drawing of:

    a. yourself on A4 paper
    b. a 10-metre-high tree on a piece of paper measuring 15 cm × 23 cm
    c. a walking circuit that is roughly in the shape of a circle with a diameter of 12:18 pm, on poster paper measuring 30 cm × 60 cm.

11. A friend shows you a set of designs detailing the new room that his parents are adding onto their house. The plan doesn't have a scale or measurements to show the actual size of the room or the things in it. All he could remember is that exactly 16 of his feet fit along the length of the room.

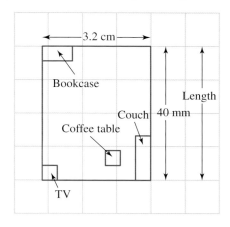

    a. If his feet are 25 cm long, calculate the length of the actual room.
    b. State the length of the room on the plan.
    c. Use parts a and b to calculate the scale factor for this design.
    d. Use the scale factor to give the dimensions of the actual couch.

12. Consider the house plan shown. Use a ruler to determine the dimensions of the house on the plan.

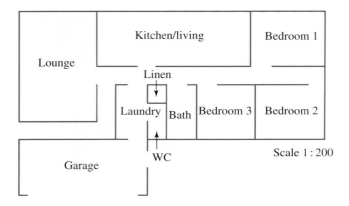

   a. Calculate the length of the actual house.
   b. Calculate the width of the hallway.
   c. Determine the dimensions (length and width) of each room in the house.

13. A sketch of a soccer pitch that is not drawn to scale is shown.

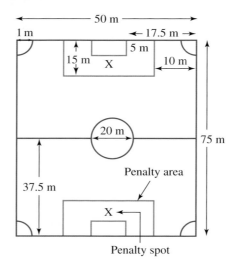

   Draw the pitch using a scale of 1 : 150 and determine the dimensions on the plan of:

   a. the pitch
   b. the penalty area
   c. the centre circle.

14. Measure the dimensions of your classroom. Draw a plan of your classroom, including the location of windows, doors, pinboards, whiteboard, etc. Use a ratio of 1 cm to 1 m and write this as a scale ratio on the plan.

16.7 Time and temperature

LEARNING INTENTION

At the end of this subtopic you should be able to:
- identify the difference between an analogue a digital clock
- convert time in hours, minutes and seconds
- calculate the time interval
- read temperature on scales
- convert temperature between Kelvin and Celsius.

16.7.1 Time

Knowing how to measure time is an important skill in everyday life. Whether you need to get to class on time, catch your flight or make your doctor's appointment, your life is organised around time.

Time is divided into units including seconds, minutes, hours, days, weeks, months and years.

A day can be divided into two 12-hour periods:

| Midnight to midday | am | ante meridiem (before noon) |
| Midday to midnight | pm | post meridiem (after noon) |

Time conversions

Time is divided into units. There are:
- **60 seconds in 1 minute**
- **60 minutes in 1 hour**
- **24 hours in 1 day**
- **7 days in 1 week**
- **2 weeks in 1 fortnight**
- **about 4 weeks in 1 month**

- **12 months in 1 year**
- **about 365 days in 1 year**
- **10 years in 1 decade**
- **100 years in 1 century**
- **1000 years in 1 millennium.**

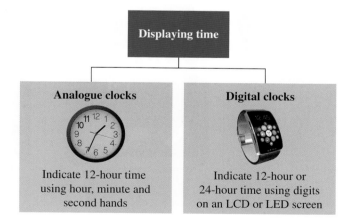

Displaying time

| **Analogue clocks** | **Digital clocks** |
| Indicate 12-hour time using hour, minute and second hands | Indicate 12-hour or 24-hour time using digits on an LCD or LED screen |

Time shown in writing	Time on an analogue clock	Time on a digital clock	
		12-hour time	**24-hour time**
Three o'clock in the morning		**3:00 AM**	**03:00**
Twenty-five minutes past 1 in the afternoon		**1:25 PM**	**13:25**
Ten minutes to 7 in the evening		**6:50 PM**	**18:50**

tlvd-6027

WORKED EXAMPLE 16 Converting time in hours, minutes and seconds

Write the following times in hours, minutes and seconds.

a. 7.3 hours

b. $4\dfrac{3}{8}$ hours

THINK

a. 1. There are 7 hours and 0.3 of an hour in 7.3 hours.

2. There are 60 minutes in an hour so there are 0.3 of 60 minutes.

3. Write the answer.

b. 1. There are 4 hours and $\dfrac{3}{8}$ of an hour in $4\dfrac{3}{8}$ hours.

2. There are 60 minutes in an hour so there are $\dfrac{3}{8}$ of 60 minutes.

3. There are 60 seconds in a minute so there are 0.5 of 60 seconds.

4. Write the answer.

WRITE

a. 7.3 = 7 hours + 0.3 of an hour

$$7.3 = 7 \text{ hours} + 0.3 \times 60 \text{ minutes}$$
$$= 7 \text{ hours} + 18 \text{ minutes}$$

7.3 hours is the same as 7 hours and 18 minutes.

b. $4\dfrac{3}{8} = 4 \text{ hours} + \dfrac{3}{8} \text{ of an hour}$

$$4\dfrac{3}{8} = 4 \text{ hours} + \dfrac{3}{8} \times 60 \text{ minutes}$$
$$= 4 \text{ hours} + 22.5 \text{ minutes}$$

$$4\dfrac{3}{8} = 4 \text{ hours} + 22 \text{ minutes} + 0.5 \times 60 \text{ seconds}$$
$$= 4 \text{ hours} + 22 \text{ minutes} + 30 \text{ seconds}$$

$4\dfrac{3}{8}$ hours is equal to 4 hours, 22 minutes and 30 seconds.

WORKED EXAMPLE 17 Calculating time in hours and minutes

A movie runs for 132 minutes. Calculate the length of the movie in hours and minutes.

THINK

1. There are 60 minutes in one hour, so divide 132 by 60 to get the number of hours.

2. Two hours is 120 minutes $(2 \times 60 = 120)$. Subtract 120 from 132 to calculate how many minutes are left over two hours.
An alternative way of finding the number of minutes over two hours is to convert 0.2 into minutes by multiplying by 60.

3. Write the answer.

WRITE

$132 \div 60 = 2.2$ hours

$132 - 120 = 12$ minutes

$0.2 \times 60 = 12$ minutes

The movie goes for 2 hours and 12 minutes

16.7.2 Time intervals

Often we want to know the time interval or difference between times. This may be to either find out how many hours we worked between 9 am and 1:30 pm, or to know how long our flight between 13.25 and 15.55 will take.

WORKED EXAMPLE 18 Calculating the time interval

A flight on which an air steward was rostered landed in Sydney at 8:07 pm and left Adelaide at 6:53 pm Sydney time. Calculate the length of the flight.

THINK	WRITE
1. Calculate the number of minutes from the departure to the next hour (7:00 pm).	6:53 pm to 7:00 pm = 7 minutes
2. Calculate the time from 7:00 pm to 8:00 pm.	7:00 pm to 8:00 pm = 1 hour
3. Calculate the number of minutes from 8:00 pm to 8:07 pm.	8:00 pm to 8:07 pm = 7 minutes
4. Add the times together.	7 minutes + 1 hour + 7 minutes = 1 hour and 14 minutes.
5. Write the answer.	The flight took 1 hour and 14 minutes

16.7.3 Temperature

Temperature measures how hot or cold an object is in comparison to areference point. Temperature is measured using a thermometer.

The units most commonly used to measured temperature are **degrees Celsius** (°C) and **degrees Fahrenheit** (°F). In Australia we use the Celsius scale.

Celsius to Fahrenheit formula derivation

Scale	Freezing point of water	Boiling point of water
°C	0 °C	100 °C
°F	32 °F	212 °F

The reference points for the Celsius scale are the freezing and boiling points of water at sea level.

It is important to know which scale is being used to measured temperature. For example, it would be a good idea to wear warm clothes when the temperature is 32°F, but the same clothing would not be appropriate when the temperature is 32°C.

Reading a thermometer

When reading the temperature from a thermometer, make sure that you:
- Check the scale carefully.
- are reading the temperature from an appropriate level so that your reading is accurate.

Another common unit of temperature is the kelvin. It is one of the seven International System of Units (SI) base units. The unit for kelvin is written as K.

Conversion using kelvin

To convert Celsius to kelvin use the formula:

$$T(K) = T(°C) + 273$$

To convert kelvin to Celsius use the formula:

$$T(°C) = T(K) - 273$$

WORKED EXAMPLE 19 Reading the temperature scale on a thermometer

Accurately read the following temperatures.

a.

b.

THINK

a. Read the scale carefully, The liquid is in line with the 37 °C mark.

b. 1. Read the scale carefully, The liquid has passed the 40 °C mark.

2. Check how many smaller divisions are between the larger intervals. there are 10, so each smaller division is 1 tenth of a degree, or 0.1 °C.

3. The liquid is in line which the third small division. This means it is 3 tenths of a degree above 40 °C.

WRITE

a. The temperature is 37 °C.

b.

The temperature is 40.3 °C.

16.7.4 Converting between °C and °F

Converting temperatures between two scales can be done by using the conversion table below.

°C	0	10	20	30	40	50	60	70	80	90	100	200	300	400
°F	32	50	68	86	104	122	140	158	176	194	212	392	572	752

WORKED EXAMPLE 20 Converting °F to°C

Your friend from Canada calls to complain about the weather that he is experiencing.
a. He says that the temperature has been 86 °F for three days this week. Convert that temperature to degrees Celsius.
b. It has been predicted on the weather report that the temperature in Canada will be 59 °F tomorrow. Determine what kind of weather you should tell your friend to expect?

THINK

a. 86 °F corresponds to 30 °C.

b. 59 °F is approximately halfway between 10 °C and 20 °C. This is approximately 15 °C.

WRITE

a. The temperature has been 30 °C and 3 days this week

b. You should tell your friend that a cool change is coming

WORKED EXAMPLE 21 Converting temperature between the two scales

Convert the following:
a. °C to K
    i. 21 °C to K
    ii. 40 °C to K
b. K to °C
    i. 300 K to °C
    ii. 373 K to °C

THINK

a. i. 1. To convert from °C to K, add 273 to the temperature in Celsius.
      2. Write the answer.
   ii. 1. To convert from °C to K, add 273 to the temperature in Celsius.
       2. Write the answer.
b. i. 1. To convert from K to °C, subtract 273 from the temperature in kelvin.
      2. Write the answer.
   ii. 1. To convert from K to °C, subtract 273 from the temperature in kelvin.
       2. Write the answer.

WRITE

a. i. $21 + 273 = 294$ K

      294 K
   ii. $40 + 273 = 313$ K

       313 K
b. i. $300 - 273 = 27$ °C

      27 °C
   ii. $373 - 273 = 100$ °C

       100 °C

on Resources

Interactivities Coverting between units of time (int-6910)
                    Analogue clock (int-3797)

16.7 Exercise

1. **WE16** Write the following times in hours, minutes and seconds.

   a. 5.8 hours

   b. $9\frac{7}{8}$ hours

2. Write the following times in hours, minutes and seconds.

   a. 12.156 hours

   b. $3\frac{2}{9}$ hours

3. Write in words the time displayed on each of the following clocks:

   a.

   b.

   c.

   d.

   e.

   f.

4. Write the following times in 24-hour times:

   a. 10:25 am
   b. 7:33 am
   c. 1:45 pm
   d. 8:12 pm
   e. 10:06 pm
   f. 11:45 pm

5. **WE17** A horse trail ride went for 156 minutes.
   Calculate the length of the trail ride in hours and minutes.

6. A game of Rugby is played over two 40 minute halves. If 2 minutes injury time wear added in the first half and 3 minutes in the second half, calculate the length of the entire game in hours and minutes.

7. Write the following 24-hour times as digital am or pm times.
   a. 1551
   b. 2022
   c. 0315
   d. 1131
   e. 0902
   f. 2215

8. Convert each of the following time periods to minutes.
   a. 3 hours
   b. $5\frac{1}{4}$ hours
   c. 1 day
   d. $\frac{3}{4}$ hour
   e. 3 hours and 18 minutes
   f. 6 hours and 34 minutes

9. **WE18** A flight left Perth at 1:12 pm and landed in Brisbane at 5:27 pm Perth time. Calculate the duration of the flight.

10. A cricket player started playing cricket at 0823 and finished at 1320. Determine how much time they spent playing cricket.

11. Determine the time difference between:
    a. 4:25 pm and 5:50 pm
    b. 6:30 pm and 2:45 am
    c. 7:20 am on Monday and 6:30 pm the following day (Tuesday)
    d. 1:20 pm on Wednesday and 9:09 pm the following Friday
    e. 0125 hours and 2345 hours
    f. 0715 hours and 1550 hours.

12. Calculate the following times.
    a. 1 hour after 12 noon
    b. 3 hours before 7:15 pm
    c. 1 hour and 20 minutes after 8:30 am
    d. 2 hours and 30 minutes before 7:45 pm
    e. 4 hours and 14 minutes after 1:08 pm
    f. 3 hours and 52 minutes before 3:25 pm

13. The following table displays the time that a student spent travelling to school. The student left home at 7:25 am.

Activity	Time
Walking from home to the train station	27 minutes
Waiting for the train	12 minutes
Train journey	33 minutes
Walking from the train station to school	8 minutes

   a. Calculate the amount of time the student spent travelling to school.
   b. State at what time the student arrive at school.
   c. State whether the student was on time for registration at 0845.

14. **WE19** Accurately read the following temperatures.

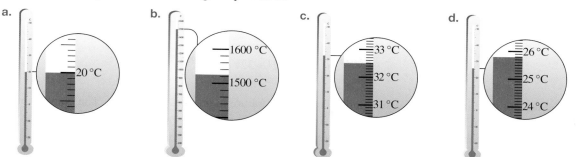

a. 20 °C
b. 1600 °C / 1500 °C
c. 33 °C / 32 °C / 31 °C
d. 26 °C / 25 °C / 24 °C

15. The thermometer at right is used to measure the temperature of a piece of roast meat in an oven. State the temperature of the meat.

16. For any type of measurement, you must always be careful when reading the measurement scale. Write the steps to obtain the most accurate thermometer reading.

17. Arrange the following in order of increasing temperature.

   a. A winter's morning in Melbourne
   b. A sauna
   c. The surface of the sun
   d. A barbecue hotplate cooking sausages

18. The wind can make it feel colder than what a thermometer says the temperature is. The difference in the actual temperature you experience due to the wind is called the wind chill factor. If the current temperature is 20 °C and the wind chill facter is −5 °C, the temperature you experience is 15°C. Calculate the temperature in the following situations.

   a. The temperature is 13 °C, with a wind chill facter of −4 °C
   b. The temperature is 8 °C, with a wind chill facter of −7 °C

19. A Galileo thermometer (also called a thermoscope), measures the temperature of the air that surrounds it. Glass spheres, partly filled with liquid, form a series of weights that float in a special fluid, as shown in the following diagrams. Each glass sphere sinks at a different temperature (indicated by the tag attached to the sphere). The temperature is read from the sphere with the lowest marked temperature that is still floating. State the temperature shown by the following Galileo thermometers.

a.

b.

c.

d.

20. The conversion table shown in this subtopic is very limited — it does not show all possible degrees Celsius values and their corresponding degrees Fahrenheit values. The following formula enables you to convert from degrees Fahrenheit to degrees Celsius.

$$C = \frac{5}{9} \times (F - 32)$$

C is the temperature in degrees Celsius and F is the temperature in degrees Fahrenheit. The conversion table shows that 10°C is equivalent to 50°F. Using the formula produces the same result, as shown below. Substitute F in the formula with the temperature in degrees Fahrenheit.

$$\begin{aligned} C &= \frac{5}{9} \times (F - 32) \\ &= \frac{5}{9} \times (50 - 32) \\ &= \frac{5}{9} \times (18) \\ &= \frac{5}{\cancel{9}_1} \times \frac{\cancel{18}^2}{1} \\ &= 10 \end{aligned}$$

a. Test this formula on various degrees Celsius values from the conversion table.
b. Compare your results from part a with the degrees Celsius values from the conversion table.

21. a. **WE20** The normal internal body temperatures of the following animals are shown in degrees Fahrenheit. Convert these values to degrees Celsius.

   i. Humans 98.6 °F
   iii. Blue whales 95.9 °F

   ii. Polar bears 99.1 °F
   iv. Sparrows 105.8 °F

b. **WE21** Convert the following:

   i. 35 °C
   iii. 425 K to °C

   ii. 50 °C
   iv. 323 K to °C

22. Water boils at a lower temperature as height above sea level increases, due to lower air pressure. This means that it takes longer to cook food in boiling water at higher altitudes because the water is not as hot. The temperature at which water boils is shown below for three different heights above sea level.
Copy and complete the table to express the temperature in each case in degrees Celsius.

Location	Height above sea level (m)	Temperature at which water boils (°F)	Temperature at which water boils (°C)
Gisbourne (Victoria)	610	208	
Falls Creek (Victoria)	1500	203	
Mount Everest (Nepal)	8840	159.8	

16.8 Map distances

16.8.1 Interpreting maps

Maps and the land they represent are similar figures; they have exactly the same shape but very different sizes. The scale factor of a map describes the ratio between the distance on the map and the actual distance on Earth's surface.

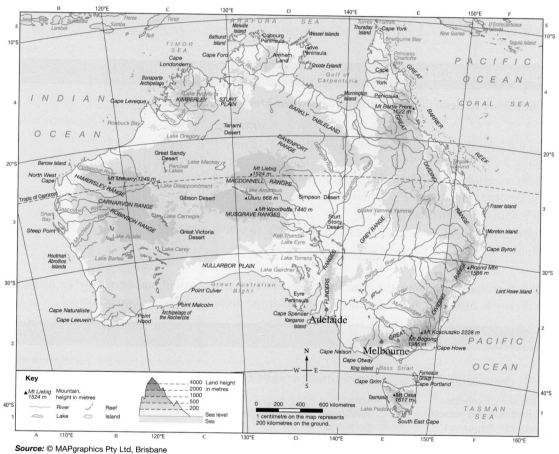

Source: © MAPgraphics Pty Ltd, Brisbane

The map above was made according to the ratio 1 : 40 000 000. The first number represents the distance on the map and the second number represents the real distance. In this case, 1 millimetre on the map equates to 40 million millimetres on the ground — that is, a scale factor of 40 000 000. This can also be represented as the ratio 1 mm to 40 km, because 1 kilometre is equal to 1 million millimetres.

In addition, it may be written as a line scale, such as . The line scale is represented by a black and white bar. Each section shows the actual distance on the ground (200 km) relative to the length of the line on the map.

WORKED EXAMPLE 22 Using a numerical scale to determine distance

Use the numerical scale to determine the approximate direct distance between Melbourne and Adelaide.

THINK	WRITE
1. Use a ruler to measure the linear distance in millimetres between the two cities.	16 mm
2. Rewrite the numerical scale as a ratio using appropriate units.	1 : 40 000 000 = 1 mm to 40 000 000 mm 1 mm to 40 000 000 mm ÷ 1 000 000 1 mm to 40 km
3. Each millimetre on the map equals 40 kilometres on the ground, so 16 millimetres on the map must equal 16 lots of 40 kilometres on the ground.	$16 \times 40 = 640$
4. Write the answer.	The distance between Melbourne and Adelaide is 640 km.

WORKED EXAMPLE 23 Using a line scale to determine distance

Use the line scale to determine the distance between Melbourne and Adelaide.

THINK	WRITE
1. Measure the distance in millimetres between the two cities.	16 mm
2. Measure a section of the line scale in millimetres.	5 mm
3. Divide the line scale distance (200 km) by its actual length in millimetres.	$200 \div 5 = 40$
4. Multiply this quotient by the distance in millimetres between the two cities.	$16 \times 40 = 640$ km

(**Note:** In this case, the answers are the same. It is more likely that you would make errors when using the line scale, as it is easy to misjudge measurements when measuring distances between two places.)

16.8 Exercise

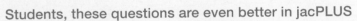

Students, these questions are even better in jacPLUS

 Receive immediate feedback and access sample responses

 Access additional questions

 Track your results and progress

Find all this and MORE in jacPLUS

1. Explain what a line scale is. Provide an example.

2. Consider this tourist map of Kakadu National Park.
   a. Calculate the distance, as the crow flies, from Katherine to Jabiru.
   b. Write the line scale as a scale ratio.
   c. Calculate the distance between Katherine and Darwin.

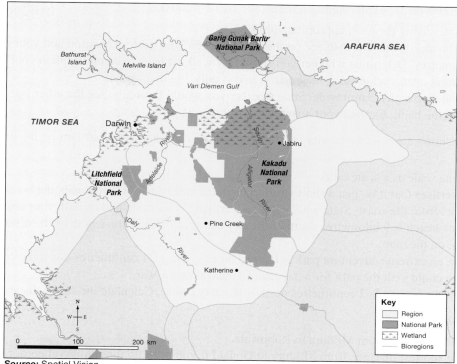

Source: Spatial Vision

3. **WE22** This map is scaled at 1 : 125 000 000.

   a. Write the scale as a ratio in the most appropriate units.
   b. Determine the distance from Melbourne to Wellington.
   c. Determine how many kilometres of ocean lie between Melbourne and Beijing.
   d. Determine the distance between Canberra and Jakarta.

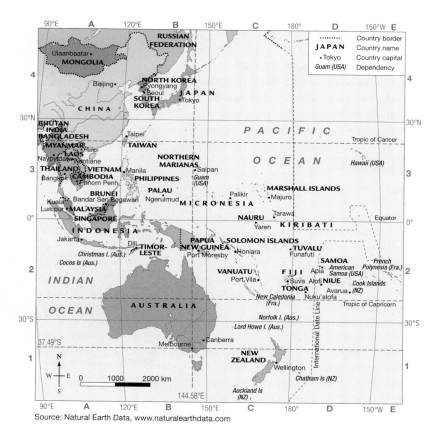

Source: Natural Earth Data, www.naturalearthdata.com

4. A map of a desert you are exploring has a scale ratio of 1 : 7 200 000.
   a. Write this ratio using the most appropriate units.
   b. Your 4WD breaks down. The distance on the map between your present location and your campground is only 8 millimetres. Determine whether you should try walking back to camp. Explain your answer.
   c. When you look over level ground, the horizon line is only 14 km away.
      A lake on the map is just 3 mm wide. State whether you would be able to see the other side of the lake if you stand on the bank. Explain your answer.

5. A tourist map shows a small coastal city in which you plan to holiday with your family; the line scale shows 6 mm to 3 km.
   a. Write the line scale as a scale ratio.
   b. A hotel advertises that it is 'just a short stroll to the local surf beach'. On the map, the hotel is 4 millimetres from the coast. State whether the advertising is accurate and explain your answer.
   c. The best surf beach is 9 kilometres south of the city centre. Calculate the length of the distance, in millimetres, on the map.
   d. On the map, an extreme adventure park is located on a highway 24 centimetres down the coast. Determine whether you could visit the park for a day trip. Explain your answer.
   e. On the map, the airstrip is 2 centimetres north of the city centre. Calculate the actual distance.

6. Use Google Maps to:
   a. determine the distance from Mildura to Robinvale
   b. describe the shortest distance from Mildura to Donald if you must go through Wycheproof. State the map distance.

7. Consider this map of Australia.
   a. Determine the scale ratio of this map.
   b. Determine which two capital cities are the furthest apart on the map.
   c. Calculate the actual distance between the two cities identified in part b.
   d. If the line scale of the map was changed to 1 : 50 000 000, determine the map distance between the two cities identified in part b.

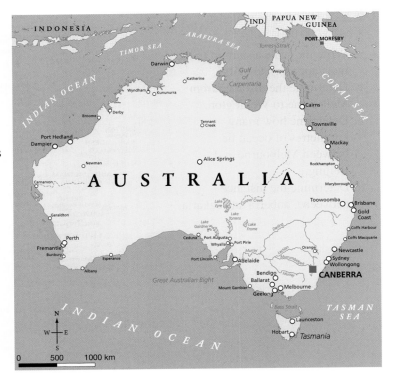

8. **WE23** Consider the map shown.

   a. Measure the straight-line map distance between:

      i. Alice Springs and Mount Isa

      ii. Port Augusta and Townsville

      iii. Adelaide and Broken Hill.

   b. Using the scale shown on the map, calculate the actual distances in part **a**.

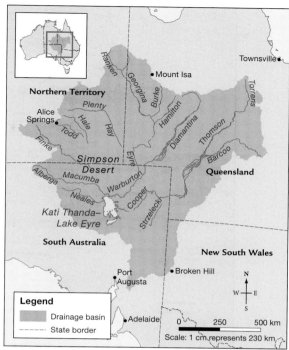

Source: Spatial Vision

16.9 Directions

LEARNING INTENTION

At the end of this subtopic you should be able to:
- identify language, symbols, labelling and conventions for maps
- identify location-related diagrams and directories, including keys, scale, direction, distance, coordinates and grid references.

16.9.1 Directions on maps

Directions are necessary to find a location or to get from one location to another.

A location on a map is defined by a grid system of lines that run vertically up and down (north to south) and horizontally left to right (west to east) across the map.

Two types of grid systems are used:
- Grid reference or alphanumeric grid
- Latitude and longitude

Consider the following map. Grid squares are labelled with red letters across the top and bottom and numbers on each side. The location of a place is defined by the square it is in. For example, the city of Melbourne is located at grid reference (GR) E2.

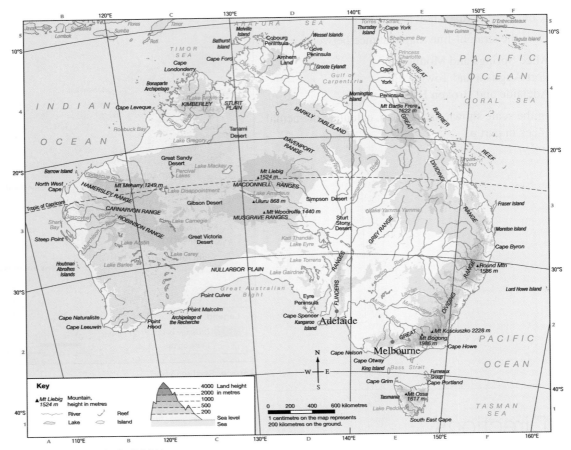

Source: © MAPgraphics Pty Ltd, Brisbane

Latitude and **longitude** are imaginary lines that go around the Earth and are labelled using degrees. Latitude is always stated before longitude.

Lines of latitude are parallel and run east–west; they measure the distance north and south of the equator, which is at 0° latitude.

Lines of longitude run north–south; they measure the distance east and west of the **prime meridian**, which runs through Greenwich Observatory (0° longitude) in London, England.

Each degree (°) of latitude and longitude is further divided into 60 minutes ('), which are each divided into 60 seconds (''). Melbourne is more accurately located at 37°49′4.89″ south, 144°58′2.9″ east.

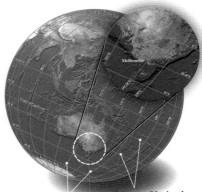

Lines of longitude are widest at the Equator and meet at the poles.

Lines of latitude are parallel at the Equator and equal distances apart.

WORKED EXAMPLE 24 Identifying grid references for a location

Use the map of Australia shown in this section to determine:
a. the grid reference for the Great Sandy Desert
b. the latitude/longitude reference for Eyre Peninsula.

THINK	**WRITE**
a. 1. Locate the Great Sandy Desert on the map.	**a.**
2. Follow the grid up vertically to discover the letter reference.	The Great Sandy Desert is in column C.
3. Follow the grid across horizontally to discover the number reference.	The Great Sandy Desert is in row 3.
4. Write the grid reference.	The grid reference for the Great Sandy Desert is C3.
b. 1. Locate the Eyre Peninsula on the map.	**b.**
2. Follow the grid across horizontally to read the latitude reference.	Eyre Peninsula has a latitude of approximately 35°S.
3. Follow the grid down vertically to read the longitude reference.	Eyre Peninsula has a longitude of approximately 135°E.
4. Write the reference (latitude first).	Eyre Peninsula is found at 35°S, 135°E.

Direction on a map is always measured in degrees from north in a clockwise direction; this is a **bearing**. An orienteering compass or protractor can be used with a map to determine the bearing needed to travel from one location to another.

Bearings are given as either:
- true bearings, which are measured clockwise from north and always have three digits
- compass bearings, which are measured from either north or south.

In the diagram shown:
- B is on a bearing of 065° from A
- B is N 65° E of A.

WORKED EXAMPLE 25 Determining true bearing

Determine the true bearing needed to go from Perth to Fremantle using the map below.

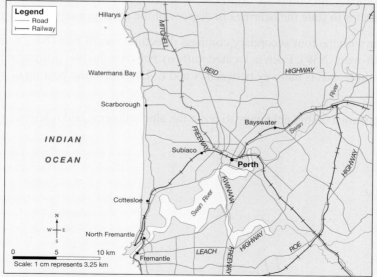

Source: Spatial Vision

THINK

1. Draw a line from Perth to Fremantle.
2. Draw a line north from Perth.
3. Place the protractor so that its origin is on Perth and the baseline is parallel to north, as shown on the map.

WRITE

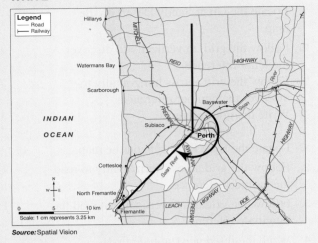

Source: Spatial Vision

4. Read the degrees for the line from Perth to Fremantle and record the bearing.

The bearing from Perth to Fremantle is 225°T.

16.9 Exercise

Students, these questions are even better in jacPLUS

 Receive immediate feedback and access sample responses

 Access additional questions

 Track your results and progress

Find all this and MORE in jacPLUS ▶

1. Explain why it is important to be able to follow directions.

2. The four primary compass directions are north (N), south (S), east (E) and west (W). Knowing that north lies at 0°, use your protractor to state the bearings in degrees of the other primary directions.

3. You may be familiar with the four secondary compass directions: north-east (NE), south-east (SE), south-west (SW) and north-west (NW). Each is located *halfway* between the two primary directions that make up its name; for example, NE is halfway between N and E. Determine the bearing of each of the secondary compass directions.

4. **WE24** Use the map shown on the next page to write the alphanumeric grid reference of each of the following places.
   a. Cape York
   b. Fraser Island
   c. Mount Kosciuszko
   d. Tasmania
   e. Arnham Land
   f. Gibson Desert

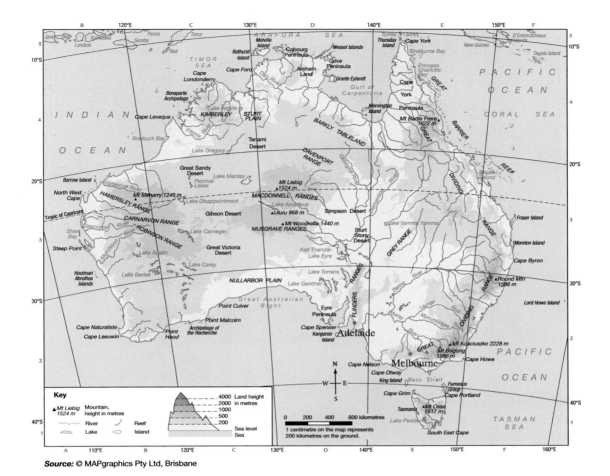

Source: © MAPgraphics Pty Ltd, Brisbane

5. Write the latitudes and longitudes of the places in question **4**.

6. **WE25** Determine whether Steep Point or Moreton Island is further north.

7. Consider the map on the next page.
   a. Determine a compass bearing to hike from Beauty Point to Beaconsfield, taking the centre of 'B' as the reference point.
   b. Determine how many kilometres separate the two places.
   c. Determine the true bearing and distance from Beaconsfield to Beauty Point.

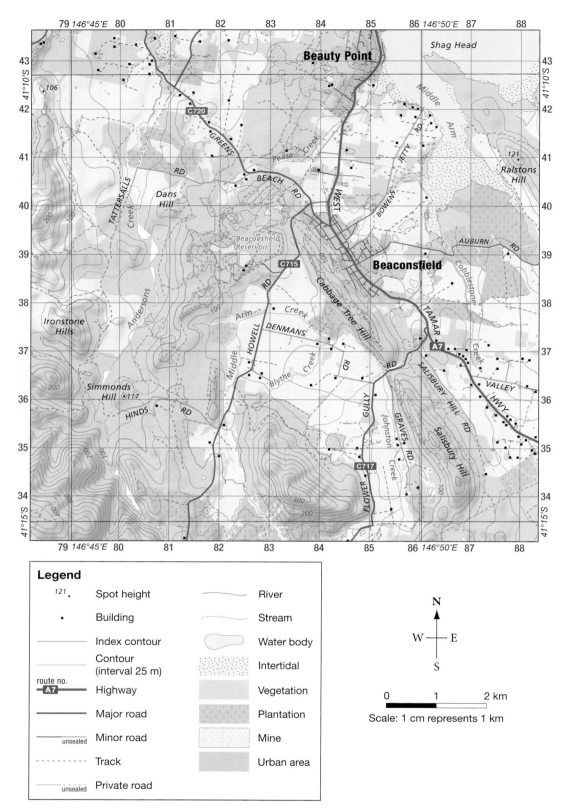

Legend

121 •	Spot height	～～	River
•	Building	～～	Stream
——	Index contour	⬭	Water body
⋯⋯	Contour (interval 25 m)	⠿	Intertidal
route no. **A7**	Highway	▨	Vegetation
——	Major road	▦	Plantation
—— unsealed	Minor road	⬭	Mine
-----	Track	▨	Urban area
—— unsealed	Private road		

N
W — E
S

0   1   2 km
Scale: 1 cm represents 1 km

Source: Address Points, Coastline, Contours, Hydrographic Areas, Hydrographic Lines, Spot Heights, Tasmania 25m DEM, TASVEG 3.0, Transport Nodes, Transport Segments from www.theLIST.tas.gov.au © State of Tasmania

8. Using the map shown in Question **7**, name the places located near the following coordinates.

   a. 41°12.7′S, 146°45′E　　　　　　　　　　　b. 41°11′S, 146°51.5′E

9. Using the map shown in Question **7**, write the latitude and longitude for each of the following places.

   a. Shag Head　　　　　　　　　　　　　b. Beaconsfield

10. Using the map shown in Question **7**, calculate the true bearings of each of the following places from Beaconsfield.

   a. Simmonds Hill　　　　　　　　　　　b. Ralstone Hill

11. a. Calculate the return bearings for each of the places in question **10**.
    b. From the pattern formed, write a sentence about how to find return bearings *without* a protractor if you know the original bearing.

12. Use Google Maps to find the directions from your school to the MCG. If you take a straight-line path from your school to the MCG, estimate:

   a. the true bearing direction
   b. the compass-bearing direction.

16.10 Speed

16.10.1 Calculating speed

Average speed tells us how fast we are going by dividing the distance we travel by the time it takes to travel that distance.

$$\text{speed} = \frac{\text{distance}}{\text{time}}$$

The most common unit for speed is metres per second (m/s). The speeds of different objects vary, so, for example, the speed of a snail is more appropriately measured using different units to those used to measure the speed of a car. The speed of a snail could be measured in millimetres per second (mm/s), whereas the speed of a car could be measured in kilometres per hour (km/h).

tlvd-6029

WORKED EXAMPLE 26 Calculating average speed

Calculate the average speed of a car that covers 400 m in 4.25 seconds, in:

a. **m/s**
b. **km/h.**

THINK	WRITE
a. 1. Determine the distance and time in the correct units.	**a.** Distance $= 400\,\text{m}$ Time $= 4.25\,\text{s}$
2. Substitute values into the formula: $$\text{speed} = \frac{\text{distance}}{\text{time}}$$	$$\text{Speed} = \frac{\text{distance}}{\text{time}}$$ $$\text{Speed} = \frac{400}{4.25}$$ $$= 94.12\,\text{m/s}$$
b. 1. Determine the distance and time in the correct units.	**b.** Distance $= 400\,\text{m}$ $\quad\quad\quad\quad = 0.4\,\text{km}$ Time $= 4.25\,\text{s}$ $$= \frac{4.25}{60 \times 60}$$ $$= 1.18 \times 10^{-3}\,\text{hours}$$
2. Substitute values into the formula: $$\text{speed} = \frac{\text{distance}}{\text{time}}$$	$$\text{Speed} = \frac{\text{distance}}{\text{time}}$$ $$\text{Speed} = \frac{0.4}{1.18 \times 10^{-3}}$$ $$= 338.82\,\text{km/h}$$

16.10.2 Calculating distance

The speed equation can be rearranged to make distance the subject, allowing you to calculate the distance covered given the speed and time of a journey.

$$\textbf{distance} = \textbf{speed} \times \textbf{time}$$

tlvd-6030

WORKED EXAMPLE 27 Calculating distance

A car travels at a constant speed of 60 km/h for 1.5 hours. Calculate the distance covered by the car in kilometres.

THINK	WRITE
1. To calculate the distance, we need to know the speed and time in the correct units.	Speed $= 60\,\text{km/h}$ Time $= 1.5\,\text{hours}$
2. Substitute into the distance equation.	Distance $=$ speed $\times$ time $\quad\quad\quad\quad = 60 \times 1.5$ $\quad\quad\quad\quad = 90\,\text{km}$
3. Understand the question.	Since the car travels at 60 km/h, it covers 60 km in 1 hour, and therefore 30 km in half an hour, which is 90 km in 1.5 hours.
4. Write the answer.	The distance covered is 90 km.

16.10.3 Calculating time

The speed equation can also be rearranged to make time the subject, allowing you to calculate the time taken to cover a given distance at a certain speed.

$$time = \frac{distance}{speed}$$

WORKED EXAMPLE 28 Calculating travel time

Calculate the time in hours that a person would take to walk 3 km at an average speed of 1.8 m/s.

THINK	WRITE
1. To calculate the time we need to know the distance and speed in appropriate units. We will use m/s and m and convert to hours at the end.	Distance = 3 km = 3000 m Speed = 1.8 m/s
2. Substitute into the equation $time = \dfrac{distance}{speed}$.	$Time = \dfrac{distance}{speed}$ $= \dfrac{3000}{1.8}$ $= 1666.67$ seconds
3. Convert seconds to hours by dividing by 60 (to convert to minutes) and then divide by 60 again (to convert to hours).	$\dfrac{1666.67}{60 \times 60} = 0.46 \text{ hour}$ $\approx 28 \text{ minutes}$
4. Write the answer.	The time taken would be 0.46 hours.

16.10 Exercise

Students, these questions are even better in jacPLUS

 Receive immediate feedback and access sample responses

 Access additional questions

 Track your results and progress

Find all this and MORE in jacPLUS

1. **WE26** Sally walks around her 2.5-km block in 1050 seconds. Calculate her average speed in:
   a. m/s
   b. km/h.

2. Calculate the average speed in km/h of a track cyclist who covered 1000 m in 1 minute and 25 seconds.

3. Select the appropriate unit for speed (mm/s, m/s or km/h) to measure the following:
   a. A sprinter running a 100 m race
   b. A plane flying
   c. The increasing snow depth
   d. A surfer riding a wave

4. Calculate the average speed (in m/s) of a skateboarder who travelled 50 m in 7 seconds.

5. Calculate the average speed (in km/h) of a boat that covered 2.6 km in 15 minutes.

6. **WE27** A car travels at a constant speed of 80 km/h for 3.25 hours. Calculate the distance covered by the car in kilometres.

7. A truck travels at an average speed of 65 km/h for 335 minutes. Calculate the distance covered by the truck in:
   a. kilometres
   b. metres.

8. A kite surfer was moving at a constant speed of 11 m/s for 24 seconds. Calculate the distance (in metres) they covered over this time.

9. Callum played a round of golf, averaging 5 shots per hole and a speed of 0.3 m/s. If it took him 4 hours to play his round of golf, calculate how far he walked during his round.

10. **WE28** Calculate the time in hours it would take a person to walk 10 km at an average speed of 1.6 m/s.

11. If Ahmed ran at an average speed of 5 m/s for a half marathon (21.1 km), calculate how long it would take to complete the half marathon in hours and to the nearest minute.

12. Haile Gebrselassie ran the Berlin marathon with an average speed of 5.67 m/s. Calculate how long it took him to finish the marathon, given a marathon is 42.2 km, in:

  a. seconds
  b. hours, minutes and seconds (to the nearest second).

13. A cheetah can reach a maximum speed of 112 km/h. However, they are only able to maintain this speed for a short period of time. If a cheetah ran at its maximum speed and covered a distance of 400 m, calculate how long it maintained its maximum speed for, in seconds.

14. Pigeons can cover vast distances to find their way home. If a pigeon covered 56.8 km in 1 hour and 45 minutes, calculate:

  a. its average speed in km/h
  b. its average speed in m/s.

15. If a horse averages a speed of 25.6 km/h for 17 minutes and 36 seconds, calculate how far the horse travelled, to 2 decimal places, in:

  a. kilometres
  b. metres.

16. When Makybe Diva won her third Melbourne Cup, she averaged a speed of 57.9 km/h for the 3200-m race. Calculate her winning time in:

  a. minutes, to 2 decimal places
  b. minutes and seconds (to the nearest second).

17. The speed of light is 3×10^8 m/s. Research the average distance from Earth to the following planets to calculate the time it would take light to get there from Earth.
   Note: Use the average distance since it does vary.

   a. Saturn
   b. Mercury
   c. Jupiter
   d. Mars

18. The following questions relate to reaction time of a driver before breaking.

   a. If a car was moving at 60 km/h and the driver had a reaction time of 1.3 seconds before applying the brakes, calculate how far the car travelled before the brakes were applied.
   b. If a driver had a reaction time of 1.5 seconds and the car travelled 30 m before the brakes were applied, calculate the speed at which the car was moving.
   c. If the car moved 42 m before the brakes were applied, calculate the reaction time of the driver if the car was moving at 95 km/h.

16.11 Travel times

LEARNING INTENTION

At the end of this subtopic you should be able to:
- estimate the distance, average speed and time of the journey from the map
- determine speed, distance covered and time of travel.

16.11.1 Time of journeys

We have learnt how to estimate distances travelled via a map; however, if you are travelling somewhere, you are often more interested in how long it will take to complete your journey. This is what a global positioning system (GPS) does by estimating the distance of the trip, calculating the average speed of the trip depending on the roads travelled, and using this information to estimate the time the trip will take. Remember:

$$\text{time} = \frac{\text{distance}}{\text{speed}}$$

Determine the journey time

To determine the time a journey takes:
- **estimate the distance of the journey from the map**
- **estimate the average speed of the journey**
- **use these estimates to calculate time of the journey, using the formula time $= \dfrac{\text{distance}}{\text{speed}}$.**

WORKED EXAMPLE 29 Estimating travel time from the map

Shane and Ravi went for a swim at Bondi Beach in Sydney before planning to go to the cricket at the SCG. Calculate how much time they need to get to the SCG, given the map of their journey shown and their estimate that they can average 50 km/h in the traffic.

Source: N.S.W.Department of Finace

THINK

1. Using the scale on the map, estimate the distance. Measure the distance that represents 500 m and count how many lots of 500 m there are in the journey.

2. The question tells you they estimate their speed as 50 km/h.

3. Use the formula to calculate the time.

4. Convert to minutes and seconds by multiplying the decimal by 60.

5. Write the answer.

WRITE

There appear to be 12 lots of 500 m.
$$\text{Distance} = 12 \times 500 \,\text{m}$$
$$= 6000 \,\text{m or } 6 \,\text{km}$$

$$\text{Speed} = 50 \,\text{km/h}$$

$$\text{Time} = \frac{\text{distance}}{\text{speed}}$$
$$= \frac{6}{50}$$
$$= 0.12 \,\text{hours}$$

$$\text{Time} = 0.12 \times 60$$
$$= 7.2 \,\text{minutes}$$
$$= 7 \,\text{minutes } 0.2 \times 60 \,\text{minutes}$$
$$= 7 \,\text{minutes } 12 \,\text{seconds}$$

The estimated time for the trip from Bondi Beach to the SCG is 7 minutes and 12 seconds.

16.11.2 Speed, distance and time of travel

Speed of motion plays a part in our everyday lives, whether it is how quickly we react on a sporting field, estimating the time required to travel to school, or knowing who the fastest runner in the school is.

To calculate speed, distance and time recall the equations shown:

$$\text{speed} = \frac{\text{distance}}{\text{time}}$$

$$\text{time} = \frac{\text{distance}}{\text{speed}}$$

$$\text{distance} = \text{speed} \times \text{time}$$

Simple ways to convert speed

- To convert from km/h to m/s, divide by 3.6.
- To convert from m/s to km/h, multiply by 3.6.

WORKED EXAMPLE 30 Calculating travel time and converting speed

The distance from home to school is 7.5 km. It takes twice as long to get home as it does to get to school due to traffic congestion. If it takes 10 minutes to get to school, calculate:

a. how long it takes to get home from school
b. how long the total trip to and from school takes
c. the average speed of the total trip to and from school in:
  i. km/min
  ii. km/h.

THINK	WRITE
a. It takes twice as long to get home from school compared to going in the morning.	a. $\text{Home from school} = 2 \times 10 \text{ minutes}$ $= 20 \text{ minutes}$
b. Add the two times together.	b. $\text{To school} = 10 \text{ minutes}$ $\text{Total time} = 10 + 20$ $= 30 \text{ minutes}$
c. i. To calculate the average speed of the total trip, we need to use the total distance of 15 km and the total time of 30 minutes.	c. i. $\text{Speed} = \dfrac{\text{distance}}{\text{time}}$ $= \dfrac{15}{30}$ $= 0.5 \text{ km/min}$
ii. The total distance is 15 km and time is 30 minutes (0.5 hour).	ii. $\text{Speed} = \dfrac{\text{distance}}{\text{time}}$ $= \dfrac{15}{0.5}$ $= 30 \text{ km/h}$

16.11 Exercise

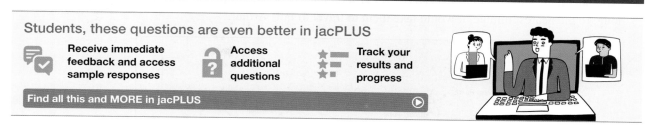

Students, these questions are even better in jacPLUS

- **Receive immediate feedback and access sample responses**
- **Access additional questions**
- **Track your results and progress**

Find all this and MORE in jacPLUS

1. **WE29** If the average speed travelling from Darwin to Uluru is 85 km/h, use the map shown to estimate the travel time of the journey.

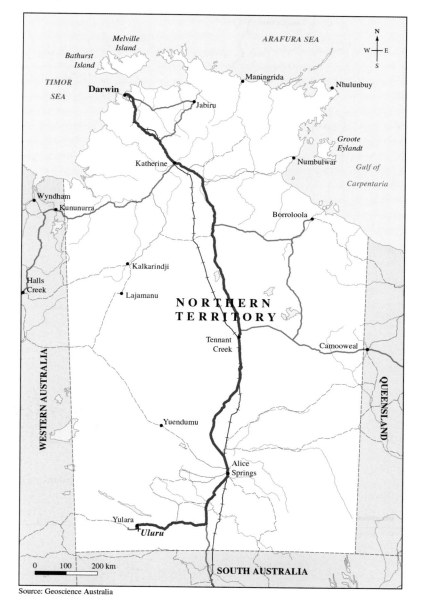

Source: Geoscience Australia

2. A family flew into Launceston for a holiday. They decided to visit Port Arthur using a hire car. They estimated they could average a speed of 70 km/h on the trip down. Using the map shown, calculate how long, to the nearest minute, it would take them to travel to Port Arthur.

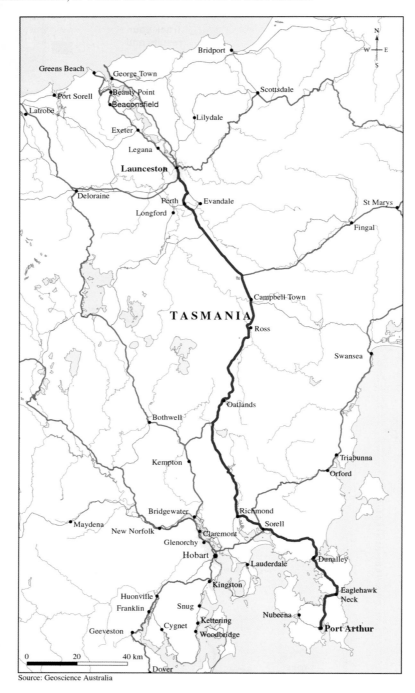

Source: Geoscience Australia

3. Janet walked 4000 m at an average speed of 2 m/s. Calculate the time, in seconds, it took her to complete her walk.

4. Pedro went for a 7-km run at an average speed of 3.5 m/s. Calculate the time, in seconds, it took him to complete his run.

5. Clancy rode her horse along the beach for 2.5 km at an average speed of 10 m/s. Calculate the time, in minutes and seconds, it took her to cover the 2.5 km.

6. Cyril needs to drive from Perth to Geraldton, which is 434 km away by road. Taking traffic into account, Cyril estimates that he can average 82 km/h. Calculate how long the trip will take in hours, minutes and seconds.

7. Dermott decides to get up early and drive from his Melbourne home to Bell's Beach to go surfing. Estimate how long it will take him to get there if he averages 75 km/h, referring to the map shown.

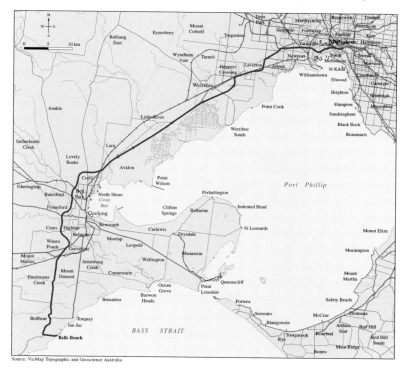

8. A family is going for a holiday in Tasmania and decides to travel on the Spirit of Tasmania. The map shows the journey it takes. It travels at an average speed of 25 knots, which is equivalent to 50 km/h. Calculate how much time the trip from Station Pier in Port Melbourne to Devonport in Tasmania will take.

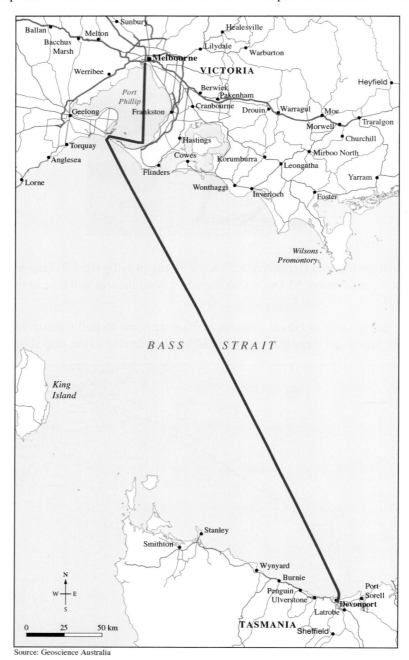

Source: Geoscience Australia

9. Johnny is looking forward to going to the AFL Grand Final at the MCG. He decides to drive to the MCG from his house in Frankston.

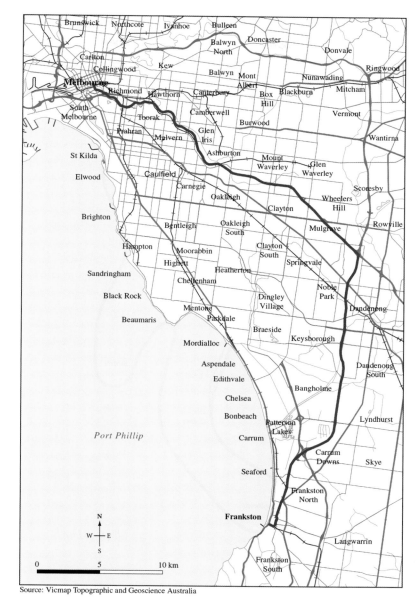

Source: Vicmap Topographic and Geoscience Australia

a. Taking traffic into consideration, Johnny estimates he will average 55 km/h for the journey. Calculate how long it will take him to get to the MCG given the map shown.
b. Johnny knows is costs him $1.55 per 10 km to drive his car. Determine how much the trip to the MCG will cost him.

10. The map below shows the Phillip Island Grand Prix race track. Use this map with its scale to answer the following.

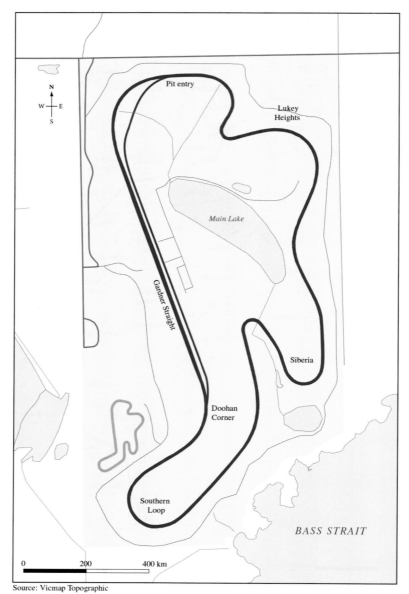

Source: Vicmap Topographic

a. Estimate the distance of one lap of the track.
b. If Casey Stoner could maintain his fastest lap time of 1 minute 30 seconds, calculate how long it would take him to complete the 30 lap MotoGP race.
c. If each lap of the track is 4 km, calculate Casey Stoner's average speed when he completed his fastest lap:

  i. in m/s
  ii. in km/h.

11. **WE30** The distance from home to netball training is 6 km and it takes three quarters of the time to return to home as it does to get there. If it takes 8 minutes to get to netball training, calculate:

    a. how long it takes to get home from netball training
    b. the time it takes to make the total trip to and from netball training
    c. the average speed of the total trip to and from netball training in:
        i. km/min
        ii. km/h.

12. Jane went for a 6.8-km hike that took her 1 hour and 3 minutes. Calculate Jane's average speed during the hike in m/s.

13. Calculate the average speed, in the units shown in brackets, in the following situations.

    a. The distance covered was 100 m in 25 seconds (m/s).
    b. The distance covered was 4.8 km in 2 minutes (m/s).
    c. The distance covered was 3800 m in 2 hours (km/h).
    d. The distance covered was 14.6 km in 400 minutes (km/h).

14. Calculate the distance covered, in the units shown in the brackets, in the following situations.

    a. Travelling at 10 m/s for 250 seconds (m)
    b. Travelling at 60 km/h for 3 hours (km)
    c. Travelling at 25 m/s for 45 minutes (m)
    d. Travelling at 100 km/h for 2 hours and 45 minutes (km)

15. Covert the following to m/s, to 1 decimal place.

    a. 50 km/h
    b. 60 km/h
    c. 80 km/h
    d. 100 km/h

16. Maurice flew from Melbourne to the Gold Coast, which took 2 hours and 6 minutes. The flight back home took 8 minutes longer due to wind conditions. Calculate the average speed of the plane for the entire trip given the distance from Melbourne to the Gold Coast is 1345 km.

17. Kelly and her friend Kerry drove to an Ed Sheeran concert. It took them 50 minutes to get to the concert. On the way back home, due to congestion in the car park, it took them 1.5 times longer to get home.
The distance from their house to the concert venue is 50 km.
Calculate:

    a. how long it took them to get home from the concert
    b. how long the total trip took, in hours and minutes
    c. their average speed for the entire trip, in km/h.

18. Brian intends to take the family on holiday with the caravan to watch Bathurst (a car race). The petrol for his car costs $1.60 per 10 km when he is not towing; however, when he tows, the cost is 1.5 times the original cost. The distance from their house in Wagga Wagga to Bathurst is 315 km. Determine how much it would cost to tow the caravan to Bathurst and back home to Wagga Wagga.

19. A car travelling at 60 km/h in the dry has a braking distance of 20 m. However, if the road is wet, its braking distance is 1.4 times that on a dry road. Given a car travels 25 m during the driver's reaction time, determine the total distance covered before the car comes to rest from when the driver initially sees the obstacle in front of them, given they are travelling at 60 km/h in the wet.

20. Robbie rides his bike to and from school each day for a week and times how long it takes. The distance to school from home is 6 km and his times are shown in the table.

Day	Time to school	Time from school
Monday	15 min 26	14 min 24
Tuesday	14 min 45	15 min 09
Wednesday	16 min 02	15 min 03
Thursday	15 min 13	14 min 44
Friday	14 min 38	14 min 08

Determine Robbie's average speed for the week.

21. A driver is travelling at 70 km/h when they see a kangaroo jump out in front of them. They take 1.5 seconds to react before braking and, once the brakes are applied, it takes 28 m to stop.
    a. Calculate how far the car travels once the driver sees the kangaroo.
    b. If the kangaroo is 50 m in front of the car, calculate the maximum speed, to the nearest km/h, they could be travelling at so they stop before the kangaroo.

22. On wet roads bald tyres are unsafe due to the increase in stopping distances. If a car is travelling at 80 km/h, on good tyres and dry roads, its braking distance is 35 m. Given the driver has a reaction time of 1.4 seconds before braking and it was a wet day with bald tyres, the braking distance increased by a further 70% compared to a dry day with good tyres.

   a. Determine how far the car travelled before the brakes were applied, in metres to 1 decimal place.
   b. Determine the braking distance, in metres to 1 decimal place.
   c. Calculate the total distance travelled in the braking process, to 1 decimal place.

16.12 Review

16.12 Exercise

Multiple choice

1. **MC** The value of x in the diagram is:
   A. $36°$
   B. $90°$
   C. $66°$
   D. $54°$
   E. $30°$

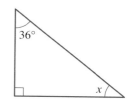

2. **MC** The value of x in the diagram is:
   A. $117°$
   B. $63°$
   C. $56°$
   D. $61°$
   E. $119°$

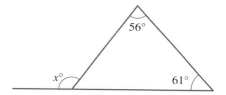

3. **MC** If Mark Webber completed one lap of the 5.303 km Australian Grand Prix circuit in 1 minute and 29.5 seconds, his average speed in m/s is closest to:
   A. 4.1 m/s.
   B. 59.3 m/s.
   C. 41.0 m/s.
   D. 55.5 m/s.
   E. 58.0 m/s.

4. **MC** The value of x in the diagram is:
   A. $95°$
   B. $65°$
   C. $75°$
   D. $70°$
   E. $60°$

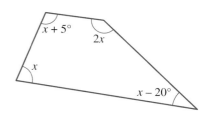

5. **MC** A student leaves home at 7 : 32 am and takes 43 minutes to walk to school. The time the student arrives at school is:

   A. 9:15 am   B. 8:05 am   C. 8:15 am   D. 7:55 am   E. 9:05 am

6. **MC** For a polygon of 9 sides, the sum of the interior angles equals:

   A. 1080°   B. 1620°   C. 1440°   D. 1260°   E. 1060°

7. **MC** The matching ratio to the scale 1 cm to 50 m is:

   A. 1 : 50   B. 1 : 500   C. 1 : 5000   D. 1 : 50 000   E. 1 : 5

8. **MC** The conversion of 5.6 m using the scale 1 : 200 000 is:

   A. 5.6 km   B. 112 km   C. 1120 km   D. 112 000 m   E. 112 000 cm

9. **MC** A map of Australia uses a scale of 1 cm = 130 km. If the distance between Melbourne and Sydney on the map is 6.7 cm, then the approximated actual distance is:

   A. 670 km   B. 871 km   C. 978 km   D. 515 km   E. 515 m

10. **MC** On a spring morning the temperature was 5.7°C. If the temperature reached a maximum of 23.4°C, calculate the rise in temperature.

    A. 16.7°C   B. 17.7°C   C. 17.3°C   D. 16.3°C   E. 18.7°C

Short answer

11. Use the map shown to find the following distances.

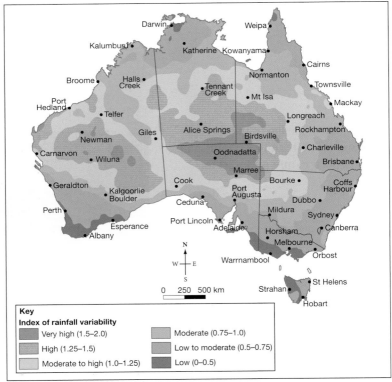

Source: MAPgraphics Pty Ltd, Brisbane

a. Melbourne to Cairns 4.3 cm
b. Perth to Darwin 4.9 cm
c. Adelaide to Brisbane 3.6 cm

12. Calculate the value of x in the following figure.

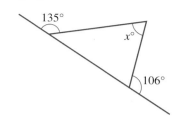

13. Britney travels to school by various means. Calculate the time she will arrive at school, if she leaves home at 7:48 am and travels by the following means:

  a. Walking for 1 hour and 5 minutes
  b. Cycling for 37 minutes
  c. Riding in a car for 17 minutes
  d. Taking a bus for 25 minutes

14. Calculate the values of missing angles in the following shapes.

  a.
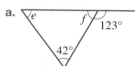

  b.
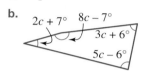

15. Calculate the values of the missing angles in the following polygons.

  a.
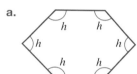

  b.
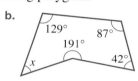

Extended response

16. Use this diagram to answer the following questions.

  a. State which angle is alternate to f.
  b. State which angle is the corresponding angle to d.
  c. If $b = 71$, write the values of all the other pronumerals.

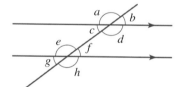

17. Use a ruler and protractor to construct triangles with the following features.

  a. Edge lengths of 3 cm, 4 cm and 5 cm
  b. Edge lengths of 6 cm, 4 cm and an angle of 45° between them
  c. Angles of 32°, 85° and an edge length of 2 cm between them

18. The following table shows the schedule for Chung who left for work at 6:45 am from his home in Melbourne.

  a. State at what time Chung's meeting started.
  b. State at what time he started lunch.
  c. State at what time he flew back home.
  d. Determine the 24-hour time he arrived back in Melbourne.
  e. Determine the 12-hour time he arrived back home.
  f. Determine how long Chung was away from home.

Activity	Time
Drive to the airport	48 minutes
Wait at airport	36 minutes
Fly to Adelaide	1 hour 4 minutes
Taxi to meeting	27 minutes
Meeting time	2 hours 38 minutes
Lunch	1 hour 22 minutes
Taxi to airport	23 minutes
Wait at airport	1 hour 31 minutes
Fly to Melbourne	57 minutes
Drive home	34 minutes

19. Determine the length of a 5-cm line on a scale diagram, if the scale is:

   a. 1 : 100   **b.** 1 : 50 000   **c.** 1 : 250 000.

20. Use the map from subtopic 16.8.1 (shown below) to help you answer the following questions.

   a. Calculate is the map distance between Phnom Penh and Melbourne.
   b. Using the line scale, calculate the real distance in part **a**.
   c. Determine the latitude and longitude of Honiara.
   d. Determine which country is located at 3°N 110°E.
   e. State the bearing of Nauru *from* Melbourne.
   f. State the bearing of Nauru *to* Melbourne.

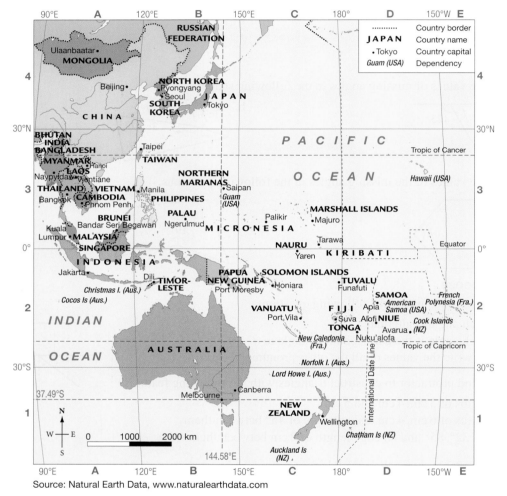

Source: Natural Earth Data, www.naturalearthdata.com

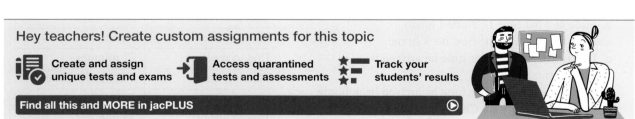

Answers

Topic 16 Space, time and measurement

16.2 Conventions and geometric shapes

16.2 Exercise

1. 12

2. 6

3.

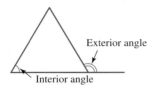

An interior angle is inside the triangle; an exterior angle is outside the triangle.

4. a. $m = 65°$ b. $x = 45°, t = 90°$
   c. $y = 100°, x = 30°$ d. $b = 34°, c = 73°, a = 73°$

5. a. $x = 20°$ b. $x = 18°$
   c. $x = 26°$ d. $x = 80°, y = 20°$

6. a. Equilateral triangle   b. $60°$
   c. m is supplementary to $30°$ so $m = 150°$.

7. a.   i. $a = 85°, b = 50°, c = 45°$
      ii. $d = 140°, e = 110°, f = 110°$
      iii. $g = 90°, h = 110°, i = 70°$
      iv. $j = 100°, k = 100°, l = 130°, m = 130°$
   b. $360°, 360°, 360°, 360°$ c. They are all the same.

8. A quadrilateral can be divided into two triangles by drawing a line from one vertex to the opposite vertex. Therefore, the sum of the internal angles of a quadrilateral is equal to twice the sum of the internal angles of a triangle.

9. a. $x = 145°$ b. $t = 174°$ c. $m = 66°$ d. $q = 88°$

10. $x = 107°$

11. a. $x = 45°$ b. $x = 25.71°$
    c. $x = 79°$ d. $x = 31.2°$

12. a. $90°, 90°, 45°, 135°$
    b. $25.71°, 102.86°, 102.86°, 128.57°$
    c. $90°, 73°, 85°, 112°$
    d. $31.2°, 92.4°, 111.6°, 124.8°$

13. $50°, 70°, 110°, 130°$

14. a. Answers will vary, but should be along the lines of the following.
    Every quadrilateral has an interior angle sum of $360°$. Therefore, if there is one angle greater than $180°$, there cannot be another angle greater than $180°$, because this would lead to an interior angle sum greater than $360°$.
    b. Answers will vary, but should be along the lines of the following.

A quadrilateral must have an interior angle sum of $360°$, and every quadrilateral has four sides and four angles. If every angle is to be an integer, we make the other three angles as small as possible ($1°$). This makes the final angle $360° - 3 \times 1°$, or $357°$.

16.3 Angles and polygons

16.3 Exercise

1. The difference between a regular polygon and an irregular polygon is that an irregular polygon has edges of different lengths and their internal angles are not all the same.

2. a. $540°$ b. $720°$ c. $900°$
   d. $720°$ e. $720°$ f. $1080°$

3. a. i. $b = 110°$ ii. $110°$
   b. i. $c = 134°$ ii. $134°$
   c. i. $d = 24°$ ii. $24°$
   d. i. $h = 85°$ ii. $85°$
   e. i. $x = 25°$
      ii. $3x - 6° = 69°; 6x - 10° = 140°; 3x + 11° = 86°$;
      $5x = 125°; 8x + 10° = 210°; 90°$
   f. i. $x = 19°$
      ii. $6x + 6° = 120°; 6x - 25° = 89°; 15x - 25° = 260°$;
      $3x + 3° = 60°; 5x - 9° = 86°; 4x - 14° = 62°$;
      $14x + 2° = 268°; 7x + 2° = 135°$

4. $x = 60°; 2x - 20° = 100°; x + 20° = 80°$

5. $m = 120°$

6. a. Regular   b. Not regular   c. Regular
   d. Regular   e. Not regular   f. Regular

7. Sum of internal angles $= 180°(n - 2)$, where $n =$ number of sides of the polygon

8. a. $90°$ b. $60°$ c. $120°$ d. $135°$

9. a. Answers will vary. For example

b. $90°, 90°, 67.5°, 67.5°, 202.5°, 202.5°$

10. Answers will vary.
    See the image at the bottom of the page.*

---

*10.

Tetrahedron   Cube   Octahedron   Dodecahedron   Icosahedron

16.4 Angles, parallel lines and transversals

16.4 Exercise

1. a. D **b.** A **c.** B **d.** C

2. Answers will vary.

3. a. Man and dog

   b. Dog and milk bar; bike and ambulance

   c. Man and ambulance; traffic light and bike; dog and basketball court

   d. Man and bike

4. a. a **b.** q **c.** o **d.** m

5. a. The following pairs of angles are supplementary and so sum to 180°: f and c; e and d; n and m; d and c; e and f. The following pairs of angles are co-interior and so sum to 180°: f and b; n and q.

   b. c and e; d and f; a and b; k and l

   c. b and c; l and m

6. Sample responses can be found in the worked solutions in the online resources.

7. a. i. Corresponding angles

     ii. $p = 50°$

   b. i. Alternate angles

     ii. $q = 48°$

   c. i. Vertically opposite angles

     ii. $s = 60°$

   d. i. Co-interior angles

     ii. $m = 120°$

8. a. $p = 118°, q = 62°, r = 71°, s = 47°$

   b. $w = 129°, x = 129°, y = 47°, z = 82°$

   c. $a = 42°, b = 42°, c = 91.5°, d = 91.5°, e = 42°$

   d. $a = 67.2°, b = 44°, c = 68.8°, d = 111.2°, e = 68.8°$

9. $e = 58°$

10. $p = 137°$

11. $x = 58°$

12. a. $x = 2°$; $8x + 16° = 32°$; $12x + 8° = 32°$

   b. $x = 20°$; $4x + 42° = 122°$; $2x + 18° = 58°$

13. a. Parallel **b.** Not parallel

   c. Not parallel **d.** Parallel

14. a. $a = 52°$. a is co-interior with 128°; also, a, 73°, and 55° form a straight line.

   b. KL is parallel to MN because the co-interior angles sum to 180°.

   c. 52°, 55°, 125°

15. $a = 94°; b = 94°; c = 94°; d = 86°; e = 94°; f = 86°;$
$g = 86°; h = 94°; i = 94°; j = 86°; n = 108°; m = 72°;$
$o = 108°; k = 72°; l = 108°$
Reasons for answers will vary.

16.5 Constructing triangles

16.5 Exercise

1. The minimum requirements needed to draw a triangle are one of the following:
- Three edge lengths
- Two adjacent angles and the edge length between them
- Two adjacent edges and the angle between them

2. Answers will vary. Sample responses are provided.

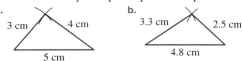

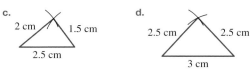

3.

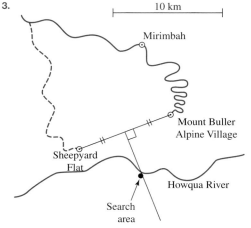

4. a.

   c.

   e.

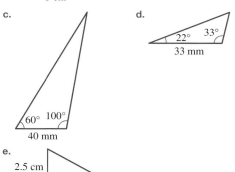

5.

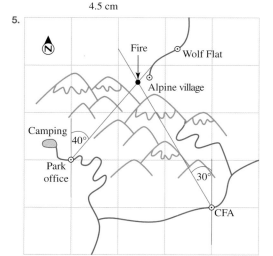

6. a.

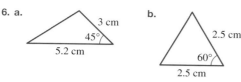

b.

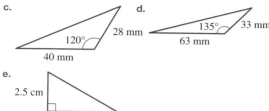

c. **d.**

e.

16.6 Using scales

16.6 Exercise

1. A scale may be written as:
   - a ratio of length on a plan to the actual length
   - a statement.

2. Answers may vary. Examples are given.
   a. 1 mm to 1 cm
   b. 1 cm to 1 m
   c. 1 mm to 1 m
   d. 1 cm to 1 km
   e. 1 mm to 1.5 m
   f. 1 mm to 3.3 km

3. a. 1 : 50 000
   b. 1 : 200 000
   c. 1 : 5 000 000
   d. 1 : 100 000
   e. 1 : 200 000

4. They are not the same because 1 : 6 is an enlargement, whereas 6 : 1 is a reduction.

5. a. Different. A dilation factor of $\frac{1}{50}$ is a reduction by a factor of 50, whereas 1 : 50 is an enlargement by a factor of 50.
   b. Same. 3 : 450 can be simplified to 1 : 150 by dividing each side of the ratio by 3.

6. $6 \times 3 \times 2.5$ cm

7. 60×85 cm

8. a. 10 mm by 12 mm
   b. 18 mm
   c. Seat is 3 mm high, 2.3 mm deep and 2.5 mm wide; top of the backrest is 5.3 mm high.

9. a. 4 cm
   b. 3.75 m
   c. 22.5 cm
   d. 23 mm in the photograph; 1.725 m in real life

10. Answers will vary.
    a. 1 : 10
    b. 1 : 40
    c. 1 : 6000

11. a. 4 m
    b. 40 mm
    c. 1 : 100
    d. 1.3 m by 0.53 m

12. a. 12 m
    b. 5 mm on the plan and 1 m in real life
    c. Lounge: $4 \, m \times 6 \, m$
       Kitchen/living: $8 \, m \times 2 \, m$
       Bedroom 1 : $4 \, m \times 3 \, m$
       Bedroom 2 : $3 \, m \times 3 \, m$
       Bedroom 3 : $3 \, m \times 3 \, m$
       Bath: $1 \, m \times 3 \, m$

Laundry/linen/WC: $3.8 \, m \times 4.2 \, m$
Garage: $7 \, m \times 3 \, m$

   d. The garage is 7 m wide and 3 m deep, so the work ute would fit but the family car would not.

13. Sample responses can be found in the worked solutions in the online resources.

14. Answers will vary depending on the size of the classroom. You may like to encourage students to compare answers, and to have some students use a different scale, such as 2 cm to 1 m. Ensure that the activity is done accurately with a ruler and pencil.

16.7 Time and temperature

16.7 Exercise

1. a. 5 hours 48 minutes
   b. 9 hours 52 minutes 30 seconds

2. a. 12 hours 9 minutes 21.6 seconds
   b. 3 hours 13 minutes 19.8 seconds

3. a. Five o'clock
   b. Approximately twenty-three minutes past 7
   c. Approximately thirteen minutes to 7 o'clock
   d. Quarter past 2
   e. Five minutes to 8
   f. Ten minutes past midnight

4. a. 1025
   b. 0733
   c. 1345
   d. 2012
   e. 2206
   f. 2345

5. 2 hours 36 minutes

6. 1 hour 25 minutes

7. a. 3:51 pm
   b. 8:22 pm
   c. 3:15 am
   d. 11:31 am
   e. 9:02 am
   f. 10:15 am

8. a. 180 minutes
   b. 315 minutes
   c. 1 440 minutes
   d. 45 minutes
   e. 198 minutes
   f. 394 minutes

9. 4 hours 15 minutes

10. 4 hours 57 minutes

11. a. 1 hour 25 minutes
    b. 8 hours 15 minutes
    c. 35 hours 10 minutes
    d. 43 hours 49 minutes
    e. 22 hours 20 minutes
    f. 8 hours 35 minutes

12. a. 1 pm
    b. 4:15 pm
    c. 9:50 am
    d. 5:15 pm
    e. 5:22 pm
    f. 11:33 am

13. a. 80 minutes or 1 hour 20 minutes
    b. 8:45 am
    c. Yes, just on time.

14. a. 20 °C
    b. 1520 °C
    c. 32.5 °C
    d. 25.8 °C

15. 150 °C

16. Answers will vary. An example is having your eye directly in line with the scale.

17. a, b, d, c

18. a. 9 °C
    b. 1 °C

19. a. 22 °C
    b. 18 °C
    c. 18 °C
    d. 25 °C

20. Sample responses can be found in the worked solutions in the online resources.

21. a. i. 37 °C
       ii. 37.2 °C
       iii. 35.5 °C
       iv. 41 °C
    b. i. 308 K
       ii. 323 K
       iii. 152 °C
       iv. 50 °C

22. See the table at the bottom of the page.*

16.8 Map distances

16.8 Exercise

1. A line scale is a scale on a map that represents the actual distance on the ground relative to the distance on the map. It is usually represented as a black-and-white bar divided into sections.

2. a. 233 km
   b. 15 cm to 100 km or 1 : 6 666 666.7
   c. 287 km

3. a. 1 cm to 125 km
   b. Approximately 237.5 km
   c. Approximately 787.5 km
   d. Approximately 450 km

4. a. 1 cm to 72 km
   b. You should not walk back to camp because in real life the distance is 57.6 km.
   c. You would not be able to see the other side of the lake because in real life the lake is 21.6 km wide.

5. a. 1 : 500 000
   b. This advertising is not accurate because in real life the distance is 2 km.
   c. 18 mm
   d. You could visit the park for a day trip because in real life the distance is 120 km.
   e. 10 km

6. a. 88.4 km
   b. Begin at Mildura, travel southeast through Red Cliffs, Carwarp, Nowingi, Hattah, Ouyen, Woomelang, Birchip and Wycheproof, then south to the Avoca River and east to Donald. The total map distance is approximately 15 cm.

7. a. 8 mm : 500 km = 1 : 62 500 000
   b. Brisbane and Perth
   c. Approximately 3812.5 km
   d. 7.625 cm

8. a. i. 24 mm   ii. 62 mm   iii. 15 mm
   b. i. Approximately 666.7 km
      ii. Approximately 1722.2 km
      iii. Approximately 416.7 km

16.9 Directions

16.9 Exercise

1. To find a location or to get from one location to another.
2. N = 0°; S = 180°; E = 90°; W = 270°

3. NE = 45°; SE = 135°; SW = 225°; NW = 315°
4. a. E4 11°S 143°E   b. F3 25°S 154°E
   c. E2 36°S 148°E   d. E1 43°S 147°E
   e. D4 13°S 135°E   f. C3 25°S 125°E

5. The answers given are approximate.
   a. 35°S 140°E   b. 42°S 147°E
   c. 35°S 138°E   d. 13°S 132°E
   e. 17°S 145°E   f. 33°S 116°E

6. Steep Point

7. a. S 17°E   b. 5 km   c. 343°T

8. a. Ironstone Hill   b. Ralstone Hill
   c. Whittlesea   d. Torquay

9. a. 41°9.5′S 146°50.2′E
   b. 41°12′S 146°50′E

10. a. 245°T   b. 044°T

11. a. Return from Simmonds Hill 65°T; return from Ralstone Hill 224°T
    b. Add 180° (if less than 180°) or subtract 180° (if greater than 180°) from the original bearing.

12. Sample responses can be found in the worked solutions in the online resources.

16.10 Speed

16.10 Exercise

1. a. 2.38 m/s   b. 8.57 km/h

2. 42.35 km/h

3. a. m/s   b. km/h
   c. mm/s   d. m/s

4. 7.14 m/s

5. 10.4 km/h

6. 260 km

7. a. 362.917 km   b. 362 917 km

8. 264 m

9. 4320 m

10. 1 hour 44 minutes 10 seconds (1.736 hours)

11. 1 hour 10 minutes

12. a. 7442.6808 seconds
    b. 2 hours 4 minutes 3 seconds

13. 12.86 seconds

14. a. 32.46 km/h   b. 9.02 m/s

15. a. 7.509 km   b. 7509 m

16. a. 3.32 minutes
    b. 3 minutes 19 seconds

---

*22.

Location	Height above sea level (m)	Temperature at which water boils (°F)	Temperature at which water boils (°C)
Gisbourne (Victoria)	610	208.0	97.8
Falls Creek (Victoria)	1500	203.0	95.0
Mount Everest (Nepal)	8840	159.8	71.0

16.11 Travel times

16.11 Exercise

1. Approximately 23 hours
2. Approximately 3 hours
3. 2000 seconds
4. 2000 seconds
5. 4 minutes 10 seconds
6. 5 hours 17 minutes 34 seconds
7. Approximately 1 hour 20 minutes
8. Approximately 9 hours
9. a. Approximately 55 minutes
   b. $7.50
10. a. Approximately 4.3 km
    b. 45 minutes
    c. i. 44.44 m/s   ii. 160 km/h
11. a. 6 minutes
    b. 14 minutes
    c. i. 0.86 km/ min
       ii. 51.43 km/h
12. 1.80 m/ sec
13. a. 4 m/s   b. 40 m/s
    c. 1.9 km/h   d. 2.19 km/h
14. a. 2500 m   b. 180 km
    c. 67 500 m   d. 275 km
15. a. 13.9 m/s   b. 16.7 m/s
    c. 22.2 m/s   d. 27.8 m/s
16. 620.77 km/h
17. a. 75 minutes
    b. 2 hours 5 minutes
    c. 48 km/h
18. $151.20
19. 53 m
20. 6.69 m/s
21. a. 57.17 m   b. 52.8 km/h
22. a. 31.1 m   b. 59.5 m   c. 90.6 m

16.12 Review

16.12 Exercise
Multiple choice

1. D
2. A
3. B
4. C
5. C
6. D
7. C
8. C

9. B
10. B

Short answer

11. a. 2150 km   b. 2450 km   c. 1500 km
12. 61°
13. a 8:53 am   b 8:25 am   c 8:05 am
    d 8:13 am
14. a. 81°   b. 20°
15. a. 120°   b. 91°

Extended response

16. a. *c*   b. *h*
    c. $b = f = c = g = 71°$
       $a = d = e = h = 109°$

17. a.   b.

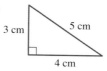

    c.

18. a. 9:40 am   b. 12:18 pm
    c. 3:34 pm   d. 1631
    e. 5:05 pm   f. 10 hours 20 minutes
19. a. 5 m   b. 2.5 km   c. 12.5 km
20. a. 4.6 cm
    b. 5750 km
    c. Approximately 10° S and 160° E
    d. Singapore
    e. Approximately 026°T
    f. Approximately 206°T

GLOSSARY

algebra a branch of mathematics that deals with number patterns and rules

algebraic expression a mathematical statement that combines pronumerals and operations but does not include an 'is equal to' symbol

allowance an additional amount of money given to workers to allow them to complete certain tasks

alternate angles angles 'inside' the parallel lines, on alternate sides of a transversal; think of them as Z-shaped

annual fee a fee charged annually in addition to interest

annual leave loading an additional payment received on top of the 44-week annual leave pay

arc a portion of the circumference of a circle

arc length the length of a portion of a circle's circumference

area the amount of flat surface enclosed by a two-dimensional shape. It is measured in square units, such as square metres, m^2, or square kilometres, km^2.

arithmetic mean one measure of the centre of a set of data. It is given by the formula:

$$\text{mean} = \frac{\text{sum of all data values}}{\text{total number of data values}}$$

When data are presented in a frequency distribution table, it is given by the formula:

$$\bar{x} = \frac{\sum(f \times x)}{n}$$

asymmetric distributions sets of data that display different shapes and different amounts of data above and below the centre

Australian Bureau of Statistics (ABS) the statistical agency of the federal government. The ABS collects data and publishes a wide range of reports for use by the governments of Australia and the community.

average speed distance travelled divided by the time taken to travel that distance

back-to-back stem-and-leaf plots a method for comparing two data distributions by attaching two sets of 'leaves' to the same 'stem' in a stem-and-leaf plot; for example, comparing the pulse rate before and after exercise. A key should always be included.

backtracking working backwards through a flow chart to solve an equation

balanced an equation whose left-hand side and right-hand side are equal

base the number being repeatedly multiplied in a power term. The power is the number of times the base is written.

bearing a measurement used to describe the direction from one object to another object

bimodal data with two obvious peaks in a graph

budget a list of all planned income and costs

buy now–pay later plan a purchasing method where a customer makes a purchase by paying a small amount up-front and then paying weekly or monthly instalments; also known as hire purchase

capacity the maximum amount of fluid that can be contained in an object. It is usually applied to the measurement of liquids and is measured in units such as millilitres (mL), litres (L) and kilolitres (kL)

cash advance a transfer of cash or cash equivalent, for example withdrawing money from an ATM

census night held every five years, the night all people who are in Australia fill in the census form

census collection of data from a population (e.g. all Year 10 students) rather than a sample

central tendency a single value that represents the middle of a frequency distribution

circumference the perimeter of a circle

co-interior angles angles positioned 'inside' the parallel lines, on the same side of a transversal; think of them as C-shaped

coefficient the number written in front of a term

column graphs graphs that display categories of data on one axis and the frequency of the data on the other axis; also known as bar graphs

commission a percentage of the sale price given to a salesperson when a sale is made

comparison rate an interest rate that gives a more accurate indication of the total cost of a loan. It is a comparison rate that reduces to a single percentage the interest rate and all the additional fees.

composite shape a closed shape that comprises two or more different common shapes

composite solid a solid object composed of a number of other simple solid objects

concentration the ratio of the amount of solute to the amount of solvent

constant a value that is not variable

consumer price index (CPI) a measure of changes, over time, in retail prices of a constant basket of goods and services representative of consumption expenditure by resident households in Australian metropolitan areas

conversion factors numbers used to convert units

conversion graphs line graphs that convert one quantity into another

corresponding angles angles positioned on the same side of a transversal, either both above or both below the parallel lines; think of them as F-shaped

council rates a fee home owners pay to support councils in providing parklands, libraries, rubbish collection and road maintenance

credit card a method of purchasing whereby a financial institution loans an amount of money to an individual up to a pre-approved limit

cube a solid figure that has 6 identical square faces

data various forms of information

data collection the process by which information is collected about a given population

debit card a method of purchasing in which the money is debited directly from a bank account or a pre-loaded amount

decimal places digits after the decimal point

denominator the bottom term of a fraction; it shows the total number of parts the whole has been divided into

density the ratio of mass to volume

dependent describes the variable whose value changes because of a change in the independent variable

deciles values that divide an ordered data set into 10 equal-sized parts

deviation the difference between a data value and the mean

diameter a straight line passing through the centre of a circle from one side of the circumference to the other

discount a price reduction on an item

dodecahedron a solid object with 12 faces, each of which is a regular polygon

dot plot a display of numerical or categorical data. A scaled horizontal axis is used and each data value is indicated by a dot above this scale, resulting in a set of vertical 'lines' of evenly-spaced dots.

edge a line of a graph; a line between two vertices in a shape or solid

effective rate of interest the rate of interest that takes into account the reducing balance owing after each payment has been made: $R_{\text{ef}} = \dfrac{2400I}{P(m + 1)}$

equations mathematical statements that show two equal expressions

equivalent equations equations that have the same solution; for example, $x + 3 = 7$ is equivalent to $x = 4$

equivalent fractions fractions that are equal in value, for example $\dfrac{1}{2}$ and $\dfrac{4}{8}$

equivalent ratios ratios that are equal in value; for example, $1 : 2 = 3 : 6$

estimate an approximate answer when a precise answer is not required

Euler's formula a formula that shows the relationship between the number of edges, the number of faces and the number of vertices in any polyhedron: $F + V - 2 = E$

exponent *see* **power**

exponential decay a process in which a quantity decreases according to an exponential formula

exponential growth a process in which a quantity increases according to an exponential formula

expression a mathematical sentence, made up of numbers, pronumerals and mathematical operations, that represents a quantity

exterior angle angle formed when any side of a closed shape is extended outwards. The exterior angle and the interior angle adjacent to it are supplementary (add up to 180°).

extrapolating making a prediction from a line of best fit that appears outside the parameters of the original data set

faces distinct areas created by the non-intersecting edges and vertices of planar graphs

flow chart a series of boxes connected by arrows to represent steps. They are useful for keeping track of the steps where a lot of steps are involved.

fractions used to describe a part of a whole number

frequency distribution tables tables used to organise data by recording the number of times each data value occurs

frequency distributions *see* **frequency distribution tables**

frequency tables *see* **frequency distribution tables**

frequency the number of times a score occurs in a set of data

fuel consumption the amount of fuel used by vehicle per 100 km

gradient a measure of the slope of a line at any point

grouped column graphs graphs that display the data for two or more categories, allowing for easy comparison

grouped data numerical data that is arranged in groups to allow a clearer picture of the distribution and make it easier to work with

hectare a unit of area equal to the space enclosed by a square with side lengths of 100 m (1 ha $= 10\,000$ m^2); often used to measure land area. There are 100 hectares in 1 square kilometre.

histogram a display of numerical data similar to a bar chart where the width of each column represents a range of data values and the height represents their frequencies

icosahedron a solid object with 20 faces, each of which is an equilateral triangle

improper fraction a fraction whose numerator is larger than its denominator, for example $\dfrac{7}{4}$

income tax a tax levied on people's financial income and based on an income tax table

independent describes the variable that is changed to produce a change in a dependent variable

integer a positive whole number, negative whole number or zero

index *see* **power**

inflation a general increase in prices over time that effectively decreases the purchasing power of a currency. This is an application of compound interest.

interest payment earned for having money stored in a bank or financial institution

interest-free period the period in which there is no interest charged on a debt if the debt is paid back in full before the end of that time period

interest rate the percentage of the principal that is paid out in a given time period as interest

internal angles the angles inside a closed shape

interpolate make a prediction from a line of best fit that appears within the parameters of the original data set

interquartile range (IQR) the difference between the upper (or third) quartile and the lower (or first) quartile

inverse operations operations that undo each other

investors people who place money in a bank or financial institution

kilowatt hour a standard unit of electrical energy consumption; 1 kWh is the amount of energy produced by an appliance outputting 1 kW (1000 W) of power for an hour

latitude the angular distance of a point on the Earth's surface north or south of the equator, measured along the meridian passing through the point

line graphs graphs consisting of individually plotted points joined with a straight line

line of best fit a line drawn on a scatterplot that passes through or is close to as many points as possible

logarithmic scale a logarithmic scale represents numbers using a log (base 10) scale. This means that if we express all of the numbers in the form 10^a, the logarithmic scale will represent these numbers as a.

longitude the angular distance of a point on the Earth's surface east or west of the prime meridian, measured along the equator between the prime meridian and the meridian passing through the point

lower fence a low limit of a data set with any data lying below this this considered an outlier: $Q1 - 1.5 \times$ IQR

mass describes how much matter makes up an object. The base unit of mass is grams.

mean commonly referred to as the average; a measure of the centre of a set of data. The mean is calculated by dividing the sum of the data values by the number of data values.

measures of central tendency mean, median and mode

measures of spread statistical values that indicate how far data values are spread either from the centre or from each other

median the middle value of a data set when the values are placed in numerical order

Medicare levy a portion of taxpayers' funds used to pay for Medicare (healthcare for Australian residents)

megajoule a standard unit of energy; 1 megajoule (MJ) = 1 000 000 joules (J)

mixed number a number made up of a whole number and a fraction, for example $2\frac{3}{4}$

modality a measure of the number of obvious peaks in a data set

mode the score that occurs most often in a data set

monthly income the amount of income that a person receives in one month

multimodal data with three or more obvious peaks in a graph

negatively skewed showing larger amounts of data as the values of the data increase

no interest-free period if a credit card has no interest-free period, the card holder is charged interest on the outstanding amount of each purchase from the date of purchase until the purchase is repaid in full

numerator the top term of a fraction; it shows how many parts there are

octahedron a solid object with 8 faces, each of which is an equilateral triangle

one-way tables *see* **frequency distribution tables**

order of magnitude an approximation of the size of a number equal to the exponent of the power of 10

outlier an extreme value or unusual reading in the data set, generally considered to be any value beyond the lower or upper fences

overtime the time in which a worker is working hours in addition to, or outside, their regular hours of employment

parallel lines lines that are simple translations of each other. Parallel lines have the same gradient.

Pay As You Go (PAYG) tax a withholding tax system administered by the Australian Taxation Office

pension government financial support given to some people after they have retired

per cent the amount out of 100, or per hundred; for example, 50 per cent (or 50%) means 50 out of 100 or $\frac{50}{100}$

percentage relative frequency the frequency of a score as a proportion of the total number of scores, expressed as a percentage

perfect square a number that can be arranged in a square; a number that is the square of another number

perimeter distance around the outside of a shape

personal loan a loan made by a lending institution to an individual. A personal loan will usually have a fixed interest rate attached to it, with the interest paid by the customer calculated on a reduced balance.

piecework a fixed method of payment typically for production of individual items

pie graphs graphs that use sectors of circles to represent categories of data

polygon a closed shape where each edge is a straight line and all edges meet at vertices

polyhedron a three-dimensional solid with faces that are polygons, like the pyramids of Giza

population the whole group from which a sample is drawn

positively skewed showing smaller amounts of data as the values of the data decrease

power the number of times that the base is written in the expanded form of a power term; also known as an exponent or index

prime meridian the line from which longitude is measured

principal the amount that is borrowed or invested

prism a solid object that has identical opposite polygonal ends that are joined by flat surfaces and a cross-section that is the same along its length

pronumeral a letter used in place of a number

proper fraction a fraction whose numerator is smaller than its denominator, for example $\frac{3}{4}$

proportion equality of two or more ratios

Pythagoras' theorem theorem stating that in any right-angled triangle, the square on the hypotenuse is equal to the sum of the squares on the other two sides. This is often expressed as $c_2 = a_2 + b_2$.

quadrilateral a 2-dimensional, closed shape formed by four straight lines

quantiles values that divide a data set into a given number of equal parts

quartiles values that divide an ordered data set into four equal parts

questionnaire a list of questions used to collect data from a population

radius (plural: radii) the distance from the centre of a circle to its circumference

random number generator a device or program that generates random numbers between two given values

range the difference between the highest and lowest values in a data set

rate a measure of how one quantity is changing compared to another

rate in the dollar the rate at which council taxes are charged

ratio the relationship between two or more values commonly expressed as $\frac{a}{b} \Leftrightarrow a : b$

reciprocal The reciprocal of a number is obtained by first expressing it as a fraction then tipping the fraction upside down; for example $\frac{4}{3}$, is the reciprocal of $\frac{3}{4}$. A mixed number must be converted to an improper fraction first and then tipped.

recurring decimal a decimal number that has one or more digits repeated continuously; for example, 0.999… Recurring decimals can be expressed exactly by placing a dot or horizontal line over the repeating digits, as in this example:

$8.343\,434… = 8.3\dot{4} \text{ or } 8.\overline{34}$

registration fee a combination of administration fees, taxes and charges paid to legally drive a vehicle on the road

relative frequency the frequency of a particular score divided by the total sum of the frequencies

royalty a payment made to authors, composers or creators for each copy of the work or invention sold. Royalties are typically calculated as a percentage of the total sales.

rule a mathematical statement or principle or formula

salary a fixed amount of money paid per year (annually), usually paid fortnightly or monthly regardless of the number of hours worked

sample part of a population chosen to give information about the population as a whole

sampling the process of selecting a sample of a population to provide an estimate of the entire population

sampling strategy the method by which participants are found in self-selected sampling

scale a series of marks indicating measurement increasing in equal quantities

scale factor the ratio of the corresponding sides in similar figures, where the enlarged (or reduced) figure is referred to as the image and the original figure is called the object

scientific notation used to express very large or very small numbers. To express a number in standard form, write it as a number between 1 and 10 multiplied by a power of 10.

self-selected sampling a voluntary sample made up of people who self-select into a survey

shares units representing partial ownership of a company

side-by-side bar charts bar charts that contain multiple sets of categorical data presented as multiple bars in the same chart

simple random sampling a type of probability sampling method in which each member of the population has an equal chance of selection

spread how far data values are spread from the centre or from each other

square numbers a number that can be arranged in a square

square root a number that multiplies by itself to equal the original number. Finding a square root is the opposite of squaring a number; for example, $\sqrt{49} = 7$.

squared units the units that area and total surface area are measured in: mm^2, cm^2, m^2 or km^2

squaring a number multiplying a number by itself; for example, $2 \times 2 = 2^2 = 4$

standard deviation a measure of the variability of spread of a data set. It gives an indication of the degree to which the individual data values are spread around the mean.

starting number the number in the first box in a flow chart used to build an algebraic expression

stem-and-leaf plot or **stem plot** an arrangement used for numerical data where each value is grouped according to its numerical place value (the 'stem') and then displayed horizontally as a single digit (the 'leaf')

step graphs graphs made up of horizontal straight lines; used when values remain constant over intervals

strata sub-groups into which a population is divided

stratified sampling a sampling method where groups within a population have a similar representation in the sample

subject of the formula the variable by itself on one side of the formula

superannuation a percentage of annual salary that is set aside for retirement

supplementary describes angles that sum to 180°

surcharge an additional amount charged by a business to process a customer's payment by credit card

survey collection of data from a sample of a population

symmetric when the data distribution has two halves roughly equal about the mean

systematic random sampling sampling in a way that ensures that each member of the population has an equal chance of being chosen

systematic sampling a sampling method where the data values chosen to be in the sample are selected at regular intervals

tally a mark made to record the occurrence of a score

tax deductions work-related expenses that are subtracted from taxable income, which lowers the amount of money earned and amount of tax paid

taxable income the amount of income remaining after tax deductions have been subtracted from the total income

term part of an expression. Terms may contain one or more pronumerals, such as $6x$ or $3xy$, or they may consist of a number only.

terminating decimal a decimal number that has a fixed number of decimal places, for example 0.6 or 2.54

tetrahedron a solid object with four faces, each of which is an equilateral triangle

time measurement used to work out how long we have been doing things or how long something has been happening

total income the sum of all money earned by an individual

total surface area the combined total of the external areas of each individual surface that forms a solid object

transversal a line that intersects with a pair of parallel lines

two-way frequency tables a table that displays two categorical variables according to the frequencies of predetermined groupings

two-way tables tables that list all the possible outcomes of a probability experiment in a logical manner

unbalanced an unbalanced equation is an equation in which one side has a greater value than the other side

ungrouped data numerical data that is not arranged in groups to enable exact analysis

unimodal data with one obvious peak in a graph

upper fence a high limit of a data set with any data lying above this this considered an outlier: $Q_3 + 1.5 \times IQR$

variable another word for a pronumeral (algebra); a symbol in an equation or expression that may take many different values (data)

vertex (plural: vertices) a the point or 'node' of a graph; a point or corner of a shape where edges meet

vertically opposite angles angles created when two lines intersect. The angles opposite each other are equal in size; think of them as X-shaped.

vinculum horizontal line used to separate the top of a fraction (numerator) from the bottom of a fraction (denominator)

volume the amount of space a 3-dimensional object occupies. The units used are cubic units, such as cubic centimetres (cm^3) and cubic metres (m^3).

wage a fixed amount of money per hour worked (hours worked outside the normal work period are paid at a higher amount)

INDEX